D1480538

Methods in Enzymology

Volume 324
BRANCHED-CHAIN AMINO ACIDS
Part B

METHODS IN ENZYMOLOGY

EDITORS-IN-CHIEF

John N. Abelson   Melvin I. Simon

DIVISION OF BIOLOGY
CALIFORNIA INSTITUTE OF TECHNOLOGY
PASADENA, CALIFORNIA

FOUNDING EDITORS

Sidney P. Colowick and Nathan O. Kaplan

Methods in Enzymology

Volume 324

Branched-Chain Amino Acids

Part B

EDITED BY

Robert A. Harris

DEPARTMENT OF BIOCHEMISTRY AND MOLECULAR BIOLOGY
INDIANA UNIVERSITY SCHOOL OF MEDICINE
INDIANAPOLIS, INDIANA

John R. Sokatch

DEPARTMENT OF BIOCHEMISTRY AND MOLECULAR BIOLOGY
THE UNIVERSITY OF OKLAHOMA
OKLAHOMA CITY, OKLAHOMA

ACADEMIC PRESS

San Diego   London   Boston   New York   Sydney   Tokyo   Toronto

Academic Press
A Harcourt Science and Technology Company
525 B Street, Suite 1900, San Diego, California 92101-4495, USA

http://www.academicpress.com

Academic Press Limited
32 Jamestown Road, London NN1 7BY, UK

International Standard Book Number: 0-12-182225-7

PRINTED IN THE UNITED STATES OF AMERICA
00 01 02 03 04 05 06 MM 9 8 7 6 5 4 3 2 1

Table of Contents

Section I. Preparation of Substrates, Assays of Intermediates and Enzymes, and Use of Enzyme Inhibitors

v

Section II. Cloning, Expression, and Purification of Enzymes of Branched-Chain Amino Acid Metabolism

Section III. Detection and Consequences of Genetic Defects in Genes Encoding Enzymes of Branched-Chain Amino Acid Metabolism

Section IV. Regulation and Expression of Enzymes of Branched-Chain Amino Acid Metabolism

Contributors to Volume 324

Article numbers are in parentheses following the names of contributors.
Affiliations listed are current.

ZE'EV BARAK (2, 10), *Department of Life Sciences, Ben-Gurion University of the Negev, Beer-Sheva 84105, Israel*

MICHAEL J. BEACH (25), *Centers for Disease Control, Atlanta, Georgia, 30341-3724*

PHILIP R. BECKETT (5), *Baylor College of Medicine, Houston, Texas 77030*

BARBARA BINZAK (24), *Department of Biochemistry and Molecular Biology, Mayo Clinic and Mayo Foundation, Rochester, Minnesota 55905*

KENNETH M. BISCHOFF (25), *Food Animal Protection Research Laboratory, U. S. Department of Agriculture, College Station, Texas 77845*

DANIEL A. BOCHAR (25), *The Wistar Institute, Philadelphia, Pennsylvania 19104-4268*

JOSEPH M. CALVO (29), *Section of Biochemistry, Molecular and Cell Biology, Cornell University, Ithaca, New York 14853*

REBECCA M. CHAN (22), *Department of Biochemistry and Molecular Biology, Indiana University School of Medicine, Indianapolis, Indiana 46202-5122*

YUHAI CHI (29), *Department of Molecular Biology and Genetics, University of Guelph, Guelph, Ontario N1G 2W1, Canada*

JEFFREY M. CHINSKY (43), *Department of Pediatrics, Johns Hopkins University School of Medicine, Baltimore, Maryland 21287*

DAVID M. CHIPMAN (2, 10), *Department of Life Sciences, Ben-Gurion University of the Negev, Beer-Sheva 84105, Israel*

DAVID T. CHUANG (18, 19, 38, 45), *Department of Biochemistry, University of Texas, Southwestern Medical Center, Dallas, Texas 75390-9038*

JACINTA L. CHUANG (18, 19, 38), *Department of Biochemistry, University of Texas, South-western Medical Center, Dallas, Texas 75390-9038*

MYRA CONWAY (33), *Department of Biochemistry, Wake Forest University School of Medicine, Winston-Salem, North Carolina 27157*

PAUL A. COSTEAS (43), *Karaiskakio Foundation, The Cyprus Bone Marrow and Platelet Donor Registry, Nicosia, Cyprus*

DAVID W. CRABB (22), *Departments of Medicine, and Biochemistry and Molecular Biology, Indiana University School of Medicine, Indianapolis, Indiana 46202-5122*

NICHOLAS P. CROUCH (32), *The Dyson Perrins Laboratory, University of Oxford, Oxford OX1 3QY, United Kingdom*

DEAN J. DANNER (31, 44), *Department of Genetics, Emory University School of Medicine, Atlanta, Georgia 30322*

BRYANT G. DARNAY (25), *UTMDA Cancer Center, Cytokine Research Section, Houston, Texas 77030*

JAMES R. DAVIE (18, 19), *Department of Biochemistry, University of Texas, Southwestern Medical Center, Dallas, Texas 75390-9038*

J. RICHARD DICKINSON (9, 36), *Cardiff School of Biosciences, Cardiff University, Cardiff CF10 3TL, United Kingdom*

CHRISTOPHER B. DOERING (44), *Department of Genetics, Emory University School of Medicine, Atlanta, Georgia 30322*

TSIONA ELKAYAM (10), *Department of Life Sciences, Ben-Gurion University of the Negev, Beer-Sheva 84105, Israel*

SABINE EPELBAUM (2), *Central Research and Development Department, E.I. duPont de Nemours and Company, Wilmington, Delaware 19880-0173*

ABDUL FAUQ (24), *Organic Synthesis Core Facility, Mayo Clinic and Mayo Foundation, Jacksonville, Florida 32224*

DEVORAH FRIEDBERG (29), *Department of Molecular Genetics and Biotechnology, The Hebrew University, Faculty of Medicine, Jerusalem 91120, Israel*

JON A. FRIESEN (25), *Department of Biochemistry and Molecular Biology, Illinois State University, Normal, Illinois 61790-4100*

SHIGEKO FUJIMOTO SAKATA (35), *Laboratory of Nutritional Chemistry, Faculty of Nutrition, Kobe Gakuin University, Kobe 651-2180, Japan*

TOSHIYUKI FUKAO (40), *Department of Pediatrics, Gifu School of Medicine, Gifu 500-8076, Japan*

K. MICHAEL GIBSON (40), *Departments of Molecular and Medical Genetics, and Pediatrics, and Biochemical Genetics Laboratory, Oregon Health Sciences University, Portland, Oregon 97201*

JOHN F. GILL (25), *Department of Protein Chemistry, Boehringer-Mannheim, Indianapolis, Indiana 46250*

GARY W. GOODWIN (21), *Division of Cardiology, Department of Internal Medicine, The University of Texas Houston Medical School, Houston, Texas 77030*

EDWIN T. HARPER (8), *Department of Biochemistry and Molecular Biology, Indiana University School of Medicine, Indianapolis, Indiana 46202-5122*

ROBERT A. HARRIS (6, 17, 20, 21, 22, 23), *Department of Biochemistry and Molecular Biology, Indiana University School of Medicine, Indianapolis, Indiana 46202-5122*

JOHN W. HAWES (8, 17, 20, 22, 23), *Department of Biochemistry and Molecular Biology, Indiana University School of Medicine, Indianapolis, Indiana 46202-5122*

HIDEYUKI HAYASHI (11), *Department of Biochemistry, Osaka Medical College, Takatsuki 569-8686, Japan*

YOKO HAYASHI-IWASAKI (28), *Department of Molecular Biology, Tokyo University of Pharmacy and Life Science, Tokyo 192-0392, Japan*

MATIJA HEDL (25), *Department of Biochemistry, Purdue University, West Lafayette, Indiana 47907-1153*

KATHRYN L. HESTER (14), *Dunlop Codding and Rogers, Oklahoma City, Oklahoma 74114*

YOUNG SOO HONG (41), *Department of Biochemistry, School of Medicine and Biomedical Sciences, State University of New York, Buffalo, New York 14214*

TOSHIMITSU HOSHINO (12, 13), *Mitsubishi Kasei Institute of Life Sciences, Tokyo 194-8511, Japan*

BOLI HUANG (23), *Department of Biochemistry and Molecular Biology, Indiana University School of Medicine, Indianapolis, Indiana 46202-5122*

YI-SHUIAN HUANG (45), *Department of Biochemistry, University of Texas Southwestern Medical Center, Dallas, Texas 75390-9038*

SUSAN HUTSON (33), *Department of Biochemistry, Wake Forest University School of Medicine, Winston-Salem, North Carolina 27157*

TERESA ITURRIAGAGOITIA-BUENO (32), *The Dyson Perrins Laboratory, University of Oxford, Oxford OX1 3QY, United Kingdom*

MARK T. JOHNSON (42), *Department of Genetics, Case Western Reserve University, Cleveland, Ohio 44106*

TUAJUANDA JORDAN-STARCK (25), *Department of Chemistry, Xavier University of New Orleans, New Orleans, Louisiana 70125*

HIROYUKI KAGAMIYAMA (11), *Department of Biochemistry, Osaka Medical College, Takatsuki 569-8686, Japan*

NATALIA Y. KEDISHVILI (21), *Department of Molecular Biology and Biochemistry, School of Biological Sciences, University of Missouri–Kansas City, Kansas City, Missouri 64110*

PETER J. KENNELLY (25), *Department of Biochemistry, Virginia Polytechnic and State University, Blacksburg, Virginia 24061-0308*

DOUGLAS S. KERR (41), *Department of Pediatrics and Center for Inherited Disorders of Energy Metabolism, Case Western Reserve University, Cleveland, Ohio 44106*

MARZIA GALLI KIENLE (7), *Department of Experimental and Environmental Medicine and Medical Biotechnology, University of Milano–Bicocca, Milan, Italy*

DONG YUL KIM (25), *Department of Biochemistry, Purdue University, West Lafayette, Indiana 47907-1153*

GYULA KISPAL (34), *Institute of Biochemistry, University Medical School of Pecs, 7624 Pecs, Hungary*

RUMI KOBAYASHI (6), *Department of Biochemistry and Molecular Biology, Indiana University School of Medicine, Indianapolis, Indiana 46202-5122*

MENG-HUEE LEE (32), *The Dyson Perrins Laboratory, University of Oxford, Oxford OX1 3QY, United Kingdom*

ROLAND LILL (34), *Institut für Zytobiologie und Zytopathologie, Philipps-Universität, 35033 Marburg, Germany*

JINHE LUO (14), *Department of Biochemistry and Molecular Biology, University of Oklahoma Health Sciences Center, Oklahoma City, Oklahoma 73190*

COLIN H. MACKINNON (32), *The Dyson Perrins Laboratory, University of Oxford, Oxford OX1 3QY, United Kingdom*

KUNAPULI T. MADHUSUDHAN (30), *Central Arkansas Veterans Healthcare System, Pathology and Laboratory Medicine, Little Rock, Arkansas 72205*

FULVIO MAGNI (7), *IRCCS San Raffaele, Mass Spectrometry Unit, Milan, Italy*

ORVAL A. MAMER (1), *The Mass Spectrometry Unit, McGill University, Montreal, Quebec H3A 1A3, Canada*

KOICHI MATSUDA (35), *Laboratory of Nutritional Chemistry, Faculty of Nutrition, Kobe Gakuin University, Kobe 651-2180, Japan*

ROWENA G. MATTHEWS (29), *Biophysics Research Division and Department of Biological Chemistry, University of Michigan, Ann Arbor, Michigan 48109-1055*

GRANT A. MITCHELL (40), *Service de Genetique Medicale, Hopital Sainte-Justine, Montreal, Quebec H3T 1C5, Canada*

HENRY M. MIZIORKO (15, 16), *Biochemistry Department, Medical College of Wisconsin, Milwaukee, Wisconsin 53226*

AL-WALID A. MOHSEN (24), *Department of Medical Genetics, Mayo Clinic and Mayo Foundation, Rochester, Minnesota 55905*

TARO MURAKAMI (23), *Department of Bioscience, Nagoya Institute of Technology, Nagoya 466-8555, Japan*

NAOYA NAKAI (6, 23), *Research Center of Health, Physical Fitness and Sports, Nagoya University, Nagoya 464-8601, Japan*

CHAKRAVARTHY NARASIMHAN (15, 16), *Lilly Research Laboratories, Indianapolis, Indiana 46285*

BASIL J. NIKOLAU (26), *Department of Biochemistry, Biophysics and Molecular Biology, Iowa State University, Ames, Iowa 50011*

TAIRO OSHIMA (28), *Department of Molecular Biology, Tokyo University of Pharmacy and Life Science, Tokyo 192-0392, Japan*

MULCHAND S. PATEL (41, 42), *Department of Biochemistry, School of Medicine and Biomedical Sciences, State University of New York, Buffalo, New York 14214*

RALPH PAXTON (22), *Laboratory of Metabolic Disorders, Departments of Anatomy, Physiology and Pharmacology, College of Veterinary Medicine, Auburn University, Auburn, Alabama 36849-5404*

KIRILL M. POPOV (6, 17, 20, 21), *Department of Molecular Biology and Biochemistry, School of Biological Sciences, University of Missouri–Kansas City, Kansas City, Missouri 64110*

CORINNA PROHL (34), *Institut für Zytobiologie und Zytopathologie, Philipps-Universität, 35033 Marburg, Germany*

JACQUELINE R. ROBERTS (16), *Chemistry Department, De Pauw University, Greencastle, Indiana 46135*

VICTOR W. RODWELL (25), *Department of Biochemistry, Purdue University, West Lafayette, Indiana 47907-1153*

CHARLES R. ROE (39), *Institute of Metabolic Disease, Baylor University Medical Center, Dallas, Texas 75226*

DIANE S. ROE (39), *Institute of Metabolic Disease, Baylor University Medical Center, Dallas, Texas 75226*

PAUL M. ROUGRAFF (22), *Naples, Florida 34102*

PETER SCHADEWALDT (3, 4), *Deutsches Diabetes Forschungsinstitut, Klinische Biochemie, D-40225 Düsseldorf, Germany*

YOSHIHARU SHIMOMURA (6, 17, 20, 23), *Department of Bioscience, Nagoya Institute of Technology, Nagoya 466-8555, Japan*

JOHN R. SOKATCH (14, 30), *Department of Biochemistry and Molecular Biology, University of Oklahoma Health Sciences Center, Oklahoma City, Oklahoma 73190*

JIU-LI SONG (18), *Department of Biochemistry, University of Texas, Southwestern Medical Center, Dallas, Texas 75390-9038*

NANAYA TAMAKI (35), *Laboratory of Nutritional Chemistry, Faculty of Nutrition, Kobe Gakuin University, Kobe 651-2180, Japan*

THOMAS W. TRAUT (37), *Department of Biochemistry and Biophysics, University of North Carolina School of Medicine, Chapel Hill, North Carolina 27599-7260*

MAGDALENA UGARTE (40), *Department Biologia Molecular, Universidad Autonoma de Madrid, 28049 Madrid, Spain*

YOSHIHIKO URATANI (12, 13), *Mitsubishi Kasei Institute of Life Sciences, Tokyo 194-8511, Japan*

JERRY VOCKLEY (24), *Department of Medical Genetics, Mayo Clinic and Mayo Foundation, Rochester, Minnesota 55905*

MARIA VYAZMENSKY (10), *Department of Life Sciences, Ben-Gurion University of the Negev, Beer-Sheva 84105 Israel*

YULI WANG (25), *Ann Arbor, Michigan 48105*

JAN WILLARD (24), *Department of Medical Genetics, Mayo Clinic and Mayo Foundation, Rochester, Minnesota 55905*

EVE SYRKIN WURTELE (26), *Department of Botany, Iowa State University, Ames, Iowa 50011*

R. MAX WYNN (18, 19), *Department of Internal Medicine, University of Texas, Southwestern Medical Center, Dallas, Texas 75390-8889*

HSIN-SHENG YANG (42), *Department of Biochemistry, School of Medicine and Biomedical Sciences, State University of New York, Buffalo, New York 14214*

YU ZHAO (20), *Bayer Corporation, Clayton, North Carolina 27520*

RAINER ZOCHER (27), *Max-Volmes-Institut für Biophysikalische Chemie, Fachgebiet Biochemie und Moleculare Biologie, Technische Universität Berlin, D-10587 Berlin, Germany*

Preface

Branched-chain amino acids were first covered in *Methods in Enzymology* Volume XVII: Metabolism of Amino Acids and Amines, Part A, edited by H. W. Tabor and C. W. Tabor. It included a section on biosynthetic pathways for branched-chain amino acids in prokaryotes, which was the state of the art at that time. In 1988, we edited *Methods in Enzymology* Volume 166, Branched-Chain Amino Acids, which included sections on the purification of several enzymes of branched-chain amino acid metabolism, specific enzyme assays, and the quantification of metabolites in branched-chain amino acid pathways. In the past twelve years, recombinant DNA technology has had a major impact on the understanding of the structure, function, regulation, and expression of enzymes of branched-chain amino acid metabolism, including genetic defects in these pathways which afflict humans.

This volume features the purification and characterization of several enzymes of branched-chain amino acid catabolism that were unknown when Volume 166 was published. Improvements on earlier methods for purification of key regulatory enzymes, the assay of various enzymes in purified form and in tissue extracts, and the determination of important metabolites of branched-chain amino acid pathways are also given. The most important feature of the volume, however, is the information resulting from the application of recombinant DNA methodology to the study of the enzymes of branched-chain amino acid metabolism. This includes sections on cloning cDNAs encoding many of the branched-chain amino acid enzymes, expression and characterization of several of these enzymes as recombinant proteins, and the identification of mutations responsible for genetic defects in enzymes of branched-chain amino acid catabolism. This volume will be of use to investigators interested in the structure, function, regulation, and expression of enzymes of branched-chain amino acid metabolism as well as to investigators who study genetic diseases caused by defects in the enzymes of catabolic pathways for these amino acids.

We would like to express our appreciation to the authors whose contributions made this volume possible.

ROBERT A. HARRIS
JOHN R. SOKATCH

METHODS IN ENZYMOLOGY

VOLUME 227. Metallobiochemistry (Part D: Physical and Spectroscopic Methods for Probing Metal Ion Environments in Metalloproteins)
Edited by JAMES F. RIORDAN AND BERT L. VALLEE

VOLUME 228. Aqueous Two-Phase Systems
Edited by HARRY WALTER AND GÖTE JOHANSSON

VOLUME 229. Cumulative Subject Index Volumes 195–198, 200–227

VOLUME 230. Guide to Techniques in Glycobiology
Edited by WILLIAM J. LENNARZ AND GERALD W. HART

VOLUME 231. Hemoglobins (Part B: Biochemical and Analytical Methods)
Edited by JOHANNES EVERSE, KIM D. VANDEGRIFF, AND ROBERT M. WINSLOW

VOLUME 232. Hemoglobins (Part C: Biophysical Methods)
Edited by JOHANNES EVERSE, KIM D. VANDEGRIFF, AND ROBERT M. WINSLOW

VOLUME 233. Oxygen Radicals in Biological Systems (Part C)
Edited by LESTER PACKER

VOLUME 234. Oxygen Radicals in Biological Systems (Part D)
Edited by LESTER PACKER

VOLUME 235. Bacterial Pathogenesis (Part A: Identification and Regulation of Virulence Factors)
Edited by VIRGINIA L. CLARK AND PATRIK M. BAVOIL

VOLUME 236. Bacterial Pathogenesis (Part B: Integration of Pathogenic Bacteria with Host Cells)
Edited by VIRGINIA L. CLARK AND PATRIK M. BAVOIL

VOLUME 237. Heterotrimeric G Proteins
Edited by RAVI IYENGAR

VOLUME 238. Heterotrimeric G-Protein Effectors
Edited by RAVI IYENGAR

VOLUME 239. Nuclear Magnetic Resonance (Part C)
Edited by THOMAS L. JAMES AND NORMAN J. OPPENHEIMER

VOLUME 240. Numerical Computer Methods (Part B)
Edited by MICHAEL L. JOHNSON AND LUDWIG BRAND

VOLUME 241. Retroviral Proteases
Edited by LAWRENCE C. KUO AND JULES A. SHAFER

VOLUME 242. Neoglycoconjugates (Part A)
Edited by Y. C. LEE AND REIKO T. LEE

VOLUME 243. Inorganic Microbial Sulfur Metabolism
Edited by HARRY D. PECK, JR., AND JEAN LEGALL

VOLUME 244. Proteolytic Enzymes: Serine and Cysteine Peptidases
Edited by ALAN J. BARRETT

Section I

Preparation of Substrates, Assays of Intermediates and Enzymes, and Use of Enzyme Inhibitors

[1] Synthesis and Gas Chromatography/Mass Spectrometry Analysis of Stereoisomers of 2-Hydroxy-3-methylpentanoic Acid

By ORVAL A. MAMER

Introduction

2-Hydroxy-3-methylpentanoic acid (HMPA) is a familiar metabolite of isoleucine commonly found in normal human urine, plasma, and other fluids. In disorders affecting branched-chain amino acid metabolism, such as branched-chain ketoaciduria, HMPA is frequently elevated in these fluids, sometimes by two orders of magnitude or more. HMPA is formed by the reduction of 2-keto-(3*S*)-methylpentanoic acid [(3*S*)-KMPA], which is the transamination product of L-isoleucine in the degradative pathway of that amino acid.[1,2] The chiral center of KMPA is spontaneously racemized in solution at high pH through keto–enol tautomerism[1] (Fig. 1).

Having two chiral centers, HMPA exists in four isomeric forms (Fig. 2): (2*R*)-hydroxy-(3*R*)-methylpentanoic acid [(2*R*,3*R*)-HMPA], (2*S*)-hydroxy-(3*S*)-methylpentanoic acid [(2*S*,3*S*)-HMPA], (2*R*)-hydroxy-(3*S*)-methylpentanoic acid [(2*R*,3*S*)-HMPA], and (2*S*)-hydroxy-(3*R*)-methylpentanoic acid [(2*S*,3*R*)-HMPA]. Each of these may be synthesized in pure form by conventional diazotization of the isomer of 2-amino-3-methylpentanoic acid having the corresponding steric configuration and trivial or common names of D-isoleucine, L-isoleucine, D-alloisoleucine, and L-alloisoleucine, respectively. All four of these amino acids are commercially available in relatively pure form.

Synthesis

General

All solutions are aqueous unless stated otherwise, reactions are carried out in a fume hood to reduce exposure to hazardous vapors, organic chemicals are from Sigma-Aldrich Canada (Oakville, ON, Canada), and inorganic materials are purchased from local suppliers and are used as received.

[1] O. A. Mamer and M. L. J. Reimer, *J. Biol. Chem.* **267,** 22141 (1992).
[2] O. A. Mamer and F. L. Lépine, *J. Mass Spectrom.* **31,** 1382 (1996).

METHODS IN ENZYMOLOGY, VOL. 324
0076-6879/00 $30.00

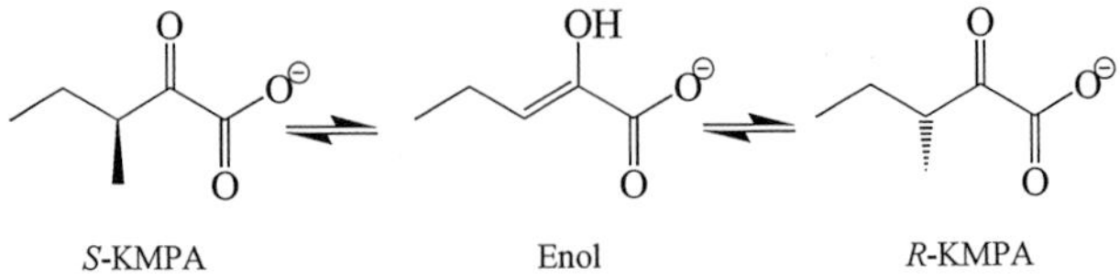

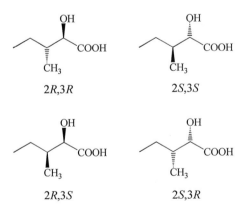

FIG. 1. Racemization of KMPA by enolization at high pH in aqueous solution.

Gas Chromatography/Mass Spectrometry Analysis

Analyses are run on a Hewlett-Packard (Palo Alto, CA) model 5988A gas chromatograph/mass spectrometer (GC/MS) equipped with a 5890A gas chromatograph fitted with a 30 m × 0.25 mm (i.d.) fused silica column having a 0.25-μm dimethylsilicone coating. The helium flow rate is set at 1 ml/min. The injector port is set at 250° and configured for splitless injection of 1-μl aliquots of the trimethylsilyl (TMS) derivative mixtures. After a 1-min hold at 70° for septum purging, the column is programmed at 5°/min to 200°. The interface and ion source temperatures are set at 270 and 200°, respectively. Ionization is by electron impact at 70 eV, and scanning is over the range 100 to 400 Da.

Racemic 2-Hydroxy-3-methylpentanoic Acid

Fully racemic HMPA may be conveniently synthesized by the sodium borohydride reduction of KMPA racemized in solution at high pH. The sodium salt of KMPA (150 mg, 1 mmol; Sigma-Aldrich) is dissolved in 5 ml of water, a few drops of 2 N sodium hydroxide are added to raise the pH to pH >12, and 50 mg of sodium borohydride is added with stirring until dissolution is complete. The resulting solution is held in a water bath

FIG. 2. The four stereoisomers of HMPA.

at 60° for 2 hr, cooled in an ice–water bath, and then cautiously acidified to pH <2 with 2 N hydrochloric acid. (*Caution:* Vigorous evolution of hydrogen gas occurs.) The acidified solution is saturated with sodium chloride and extracted three times with volumes of diethyl ether equal to the aqueous phase. The ether extracts are combined, treated with 200 mg of anhydrous sodium sulfate to remove water, and evaporated to an oil (approximately 100-mg yield) under a warm, dry nitrogen stream. The oil may be crystallized from an ethyl acetate–petroleum ether solvent pair if desired, although this is not necessary for further work.

A few micrograms of the product oil is converted to the TMS derivative in a mixture of 25-μl volumes of anhydrous pyridine and *N,O*-bis(trimethylsily)trifluoroacetamide (BSTFA) in a capped autoinjector vial, heated at 60° for 20 min, and analyzed by GC/MS as described above. The relevant portion of the resulting chromatogram is reproduced in Fig. 3A.

Two unequal peaks are found for racemic HMPA made from fully racemized KMPA. Although it is not possible to separate by chromatography enantiomers in achiral stationary phases, diastereomers that are not mirror images of each other are frequently easily separable, as they have different chemical and physical properties. Sodium borohydride reduction is expected to produce less of a racemic (2*S*,3*R*)- and (2*R*,3*S*)-HMPA enantiomeric pair than of a (2*S*,3*S*)-HMPA, (2*R*,3*R*)-HMPA pair.[1] This is borne out in the resulting chromatogram. The mass spectra of the two peaks are reproduced in Fig. 4A and B, and cannot be distinguished reliably.

L-2-Hydroxy-3-methylpentanoic Acid

Sodium L-2-hydroxy-3-methylvalerate (50 mg, 0.3 mmol; Sigma) is dissolved in 2 ml of water, which is then acidified, extracted, isolated, and derivatized as described above. The resulting mixture when analyzed by GC/MS produces a single peak (Fig. 3B) having the same retention time as the second eluting peak in Fig. 3A and the mass spectrum shown in Fig. 4C. Again, the mass spectrum of this isomer is indistinguishable from those of the racemate. Because (3*S*)-KMPA is probably not racemized after its biosynthesis from L-isoleucine, reduction to HMPA, presumably by a lactate dehydrogenase (LDH) isozyme, will leave the *S*-chirality at the 3-carbon unaffected. LDH reductions of 2-keto acids produce *S*-chirality at the 2-carbon. This assignment of "L-HMPA" as (2*S*,3*S*)-HMPA is consistent then with its elution with the retention time of peak 2 in the racemate (Fig. 3).

2-Hydroxy-3-methylpentanoic Acid Stereoisomers

HMPA isomers may be conveniently synthesized in pure form by diazotization of the corresponding amino acid in dilute perchloric acid. Replace-

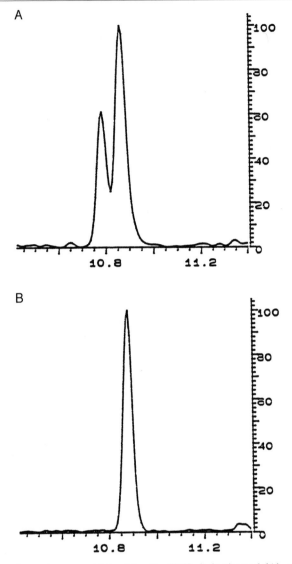

FIG. 3. Gas chromatograms obtained for the TMS derivatives of (A) racemic HMPA synthesized by sodium borohydride reduction of racemic KMPA, and (B) commercially available, isomerically pure (2S,3S)-HMPA.

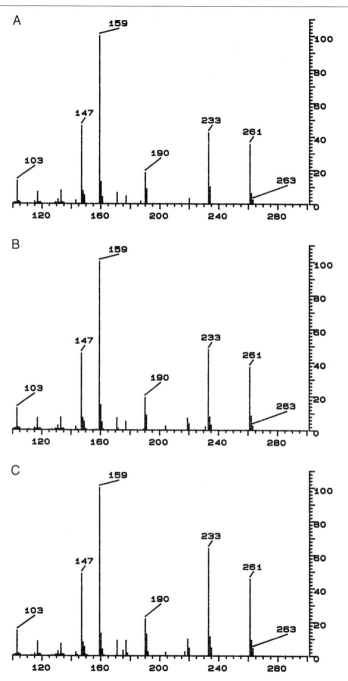

FIG. 4. Mass spectra obtained for (A) peak 1 in Fig. 3A, (B) peak 2 in Fig. 3A, and (C) the major peak in Fig. 3B.

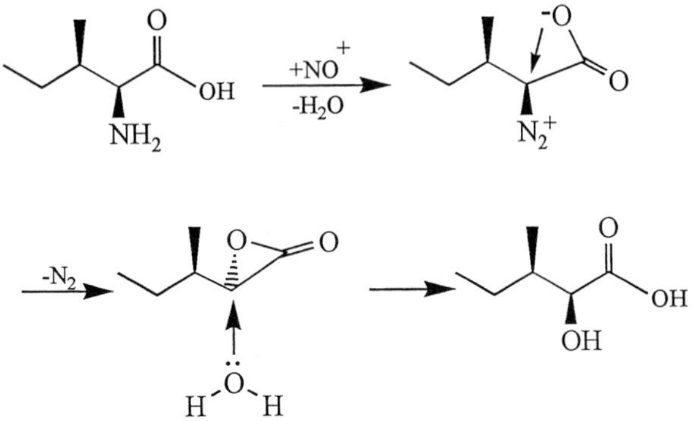

FIG. 5. Retention of configuration by anchiomeric participation of the carboxylate group in the diazotization of L-isoleucine.

ment of the amino group by a hydroxyl group is known to occur with retention of configuration about the 2-carbon atom, owing to anchiomeric participation of the neighboring carboxyl function[1] (Fig. 5).

L-Isoleucine (100 mg, 0.75 mmol is dissolved in 50 ml of cold (0°) 0.2 N perchloric acid. To this is added a cold solution (0°) of sodium nitrite (1.4 g, 20 mmol) in 20 ml of water with rapid stirring. Stirring is continued while the mixture is allowed to warm to room temperature and the evolution of gas subsides (approximately 0.5 hr). The resulting solution is raised to the boiling point for a few minutes, cooled to room temperature, and saturated with sodium chloride. Ether extracts are made and dried as described above, and the product is isolated similarly to yield 75 mg of (2S,3S)-HMPA as an oil. This product is also analyzed as the TMS derivative by GC/MS.

The three other stereoisomers, (2R,3R)-HMPA, (2R,3S)-HMPA, and (2S,3R)-HMPA, are synthesized in a similar manner from D-isoleucine, D-alloisoleucine, and L-alloisoleucine, respectively.

All four of the stereoisomers have mass spectra indistinguishable from each other.[1] Relative retention times are: (2S,3S)- and (2R,3R)-HMPA coelute after (2R,3S)- and (2S,3R)-HMPA, which also coelute.

Resolution of Four Stereoisomers in Racemic Mixture

The four stereoisomers may be resolved by introducing a third chiral center (Fig. 6). For this purpose, 20 mg (0.2 mmol) of (S)-(+)-2-methylbutanoic acid is dissolved in 10 ml of toluene, 75 μl of thionyl chloride (1 mmol) is added, and the resulting solution is heated at reflux for 2 hr. While still

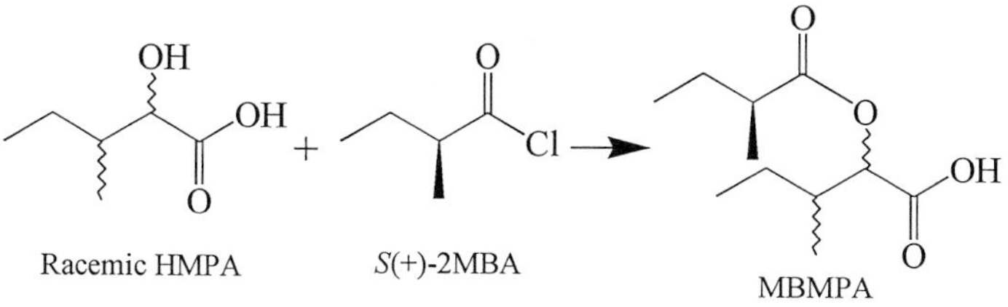

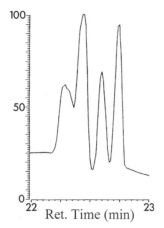

FIG. 6. Introduction of a third chiral center by esterification of racemic HMPA with (S)-(+)-2MBA.

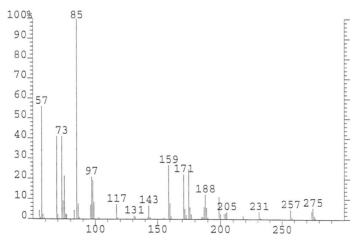

FIG. 7. Gas chromatogram obtained for the TMS derivative of the (S)-(+)-2-methylbutanoyl ester of racemic HMPA.

FIG. 8. The mass spectrum of the fourth peak in the gas chromatogram of the four stereoisomers of MBMPA.

warm, the toluene solution is reduced in volume under reduced pressure to approximately 3 ml and then added to 10 mg of racemic HMPA produced as described above and dissolved in 15 ml of toluene. A few drops of pyridine are added, and the resulting mixture is heated at reflux overnight.

The toluene solution is washed with 10 ml of 2 N hydrochloric acid and then extracted with two 10-ml volumes of 1 N sodium bicarbonate. The bicarbonate extracts are combined, acidified with 2 N hydrochloric acid to pH <2 (*caution:* carbon dioxide evolution), saturated with sodium chloride, extracted with diethyl ether, and isolated as described above. The product ester, 2-(2-methylbutanoyloxy)-3-methylpentanoic acid (MBMPA), is isolated as 11 mg of crude product. A small portion of this was analyzed as the TMS derivative by GC/MS (Fig. 7). Four peaks are detected, all with identical spectra,[3] one of which is reproduced in Fig. 8. Introduction of a third chiral center [(2S)-MB] esterified to the four isomers of HMPA will produce (2R)-[(2S)-MB]-(3R)-MPA, (2S)-[(2S)-MB]-(3S)-MPA, (2R)-[(2S)-MB]-(3S)-MPA, and (2S)-[(2S)-MB]-(3R)-MPA.

---

[3] M. L. J. Reimer and O. A. Mamer, *in* "Advances in Chemical Diagnosis and Treatment of Metabolic Disorders" (I. Matsumoto, O. A. Mamer, D. S. Millington, and T. Shinka, eds.), Vol. I, p. 19. John Wiley & Sons, New York, 1992.

[2] Analysis of Intracellular Metabolites as Tool for Studying Branched-Chain Amino Acid Biosynthesis and Its Inhibition in Bacteria

By Sabine Epelbaum, David M. Chipman, and Ze'ev Barak

Introduction

Despite the great interest in the pathway for biosynthesis of the branched-chain amino acids (BCAA pathway; Fig. 1), reinforced by the economic importance of herbicides that inhibit the pathway, the function of its components as they act together is still incompletely understood. The pathway includes a sequence of three enzymes with dual specificity, leading to 2-ketoisovalerate (the precursor of valine, leucine, and pantoic acid), on the one hand, and to 2-keto-3-methylvalerate (the precursor of isoleucine),

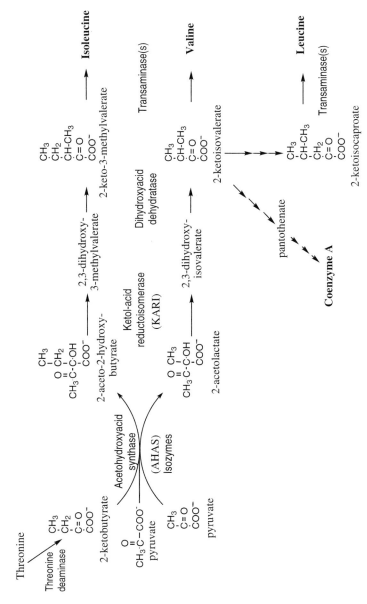

Fig. 1. Pathway for the biosynthesis of the branched-chain amino acids. Structures are shown for the intermediates whose determination is described here, and the enzymes of the main pathway to isoleucine and valine are named.

on the other.[1] The pathway has many regulatory elements: (1) allosteric inhibition of the first step in each branch of the pathway by its end product; e.g., threonine deaminase (threonine dehydratase) by isoleucine, acetohydroxy acid synthase (AHAS) by valine, and isopropylmalate synthase by leucine; and (2) regulation of gene expression. However, in any metabolic pathway the product of one enzyme is the substrate of another, so that all the enzymes are kinetically coupled to and affect one another. Control of a metabolic pathway is a property of the system as a whole, not of its isolated elements.[2]

Thus, the effect of an enzyme inhibitor on a target organism is critically dependent on the role the enzymatic reaction plays in the dynamics of the metabolic pathway involved,[2] so that an understanding of the control of the pathway is as important to the design of biologically active compounds as is the understanding of the mechanism of individual enzymes. The effect on an organism of inhibition of an enzyme in a pathway may result either from starvation for its end product(s) or from toxic effects of an abnormal accumulation of an intermediate proximal to the inhibited enzyme.

Several groups of herbicides and candidate lead compounds are inhibitors of enzymes of the BCAA pathway. The sulfonylurea and imidazolinone herbicides are specific inhibitors of AHAS[3–11] and at least two mechanism-based inhibitors of ketol-acid reductoisomerase (KARI) have been identified: N-isopropyloxalylhydroxamate[7,12] (IpOHA) and HOE-704.[13] In addi-

[1] H. E. Umbarger, in *"Escherichia coli* and *Salmonella typhimurium:* Cellular and Molecular Biology" (F. C. Neidhardt, J. L. Ingraham, B. L. Low, B. Magasanik, M. Schaechter, and H. E. Umbarger, eds.), Vol. 1, pp. 352–367. American Society for Microbiology, Washington, D.C., 1987.

[2] H. Kacser, in "Physiological Models in Microbiology" (M. J. Bazin and J. I. Prosser, eds.), Vol. 1, pp. 1–23. CRC Press, Boca Raton, Florida, 1988.

[3] T. R. Hawkes, J. L. Howard, and S. E. Pontin, in "Herbicides and Plant Metabolism" (A. D. Dodge, ed.), pp. 113–137. Cambridge University Press, Cambridge, 1989.

[4] V. A. Wittenbach, D. R. Rayner, and J. V. Schloss, in "Biosynthesis and Molecular Regulation of Amino Acids in Plants" (B. K. Singh, H. E. Flores, and J. C. Shannon, eds.), pp. 69–88. American Society of Plant Physiologists, Rockville, Maryland, 1992.

[5] B. K. Singh and D. L. Shaner, *Plant Cell* **7**, 935 (1995).

[6] V. A. Wittenbach, P. W. Teaney, W. S. Hanna, D. R. Rayner, and J. V. Schloss, *Plant Physiol.* **106**, 321 (1994).

[7] J. V. Schloss and A. Aulabaugh, *Z. Naturforsch.* **45c**, 544 (1990).

[8] A. Aulabaugh and J. V. Schloss, *Biochemistry* **29**, 2824 (1990).

[9] M. J. Muhitch, D. L. Shaner, and M. A. Stidham, *Plant Physiol.* **83**, 451 (1987).

[10] T. B. Ray, *Plant Physiol.* **75**, 827 (1984).

[11] W. K. Moberg and B. Cross, *Pest. Sci.* **29**, 241 (1990).

[12] V. A. Wittenbach, A. Aulabaugh, and J. V. Schloss, in "Pesticide Chemistry" (H. Frehse, ed.), pp. 151–160. VCH, Weinheim, Germany, 1991.

[13] A. Schulz, P. Sponemann, H. Kocher, and F. Wengenmayer, *FEBS Lett.* **238**, 375 (1988).

tion, an understanding of the mechanisms of several other enzymes in the pathway has made it possible to design inhibitors for them.

The study of the dynamics of the BCAA pathway, as well as an understanding of the physiological effects of inhibition of a given enzyme, require the simultaneous determination of as many intracellular parameters as possible under changing physiological conditions. This chapter describes methods for determination of intracellular concentrations of a number of keto acids and of the two relevant acetohydroxy acids of the pathway. These methods have been used successfully to understand the differential metabolic effects of two inhibitors[14] and the roles of two AHAS isozymes[15] in the enterobacterium *Salmonella typhimurium* and in other organisms.[15–17]

Determination of Intracellular Concentrations of 2-Keto Acids

Principle

There are fundamental problems involved in the analysis of cellular concentrations of intermediates in biosynthetic pathways. One difficulty is the specific and quantitative detection of compounds that occur at low concentrations in a complex mixture containing many other compounds at much higher levels. The method described here is based on conversion of the 2-keto acids to fluorescent derivatives of 1,2-diamino-4,5-methylenedioxybenzene dihydrochloride (DMB).[18]

2-Keto acid   **DMB**   **Fluorescent derivative**

This reagent is the best of several derivatives of *o*-phenylenediamine tested by Ohkura and co-workers,[19] as far as completeness of the conver-

[14] S. Epelbaum, D. M. Chipman, and Z. Barak, *J. Bacteriol.* **178**, 1187 (1996).

[15] S. Epelbaum, R. A. LaRossa, T. K. VanDyk, T. Elkayam, D. M. Chipman, and Z. Barak, *J. Bacteriol.* **180**, 4056 (1998).

[16] S. Epelbaum, The Dynamics of the Branched Chain Amino Acid Pathway in Enterobacteria. Ben-Gurion University, Beer-Sheva, Israel, 1995.

[17] D. Landstein, S. Epelbaum, S. Arad, Z. Barak, and D. M. Chipman, *Planta* **197**, 219 (1995).

[18] Z.-J. Wang, K. Zaitsu, and Y. Ohkura, *J. Chromatogr.* **430**, 223 (1988).

[19] M. Nakamura, S. Hara, M. Yamaguchi, Y. Takemori, and Y. Ohkura, *Chem. Pharm. Bull.* **35**, 687 (1987).

sion of keto acids to their derivatives and sensitivity are concerned. After reaction, the mixture is separated by reversed-phase high-performance liquid chromatography (HPLC) on a C_{18} column and the fluorescent 3-alkyl-6,7-dimethoxy-2(1H)-quinoxalinone derivatives detected with a dual-monochromator detector. The use of a dual-monochromator detector (excitation, 365 nm; emission, 445 nm) makes the method highly specific; only a few peaks are observed that cannot be identified, by use of authentic standards, as derivatives of 2-keto acids, and these are of low intensity. An internal standard of 2-ketovalerate is added to the samples; the endogenous level of this keto acid in bacteria is generally below the limits of detection by this method. The method, originally developed for analysis of keto acids in human serum and urine,[18] has been further adapted by us for the more complex cellular extracts by use of a gradient.[14]

We have found that phosphoenolpyruvate (PEP), a major cellular metabolite under some conditions, is labile and is converted to 3-methyl-6,7-dimethoxy-2(1H)-quinoxalinone (the DMB derivative of pyruvate) in this assay. Analysis of PEP standards after DMB derivatization gives a peak whose area is 0.7 times that of the equivalent molar amount of pyruvate. Therefore, pyruvate must be determined independently by the spectrophotometric lactate dehydrogenase (LDH) assay.[20]

Another problem in such an analysis is the efficient extraction of metabolites from cells in a manner that allows only a minimal change in the physiological quantities of the analyte. We adopted the method proposed by Payne and Ames[21] with some modifications. This method involves rapid vacuum filtration through a large-diameter polycarbonate filter, flash freezing, and extraction of the frozen cells with perchloric acid. Some other methods proved unsatisfactory; e.g., the use of centrifugal sedimentation of cells through a silicone fluid into an acidic layer[22–24] leads to low recoveries of most of the keto acids.

Reagents

1,2-Diamino-4,5-methylenedioxybenzene   dihydrochloride   (DMB) (available from Sigma-Aldrich Chemicals, St. Louis, MO)
Sodium salts of pyruvic acid, glyoxylic acid, 2-ketobutyric acid, 2-ketovaleric acid, 2-ketoisovaleric acid, 2-keto-4-methylvaleric acid (2-ketoisocaproic acid), DL-2-keto-3-methylvaleric acid, and phos-

---

[20] W. Lamprecht and F. Heinz, in "Methods of Enzymatic Analysis" (H. U. Bergmeyer, ed.), Vol. 6, 3rd Ed., pp. 570–577. Verlag Chemie, Weinheim, Germany, 1984.

[21] S. M. Payne and B. N. Ames, *Anal. Biochem.* **123**, 151 (1982).

[22] M. Klingenberg and E. Pfaff, *Methods Enzymol.* **10**, 680 (1967).

[23] F. Palmieri and M. Klingenberg, *Methods Enzymol.* **56**, 283 (1979).

[24] J. Eggeling, C. Cordes, E. Eggeling, and H. Sahm, *Appl. Microbiol. Biotechnol.* **25**, 346 (1987).

phoenolpyruvic acid, as standards for the intermediates of interest with regard to BCAA biosynthesis
HPLC-grade acetonitrile and methanol
Filtered, double-distilled water
Lactate dehydrogenase [crystalline rabbit muscle enzyme in $(NH_4)_2SO_4$ suspension, Sigma type II]
β-NADH

Apparatus

Filtration. A 142-mm-diameter vacuum filtration apparatus, from which the filter is rapidly demountable, is required. Appropriate apparatus are the Selectron pressure filter holder MD (Schleicher & Schuell, Dassel, Germany) or the 142 MM pressure holder (Sartorius, Göttingen, Germany). Other apparatus with appropriate filter supports can be adapted for this purpose. The apparatus is fitted with a 142-mm-diameter polycarbonate membrane filter with 0.2-μm pores [ME24 membrane filter (Schleicher & Schuell) or membrane filter 166130 (Poretics, Livermore, CA)].

High-Performance Liquid Chromatography. An appropriate high-performance liquid chromatography system includes a Waters workstation and 600E controller controlling a Waters M-45 pump (Waters-Millipore, Milford, MA) equipped with a Rheodyne 7125 syringe-loading sample injector valve (50-μl loop), as well as a Jasco 821-FP dual-monochromator fluorescence detector (Japan Spectroscopic, Tokyo, Japan). The separation is carried out on a 250 mm × 4.6 mm (i.d.) column of TSKgel ODS-80TM (particle size, 5 μm; Tosoh, Tokyo, Japan).

Procedure

Bacterial Extracts. For analysis of metabolites in steady state cultures of *S. typhimurium,* 250 ml of prewarmed medium in a 1-liter Erlenmeyer flask is inoculated (1% by volume) with an exponential-phase culture and incubated with shaking (200 rpm) in a gyratory water bath at 37°. Growth is monitored with a Klett–Summerson colorimeter equipped with a no. 66 filter or with a spectrophotometer at 660 mm. The entire culture of exponentially growing bacteria [~40 Klett units (KU), about 10^{11} cells] is filtered through the large-diameter polycarbonate membrane filter in <1 min and the filter immediately demounted and frozen in liquid nitrogen. The frozen filter is broken up with forceps in a 50-ml centrifuge tube and 4 ml of $HClO_4$ (0.3 M) containing 1 mM EDTA is added. As an internal standard, 25 μl of a 4 mM solution of 2-ketovalerate (100 nmol) is added to the extract. The tube is vortexed for 1 min and then centrifuged at

10,000g for 15 min. The supernatant is kept at $-70°$ until further analysis. To test whether 2-ketovalerate occurs at a significant level in the cells of interest, the extract can be prepared without the standard, and separate aliquots derivatized with and without addition of 2-ketovalerate.

Derivatization. A 5 mM solution of DMB is prepared in 0.4 M HClO$_4$, containing 1.0 M 2-mercaptoethanol and 28 mM sodium hydrosulfite as stabilizers for the DMB,[18] and filtered. The solution can be used for 2 days, and can be used for at least 1 month if frozen in portions and stored at $-20°$. The derivatization mixture is prepared in a screw-topped glass vial and should include an aliquot (0.05 to 0.2 ml) of the extract prepared as described above, 0.25 ml of the DMB solution, and sufficient 0.4 M HClO$_4$ to bring it to a total volume of 0.5 ml. It may also contain 1 nmol of 2-ketovalerate added separately as an internal standard when this compound is not added directly to cell extracts. The tubes are capped and incubated for 50 min at 100°. The samples are cooled on ice, diluted fivefold in the starting elution buffer (see below), and passed through a 0.2-μm Prep-Disc membrane filter (Bio-Rad, Hercules, CA) before injection into the HPLC instrument. A mixture of all the 2-keto acids of interest (0.2 to 2 nmol each) is derivatized in the same manner for determination of retention times and calibration of the response parameters relative to 2-ketovalerate.

Analysis. The standard separation uses as starting elution buffer 40 mM potassium phosphate buffer (pH 7.0)–doubly distilled water–acetonitrile (12:8:5, v/v/v) and as final elution buffer 40 mM phosphate–methanol–acetonitrile (12:8:5, v/v/v). The separation of the DMB derivatives is performed at ambient temperature at a flow rate of 0.8 ml/min, using a linear gradient from the starting elution buffer to the final buffer over 100 min, followed by elution with the final buffer for 20 min more. The eluate is monitored at an excitation wavelength of 365 nm and an emission wavelength of 445 nm. A typical chromatogram is shown in Fig. 2. Table I gives the HPLC retention times for several of the keto acids, using the standard gradient. For more accurate analysis of the keto acids that elute early, such as glyoxylate, a more gradual gradient can be used, e.g., over 200 min. The areas of the peaks of interest and of the 2-ketovalerate peak are determined by integration of the chromatogram. The relative area responses of the standards determined in the calibration chromatograms are used to determine the molar quantity of each keto acid in the sample of cells extracted relative to the amount of 2-ketovalerate added to the extract.

Pyruvate Assay Using Lactate Dehydrogenase. Bacterial extracts are prepared as described above and neutralized with K$_2$CO$_3$ to pH 7.0–7.2 (approximately 65 mg of K$_2$CO$_3$ for 3 ml of extract). The resulting precipitate is separated by centrifugation and 0.5–1.5 ml of the neutralized supernatant taken for reaction with LDH.[20] For the assay of such bacterial extracts,

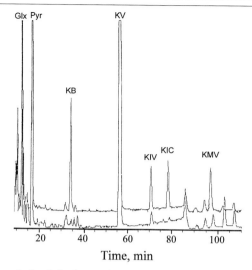

FIG. 2. HPLC analysis of the keto acid content of an extract of *S. typhimurium* LT2. Bacteria were grown in minimal medium with glucose and harvested rapidly by the method described here. The frozen cells were extracted with HClO₄ containing 2-ketovalerate (KV) as internal standard, and 1/40 of the extract derivatized with DMB. The mixture was diluted fivefold into the eluent and a 50-μl aliquot injected onto the 250 × 4.6 mm TSKgel ODS-80TM column and analyzed as described (lower trace). To another aliquot of the diluted mixture, 0.024 nmol (1.2 pmol per injection) of each of the identified keto acid derivatives was added and a 50-μl sample injected into the HPLC and reanalyzed (upper trace). The keto acids whose derivatives are identified on the trace are as follows: Glx, glyoxylate; Pyr, pyruvate; KB, 2-ketobutyrate; KIV, 2-ketoisovalerate; KIC, 2-ketoisocaproate; and KMV, 2-keto-3-methylvalerate. The ordinate represents an arbitrary relative detector voltage.

the reaction mixture contains, in a 2.5-ml total reaction volume, 0.1 M sodium phosphate buffer (pH 7.0) and 0.06 mM NADH. LDH (50 μl, 2.7 units) is added to start the reaction and the decrease in absorbance of NADH at 340 nm is monitored. A slow continued decrease may be observed in some cases, due, e.g., to the slow reduction of other keto acids by LDH. This decrease is extrapolated back to time of addition of the enzyme and the difference between the initial absorbance and the extrapolated final absorbance is used to calculate the molar quantity of pyruvate present in the assay from ε_{340}(NADH) = 6220 M^{-1} cm^{-1}. The molar quantity of pyruvate in the sample of cells extracted is obtained by dividing by the fraction of the extract volume taken for assay.

Calculation of Intracellular Concentrations. Intracellular metabolite concentrations are calculated by dividing the molar quantities detected in a bacterial extract by the cytoplasmic volume the extract represents. Measurements with a Coulter counter verified that 1 Klett unit with a no. 66 filter

TABLE I

CONCENTRATION OF KETO ACIDS IN *Salmonella typhimurium* LT2 DURING
STEADY STATE GROWTH[a]

Metabolite	Method[b]	Retention time (min)[c]	Concentration (μM) on:	
			Glucose	Acetate
Pyruvate	LDH	n.r.	1350	660
Pyruvate + PEP[d]	DMB–HPLC	17.5	1650	1200
Glyoxylate	DMB–HPLC	12.4	11	120
2-Ketobutyrate	DMB–HPLC	36.0	17	17
2-Ketoisovalerate	DMB–HPLC	72.6	70	10
3-Methyl-2-ketovalerate	DMB–HPLC	98.1	27	29
4-Methyl-2-ketovalerate	DMB–HPLC	80.5	7	7
2-Ketoglutarate	DMB–HPLC	5.32	670	160
2-Ketovalerate	DMB–HPLC	58.3	(Standard)	(Standard)

[a] Cells were grown to a density of about 40 KU, harvested rapidly, and extracted as described in this chapter. The concentrations measured in extracts from different cultures were reproducible within 15% for pyruvate, and within about 25% for the other metabolites. The bacterial strain used was a derivative of *S. typhimurium* LT2 that carries the episome F'*pro-lac zzf1836*::Tn*10* Cm, obtained from R. A. LaRossa (Central Research and Development Department, E.I. duPont de Nemours, Wilmington, DE). The episome encodes an isopropylthiogalactoside (IPTG)-inducible β-galactosidase that is irrelevant to the analysis. Cultures were grown in MOPS minimal medium [F. C. Neidhardt, P. L. Bloch, and D. F. Smith, *J. Bacteriol.* **119**, 736 (1974)] with glucose added to a final concentration of 0.4% (w/v) or acetate to 0.8% (w/v).

[b] The methods used were either (1) DMB–HPLC, HPLC analysis of the fluorescent 1,2-diamino-4,5-methylenedioxybenzene derivatives of the 2-keto acids on a 250 × 4.6 mm TSKgel ODS-80TM column or (2) LDH, lactate dehydrogenase-linked disappearance of NADH.

[c] The retention times are given for the standard linear gradient described. n.r., Not relevant.

[d] Phosphoenolypyruvate reacts under the derivatization conditions of the assay to yield the pyruvate–DMB derivative.

represents 10^7 *S. typhimurium* cells ml^{-1}. The cytoplasmic volume (excluding the periplasmic space) of 10^{11} cells is taken as 0.06 ml.[25,26] Table I gives typical intracellular concentrations determined in *S. typhimurium* cells grown in minimal medium with glucose or acetate as carbon sources.

[25] D. S. Goodsell, *Trends Biochem. Sci.* **16**, 203 (1991).

[26] F. C. Neidhardt, in "*Escherichia coli* and *Salmonella typhimurium*: Cellular and Molecular Biology" (F. C. Neidhardt, J. L. Ingraham, B. L. Low, B. Magasanik, M. Schaechter, and H. E. Umbarger, eds.), Vol. 1, pp. 3–6. American Society for Microbiology, Washington, D.C., 1987.

Determination of Intracellular Concentrations of Acetohydroxy Acids

Principle

We have previously described a method for determining the rates of production of different acetohydroxy acids by acetolactate synthase.[27] The method is based on the catalyzed air oxidation of these compounds to the respective diketones, distillation of the diketones into a methanol solution, and analysis of the diketones by gas–liquid chromatography (GLC) with an electron-capture detector. However, to prevent conversion of the aceto-hydroxy acids to acetoin at strongly acid pH, the measurement of intracellu-lar levels of these compounds requires a different method for the disruption and extraction of the cells from that described above. The acetohydroxy acids can be successfully extracted from rapidly filtered and frozen bacterial cells by using a citrate buffer at pH 4.0.

Reagents

2,3-Pentanedione and 2,3-butanedione (available from Aldrich, Mil-waukee, WI)
Citric acid
$FeCl_3$, $FeSO_4$
Methanol

Procedure

Cultures are grown as described above, and rapidly filtered and frozen in the same manner. The frozen cells are broken by the addition of 2 ml of sodium citrate buffer (1 M, pH 4.0). The tube is vortexed for 1 min and then centrifuged at 10,000g for 15 min at 4°. The reaction volume is brought to 10 ml with distilled water and $FeCl_3$ and $FeSO_4$ added to a final concentra-tion of 0.15 mM each. This mixture, in a stoppered glass tube, is incubated at 80° for 10 min to convert acetolactate and acetohydroxybutyrate to 2,3-butanedione and 2,3-pentanedione, respectively. The separation and identification of the diketones is performed by gas chromatography, as previously described in this series.[27] Commercially available 2,3-butanedi-one and 2,3-pentanedione serve as standards, but caution must be taken because the latter compound may slowly deteriorate at 4°. A number of controls described by Gollop *et al.*[28] demonstrate that with fresh standards, the calibration of the method is accurate. Addition of acetolactate to dupli-cate extracts leads to a recovery of 95%. Calculated intracellular acetolactate

[27] N. Gollop, Z. Barak, and D. M. Chipman, *Methods Enzymol.* **166,** 234 (1988).
[28] N. Gollop, B. Damri, Z. Barak, and D. M. Chipman, *Biochemistry* **28,** 6310 (1989).

and acetohydroxybutyrate concentrations in *S. typhimurium* grown in a minimal medium with glucose as carbon source are typically 300–400 and about 50 μM, respectively.

Discussion

The method described here for analysis of the intracellular concentrations of 2-keto acids may be used to detect a wide variety of these compounds in bacteria and other organisms, including all those that are intermediates in the biosynthesis of the BCAAs. The intracellular concentrations of the latter in the enterobacteria are one to two orders of magnitude lower than those of more "central" metabolites such as pyruvate (Table I). The sensitivity of the method allows detection of these compounds down to (calculated) intracellular concentrations of about 1 μM. Essentially all the uncertainty in the method (±15–20%) seems to be derived from the variation among extracts prepared under ostensibly similar physiological conditions. The GLC analysis of the intracellular levels of acetohydroxy acids, as described here, is more than an order of magnitude less sensitive. It can quantitatively detect acetolactate and acetohydroxybutyrate at levels above 10 and 20 μM, respectively.

The analysis of intracellular 2-keto acid concentrations can be used to clarify important aspects of the physiology of the BCAA pathway. The enterobacteria, which are capable of growing in minimal medium with a variety of carbon sources, encode three types of biosynthetic AHAS.[1,29–32] The expression of AHAS isozyme I is induced about four- to fivefold when the bacteria are grown in so-called poor carbon sources such as acetate[33,34] and mutants that cannot express AHAS I grow poorly, if at all, in acetate.[15,33,35] On the other hand, mutants that are capable of expressing AHAS I only grow poorly in glucose minimal medium.[15,36] The 2-keto acid levels

[29] Z. Barak, N. Kogan, N. Gollop, and D. M. Chipman, *in* "Biosynthesis of Branched Chain Amino Acids" (Z. Barak, D. M. Chipman, and J. V. Schloss, eds.), pp. 91–107. VCH, Weinheim, Germany, 1990.

[30] M. DeFelice, C. T. Lago, C. H. Squires, and J. M. Calvo, *Ann. Microbiol. (Paris)* **133A**, 251 (1982).

[31] M. DeFelice, G. Griffo, C. T. Lago, D. Limauro, and E. Ricca, *Methods Enzymol.* **166**, 241 (1988).

[32] H. E. Umbarger, *in* "Biosynthesis of Branched Chain Amino Acids" (Z. Barak, D. M. Chipman, and J. V. Schloss, eds.), pp. 1–24. VCH, Weinheim, Germany, 1990.

[33] F. E. Dailey, J. E. Cronan, Jr., and S. R. Maloy, *J. Bacteriol.* **169**, 917 (1987).

[34] M. Freundlich, R. O. Burns, and H. E. Umbarger, *Proc. Natl. Acad. Sci. U.S.A.* **48**, 1804 (1962).

[35] F. E. Dailey and J. E. Cronan, Jr., *J. Bacteriol.* **165**, 453 (1986).

[36] K. J. Shaw, C. M. Berg, and T. J. Sobol, *J. Bacteriol.* **141**, 1258 (1980).

in wild-type *S. typhimurium* LT2 grown in glucose and acetate minimal medium (Table II) reveal an important difference that we believe is central to the requirement for AHAS I when acetate is the sole carbon source.[15] During growth on acetate, the intracellular glyoxylate concentration is raised by at least an order of magnitude, and the glyoxylate/pyruvate ratio increases more than 20-fold (Table II; compare first two entries). The 2-ketobutyrate/pyruvate ratio differs by only a factor of two in these two media. This suggests that the important functional advantage of AHAS I

TABLE II

EFFECTS OF MEDIUM, ENZYME EXPRESSION, AND INHIBITORS ON INTRACELLULAR CONCENTRATION OF KETO ACIDS IN *Salmonella typhimurium*

Strain[a]	Enzyme expressed[a]	Medium[b]	Intracellular concentration (μM)[c]				Ratio		
			Pyruvate	2KB	Glx	KIV	2KB/Pyr	KIV/2KB	Glx/Pyr
LT2	AHAS I and II	GLC	1350	17	11	70	0.013	3.9	0.008
	AHAS I and II	Acetate	660	17	120	10	0.026	0.6	0.18
TV496	AHAS I	GLC + pan	900	60		17	0.07	0.27	
TV497	AHAS I, mut TD (*ilvA219*)	GLC + pan	990	140		23	0.14	0.17	
TV105	AHAS II	GLC	930	9		30	0.01	2.2	
	AHAS II	GLC + 0.2 μM SMM[d]	2200	71		nd	0.03	≤0.02	
	AHAS II	GLC + 10 μM SMM[e]	600	12		nd	0.02	≤0.08	
	AHAS II	GLC + 1 μM IpOHA[e]	800	74		nd	0.02	≤0.08	

[a] The bacterial strains used here are isogenic derivatives of the *S. typhimurium* LT2 strain described in Table I, and were also obtained from R. A. LaRossa (Central Research and Development Department, E.I. duPont de Nemours, Wilmington, DE). LT2 expresses AHAS isozymes I and II. The other strains express only a single AHAS isozyme and some contain the *ilvA219* mutation leading to a lack of feedback inhibition in threonine deaminase. (See Ref. 15.)

[b] Cells growing at steady state, unless otherwise indicated, in MOPS minimal medium [F. C. Neidhardt, P. L. Bloch, and D. F. Smith, *J. Bacteriol.* **119**, 736 (1974)] with 0.4% (w/v) glucose (GLC) or 0.8% (w/v) acetate. Where indicated, 0.34 m*M* pantothenate (pan) was added. Sulfometuron methyl (a gift from E.I. duPont de Nemours, Wilmington, DE) was added as a freshly prepared solution in dimethyl sulfoxide. IpOHA was prepared as described in A. Aulabaugh and J. V. Schloss, *Biochemistry* **29**, 2824 (1990).

[c] nd, Not detected; ≤1 μM. Where entries are absent for glyoxylate these data were not collected. The keto acids whose concentrations are listed are as follows: Glx, glyoxylate; Pyr, pyruvate; 2KB, 2-ketobutyrate; KIV, 2-ketoisovalerate.

[d] At this low concentration of inhibitor, the cells continue to grow exponentially, but with a doubling time of about 90 min (compared with 60 min in the absence of SMM). Analyses were made after 10 generations under these partially inhibited conditions.

[e] The inhibitor levels are above the minimal inhibitory concentrations in these cases and there is no steady state growth. The analyses were made 20 min after addition of inhibitor.

relative to the other isozymes, during growth on acetate as carbon source, is the relative resistance of AHAS I to inhibition by glyoxylate.[15]

On the other hand, the low specificity of AHAS I for reaction with 2-ketobutyrate has long been known to make it less than optimal for prototrophic growth on glucose as carbon source.[37] Analysis of the 2-keto acid levels in strain TV496 (which expresses only AHAS I) shows how the metabolic network (Fig. 1) adjusts to the absence of an AHAS with high specificity for reaction with 2-ketobutyrate (Table II). Isoleucine limitation causes activation of threonine deaminase, which leads to an increase in steady state levels of 2-ketobutyrate and, because of the competition between 2-ketobutyrate and pyruvate as substrates for AHAS,[14] to a decrease in 2-ketoisovalerate. The low ratio of 2-ketoisovalerate to 2-ketobutyrate appears to be responsible for the slow growth of this strain in glucose, which can be relieved by pantothenate as well as isoleucine.[15] TV497, in which threonine deaminase is not inhibited by isoleucine, shows more extreme behavior and does not grow at all without appropriate additions to the medium.[15]

The analysis of intracellular metabolites can also help to understand the effects of inhibitors of two of the enzymes of the BCAA pathway.[14] The sulfonylurea herbicide sulfometuron methyl (SMM) is a tight, slow-binding inhibitor of AHAS. It has been proposed that the toxicity of SMM is due to the accumulation of the AHAS substrate, 2-ketobutyrate,[38–40] although experiments in plant[41] and algal[17] cultures failed to show a clear correlation between inhibition and 2-ketobutyrate levels. The minimal inhibitory concentration (MIC) for SMM is about 2 μM in *S. typhimurium* TV105, which expresses only AHAS II, the most sensitive of the enterobacterial AHAS isozymes. At a level 10-fold below the MIC, the inhibition of AHAS leads to accumulation of its substrate, 2-ketobutyrate, to a moderate level. The level of the valine precursor 2-ketoisovalerate drops more than 30-fold, however (Table II), and it can be shown[14] that this is a result of the combined effect of direct inhibition of AHAS by SMM and the diversion of AHAS away from the valine pathway by the excess of 2-ketobutyrate. At higher levels of SMM, at which bacterial replication ceases, the cells do not reach a physiological steady state and the 2-ketobutyrate levels at the arbitrary time of 20 min after SMM addition are actually lower than those observed in the previous case (Table II).

[37] Z. Barak, D. M. Chipman, and N. Gollop, *J. Bacteriol.* **169**, 3750 (1987).
[38] R. A. LaRossa, T. K. VanDyk, and D. R. Smulski, *J. Bacteriol.* **169**, 1372 (1987).
[39] P. A. Singer, M. Levinthal, and L. S. Williams, *J. Mol. Biol.* **175**, 39 (1984).
[40] T. K. VanDyk and R. A. LaRossa, *J. Bacteriol.* **165**, 386 (1986).
[41] D. L. Shaner and B. K. Singh, *Plant Physiol.* **103**, 1221 (1993).

IpOHA is a potent inhibitor of KARI.[42,43] At the MIC in *S. typhimurium* TV105 (1 μM), the cellular levels of the keto acid precursors of the branched-chain amino acids drop below detectable levels (Table II and data not shown[14]). The intracellular concentrations of both substrates of KARI, acetolactate and acetohydroxybutyrate. increase by about 10-fold within 20 min, from 0.39 to 4.3 mM and from 0.05 to 0.43 mM, respectively.[14] In this case, too, the feedback controls in the metabolic network lead to an increase in the levels of 2-ketobutyrate (Table II), which may also play a role in the toxicity of IpOHA.[14]

The observations described above, some of which were quite unexpected, have led to reassessment of several assumptions concerning the branched-chain pathway. The analyses of intracellular metabolites described here have proved themselves to be valuable tools for opening new insights into metabolic function.

Acknowledgment

This work was supported in part by Grant 338/92 from the Israel Science Foundation.

[42] A. Aulabaugh and J. V. Schloss, *J. Cell Biol.* **107**, 402A (1988).
[43] J. V. Schloss and A. Aulabaugh, *in* "Biosynthesis of Branched Chain Amino Acids" (Z. Barak, D. M. Chipman, and J. V. Schloss, eds.), pp. 329–356. VCH, Weinheim, Germany, 1990.

[3] Determination of Branched-Chain L-Amino-Acid Aminotransferase Activity

By PETER SCHADEWALDT

Introduction

The first step in the catabolism of branched-chain L-amino acids is reversible transamination, which is catalyzed by branched-chain L-amino-acid aminotransferase(s) (BCAA-AT; EC 2.6.1.42). Activity and (subcellular) distribution of the enzyme has been most extensively studied in the rat. Reportedly, (iso)enzyme I (mBCAA-AT) prevails in the mitochondria of most organs and tissues. (Iso)enzyme II (leucine–methionine specific) is found in rodent liver only, and the cytosolic (iso)enzyme (cBCAA-AT) is expressed mainly in brain, placenta, and ovary. Information on the enzymes from other mammals is less abundant.[1-4] The early findings of Goto et al.[5] on (iso)enzyme distribution in human tissues have been extended. Suryawan et al.[6] performed enzyme activity and mRNA analyses and revealed considerable differences in organ distribution and established the clear predominance of mBCAA-AT in all human tissues studied with the exception of brain.

Aminotransferases play an important role in the partitioning of amino acid carbon skeletons into anabolic and catabolic processes. Unlike the rather specific L-alanine (EC 2.6.1.2) and L-aspartate (EC 2.6.1.1) aminotransferases, BCAA-AT exhibits a rather broad substrate specificity. Under physiological conditions, four branched-chain L-amino acids, including L-alloisoleucine, L-glutamate, as well as the derived 2-oxo acids, may be used as substrates. Thus, competition among the various compounds may be of physiological significance for the regulation of the overall metabolism of branched-chain compounds.[1,2,4,6]

For measurement of BCAA-AT activity, radiochemical, colorimetric, and fluorimetric end-point methods have been described.[7-10] All these methods are reliable, but somewhat laborious, and none allows monitoring of the course of the transamination reaction. The coupled enzymatic assay procedures described below were developed in order to allow continuous and sensitive spectrophotometric determination of BCAA-AT activity.[11,12]

Assay Using Branched-Chain L-Amino Acid Substrate

Principle of Method

This assay is based on the transamination of L-leucine in the presence of 2-oxoglutarate by branched-chain L-amino-acid aminotransferase (BCAA-

[1] A. Ichihara, in "Transaminases" (P. Christen and E. E. Metzler, eds.), pp. 433–439. John Wiley & Sons, New York, 1985.

[2] A. E. Harper, R. H. Miller, and K. P. Block, Annu. Rev. Nutr. **3,** 409 (1984).

[3] S. M. Hutson and T. R. Hall, J. Biol. Chem. **268,** 3084 (1993).

[4] T. R. Hall, R. Wallin, G. D. Reinhard, and S. M. Hutson, J. Biol. Chem. **268,** 3092 (1993).

[5] M. Goto, H. Shinno, and A. Ichihara, Gann **68,** 663 (1977).

[6] A. Suryawan, J. W. Hayes, R. A. Harris, Y. Shimomura, A. E. Jenkins, and S. Hutson, Am. J. Clin. Nutr. **68,** 72 (1998).

[7] A. Ichihara and E. Koyama, J. Biochem. **59,** 160 (1966).

[8] C. S. Hintz, W. R. Turk, N. Chambon, H. B. Bruch, P. M. Nemeth, and O. H. Lowry, Anal. Biochem. **146,** 418 (1970).

[9] A. Akabaysahi and T. Kato, Anal. Biochem. **182,** 129 (1989).

[10] R. T. Taylor and W. T. Jenkins, J. Biol. Chem. **241,** 4391 (1966).

[11] P. Schadewaldt, W. Hummel, U. Wendel, and F. Adelmeyer, Anal. Biochem. **230,** 199 (1995).

[12] P. Schadewaldt and F. Adelmeyer, Anal. Biochem. **238,** 65 (1996).

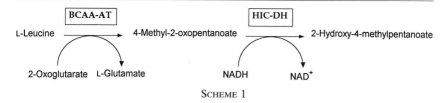

SCHEME 1

AT), yielding 4-methyl-2-oxopentanoate. The rate of formation of the 2-oxo acid is specifically monitored in a coupled enzymatic reaction. NAD^+-dependent D-2-hydroxyisocaproate dehydrogenase (HIC-DH) is used as a coupling enzyme and the rate of decrease in NADH absorbance is continuously measured spectrophotometrically at 334 nm (Scheme 1).

Materials and Methods

Reagents and Solutions

Tris buffer: Prepare Tris-HCl buffer, pH 8.4, containing 0.2 M Tris; adjust the pH with HCl solution (2 M); at $-20°$ this buffer is stable for months

Leu solution: 85 mM L-leucine

2-OG solution: 250 mM 2-oxoglutarate, sodium salt (Merck, Rahway, NJ); freshly prepared

NADH solution: 4 mM, prepare daily

PLP solution: 2 mM pyridoxal phosphate (Boehringer Mannheim, Indianapolis, IN); store protected from light in suitable portions at $-20°$ (stable for months)

KIC solution: 42 mM 4-methyl-2-oxopentanoate, sodium salt (Sigma, St. Louis, MO)

HIC-DH working solution: D-2-Hydroxyisocaproate dehydrogenase is not commercially available. A suitable ~40-fold purified preparation in 50% glycerol (w/w) can be obtained from disrupted cells of *Lactobacillus casei* ssp. *pseudoplantarum* by a two-step liquid–liquid extraction using polyethylene glycol 1540 and polyethylene glycol 10000 followed by DEAE-cellulose chromatography (DE52; Whatman, Clifton, NJ) described by Hummel *et al.*[13] Adjust the enzyme activity to give a final activity concentration of 10 mM min^{-1} (see below); store in suitable portions at $-20°$ (stable for months)

Enzyme dissolution buffer: Potassium phosphate buffer, 50 mM, pH 7.4

[13] W. Hummel, H. Schütte, and M.-R. Kula, *J. Appl. Microbiol. Biotechnol.* **21,** 7 (1985).

Check of Coupling Enzyme Activity. The activity of HIC-DH decreases considerably at pH >8.[11] Therefore, the enzyme working solution should be checked as follows. Dilute the working solution 1:100 with enzyme dissolution buffer. Enzyme activity is assessed in a 10-mm cuvette at 25° by measuring spectrophotometrically the NADH consumption at 334 nm against time. The assays comprise (1-ml final volume) the following: 0.5 ml of Tris buffer, 0.05 ml each of NADH, PLP, and KIC solution, 0.1 ml of Leu solution, and 0.25 ml of H_2O. The reaction is started by addition of 0.05 ml of diluted HIC-DH. The test should yield a linear decrease of absorbance of 0.031 ΔA min^{-1} ($\pm\sim$10%).

Analysis of Branched-Chain L-Amino-Acid Aminotransferase Activity. Enzyme activity is assessed at 25° in a 10-mm cuvette and a reaction volume of 1 ml by measuring spectrophotometrically the decrease in NADH absorption at 334 nm against time. The cuvette contains 0.5 ml of Tris buffer, 0.05 ml each of NADH, PLP, and HIC-DH working solution, and up to 0.25 ml of a (crude) BCAA-AT preparation. When necessary, an appropriate volume of H_2O is added to give 0.90 ml. After recording the baseline, the reaction is started by addition of 0.1 ml of Leu solution.

Coupling enzyme activity in the assay should be in at least 25-fold excess. Therefore, either the sample volume should be reduced or, alternatively, the BCAA-AT preparation diluted appropriately when the linear change in absorbance exceeds 0.12 ΔA min^{-1}. For assessment of reagent blanks, BCAA-AT in the assay is replaced by an appropriate volume of H_2O. After correction of the measured absorbance change per minute for baseline and reagent blank, BCAA-AT activity concentration of the preparation is calculated by the formula

$$[\text{BCAA-AT}] \ (\text{in } \mu M \ \text{min}^{-1}) = \left(\frac{\Delta A \ \text{min}^{-1}}{6.22}\right)\left(\frac{1}{V_S}\right) \times 10^3$$

with V_S the sample volume in the cuvette in milliliters.

Performance of Method

Precision and Accuracy. In this assay, the rates of NADH consumption (equivalent to 4-methyl-2-oxopentanoate production from L-leucine) and of L-glutamate formation (from 2-oxoglutarate) are equivalent, indicating that stoichiometry is fully met under these conditions. Furthermore, practically identical results were obtained when enzymatic measurements were compared with data from an independent radiochemical assay.[11]

Absorbance changes of <0.001 ΔA min^{-1} are reliably measurable. Thus, the sensitivity of the method for detection of BCAA-AT activity is ≤0.1 μM min^{-1}. The assay is linear up to a BCAA-AT activity concentration of about 20 μM min^{-1}. When determined with rat heart BCAA-AT activities of 1, 7.5, and 14.5 μM min^{-1} in the assay, the coefficient of variation within the run amounts to 7.9, 2.1, and 2.2%, respectively ($n = 7$).[11]

The present test conditions have been optimized for analysis of rat heart mitochondrial BCAA-AT and allow measurement of the reaction rate at $0.9V_{max}$ for this enzyme. It may be nescessary to introduce some modifications in order to optimize the assay for (iso)enzymes from other species.

Temperature Conversion Factors. When related to the rates obtained with partially purified BCAA-AT preparations from rat heart at 25°, estimated temperature conversion factors for 20, 30, and 37° amounted to 0.7, 1.4, and 2.0, respectively ($n = 3$).[11]

Comments. Because of the kinetic properties of HIC-DH from *L. casei* ssp. *pseudoplantarum*,[13] the present assay is largely limited to L-leucine as a substrate. However, studies including the other branched-chain L-amino acids may be performed with HIC-DH activities from other sources that exhibit the desired substrate properties, e.g., the enzyme from *Lactobacillus curvatus*.[14] Based on our experience, no valid measurements in tissue and cell homogenates can be performed in the presence of ammonia. Presumably, these interferences are due to the presence of L-glutamate dehydrogenase (EC 1.4.1.3), which also exhibits some L-leucine dehydrogenase activity.[15] Therefore, it is essential to remove NH_4^+ from the samples when necessary, e.g., by dialysis.

Assays Using Branched-Chain 2-Oxo Acid Substrates

Principle of Methods

These methods are based on the branched-chain L-amino-acid aminotransferase (BCAA-AT)-dependent transamination of 4-methyl-2-oxopentanoate in the presence of L-glutamate. The formation of 2-oxoglutarate is coupled either to L-aspartate aminotransferase (ASAT) plus L-malate dehydrogenase (MDH) (BAM assay procedure) or, alternatively, to L-alanine aminotransferase (ALAT) plus L-lactate dehydrogenase (LDH) (BAL assay procedure) as indicator systems. The transamination rate is continuously monitored by spectrophotometric measurement of the decrease in NADH absorbance at 334 nm over time (Scheme 2).

[14] W. Hummel, H. Schütte, and M.-R. Kula, *J. Appl. Microbiol. Biotechnol.* **28,** 433 (1988).
[15] E. L. Smith, B. M. Austen, K. M. Blumenthal, and J. F. Nyc, *in* "The Enzymes" (P. D. Boyer, ed.), Vol. 11, 3rd Ed., pp. 293–367. Academic Press, New York, 1975.

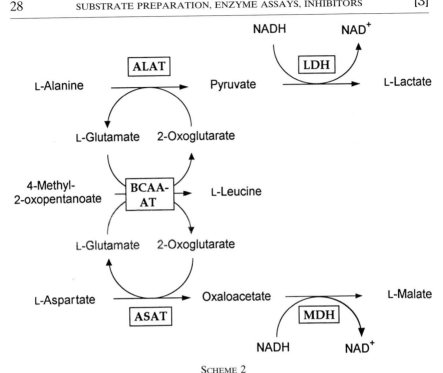

SCHEME 2

Materials and Methods

Reagents and Solutions

The following are required for the BAM assay procedure.

BAM buffer: Prepare Tris-Glu-Asp-NaOH buffer, pH 8.3, containing 0.2 M Tris, 0.6 M L-glutamic acid, 0.4 M L-aspartic acid; adjust pH with NaOH solution (2 M); stable for months at $-20°$

KIC solution: 42 mM 4-methyl-2-oxopentanoate, sodium salt (Sigma); store in suitable portions at $-20°$; stable for months

NADH solution: 4 mM, prepare daily

PLP solution: 2 mM pyridoxal phosphate (Boehringer Mannheim); store protected from light in suitable portions at $-20°$; stable for months

Enzyme dissolution buffer: potassium phosphate buffer, 100 mM, pH 7.4

ASAT working solution: Dissolve lyophilized L-aspartate aminotransferase (EC 2.6.1.1, from porcine heart; Sigma) in enzyme dissolution buffer to give a final activity concentration of 10 mM min⁻¹; store in suitable portions at −20°; stable for weeks

MDH working solution: Dilute L-malate dehydrogenase preparation in 50% glycerol (EC 1.1.1.37, from pig heart; Boehringer Mannheim) with enzyme dissolution buffer to give a final activity concentration of 20 mM min⁻¹; prepare daily

2-OG solution: 250 mM 2-oxoglutarate, sodium salt (Merck), freshly prepared

OAA solution: 20 mM oxaloacetate, sodium salt (Merck), freshly prepared

The following are required for the BAL assay procedure.

BAL buffer: Prepare Tris-Glu-Ala-NaOH buffer, pH 8.3, containing 0.2 M Tris, 0.6 M L-glutamic acid, and 0.4 M L-alanine, and adjust the pH with NaOH solution (2 M); stable for months at −20°

ALAT working solution: Dissolve lyophilized L-alanine aminotransferase (EC 2.6.1.2, from porcine heart; Sigma) in enzyme dissolution buffer to give a final activity concentration of 10 mM min⁻¹; store in suitable portions at −20° (stable for weeks)

LDH working solution: Dilute L-lactate dehydrogenase preparation in 50% glycerol (EC 1.1.1.27, from pig heart; Boehringer Mannheim) with enzyme dissolution buffer to give a final activity concentration of 20 mM min⁻¹; prepare daily

Pyruvate solution: 40 mM pyruvate, sodium salt (Merck), freshly prepared

Solutions of KIC, NADH, PLP, 2-OG, and enzyme dilution buffer: Prepare solutions as described above for the BAM assay procedure

Check of Coupling Enzyme Activities. The coupling enzyme working solutions may be checked as follows. Dilute the working solutions 1:100 with enzyme dissolution buffer. Enzyme activities are then assessed in a 10-mm cuvette at 25° by measuring spectrophotometrically the NADH consumption at 334 nm against time. The assays comprise (1-ml final volume) the following: for ASAT (specifications for ALAT in parentheses): 0.5 ml of BAM buffer (BAL buffer), 0.05 ml each of KIC, NADH, PLP, 2-OG, and MDH (LDH) working solution, and 0.2 ml of H₂O. The reaction is started by addition of 0.05 ml of ASAT (ALAT); for MDH (specifications for LDH in parentheses): 0.5 ml of BAM buffer (BAL buffer), 0.05 ml each of KIC, NADH, PLP, and OAA (pyruvate) solution, and 0.25 ml of H₂O. The reaction is started by addition of 0.05 ml of diluted MDD (LDH) solution. Under these conditions, the diluted ASAT (ALAT) preparation should yield a linear decrease in absorbance of 0.031 ΔA min⁻¹, and the diluted MDH (LDH) preparation should yield an absorbance change of 0.062 ΔA min⁻¹ (±∼10%).

Branched-Chain L-Amino-Acid Aminotransferase Measurement Procedure. Enzyme activity is assessed at 25° in a 10-mm cuvette and a reaction volume of 1 ml by measuring spectrophotometrically the linear decrease in NADH absorption at 334 nm against time. For the BAM assay procedure (specifications for the BAL assay in parentheses), the cuvette contains 0.5 ml of BAM buffer (BAL buffer), 0.05 ml each of NADH, PLP, ASAT (ALAT), and MDH (LDH) working solution, and up to 0.25 ml of a (crude) BCAA-AT preparation. When necessary, an appropriate volume of H_2O is added to give 0.95 ml. After recording the baseline, the reaction is started by addition of 0.05 ml of KIC solution.

When the linear decrease of absorbance exceeds 0.15 ΔA min^{-1}, it is advisable to reduce the sample volume or to dilute the BCAA-AT preparation appropriately. For the preparation of additional reagent blanks, the BCAA-AT in the assay is replaced by an appropriate volume of H_2O. The latter determination is essential in the case of the BAL assay procedure, because 4-methyl-2-oxopentanoate is also a (although poor) substrate for L-lactate dehydrogenase.

After correction of the measured linear change of absorbance per minute for baseline and reagent blank, the BCAA-AT activity concentration of the preparation is calculated by the formula given above.

Performance of Methods

Precision and Accuracy. When stoichiometry was checked in the BAM assay, the rates of NADH consumption and L-leucine formation were found to be equivalent. When comparative BCAA-AT activity measurements were performed with the BAM and BAL assays, practically identical results were obtained (linear regression analysis: $y = 0.41 + 0.93x$, $r = 0.998$, $n = 10$). In addition, a good linear correlation was found between the BAM assay and a procedure using L-leucine as substrate and D-2-hydroxyiso-caproate dehydrogenase as a coupling enzyme (see above).[12] As already noted, the sensitivity of the spectrophotometric assay methods for detection of BCAA-AT activity is better than 0.1 μM min^{-1}. The assays are linear up to a BCAA-AT activity of about 25 μM min^{-1}. When determined with rat heart BCAA-AT activities of 4.5, 9.7, and 23.0 μM min^{-1} in the BAM assays, the coefficient of variation (within run) amounted to 2.3, 1.0, and 1.4%, respectively ($n = 5$).[12]

Temperature Conversion Factors. When related to the rates obtained with partially purified BCAA-AT preparations from rat heart in the BAM assay at 25°, estimated temperature conversion factors for 30 and 37° were 1.4 and 2.3, respectively ($n = 6$).[12]

Comments. The present assays are not restricted to 4-methyl-2-oxopentanoate as substrate. Other branched-chain 2-oxo acids as well as straight-chain 2-oxo acids can be used as well. Thus, the assays are of general usefulness when studies of the substrate properties of BCAA-AT (iso)enzymes are to be conducted.[12]

The present assay conditions have been optimized for analysis of rat heart mitochondrial BCAA-AT. According to the available kinetic data, the reaction rate should amount to 0.9 V_{max} for this enzyme. Some modifications may be needed when the test conditions are to be optimized for analysis of (iso)enzymes from other sources.

As with the other NADH-dependent assay procedure using L-leucine as substrate (described above), NH_4^+ may cause interferences in the BAM and BAL assay procedures and must be omitted from the assay.

Sample Preparation

Procedures for the extraction and purification of branched-chain L-amino-acid aminotransferase activity have been described (see Refs. 16 and 17). When the procedure described here is to be used in crude tissue extracts, laboratory animals should be heparinized and the organs thoroughly rinsed, e.g., by perfusion with ice-cold saline (0.154 M NaCl), prior to extraction in order to remove interfering hemoglobin.

With rat hearts, the following extraction procedure is applied: After rinsing by retrograde perfusion, the blood-free organ is frozen with liquid N_2 and pulverized in a percussion mortar, and the powder is extracted by treatment with a 10-fold volume (w/v) of potassium phosphate buffer (200 mM, pH 6.0, 5 mM EDTA) and centrifugation. For extraction of bovine aorta endothelial cells, microcarrier cultures at confluency are washed three times with a twofold volume of ice-cold phosphate-buffered saline, and cell-free extracts are then prepared by addition of an equal volume of extraction buffer [50 mM Tris-HCl (pH 7.4), 2mM EDTA, 0.2% (v/v) Triton X-100] and centrifugation.[11,12] The latter extraction procedure has also been applied to cultured human skin fibroblasts.

Tissue Activities and Kinetic Properties

Branched-chain L-amino-acid aminotransferase activities have been reported for a number of tissues from various mammalian species[1–4] and for

[16] T. K. Korpela, *Methods Enzymol.* **166,** 269 (1988).
[17] R. Kido, *Methods Enzymol.* **166,** 275 (1988).

cultured cells.[12,18] One study compared enzyme activities in rat, human, and monkey tissue and found considerable tissue and species differences.[6] Because numerous, quite different assay conditions have been applied it is hardly possible, however, to compare the enzyme activity data obtained in different laboratories (see Refs. 12 and 18 for discussion and references).

The human mitochondrial (iso)enzyme, quite comparable to enzymes from other mammalian sources, exhibits a rather broad substrate specificity.[1,18] With L-leucine (1 mM), the activity against amino group acceptors (forward reaction) was in the order 2-oxoglutarate $\geq$ branched-chain > straight-chain 2-oxo acids (C_3–C_8). With 4-methyl-2-oxopentanoate (1 mM) the activity against amino group donors (reverse reaction) was in the order L-glutamate $\geq$ branched-chain > straight chain (C_3–C_6) > other L-amino acids.[18]

With respect to the kinetic properties of the enzyme, systematic studies using adequate methodology[19] appear not to have been published. Our studies of the Michaelis constants of the rat heart mitochondrial (iso)enzyme yielded the following results ($K_{m,app.}$, means $\pm$ asymptotic standard errors): with L-leucine and 2-oxoglutarate as substrates, 0.30 ± 0.02 and 0.68 ± 0.03 mM, respectively; with branched-chain 2-oxo acid substrates and L-glutamate ($K_{m,app.}$ for the amino group donor in parentheses): 4-methyl-2-oxopentanoate (KIC), 0.14 ± 0.01 (6.65 ± 0.34) mM; 3-methyl-2-oxobutanoate (KIV), 0.11 ± 0.01 (3.60 ± 0.28) mM; (R,S)-3-methyl-2-oxopentanoate [(R,S)-KMV], 0.07 ± 0.01 (2.45 ± 0.19); $K_{i,app.}$ values amounted to 2.10 ± 0.18, 4.19 ± 0.74, and 1.57 ± 0.21 mM, respectively.[11,12] Using cell-free extracts from cultured fibroblasts, the following $K_{m,app.}$ values were found for the human mitochondrial (iso)enzyme[20]: forward reaction with 2-oxoglutarate ($K_{m,app.}$ for the common amino group acceptor in parentheses): L-Leu, 0.62 ± 0.04 (3.8 ± 0.2) mM; L-Val, 2.96 ± 0.11 (4.0 ± 0.1) mM; L-Ile, 0.56 ± 0.03 (5.5 ± 0.2) mM; L-alloisoleucine, 1.54 ± 0.07 (2.4 ± 0.1) mM. For 2-oxoglutarate, the estimated $K_{i,app.}$ was >60 mM. For the reverse reaction with L-glutamate ($K_{m,app.}$ for the common amino group donor in parentheses) $K_{m,app.}$ estimates were as follows: KIC, 0.53 ± 0.03 (28.3 ± 1.6) mM; KIV, 0.61 ± 0.03 (10.4 ± 0.5) mM; (S)-KMV, 0.17 ± 0.02 (21.3 ± 2.3) mM; (R)-KMV, 0.06 ± 0.01 (4.5 ± 0.5) mM. Estimated $K_{i,app.}$ values with branched-chain 2-oxo acid substrates were $\geq$~5 mM. No inhibitory effects of L-amino acids were noted.

[18] P. Schadewaldt, U. Wendel, and H.-W. Hammen, *Amino Acids* **9,** 147 (1995).
[19] R. D. Allison and D. L. Purich, *Methods Enzymol.* **63,** 3 (1979).
[20] P. Schadewaldt and H.-W. Hammen, unpublished data, 1999.

[4] Analysis of (S)- and (R)-3-Methyl-2-oxopentanoate Enantiomorphs in Body Fluids

By PETER SCHADEWALDT

Introduction

The branched-chain amino acids L-leucine, L-valine, L-isoleucine and its diastereomer L-alloisoleucine, and their corresponding branched-chain 2-oxo acids, 4-methyl-2-oxopentanoate, 3-methyl-2-oxobutyrate, and (S)- and (R)-3-methyl-2-oxopentanoate are normal constituents of human plasma.[1] L-Alloisoleucine is derived from L-isoleucine *in vivo,* most probably via retransamination of (R)-3-methyl-2-oxopentanoate, which is an apparently inevitable by-product formed during normal L-isoleucine transamination.[2,3] Concentrations of L-alloisoleucine and its transamination product, (R)-3-methyl-2-oxopentanoate, are low in healthy subjects. Increased concentrations are regularly found in patients with maple syrup urine disease.[4,5] The diastereomeric amino acids L-isoleucine and L-alloisoleucine can easily be determined by standard amino acid analysis. Physiological studies on the L-isoleucine–L-alloisoleucine interrelationship, however, are complicated by the fact that the procedures worked out for branched-chain 2-oxo acid analysis in body fluids[6–9] do not allow a differential quantification of the related 2-oxo acid enantiomorphs.

The present method was developed for reliable and sensitive quantitation of (R)-3-methyl-2-oxopentanoate in body fluids in the presence of excess (S)-3-methyl-2-oxopentanoate. The procedure is also valuable when measurements of stable isotope enrichments in (S)- and (R)-3-methyl-2-

[1] P. Schadewaldt, U. Wendel, and H.-W. Hammen, *J. Chromatogr. B* **682,** 209 (1997).

[2] O. A. Mamer and M. L. J. Reimer, *J. Biol. Chem.* **267,** 22141 (1992).

[3] P. Schadewaldt, H.-W. Hammen, A. Bodner, and U. Wendel, *J. Inher. Metab. Dis.* **21** (Suppl. 1), 19 (1997).

[4] D. T. Chuang and V. E. Shih, *in* "The Metabolic and Molecular Bases of Inherited Disease" (C. R. Scriver, A. L. Beaudet, W. S. Sly, and D. Valle, eds.), pp. 1239–1277. McGraw-Hill, New York, 1995.

[5] P. Schadewaldt, H.-W. Hammen, A.-C. Ott, and U. Wendel, *J. Inher. Metab. Dis.* **22,** 706 (1999).

[6] G. Livesey and P. Lund, *Methods Enzymol.* **166,** 3 (1988).

[7] A. L. Gasking, W. T. E. Edwards, A. Hobson-Frohock, E. Elia, and G. Livesey, *Methods Enzymol.* **166,** 20 (1988).

[8] O. A. Mamer and J. A. Montgomery, *Methods Enzymol.* **166,** 27 (1988).

[9] P. L. Crowell, R. H. Miller, and A. E. Harper, *Methods Enzymol.* **166,** 39 (1988).

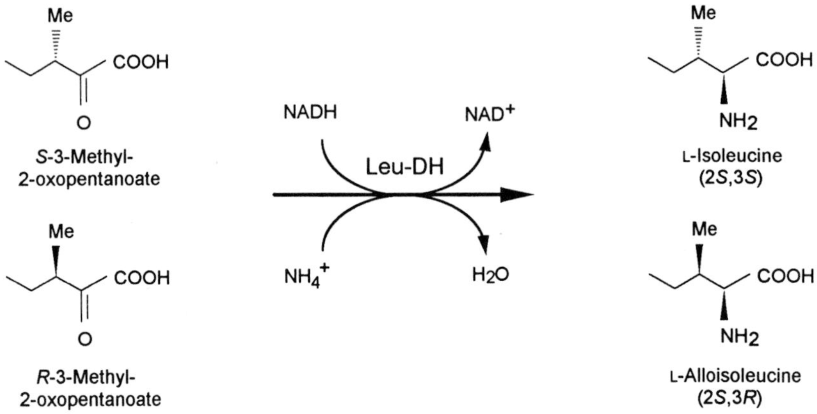

SCHEME 1

oxopentanoate by gas chromatography–mass spectrometry are to be performed.[1,3]

Principle of Method

(S)- and (R)-3-Methyl-2-oxopentanoate enantiomorphs are extracted from body fluids by acid, separated from the L-amino acid fraction by cation-exchange chromatography, and then specifically converted to the corresponding diastereomeric L-amino acids by L-leucine dehydrogenase-catalyzed reductive amination[6] (see Scheme 1). The products are purified by cation-exchange chromatography and finally quantitated by standard amino acid analysis. In principle, the procedure is suitable for the simultaneous and sensitive measurement of all four naturally occurring branched-chain 2-oxo acids.

Materials and Methods

Reagents and Solutions

5-Sulfosalicylic acid (Merck, Rahway, NJ) solutions: 3 and 60% (w/v)

Sample diluent: NaCl, 0.154 M

Internal standard solution: 2-Oxohexanoate, sodium salt (Sigma, St. Louis, MO), 0.5 mM; store in suitable portions at $-20°$; stable for months

(R,S)-3-Methyl-2-oxopentanoate solution: (R,S)-3-Methyl-2-oxopentanoate, sodium salt (Fluka, Ronkonkoma, NY), 50 mM; store in suitable portions at −20°, stable for months. Prior to use, the degree of racemization of the preparation should be checked, e.g., by optical rotation analysis.

Control samples: Spike reconstituted serum/plasma (e.g., Precinorm; Boehringer Mannheim, Indianapolis, IN) and/or normal spot urine with racemic (R,S)-3-methyl-2-oxopentanoate solution to give the following final concentrations: 0.0 mM (blank), 0.1 mM, and 0.5 mM, prepared daily

NH₄Cl–NH₄OH buffer: 5 M, pH 8.35; adjust pH in 5 M NH₄Cl by adding about 30 ml of NH₄OH (25%, w/w) per liter

NH₄OH solution: 2.5 M

NADH solution: 0.1 M; prepare daily

Enzyme dissolution buffer: Potassium phosphate buffer, 50 mM, pH 7.2; mix with glycerol (Merck) to give 50% (w/w)

L-Leucine dehydrogenase working solution: Dissolve lyophilizate (Leu-DH, EC 1.4.1.9, from *Bacillus* sp., 150–300 U of lyophilizate per milligram; Sigma) in enzyme dissolution buffer to give 12.5 U/ml; store in suitable portions at −20°, stable for months

Extraction columns: Disposable 1-ml SPE columns (Baker, Phillipsburg, NJ) containing 0.5 ml of Dowex 50W-X8, (200–400 mesh, H⁺ form; Serva, Heidelberg, Germany) equilibrated with 3% 5-sulfosalicylic acid solution

Preparation of Body Fluids and Control Samples

The following extraction procedure is applicable to body fluids such as EDTA-plasma, fresh urine, and liquor.

One milliliter of the sample fluid is transferred into a 1.2-ml reaction cup. 2-Oxohexanoate solution is added (0.1 ml) for internal standardization. For deproteinization, 60% 5-sulfosalicylic acid solution (0.06 ml) is added. The sample is thoroughly mixed and kept on ice for 15 min. After removal of the protein precipitate by centrifugation (10,000g, 10 min, 4°), the acid supernatant is used for analysis.

In samples from patients with maple syrup urine disease, concentrations of an individual branched-chain 2-oxo acid in plasma and urine may exceed 1 and 20 mM, respectively. Therefore, the following dilution scheme as used in our laboratory is recommended: Plasma from maple syrup urine disease patients is generally diluted 1:5 (v/v) with NaCl solution prior to use. For determinations in urine specimens, three samples are prepared and analyzed in parallel. One sample without prior dilution is used to cover the concentration range up to 0.2 mM. Two dilutions of 1:10 and 1:50 (v/v) with NaCl solution are prepared to cover the concentration range up to about 2 and 20 mM, respectively.

The appropriate set of control samples is run in parallel with the physiological samples in order to allow an overall control of the analytical procedure and especially the completion of the enzymatic reductive amination of the 2-oxo acids.

Separation and Amination of 2-Oxo Acids

2-Oxo acids are separated from the amino acids in the acid extract by the following procedure: 0.5 ml of the acid extract is allowed to penetrate slowly into the resin of an extraction column. The eluate is discarded. Subsequently, 3% 5-sulfosalicylic acid solution (0.5 ml) is applied to the column. The ensuing eluate (0.5 ml) is collected in a 2-ml reaction cup. NH_4Cl–NH_4OH buffer (0.1 ml) and NH_4OH solution (0.05 ml) are added. After thorough mixing, NADH solution (0.04 ml) and L-leucine dehydrogenase solution (0.01 ml) are added. The reaction mixture is shaken well and incubated in a water bath at 37° for 2 hr. Thereafter, 60% 5-sulfosalicylic acid (0.3 ml) and H_2O (0.8 ml) are added.

Purification of Products

After centrifugation (10,000g, 10 min, 4°), the above mixture (1.8 ml, three portions) is applied to an extraction column and allowed to penetrate slowly as described above. The column is carefully washed with H_2O (about 2 ml, eluate pH >3). Thereafter, NH_4OH solution (0.4 ml) is added and the eluate is discarded. For elution of the amino acids, two 0.5-ml volumes of NH_4OH solution are applied to the column. The eluate is collected and evaporated to dryness by freeze drying, in a desiccator or under a stream of nitrogen at 60°. The residue is dissolved in an appropriate volume of a suitable loading buffer depending on the method for quantitative amino acid analysis established in the laboratory.

In general, the final yield of (R)-3-methyl-2-oxopentanoate-derived L-alloisoleucine is ≥50 pmol, together with significantly higher yields of all other branched-chain 2-oxo acid-derived L-amino acids. Thus, the sensitivity of the amino acid analyzer used should allow quantitation of L-alloisoleucine down to about 50 pmol per run. If the available sensitivity is lower, it may be necessary to combine residues of repeated sample work-up or, alternatively, to scale up the above procedure appropriately.

Quantitation

Standard methods for amino acid analysis have been extensively reviewed and need not be repeated here.[10] In general, quantitation poses no

---

[10] R. H. Slocum and J. G. Cummings, in "Techniques in Diagnostic Human Biochemical Genetics" (F. A. Hommes, ed.), pp. 87–126. Wiley-Liss, New York, 1991.

special problems as the branched-chain 2-oxo acid-derived amino acids can be determined on any modern commercial amino acid analyzer using an ion-exchange column for separation and postcolumn derivatization with ninhydrin or o-phthaldialdehyde followed by spectrophotometric and fluorimetric detection, respectively. Availability of an appropriate peak integration software generally increases reproducibility and accuracy. To speed up the analysis, a modification of the short program for branched-chain amino acid analysis described by Benson et al.[11] is used in our laboratory.[12]

Various sensitive procedures using reversed-phase high-performance liquid chromatography (HPLC), gas chromatography, or capillary electrophoresis for separation of the amino acids have been described and may be applicable as well.[13–15]

Considering the generally quite different concentrations of (S)- and (R)-3-methyl-2-oxopentanoate in body fluids and, moreover, the wide range of concentrations occurring in patients with maple syrup urine disease,[4,5] it is imperative to check carefully the performance of the amino acid analysis method with respect to sensitivity, imprecision, linearity, and accuracy. For calibration, either homemade standard solutions based on high-purity L-amino acid preparations (from Merck or Bachem [Bubendorf, Switzerland]) or checked commercial calibrators (Benson) appropriately spiked with L-norleucine and L-alloisoleucine (Bachem) may be used.

Calculations

The micromolar concentration of (S)-3-methyl-2-oxopentanoate, (S)-KMV [or (R)-3-methyl-2-oxopenanoate, (R)-KMV], in a sample is calculated from the peak areas of isoleucine (or alloisoleucine) and norleucine (Nleu) from analysis runs of the sample (S) and of the external standard (ES) with known concentrations (conc) of components as follows:

$$ C_{(S)\text{-KMV}} = \frac{\text{area}_{\text{Ile, S}}}{\text{area}_{\text{Nleu, S}}} \left(\frac{\text{area}_{\text{Nleu, ES}}}{\text{area}_{\text{Ile, ES}}} \times \frac{\text{conc}_{\text{Ile, ES}}}{\text{conc}_{\text{Nleu, ES}}} \right) \times 50 $$

The two control samples should yield concentrations of 50 and 250 μmol/liter ($\pm5\%$).

[11] J. V. Benson, J. Cormick, and J. A. Patterson, *Anal. Biochem.* **18**, 481 (1967).
[12] P. Schadewaldt, H.-W. Hammen, C. Dalle-Feste, and U. Wendel, *J. Inher. Metab. Dis.* **13**, 137 (1991).
[13] G. B. Irvine, *Methods Mol. Biol.* **32**, 257 (1994).
[14] V. Walker and G. A. Mills, *Ann. Clin. Biochem.* **32**, 28 (1995).
[15] J. T. Smith, *Electrophoresis* **18**, 2377 (1997).

Performance of Method

Precision and Accuracy

The results of the present method have been shown to be in excellent agreement with data obtained by standard HPLC procedures (r = 0.993, n = 30).[1] When related to the 2-oxo acid content in original plasma samples, the final yields of the L-amino acids derived from (*S*)- plus (*R*)-3-methyl-2-oxopentanoate amounted to 47.7 ± 3.6% (n = 10). When checked with (*S*)-3-methyl-2-oxopentanoate at plasma concentrations of 24.5 and 262 μM, the method yielded coefficients of variation of 2.9 and 1.9% within run (n = 10) and of 2.1 and 2.0% between run (n = 9), respectively. With (*R*)-3-methyl-2-oxopentanoate at plasma concentrations of 0.9 and 243 μM, the coefficients of variation amounted to 7.5 and 2.1% within run (n = 10) and to 6.2 and 1.8% between run (n = 9), respectively.[1]

Stability of 2-Oxo Acids

The 2-oxo acids are fairly stable in plasma specimens when stored at $-20°$. This is probably due to the presence of sulfhydryl groups that protect 2-oxo acids from oxidative damage.[7,16] According to our own experience[5] and those of others,[17] however, the 2-oxo acids in urine specimens decompose rather rapidly and in an apparently unpredictable manner. Appropriate data for the 2-oxo acid stability in liquor are not yet available.

To overcome potential errors arising from 2-oxo acid decomposition, the analyses should be performed directly after collection of samples, at least in urine specimens. When samples are to be stored, the hydrazide gel column treatment for stabilization of the 2-oxo acids developed by Hayashi and co-workers may be used.[7,17] Alternatively, the following procedure may be applied: The total concentration of (*R*)- plus (*S*)-3-methyl-2-oxopentanoate is determined in the fresh samples. For this purpose, the 2-oxo acids are reacted with *o*-phenylenediamine to yield the quinoxaline derivatives and then quantitated by reversed-phase HPLC and fluorimetric detection.[7,18] The samples are then stored at $-20°$, and the relation of (*S*)- and (*R*)-methyl-2-oxopentanoate is later determined by the method described above. The resulting ratio is used for calculation of the original concentrations of the enantiomorphs in the fresh specimens. It is noteworthy that quinoxaline analysis also provides an excellent overall control method for

[16] G. Livesey and W. T. E. Edwards, *J. Chromatogr.* **337**, 98 (1985).

[17] T. Hayashi, H. Tsuchiya, and H. Naruse, *Clin. Chim. Acta* **132**, 321 (1983).

[18] P. Schadewaldt, W. Hummel, U. Trautvetter, and U. Wendel, *Clin. Chim. Acta* **183**, 171 (1989).

the determination of branched-chain 2-oxo acids via the corresponding L-amino acids.

Concentrations in Human Body Fluids

In the plasma of healthy subjects (n = 15), the mean (±SD) (S)- and (R)-3-methyl-2-oxopentanoate concentrations amounted to 22.4 ± 5.7 and to 0.7 ± 0.2 μM, respectively. The (S)-3-methyl-2-oxopentanoate: (R)-3-methyl-2-oxopentanoate ratio was 32 ± 4. In the plasma of patients with diabetes mellitus (n = 15), (S)- and (R)-3-methyl-2-oxopentanoate concentrations were 24.4 ± 5.4 and 0.8 ± 0.3 μM, respectively, and the ratio was 31 ± 5.

In patients with the classic form of maple syrup urine disease, the plasma levels of (S)- and (R)-3-methyl-2-oxopentanoate (n = 102) were highly variable and ranged (in parentheses, mean ± SD) from 17.2 to 406 (125 ± 76) μM and from 3.8 to 112 (43.8 ± 25.4) μM, respectively. The ratio of the (S)- to (R)-form was 3.0 ± 1.1 (Fig. 1). In spot urine specimens (n = 63) from these patients, concentrations of (S)- and (R)-3-methyl-2-oxopentanoate ranged from 9.0 to 5386 μM and from 0.0 to 2853 μM, respectively. Related to urinary creatinine, the ranges were 2.1–3174 and 0.0–865 $\mu mol/mmol$ of creatinine, respectively (Fig. 1). The mean ratio of the (S)- to (R)-enantiomorph in urine was 4.7 ± 2.3.

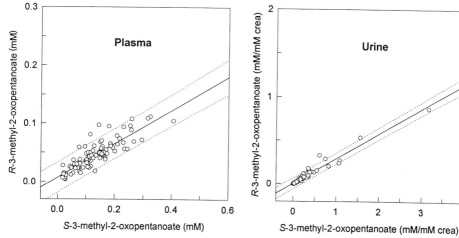

FIG. 1. Relation of (S)- and (R)-3-methyl-2-oxopentanoate in plasma and urine specimens of patients with the classic form of maple syrup urine disease. Plasma: solid line, regression line (y = 0.008 + 0.288x, r = 0.864, n = 102); urine: solid line, regression line (y = 0.005 + 0.285x, r = 0.975, n = 63). The 95% prediction intervals are indicated by dotted lines.

[5] Spectrophotometric Assay for Measuring Branched-Chain Amino Acids

By PHILIP R. BECKETT

Introduction

The branched-chain amino acids isoleucine, leucine, and valine are essential substrates for protein synthesis, growth, and maintenance of lean body mass in animals and humans. Loss of lean body mass because of undernutrition, disease, or trauma increases mortality. Endocrine factors that regulate protein metabolism have been researched extensively. However, amino acids, especially the branched-chain amino acids, also appear to regulate protein metabolism independent of their role as a supply of substrate.[1] To investigate acute regulation of protein metabolism by amino acids, amino acid concentrations must be manipulated predictably and reproducibly. This can be achieved with an indexed infusion of a solution of amino acids with the infusion rate based on real-time plasma concentrations. Branched-chain amino acid concentrations can be measured accurately by liquid chromatography; however, the chromatography techniques generally require several hours to complete an analysis. Brown *et al.*[2] published a chromatography technique that requires 5–7 min to measure leucine concentrations.

The practical difficulties of setting up a chromatography technique at the patient's bedside and the possibility of substantially reducing the analysis time led us to explore the possibility of using an enzyme assay to determine branched-chain amino acid concentration.[3] This assay was modified from previously published assays[4–7] for the explicit purpose of measuring branched-chain amino acids at the bedside.

[1] P. J. Garlick and I. Grant, *Biochem. J.* **254**, 579 (1988).
[2] L. L. Brown, P. E. Williams, T. A. Becker, R. J. Ensley, M. E. May, and N. N. Abumrad, *J. Chromatogr.* **426**, 370 (1988).
[3] P. R. Beckett, D. S. Hardin, T. A. Davis, H. V. Nguyen, D. Wray-Cahen, and K. C. Copeland, *Anal. Biochem.* **240**, 48 (1996).
[4] G. Livesey and P. Lund, *Biochem. J.* **188**, 705 (1980).
[5] J. M. Burrin, J. L. Paterson, and G. M. Hall, *Clin. Chim. Acta* **153**, 37 (1985).
[6] T. Ohshima, H. Misono, and K. Soda, *J. Biol. Chem.* **253**, 5719 (1978).
[7] K. Soda, H. Misono, K. Mori, and H. Sakato, *Biochem. Biophys. Res. Commun.* **44**, 931 (1971).

Reaction

Leucine dehydrogenase (BCKDH, EC 1.4.1.9) from *Bacillus* spp. (Toyobo, New York, NY) catalyzes the following reversible reaction:

$$RCNH_2COO^- + NAD^+ + H_2O \leftrightharpoons RCOCOO^- + NH_3 + NADH + H^+$$

where R is the side chain of the branched-chain amino acids isoleucine, leucine, and valine.

The generation of NADH is stochiometric with the oxidation of the amino acid and can be measured by the increase in absorbance at 340 nm, using a spectrophotometer.

Reagents

The following solutions are prepared fresh each day:
Glycine–KCl–KOH buffer: 0.1 M, pH 10.5; 2 mM EDTA
NAD (120 mM) diluted in 0.1 M sodium carbonate buffer, pH 10.7
Leucine dehydrogenase: 200 units/ml in 25 mM sodium phosphate
    buffer, pH 7.2, with bovine serum albumin (BSA, 1 mg/ml)

The units of enzyme activity described here for the preparation of the enzyme solution are those of the manufacturer, i.e., 1 unit of enzyme activity causes the formation of 1 μmol of NADH per minute in the presence of 18 mM L-leucine, 1.1 mM NAD$^+$, and 0.18 M glycine–KCl–KOH buffer, pH 10.5, at 37°. On receipt from the manufacturer, we diluted lyophilized leucine dehydrogenase in 25 mM sodium phosphate buffer, pH 7.2, to a concentration of 1 mg of protein per milliliter, divided the diluted enzyme into convenient aliquots, and stored it at −20°. After further dilution of the enzyme for the assay, activity is stable for at least 24 hr when stored on ice and is, therefore, suited for prolonged studies. The NAD is also maintained on ice and the glycine–KCl buffer is maintained at room temperature.

Procedure

All spectrophotometric measurements are made at room temperature with a UV spectrophotometer (Shimadzu, Columbia, MD) equipped with an adjustable microcuvette cell holder and 1-ml quartz microcuvettes. The adjustable cell holder enables the reaction volume to be decreased to 300 μl, which increases the sensitivity of the assay. The reaction mixture contains 270 μl of assay buffer, 4 mM NAD (10 μl), and 10 μl of sample (plasma). The optical density in the absence of enzyme is about 0.22. The absorbance before the addition of enzyme should be stable (± 0.002); however, this is confirmed for every sample by recording the absorbance for 30 sec before addition of the enzyme.

Background noise of 0.002 optical density (OD) units limits measurements to about 0.02 OD units, equivalent to 100 μM branched-chain amino acids in a sample. The reaction is initiated with 2 units of leucine dehydrogenase (10 μl) and the optical density is measured after 1 min. The time course of the reaction is shown in Fig. 1. The addition of the enzyme itself does not cause any change in absorption in the absence of branched-chain amino acids. The concentration of amino acid in the sample is calculated

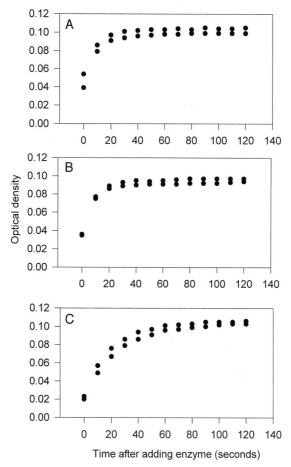

Fig. 1. Time course of the change in absorbance at 340 nm measured at room temperature with a reaction mixture containing 0.2 M glycine–KCl–KH$_2$PO$_4$–KOH buffer (pH 10.5), 4 mM NAD, 5 nmol of leucine (A), isoleucine (B), or valine (C), and 2 units of leucine dehydrogenase. Individual points are shown for duplicate measurements.

from the extinction coefficient of NADH, corrected for dilution and the assay volume as follows:

$$(OD_{t1} - OD_{t0})/(((\varepsilon/1000)/0.3) \times 0.01)$$

where OD_{t1} is optical density at 1 min after adding leucine dehydrogenase, OD_{t0} is optical density before adding enzyme, ε is the extinction coefficient of NADH (6.22), 1000 converts millimolar to micromolar units, 0.3 is the assay volume in milliliters, and 0.01 is the sample volume in milliliters. As an internal test of the assay when analyzing unknown samples we always compare the change in optical density for a known standard (500 μM leucine) with the calculated value at frequent intervals (approximately hourly).

Standard Curve

The production of NADH (nanomoles) is determined from 10-μl standards of leucine, isoleucine, and valine individually at concentrations of 100, 200, 400, 600, 800, and 1000 μM (corresponding to 1, 2, 4, 6, 8, and 10 nmol of amino acid in the assay). The relationships (slopes) between the calculated values and the measured values, using regression analysis, are 1.002, 1.0126, and 0.938 for leucine, isoleucine, and valine, respectively. The corresponding correlation coefficients are 0.9996, 0.9988, and 0.996 for leucine, isoleucine, and valine, respectively (Fig. 2). A more critical method of analyzing the data is to use a Bland–Altman plot to compare the differences (nanomoles per milliliter) of the assay from the known values at each concentration. The plots for the three amino acids are shown in Fig. 3. The difference between the assay and the mean concentration for leucine is 3.3 ± 11 (95% confidence interval, −4.0 to 10.6); for isoleucine it is −7.3 ± 14.7 (95% confidence interval, −16.6 to 2.1). For valine, the error is greater because, as can be seen from Fig. 3, the assay underestimates the 1000 μM standard, resulting in a mean difference of −21 ± 29. Omission of the 1000 μM standard results in a mean difference of −9.6 ± 9.5 (95% confidence interval, −2.8 to −16.0). The correct value for 1000 μM valine can be obtained by running the assay for 90 sec rather than 60 sec, in order to reach the end point. However, a concentration of 1000 μM valine is unlikely to be encountered in physiologic samples.

Comparison with High-Performance Liquid Chromatography

Plasma is prepared by obtaining 20 ml of blood in heparinized vacutainers. The blood is centrifuged at 3000g for 10 min and the plasma split into 1-ml aliquots. The plasma is spiked by the addition of 0, 20, 40, 60, or 80 nmol each of leucine, isoleucine, and valine, to give branched-chain amino

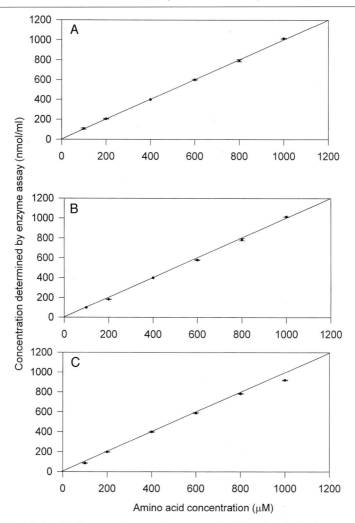

FIG. 2. Relationship between the actual concentration of leucine, isoleucine, or valine in the reaction mixture and that determined from the change in absorbance at 340 nm (mean ± SD). The reaction mixture contained 0.2 M glycine–KCl–KH$_2$PO$_4$–KOH buffer (pH 10.5), 4 mM NAD, amino acid standard, and 2 units of leucine dehydrogenase. The amino acid standards were prepared from crystalline leucine, isoleucine, and valine at concentrations of 100, 200, 400, 600, 800, and 1000 μM. Each branched-chain amino acid was analyzed separately: (A) leucine; (B) isoleucine; (C) valine. The lines drawn on the plots are lines of unity. The actual slopes were 1.002, 1.0126, and 0.938 for leucine, isoleucine, and valine, respectively.

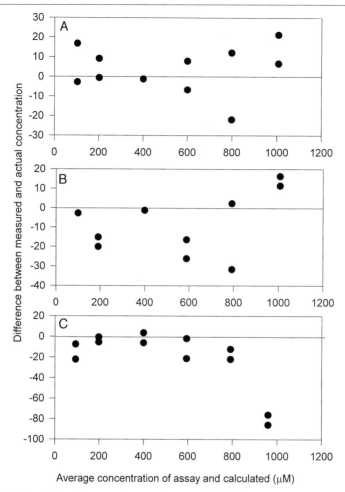

Fig. 3. Plot of the difference (nanomoles per milliliter) between the concentration of branched-chain amino acid in the standards and that determined from the enzyme assay versus the concentration in the standard for leucine (A), isoleucine (B), and valine (C). A significant correlation indicates a methodological difference. The correlation coefficients were 0.4, 8.2, and 50% for leucine, isoleucine, and valine, respectively. Only the correlation coefficient for valine was significant ($p < 0.01$), which occurred because of the underestimation of the highest standard. On removal of the 1000 μM standard from the analysis, there were no differences between the actual concentration and the measured concentration.

acid concentrations of the unadulterated plasma plus 0, 60, 120, 180, and 240 nmol. Samples are analyzed in duplicate both with an ion-exchange amino acid analyzer (Beckman, Fullerton, CA) and the enzyme assay (Fig. 4). The concentration in the unadulterated plasma is $420 \pm 9 \ \mu M$ using the amino acid analyzer and $436 \pm 3.5 \ \mu M$ using the enzyme assay. In both assays, the concentration increases in proportion to the spike concentration added. After subtracting the concentration added as a spike, the remainder is not different from the concentration in the unadulterated sample, irrespective of the spike concentration added. Thus, the enzyme assay is sensitive to total branched-chain amino acid concentrations in serum. The enzyme assay overestimates concentrations measured with the amino acid analyzer by $23 \pm 19 \ \mu M$ ($p < 0.05$), using a paired t test (95% confidence interval, -9.6 to -36.5), an error of about 5% at a concentration of 420 μM. The reason for the difference is unclear, but could be caused by errors

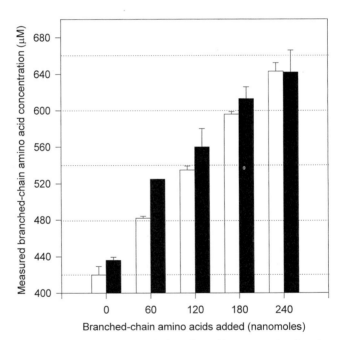

Fig. 4. Comparison of total branched-chain amino acid concentration from human plasma determined by ion-exchange amino acid analysis and the enzyme assay. The plasma was measured unadulterated and after the addition of 60, 120, 180, or 240 nmol of branched-chain amino acids (one-third each leucine, isoleucine, and valine). The open bars represent the results of the ion-exchange amino acid analysis, and the solid bars the results of the enzyme assay. The dashed lines represent the calculated concentration of the unadulterated plasma measured by ion-exchange amino acid analysis plus the added amino acids.

in the concentrations of internal standards used in the assays, or by pipetting errors. The intraassay variability, determined with 5 blood samples measured 15 times by the enzyme assay, was 4.1%.

Perspectives

Leucine dehydrogenase has four features that render it applicable to a rapid bedside assay: (1) The assay is highly substrate specific for the branched-chain amino acids, which enables its application in analyzing physiologic samples containing other amino acids and compounds; (2) during the reaction, NAD is reduced stoichiometrically with oxidation of the branched-chain amino acid. Reduction of NAD can be detected with a spectrophotometer, which can be transported readily to the bedside; (3) the sample volume is only 10 μl, and thus the assay is suitable for multiple sampling, or for premature infants. For multiple sampling, we use a closed system with two syringes on an intravenous line so that blood taken as a discard from the catheter is returned to the patient. Using this technique, we avoid the loss of 5 ml of blood to obtain a 50-μl sample; and (4) sample turnaround time, from blood sampling to result, can be achieved in less than 2 min with practice and an efficient set-up. These real-time data can be used to modify the infusion rate of an amino acid mixture in order to maintain a stable plasma concentration of branched-chain amino acids, in a manner equivalent to the glucose clamp.[8] This assay may be clinically applicable for maintaining plasma branched-chain amino acid concentrations during physiologic studies and to monitor therapy for clinical conditions such as diabetic ketoacidosis or during the use of total parenteral nutrition.

Acknowledgments

The author thanks Kenneth Copeland, M.D., and Teresa Davis, Ph.D., for their contribution to the development of the assay, and Morey Haymond, M.D., for ion-exchange amino acid analysis.

[8] R. A. DeFronzo, J. D. Tobin, and R. Andres, *Am. J. Physiol.* **237,** E214 (1979).

[6] Determination of Branched-Chain α-Keto Acid Dehydrogenase Activity State and Branched-Chain α-Keto Acid Dehydrogenase Kinase Activity and Protein in Mammalian Tissues

By Naoya Nakai, Rumi Kobayashi, Kirill M. Popov, Robert A. Harris, and Yoshiharu Shimomura

Introduction

The branched-chain α-keto acid dehydrogenase (BCKDH) complex, a large multienzyme complex located in the mitochondrial matrix space, catalyzes oxidative decarboxylation of branched-chain α-keto acids (α-ketoisocaproate, α-keto-β-methylvalerate, and α-ketoisovalerate) derived from transamination of the corresponding branched-chain amino acids (leucine, isoleucine, and valine). The reaction catalyzed by the BCKDH complex is considered the rate-limiting step for the catabolism of all three of the branched-chain amino acids.[1,2] The activity of the BCKDH complex is regulated by reversible covalent modification catalyzed by a specific BCKDH kinase that phosphorylates and inactivates the complex and by a specific BCKDH phosphatase that dephosphorylates and activates the complex.[3] The activity state of the BCKDH complex is defined as the proportion of the enzyme in the active, dephosphorylated state in a tissue in a particular physiological or pathological condition of an animal. It provides an estimation of the *in vivo* phosphorylation state of the enzyme, which in turn provides insight with respect to the relative importance of covalent modification of the complex in regulating flux through the catabolic pathways for branched-chain amino acids. The activity state of the complex in rat liver and muscle is affected by hormones,[4,5] diabetes,[6,7]

[1] A. E. Harper, R. H. Miller, and K. P. Block, *Annu. Rev. Nutr.* **4,** 409 (1984).

[2] R. A. Harris, B. Zhang, G. W. Goodwin, M. J. Kuntz, Y. Shimomura, P. Rougraff, P. Dexter, Y. Zhao, R. Gibson, and D. W. Crabb, *Adv. Enzyme Regul.* **30,** 245 (1990).

[3] S. J. Yeaman, *Biochem. J.* **257,** 625 (1989).

[4] K. P. Block, W. B. Richmond, W. B. Mehard, and M. G. Buse, *Am. J. Physiol.* **252,** E396 (1987).

[5] R. Kobayashi, Y. Shimomura, T. Murakami, N. Nakai, N. Fujitsuka, M. Otsuka, N. Arakawa, K. M. Popov, and R. A. Harris, *Biochem. J.* **327,** 449 (1997).

[6] R. P. Aftring, W. J. Miller, and M. G. Buse, *Am. J. Physiol.* **254,** 292 (1988).

[7] R. Gibson, Y. Zhao, J. Jaskiewicz, S. E. Fineberg, and R. A. Harris, *Arch. Biochem. Biophys.* **306,** 22 (1993).

exercise,[8-10] starvation,[2,11,12] and low dietary protein feeding.[2,11,12] Spectrophotometric and radiochemical assays for the estimation of BCKDH activity and its activity state in mammalian tissues are presented in this chapter. Related methods have been described and used by other investigators.[13-16]

The BCKDH kinase exists tightly bound to the BCKDH complex, with perhaps little or no free kinase in the mitochondrial matrix space. The relative activity of the kinase is affected by various treatments of animals[2,17-20] and is therefore believed to be important in setting the activity state of the BCKDH complex. Although not easy to assay for technical reasons, the determination of BCKDH kinase activity in tissue extracts usually reveals an inverse relationship between this activity and the activity state of the BCKDH complex. Changes in kinase activity induced by low dietary protein feeding have been shown to correlate with increased amounts of BCKDH kinase protein.[20] The activity of BCKDH phosphatase, the other major player in this regulatory mechanism, cannot be readily measured for want of an assay suitable for crude tissue extracts.

Principle

The level at which the BCKDH complex is expressed in a given tissue determines whether a spectrophotometric or radiochemical assay should be used. Tissues with relatively high activity of the complex (rat liver, kidney, and heart) can be readily assayed spectrophotometrically with one of the branched-chain α-keto acids as substrate, whereas tissues with rela-

[8] A. J. M. Wagenmakers, J. H. Brookes, J. H. Coakley, T. Reilly, and R. H. Edwards, *J. Appl. Physiol.* **59,** 159 (1989).
[9] Y. Shimomura, T. Suzuki, S. Saitoh, Y. Tasaki, R. A. Harris, and M. Suzuki, *J. Appl. Physiol.* **68,** 161 (1990).
[10] Y. Shimomura, H. Fujii, M. Suzuki, T. Murakami, N. Fujitsuka, and N. Nakai, *J. Nutr.* **125,** 1762S (1995).
[11] P. A. Patston, J. Espinal, J. M. Shaw, and P. J. Randle, *Biochem. J.* **235,** 429 (1986).
[12] R. A. Harris, G. W. Goodwin, R. Paxton, P. Dexter, S. M. Powell, B. Zhang, A. Han, Y. Shimomura, and R. Gibson, *Ann. N.Y. Acad. Sci.* **573,** 306 (1989).
[13] P. A. Patston, J. Espinal, M. Beggs, and P. J. Randle, *Methods Enzymol.* **166,** 175 (1988).
[14] G. W. Goodwin, B. Zhang, R. Paxton, and R. A. Harris, *Methods Enzymol.* **166,** 189 (1988).
[15] K. P. Block, R. P. Aftring, M. G. Buse, and A. E. Harper, *Methods Enzymol.* **166,** 201 (1988).
[16] J. Espinal, M. Beggs, H. Patel, and P. J. Randle, *Biochem. J.* **237,** 285 (1986).
[17] H. Fujii, Y. Shimomura, T. Murakami, N. Nakai, T. Sato, M. Suzuki, and R. A. Harris, *Biochem. Mol. Biol. Int.* **44,** 1211 (1998).
[18] R. A. Harris, J. W. Hawes, K. M. Popov, Y. Zhao, Y. Shimomura, J. Sato, J. Jaskiewicz, and T. D. Hurley, *Adv. Enzyme Regul.* **37,** 271 (1997).
[19] H. S. Paul, W. Q. Liu, and S. A. Adibi, *Biochem. J.* **317,** 411 (1996).
[20] K. M. Popov, Y. Zhao, Y. Shimomura, J. Jaskiewicz, N. Y. Kedishvili, J. Irwin, G. W. Goodwin, and R. A. Harris, *Arch. Biochem. Biophys.* **316,** 148 (1995).

tively low activity are best assayed by radiochemical measurement of $^{14}CO_2$ produced from a 1-^{14}C-labeled branched-chain α-keto acid. One unit of BCKDH complex activity catalyzes the formation of 1 μmol of NADH per minute in the spectrophotometric assay and the production of 1 μmol of CO_2 per minute in the radiochemical assay.

The actual activity of the BCKDH complex refers to the activity as it occurs *in vivo* as a consequence of the complex existing partially in its active, dephosphorylated state. Accurate determination of the actual activity requires preservation of the *in vivo* phosphorylation state of the complex during tissue removal, extraction, and enzyme assay. This demands rapid removal and freeze-clamping of the tissue followed by extraction and assay with reagents that minimize kinase, phosphatase, and protease activities. Determination of the total activity of the BCKDH complex in tissue extracts requires complete activation of the enzyme, which can be accomplished only by dephosphorylation with a suitable phosphoprotein phosphatase.

BCKDH kinase activity can be measured either by determining the rate at which radioactive phosphate of [γ-^{32}P]ATP is incorporated into the E1α component of the complex or by determining the rate at which overall BCKDH complex activity is lost as a consequence of phosphorylation.[21] Although either approach can be used to assay kinase activity in crude tissue extracts,[21] only the latter assay has been used extensively and is described in detail here. The BCKDH complex frequently exists in tissue extracts in its phosphorylated, inactive form. Determination of BCKDH kinase activity therefore requires prior activation of the complex by dephosphorylation. This extra step can prove problematic for this assay as discussed below.

Materials

Reagents are obtained as described previously[9] or from Sigma (St. Louis, MO) and Wako Pure Chemical Industries (Osaka, Japan). α-Keto [1-^{14}C]isocaproate is obtained from Amersham Japan (Tokyo, Japan). α-Chloroisocaproate is prepared as described previously.[22] The broad-specificity protein phosphatase is prepared from rat livers or bovine hearts as described previously.[23] Lambda protein phosphatase is purchased from New England BioLabs (Beverly, MA). The E3 component of the BCKDH complex (dihydrolipoamide dehydrogenase from porcine heart) is obtained

[21] J. Espinal, M. Beggs, and P. J. Randle, *Methods Enzymol.* **166**, 166 (1988).
[22] R. A. Harris, M. J. Kuntz, and R. Simpson, *Methods Enzymol.* **166**, 114 (1988).
[23] R. A. Harris, R. Paxton, and R. A. Parker, *Biochem. Biophys. Res. Commun.* **107**, 1497 (1982).

from Sigma. Xylene-base scintillation fluid (Scintisol EX-H) is purchased from Wako.

Branched-Chain α-Keto Acid Dehydrogenase Complex Activity

Buffers and Reagents

Extraction buffer: 50 mM N-2-hydroxyethylpiperazine-N'-2-ethane-sulfonic acid (HEPES), 3% (w/v) Triton X-100, 2 mM EDTA, 5 mM dithiothretiol (DTT), 0.5 mM thiamine pyrophosphate, 1 mM α-chloroisocaproate, 50 mM potassium fluoride, 2% (v/v) bovine serum, 0.1 mM N-tosyl-L-phenylalanine chloromethyl ketone (TPCK), trypsin inhibitor (0.1 mg/ml), leupeptin (0.02 mg/ml; pH 7.4 at 4°, adjusted with KOH). Triton X-100 and bovine serum are added to the buffer before addition of TPCK (prepared in ethanol as a 20 mM stock solution). The extraction buffer is divided into aliquots and stored at −40°

Suspending buffer: 25 mM HEPES, 0.1% (w/v) Triton X-100, 0.2 mM EDTA, 0.4 mM thiamine pyrophosphate, 1 mM DTT, 50 mM KCI, and leupeptin (0.02 mg/ml; pH 7.4 at 37°, adjusted with KOH). The suspending buffer can be stored at −40°

Assay buffer (2×): 60 mM potassium phosphate, 4 mM MgCl$_2$, 0.8 mM thiamine pyrophosphate, 0.8 mM CoA, 2 mM NAD$^+$, 0.2% (w/v) Triton X-100, 4 mM DTT, and pig heart dihydrolipoamide dehydrogenase (10 units/ml; pH 7.3 at 30°). The 2× assay buffer is prepared as a stock solution and stored at −80° without MgCl$_2$ and dihydrolipoamide dehydrogenase

Substrate solution: A 50 mM sodium α-ketoisovalerate solution is prepared and stored in aliquots at −80° for the spectrophotometric assay. A 4 mM α-keto[1-^{14}C]isocaproate solution with a specific activity of 600–1000 cpm/nmol is prepared and stored in aliquots at −80° for the radiochemical assay. Freezing and thawing of these solutions is avoided

Stop solution: 2 M acetic acid made 1% (w/v) in sodium dodecyl sulfate (SDS)

Polyethylene glycol (PEG) 6000 solution, 27% (w/v)

Phenethylamine, 33% (v/v) in methanol

Preparation of Tissue Extracts

Frozen tissue is pulverized to a fine powder under liquid nitrogen with a precooled mortar and pestle. This step is critically important for quantitative extraction of the BCKDH complex from hard tissues such as skeletal muscle and heart. A stainless steel mortar and pestle has been found to work particularly well. Approximately 0.25 g of powdered tissue is transferred with a precooled spatula into a tared, precooled plastic tube. The tube is reweighed rapidly to establish the wet weight and the tissue is immediately homogenized in 2 ml of ice-cold extraction buffer with a motor-driven Teflon pestle. Insoluble material is removed by centrifugation at 20,000g for 5 min at 4°, and the supernatant is made 9% in PEG from the 27% solution. After standing for 20 min on ice the mixture is centrifuged at 12,000g for 10 min at 4° and the pellet obtained is suspended in 1 ml of ice-cold suspending buffer. Because freezing and thawing as well as storage of tissue extracts on ice have detrimental effects, assays of enzyme activity should be conducted as soon as possible after preparation of tissue extracts.

Spectrophotometric Assay

The E3 component (dihydrolipoamide dehydrogenase) of the BCKDH complex dissociates from the complex during extraction, is not precipitated by PEG, and consequently is rate limiting for the assay of BCKDH complex activity in the absence of added E3.[24] Excess E3 as well as excess Mg^{2+} to offset chelation by EDTA (present in the extraction and suspension buffers to help minimize kinase, phosphatase, and protease activities) must be present in the assay cocktail to obtain maximum BCKDH complex activity. E3 and $MgCl_2$ are best added to the otherwise complete assay mixture just prior to assay. The total volume of the assay mixture for the spectrophotometric assay is 1 ml. The 2× assay buffer, 0.5 ml, is diluted with 0.46 ml of H_2O and prewarmed to 30° in a cuvette. The tissue extract, 20 μl, is added to the assay buffer and fully mixed. The absorbance at 340 nm is recorded for several minutes to establish a baseline and the reaction is initiated with 20 μl of 50 mM α-ketoisovalerate (final concentration of 1 mM) prewarmed to 30°.

Radiochemical Assay

For the decarboxylation assay,[9] a 2-ml Eppendorf tube containing the mixture of 200 μl of 2× assay buffer and 100 μl of the tissue extract is placed in a 20-ml glass (transparent) scintillation vial that contains 0.35 ml of 1.2 M KOH. After preincubation for 5 min at 30°, the reaction is initiated by addition of 100 μl of 4 mM α-keto[1-^{14}C]isocaproate prewarmed to 30°. (α-Keto[1-^{14}C]isovalerate can be used as substrate in place of α-keto[1-^{14}C]isocaproate. Indeed, the latter compound is a better substrate for assay of the BCKDH complex in liver, which is known to express significant

[24] R. Odessey, *Biochem. J.* **192**, 155 (1980).

amounts of α-ketoisocaproate oxidase.[25] α-Keto[1-^{14}C]isovalerate is not commercially available but can be produced from [1-^{14}C]valine.) The vial is immediately sealed with a rubber serum cap that has a short inner plug. (Caps with long inner plugs may interfere with the diffusion of CO_2 from the reaction mixture to the KOH solution.) After 10 min of incubation at 30° with gentle shaking, 0.8 ml of the stop solution is injected through the serum cap with a disposable syringe to terminate the reaction. The syringe is used to fully mix the stop solution with the reaction mixture to ensure complete and immediate termination of the reaction. A blank is prepared by adding radioactive substrate to a tube that already contains stop solution, assay buffer, and tissue extract. CO_2 is collected by incubation at 30° for 1 hr with gentle shaking. The Eppendorf tube is removed, and 1 ml of phenethylamine in methanol is used to wash KOH from the outer surface of the Eppendorf tube, and 10 ml of xylene-base scintillation fluid is added to the vial for measuring radioactivity. The mixture of scintillation fluid, KOH, phenethylamine, and methanol should be clear. If not, small amounts of the phenethylamine solution should be added. The radioactivity of the blank (which should be no greater than 600 cpm) is subtracted from those of sample vials. Although the assay as described has proved linear with respect to both time and quantity of tissue extract, it is important to validate the assay by these criteria each time experimental conditions are modified and new reagents are used.

Actual Activity of Branched-Chain α-Keto Acid Dehydrogenase Complex

The assay carried out as described is designed to give the actual activity of the BCKDH complex. However, to preserve the *in vivo* phosphorylation state and therefore the actual activity *in vivo,* tissues must be removed quickly from the animal and freeze-clamped (<20 sec) with aluminum tongs cooled to the temperature of liquid nitrogen. This is particularly important for muscle because this tissue contains high BCKDH kinase activity. Although immediate processing of the tissue is considered best, tissue has been stored at −80° for several days prior to extraction with good results. The extraction buffer contains α-chloroisocaproate, thiamine pyrophosphate, and EDTA, all of which inhibit BCKDH kinase.[26] Thiamine pyrophosphate may also stabilize the BCKDH complex against denaturation.[27] The extraction buffer also contains potassium fluoride and EDTA, both

[25] P. J. Sabourin and L. L. Bieber, *J. Biol. Chem.* **257,** 7460 (1982).
[26] Y. Shimomura, N. Nanaumi, M. Suzuki, K. M. Popov, and R. A. Harris, *Arch. Biochem. Biophys.* **283,** 293 (1990).
[27] Y. Shimomura, M. J. Kuntz, M. Suzuki, T. Ozawa, and R. A. Harris, *Arch. Biochem. Biophys.* **266,** 210 (1988).

of which inhibit phosphoprotein phosphatases that activate the BCKDH complex, and it is further fortified with protease inhibitors and bovine serum in an attempt to minimize inactivation of the complex by proteolysis. In spite of all these additions designed to stabilize the enzyme, it has not been found possible to store tissue extracts for any extended period of time prior to assay. Reliable data are obtained only when assays are carried out immediately after preparation of tissue extracts.

Total Activity of Branched-Chain α-Keto Acid Dehydrogenase Complex

For measurements of total enzyme activity the BCKDH complex must be completely activated by dephosphorylation before assay of its activity. Three methods can be used to achieve this in crude tissue extracts:

Method 1: Incubation with Mg^{2+}. Incubation with Mg^{2+} is applicable only in tissues that express endogenous protein phosphatase activity high enough to effect complete activation of the BCKDH complex. From a practical standpoint this restricts this method to rat heart and rat liver. Other tissues examined in this laboratory (rat skeletal muscle and digestive tract tissue; dog liver, heart, kidney, skeletal muscle, and digestive tract tissue; bovine liver and heart; and human liver and skeletal muscle) require supplementation with an exogenous protein phosphatase to achieve full activation of the complex.

The tissue extract is diluted to 2 volumes with suspending buffer supplemented with $MgSO_4$ (final concentration 15 mM) and incubated at 37° for approximately 50 min. Partial inactivation of the BCKDH complex sometimes occurs during this long incubation period. Whether this is a consequence of inactivation by denaturation or proteolysis has not been established. This problem can be partly avoided by supplementing the medium with an exogenous phosphatase to shorten the incubation time as described below.

Method 2: Incubation with Broad-Specificity Protein Phosphatase and Mg^{2+}. The tissue extract is diluted to 2 volumes with suspending buffer supplemented with a broad-specificity protein phosphatase and $MgSO_4$ (final concentration, 15 mM). The mixture is incubated at 37° until full activation of the BCKDH complex is achieved. The amount of the phosphatase and the incubation time required must be established empirically for each phosphatase preparation. The amount of phosphatase should be adjusted so that an incubation time of no more than 30 min is required. The time required to reach maximal activity is also affected by the *in vivo* phosphorylation state of the BCKDH complex. Thus, the time course of activation should be examined for each type of tissue extract, with particular attention given to determining whether the BCKDH complex enzyme activity is stable after reaching a plateau value. Loss of activity once a plateau has been reached is sometimes observed. Supplementation with additional protease inhibitors or using less of the extract or even less of the phosphatase preparation may stabilize the activated enzyme. This particular problem does not happen with all phosphatase preparations but can be avoided by using commercially available lambda protein phosphatase as described below.

Method 3: Incubation with Lambda Protein Phosphatase and Mn^{2+}. The tissue extract is diluted to 2 volumes with suspending buffer supplemented with lambda protein phosphatase (final concentration, 2000 units/ml) and $MnCl_2$ (final concentration, 2 mM). The mixture is incubated at 37° for 15–20 min. The time course of the activation must be followed with a representative sample to ensure that complete activation occurs under the chosen incubation conditions.

Activity State of Branched-Chain α-Keto Acid Dehydrogenase Complex

The activity state of the BCKDH complex is calculated as a percentage by multiplying the ratio of the actual activity to the total activity by 100.

Branched-Chain α-Keto Acid Dehydrogenase Kinase Activity

See also Popov *et al.* ([17] in this volume[28]).

Buffers and Reagent

Extraction buffer: The composition of this buffer is the same as that used for the BCKDH complex assay, except that kinase inhibitors (α-chloroisocaproate and thiamine pyrophosphate) are omitted

Suspending buffer: The composition of this buffer is the same as that used for the BCKDH complex assay, except that thiamine pyrophosphate is omitted

Kinase assay buffer (1.25×): 25 mM HEPES, 1.875 mM MgCl$_2$, 2.5 mM DTT, 31.2 mM potassium phosphate (K$_2$HPO$_4$), 62.5 mM KF, and 25% (v/v) glycerol. This solution is adjusted to pH 7.35 at 30° with HCl and stored in aliquots at −40°

ATP solution: 2.5 mM (neutralized with NaOH). This solution is stored in aliquots at −40°

---

[28] See also K. M. Popov, Y. Shimomura, J. W. Hawes, and R. A. Harris, *Methods Enzymol.* **324,** Chap. 17, 2000 (this volume).

Preparation of Tissue Extract

The BCKDH kinase is quantitatively precipitated by 9% (w/v) PEG as a consequence of its tight association with the BCKDH complex. Therefore tissue extracts prepared by the method described for the BCKDH complex activity assay can be used. Omission of kinase inhibitors from the extraction and suspension buffers is the only necessary modification.

Activation of Branched-Chain α-Keto Acid Dehydrogenase Complex in Tissue Extract

The BCKDH complex present in tissue extracts is completely activated by dephosphorylation with a broad-specificity protein phosphatase (method 2) and then precipitated a second time with 9% (w/v) PEG to remove most of the added phosphatase. The precipitate is resuspended in a 1.25× kinase assay buffer to give a final BCKDH complex activity of 0.5–1.0 units/ml as measured by the spectrophotometric assay.

The lambda protein phosphatase cannot be used to activate the BCKDH complex for the BCKDH kinase assay. For unknown reasons, incubation of the BCKDH complex with lambda protein phosphatase plus Mn^{2+} causes partial inactivation of BCKDH kinase. Attempts to prevent this loss of activity have not been successful.

When the kinase activity is measured in tissue extracts with high ATPase activity, such as rat heart, the kinase assay buffer is further supplemented with oligomycin at a final concentration of 25 μg/ml (added from a 5-mg/ml stock solution in ethanol) in order to conserve ATP during the BCKDH kinase assay.

Assay of Branched-Chain α-Keto Acid Dehydrogenase Kinase

A 160-μl volume of the final suspension of the tissue extract is placed in an Eppendorf tube and prewarmed at 30° for 2 min. The kinase reaction is initiated by addition of 40 μl of 2.5 m*M* ATP, prewarmed to 30°. Aliquots (20 μl) of the mixture are transferred to the assay buffer for the measurement of BCKDH complex activity at four time points, and the remaining BCKDH activity is immediately assayed spectrophotometrically as described above. Sampling times depend on the level of kinase activity in the extract. Glycerol is added to the kinase assay buffer to partially inhibit kinase activity, presumably as a consequence of greater viscosity of the assay mixture. This has been found useful because it gives a ratio of kinase activity to BCKDH complex activity that makes it easy to follow loss of BCKDH complex activity over a 3- to 4-min period of time. For example, four samples are taken over a 3-min time period during the assay of rat liver BCKDH kinase under the described assay conditions. Apparent first-order rate constants for the rate of BCKDH complex inactivation are calculated by least-squares linear regression analysis.[21]

Measurement of Amount of Branched-Chain α-Keto Acid Dehydrogenase Kinase Protein by Immunoaffinity Chromatography and Western Blot Analysis

Principle

The BCKDH complex is isolated from tissue extracts by immunoaffinity chromatography and the amount of BCKDH kinase associated with the complex is quantified by Western blot analysis with antibodies against the kinase. Because BCKDH kinase is tightly bound to the BCKDH complex, the procedure allows quantitative measurement of the amount of the kinase relative to the other subunits of the complex.

Materials. ^{125}I-Labeled protein A is obtained from ICN Pharmaceuticals (Costa Mesa, CA). Centricon-10 is purchased from Amicon (Beverly, MA). All other chemicals are from Sigma.

Buffers and Reagents

Goat anti-BCKDH complex IgG purified on DEAE-Sepharose[20]
CNBr-activated Sepharose 4B (Sigma)
Rabbit BCKDH E1/E2 antisera[20]
Purified rabbit BCKDH kinase antibodies generated against a 10-residue peptide from the amino-terminal end of BCKDH kinase[20]
Dialysis buffer: 0.5 M NaHCO$_3$, pH 8.2
2-Aminoethanol (pH 8.5), 100 mM
Extraction buffer: 10 mM Tris-HCl (pH 7.5), 1 mM EDTA, 5 mM 2-mercaptoethanol, 0.1 mM TPCK, trypsin inhibitor (0.1 mg/ml), leupeptin (1 mg/ml), pepstatin A (1 mg/ml), aprotinin (1 mg/ml), and 0.5% (w/v) Triton X-100, adjusted to pH 7.5
Washing buffer 1: 10 mM Tris-HCl (pH 7.5), 1 mM EDTA, 3 mM DTT, 0.1 mM TPCK, trypsin inhibitor (0.1 mg/ml), leupeptin (1 mg/ml), pepstatin A (1 mg/ml), aprotinin (1 mg/ml), 0.1% (w/v) Triton X-100, and 100 mM NaCl, adjusted to pH 7.5
Washing buffer 2: 10 mM Tris-HCl (pH 7.5) and 1 mM EDTA
Eluting solution: 2% (w/v) sodium dodecyl sulfate (SDS)
Loading buffer: 2% (w/v) SDS, 62.4 mM Tris-HCl (pH 6.8), 10% (v/v) glycerol, and 5% (v/v) 2-mercaptoethanol

Immobilization of Goat Anti-BCKDH IgG on CNBr-Activated Sepharose

Purified goat anti-BCKDH IgG (approximately 180 mg of protein) is dialyzed overnight against 2 liters of dialysis buffer at 4°. Insoluble material is removed by centrifugation at 14,000g for 10 min at 4°, and the supernatant is brought to 15 ml with 0.5 M NaHCO₃. Approximately 5 g of CNBr-activated Sepharose 4B is reconstructed in 50 ml of distilled water for 3 to 5 min and washed first with 100 ml of distilled water and then with 50 ml of 0.5 M NaHCO₃ through a funnel to remove lactose and dextran. The washed Sepharose is added directly to the IgG solution and rocked for 8 hr at room temperature. After the immobilization of IgG, the buffer is removed from the resin through a funnel. The resin is further washed twice with 50 ml of 0.5 M NaHCO₃ and twice with 100 ml of 0.5 M NaHCO₃ including 1 M NaCl. The resin is mixed with 40 ml of 100 mM 2-aminoethanol (pH 8.5) and rocked overnight at room temperature. Finally, the resin is washed and resuspended in phosphate-buffered saline (1 : 2, v/v). Approximately 16 ml of resin is obtained, with about 10 mg of anti-BCKDH complex IgG immobilized per 1 ml. Approximately 0.5 mg of purified BCKDH complex is bound by 1 ml of the immunoadsorbent. Adsorption of the complex does not release bound BCKDH kinase. The immunoadsorbent is stable when stored at 4°.

Tissue Extraction and Immunoaffinity Chromatography of Branched-Chain α-Keto Acid Dehydrogenase–Kinase Complex

Freeze-clamped liver (~0.5 g) is weighed into a plastic tube precooled in liquid nitrogen and homogenized in 5 volumes of extraction buffer. After centrifugation at 60,000g for 30 min, the supernatant is carefully removed without lipid and incubated for 4 hr with 0.5 ml of BCKDH immunoadsorbent at 4°. The gel is washed four times in 10 volumes of washing buffer 1 and one time with 5 volumes of washing buffer 2. BCKDH–kinase complex bound to the gel is eluted by vigorous mixing with 0.5 ml of eluting solution at room temperature. The elution fractions are concentrated on a Centricon filter, diluted with loading buffer, and boiled for 3 min.

Western Blot Analysis

The protein components of the immunoaffinity-purified BCKDH–kinase complex are separated by SDS–polyacrylamide gel electrophoresis (PAGE) on a 10% gel. The proteins are transferred to nitrocellulose membranes and blocked in 3% (w/v) bovine serum albumin (BSA) in PBS-T [phosphate-buffered saline, pH 7.4, with 0.02% (w/v) Tween 20] at 37° for 1 hr. The membrane is incubated with a mixture of rabbit polyclonal antisera specific for the E1 and E2 subunits of BCKDH complex [diluted 1:1000 in 3% (w/v) BSA in PBS-T] at 4° overnight. After washing the membrane three times for 10 min in PBS-T at room temperature, it is incubated with ^{125}I-labeled protein A (4.5 μCi, 5.8 μCi/μg) in BSA (3%, w/v) in PBS-T at room temperature for 1 hr, and then washed as described above. Based on quantification with an Ambis β scanner (AMBIS, Inc., San Diego, CA), the suspensions are adjusted to load identical amounts of the BCKDH complex on a second gel for SDS–PAGE. The proteins resolved by SDS–PAGE are transferred to a nitrocellulose membrane. This membrane in turn is blocked as described above, incubated with purified rabbit BCKDH kinase antibodies (1:50), washed as described above, and finally incubated with ^{125}I-labeled protein A. The amount of BCKDH kinase protein is quantified by autoradiography.

Activities of Branched-Chain α-Keto Acid Dehydrogenase Complex and Kinase in Various Mammalian Tissues

There have been many reports concerning the regulation of BCKDH complex in rat liver and muscle. Rat liver has the highest activity of BCKDH complex among the tissues that have been examined, suggesting that liver plays as important role in the disposal of the branched-chain amino acids in the rat. A remarkable sex-based difference occurs in the activity state of the liver BCKDH complex in the rat (Table I).[5] The activity of BCKDH

TABLE I
LIVER BCKDH COMPLEX ACTIVITY AND ACTIVITY STATE AND BCKDH KINASE ACTIVITY[a]
IN MALE AND FEMALE RATS[b]

| Sex | BCKDH complex activity (nmol/min/g wet weight) | | Activity state (%) | BCKDH kinase (min^{-1}) |
	Actual	Total		
Male	1669 ± 70	1847 ± 57	90 ± 2	0.25 ± 0.02
Female	213 ± 85^c	1246 ± 142^c	15 ± 5^c	1.25 ± 0.18^c

[a] Groups of five or six rats were maintained on chow diet for 3 weeks. Liver BCKDH complex activity was assayed spectrophotometrically. Total activity of the BCKDH complex was obtained by incubation with the broad-specificity protein phosphatase and MgSO$_4$. BCKDH kinase activity was assayed after dephosphorylation of the complex by incubation with the broad-specificity protein phosphatase and MgSO$_4$. All values represent means ± SEM.

[b] Adapted with permission from R. Kobayashi, Y. Shimomura, T. Murakami, N. Nakai, N. Fujitsuka, M. Otsuka, N. Arakawa, K. M. Popov, and R. A. Harris, *Biochem. J.* **237**, 449 (1997).

[c] Significantly different from male rats ($p < 0.05$).

kinase is three- to fivefold higher in females than in males (Table I), giving an inverse relationship between BCKDH kinase activity and the activity state of the BCKDH complex. The increase in BCKDH kinase activity caused by low dietary protein feeding correlates with an increase in the amount of BCKDH kinase bound to the BCKDH complex.[20] A concurrent increase in the abundance of the message[20] encoding BDKDH kinase suggests expression of BCKDH kinase may be regulated at the level of expression of its gene. Exercise training and starvation have significant effects on BCKDH complex activity in rat skeletal muscle.[10] Although BCKDH activity in skeletal muscle is much lower than in liver, exercise training increases the actual activity of the complex in skeletal muscle of resting rats in the fed condition. Starvation for 24 hr increases the actual activity in both untrained and trained rats (Table II). Exercise training also increases the total activity of BCKDH complex in the muscle (Table II). The amount of BCKDH kinase protein is decreased in skeletal muscle by exercise

TABLE II
ACTUAL AND TOTAL ACTIVITIES OF BCKDH COMPLEX IN SKELETAL
MUSCLE OF TRAINED AND UNTRAINED RATS[a,b]

Group	BCKDH complex activity in muscle (nmol/min/g wet weight)		Activity state (%)
	Actual	Total	
Untrained			
Fed	1.0 ± 0.2	32.8 ± 2.9	3.5 ± 0.7
Starved	2.5 ± 0.2^c	30.5 ± 2.1	8.5 ± 0.9^c
Trained			
Fed	1.8 ± 0.2^d	45.3 ± 3.5^d	4.1 ± 0.5
Starved	5.0 ± 0.7^c	42.6 ± 1.8^d	11.8 ± 1.5^c

[a] Rats were divided into untrained and trained groups ($n = 16$). Rats in the latter group were run on the treadmill for 30 min/day in the initial 5 weeks and for 60 min/day in the following 7 weeks. Rats were exercised for 5 days during each week at a running speed of 30–35 m/min. Rats were anesthetized with pentobarbital 24 hr after the final exercise and the gastrocnemius muscle was rapidly removed and freeze-clamped. Before anesthetization, half the rats in each group were starved for 24 hr. BCKDH complex was assayed radiochemically. The total activity of BCKDH complex was obtained by incubation with broad-specificity protein phosphatase. Values represent means $\pm$ SEM.
[b] Adapted with permission from Y. Shimomura, H. Fujii, M. Suzuki, T. Murakami, N. Fujitsuka, and N. Nakai, *J. Nutr.* **125,** 1762S (1995).
[c] Significantly different from the corresponding fed group ($p < 0.05$).
[d] Significantly different from corresponding untrained rats ($p < 0.05$).

training,[17] suggesting that decreased BCKDH kinase activity may be partially responsible for the increased BCKDH activity state in this condition.

In contrast to rats, the BCKDH complex activity of human and monkey livers is not nearly as high as that found in other tissues of these animals. The total activity in human liver is less than 2% of that in rat liver.[29,30] Kidney has the highest activity in both humans and monkeys.[29] It is likely that skeletal muscle plays a quantitatively more important role in branched-chain amino acid catabolism in the primates than in rodents.[29,30]

Remarks

Spectrophotometric and radiochemical assays are described here for the determination of BCKDH complex activity in crude tissue extracts. These methods are modifications of those reported previously.[14] Three methods for dephosphorylation of the BCKDH complex are given for the determination of total activity of the complex in tissue extracts. The use of commercially available lambda phosphatase for this purpose circumvents the need to purify broad-specificity protein phosphatase, which sometimes is found to be unstable or contaminated with apparent protease activity.

The radiochemical method for the assay of BCKDH complex activity is about 100 times more sensitive than the spectrophotometric method. The enzyme activities in rat liver, heart, and kidney are usually assayed by the spectrophotometric method whereas those in other rat and human tissues are assayed by the radiochemical method. The latter can also be used to determine pyruvate dehydrogenase activity by substituting [1-^{14}C]pyruvate for the labeled branched-chain α-keto acid.[31,32]

The activity of BCKDH complex in rat liver exhibits a marked sex difference (Table I) and a marked diurnal variation in females.[5] Therefore, the sex of the animals and the time of day that animals are killed should be carefully considered. Wide variability in the activity state of the BCKDH complex in rat liver has been reported,[15] which may in part be explained by the diurnal variation in BCKDH kinase activity. On the other hand, most studies so far conducted have used male rats. Because female rats have a high BCKDH complex activity state and a low kinase activity at

[29] A. Suryawan, J. W. Hawes, R. A. Harris, Y. Shimomura, A. E. Jenkins, and S. M. Hutson, *Am. J. Clin. Nutr.* **68**, 72 (1998).

[30] K. Taniguchi, T. Nonami, A. Nakao, A. Harada, T. Kurokawa, S. Sugiyama, N. Fujitsuka, Y. Shimomura, S. M. Hutson, R. A. Harris, and H. Takagi, *Hepatology* **24**, 1395 (1996).

[31] N. Nakai, G. R. Collier, Y. Sato, Y. Oshida, N. Fujitsuka, and Y. Shimomura, *Life Sci.* **60**, 51 (1997).

[32] N. Nakai, Y. Sato, Y. Oshida, A. Yoshimura, N. Fujitsuka, S. Sugiyama, and Y. Shimomura, *Life Sci.* **60**, 2309 (1997).

the beginning of the light period, and a low activity state and a high kinase activity at the end of the light period, they provide a good model system for the study of hormonal and nutritional control of branched-chain amino acid metabolism.

Acknowledgments

This work was supported by grants from the U.S. Public Health Services (NIH DK19259 and DK40441 to R.A.H., GM51262 to K.M.P.), the Grace M. Showalter Trust (to R.A.H.), the Uehara Memorial Foundation (to Y.S.), and the University of Tsukuba Project Research Fund (to Y.S.).

[7] Simultaneous Quantification of Plasma Levels of α-Ketoisocaproate and Leucine by Gas Chromatography–Mass Spectrometry

By Marzia Galli Kienle and Fulvio Magni

Introduction

Metabolism of the branched-chain amino acids (BCAAs) leucine, isoleucine, and valine is disturbed in a number of hereditary and acquired diseases. Therefore quantification of both the amino acids and their closely related keto analogs is often required in investigation of amino acid metabolism. Leucine and α-ketoisocaproic acid (KIC) are the most extensively investigated components in this respect. Variably labeled leucines are infused in humans and both the concentration and the isotopic enrichment in plasma are determined in order to study muscle protein synthesis. Because KIC is not recycled directly to amino acids,[1,2] its isotopic enrichment is a better index of metabolism and of intra- and extracellular exchange of amino acids than that of leucine. Experiments carried out for this purpose involve multilabeled tracers and complex schemes.

Plasma levels and enrichment of leucine have been measured by gas chromatography–mass spectrometry (GC–MS) after derivatization with various reagents.[3,4] Keto acid transformation into derivatives has been also

[1] S. M. Hutson, T. C. Cree, and A. E. Harper, *J. Biol. Chem.* **253**, 8126 (1978).

[2] M. J. Rennie, R. H. T. Edwards, D. Halliday, D. E. Matthews, S. L. Wolman, and D. J. Millward, *Clin. Sci.* **63**, 519 (1982).

[3] G. C. Ford, K. N. Cheng, and D. Halliday, *Biomed. Mass Spectrom.* **12**, 432 (1985).

[4] G. N. Thompson, P. J. Pacy, G. C. Ford, H. Merritt, and D. Halliday, *Europ. J. Clin. Invest.* **18**, 639 (1988).

described, and the most widely used procedures are based on the conversion into O-(trimethylsilyl)quinoxalinols,[3–5] pentafluorobenzyl (PFB) esters,[6] or silylated oximes.[7] Formation of quinoxalinols is tedious and requires carefully prepared o-phenylenediamine-based reagents, which are carcinogenic and sensitive to oxygen.[8] The high background due to the natural abundance of ^{29}Si and ^{30}Si introduced into the derivative can be avoided by the use of N- and O-alkylated quinoxalinols as reported by Fernandez et al.[9] Nevertheless, o-phenylenediamine is still used. Higher sensitivity is achieved with the formation of PFB esters followed by analysis in the negative ion chemical ionization (NICI) mode. Nevertheless, electron ionization (EI) is a more widely diffused technique, and thus methods based on this type of ionization represent a useful tool for many mass spectrometry laboratories. Derivatization with trimethylsilylating agents is the most common procedure for GC analysis of these compounds. Amino acids are converted into their trimethylsilyl (TMS) derivatives at both the carboxylic and amino groups. However, protection of the keto group of keto acids is required before silylation.[10] N-Methyl-N-($tert$-butyldimethylsilyl)trifluoroacetamide (MTBSTFA) has been used as silylating agent, being considerably more stable than BSTFA and other reagents utilized for TMS preparation. Analyses of amino acids[11,12] and studies of leucine turnover[13,14] that employ $tert$-butyldimethylsilyl derivatives have already been reported and MTBSTFA has also been shown to be able to react with keto acids.[12]

The purpose of this chapter is to describe a method implying EI–GC–MS measurement of $tert$-butyldimethylsilyl and methyloxime $tert$-butyldimethylsilyl derivatives of leucine and KIC, respectively. Plasma levels of the two compounds are determined accurately with norleucine and α-ketovaleric acid (KVA) as internal standards. [1-^{13}C]Leucine and

[5] H. P. Schwarz, I. E. Karl, and D. M. Bier, *Anal. Biochem.* **108**, 360 (1980).

[6] W. Kulik, L. van Toledo-Eppinga, R. M. Kov, W. S. Guèrand, and H. N. Lafeber, *J. Mass Spectrom.* **30**, 1260 (1995).

[7] C. Aussel, L. Cynober, and J. Giboudeau, *J. Chromatogr.* **423**, 270 (1987).

[8] G. Liversey and W. T. E. Edwards, *J. Chromatogr.* **377**, 98 (1985).

[9] A. A. Fernandez, S. C. Kalhan, N. G. Njoroge, and G. S. Matousek, *Biomed. Mass Spectrom.* **13**, 569 (1985).

[10] T. Niwa, *J. Chromatogr. Biomed. Appl.* **379**, 313 (1986).

[11] T. P Mawhinney, R. S. R. Robinett, A. Atalay, and M. A. Madson, *J. Chromatogr.* **358**, 231 (1986).

[12] W. F. Schwenk, P. J. Berg, B. Beaufrere, J. M. Miles, and M. W. Haymond, *Anal. Biochem.* **141**, 101 (1984).

[13] W. M. Bennet, A. A. Connacher, C. M. Scrimgeour, K. Smith, and M. J. Rennie, *Clin. Sci.* **76**, 447 (1989).

[14] G. L. Loy, A. N. Quick, Jr., C. C. Teng, W. W. Hay, Jr., and P. V. Fennessey, *Anal. Biochem.* **185**, 1 (1990).

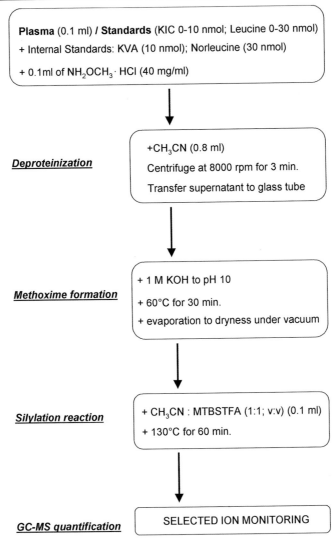

Plasma (0.1 ml) / **Standards** (KIC 0-10 nmol; Leucine 0-30 nmol)
+ Internal Standards: KVA (10 nmol); Norleucine (30 nmol)
+ 0.1ml of $NH_2OCH_3 \cdot HCl$ (40 mg/ml)

Deproteinization

+CH_3CN (0.8 ml)
Centrifuge at 8000 rpm for 3 min.
Transfer supernatant to glass tube

Methoxime formation

+ 1 M KOH to pH 10
+ 60°C for 30 min.
+ evaporation to dryness under vacuum

Silylation reaction

+ CH_3CN : MTBSTFA (1:1; v:v) (0.1 ml)
+ 130°C for 60 min.

GC-MS quantification

SELECTED ION MONITORING

SCHEME 1. Outline of the procedures used for the extraction, purification, derivatization, and mass spectrometry analysis of leucine and α-ketoisocaproic acid in biological samples.

[1-^{13}C]KIC enrichment can also be evaluated by the present method in plasma samples treated with the labeled ^{13}C isotopomers (see Scheme 1). Both sample preparation and GC–MS analysis are simple and rapid, so that several samples can be analyzed daily.

Reagents

L-Leucine (99% pure) and L-norleucine (99% pure): Purchase from Sigma (St. Louis, MO)

Sodium α-ketoisocaproate and α-ketovalerate: Obtain from Fluka Chemie (Buchs, Switzerland)

L-[1-^{13}C]Leucine (99 atom% excess) and sodium [1-^{13}C]α-ketoisocaproate (99 atom% excess): Obtain from Tracer Technologies (Sommerville, NJ)

N-Methyl-N-(*tert*-butyldimethylsilyl)trifluoroacetamide (MTBSTFA): Obtain from Regis Chemical (Morton Grove, IL)

Methoxyamine hydrochloride: Obtain from Deside Industrial Park (Cluyd, UK)

Methoxyamine hydrochloride solution: Make in water at a concentration of 40 mg/ml when needed

All other reagents and solvents: Analytical grade

Standard Solutions

Stock solutions of leucine, [1-^{13}C]leucine, and norleucine are prepared by dissolving a weighed amount of about 10–15 mg in 1 M HCl to a 30 mM concentration. Working solutions are prepared by 1:20 (v/v) dilution of stock solutions with 1 M HCl to reach a final concentration of 1.5 mM. All solutions can be stored at 4° and are stable for at least 30 days.

Solutions of the sodium salts of α-ketoisocaproic acid, α-ketovaleric acid, and [1-^{13}C]KIC are freshly prepared at a concentration of 15 mM by dissolving about 10 mg of each compound in distilled water. Working solutions are prepared by dilution of stock solutions with water to reach the final concentration of 0.4 mM.

Procedures

Preparation and Derivatization of Plasma Extracts

Plasma aliquots of (0.1 ml) are spiked with 30 nmol of norleucine and with 10 nmol of KVA, to be used as internal standards for leucine and KIC quantification, respectively. Samples are vortexed for 30 sec to allow equilibration. After treatment with 0.1 ml of methoxyamine hydrochloride, deproteinization is carried out by addition of 0.8 ml of CH$_3$CN followed by centrifugation at 8000 rpm for 3 min in a Beckman (Fullerton, CA) Microfuge 11. The supernatant is transferred to a 10-ml glass tube and about 50 μl of 1 M potassium hydroxide is added to reach pH 10. The mixture is then heated at 60° for 30 min to form the methyloxime derivative of keto acids. Particular attention must be paid to the pH, which represents the most critical parameter of the reaction. After evaporation to dryness under vacuum the residue is reacted with 0.1 ml of a freshly prepared derivatizing solution [CH$_3$CN–MTBSTFA (2:1, v/v)] at 130° for 1 hr and 0.5- to 1-μl volumes of the resulting solutions are injected for GC–MS analysis. Derivatives of both leucine and KIC are stable for at least 3 days.

Mass Spectra of Derivatives

Mass spectra of methyloxime–*tert*-butyldimethylsilyl (TBDMS) derivatives of pure KIC and KVA are shown in Fig. 1. Under the EI conditions used, molecular ions (at m/z 273 and 259, respectively) are not detectable. A major signal resulting from the loss of the butyl group from the ionized molecule ([M-57]$^+$) is present at m/z 216 and 202 in the spectra of KIC and KVA derivatives, respectively. The spectrum of [^{13}C]KIC shows the corresponding ion at m/z 217.[15] Mass spectra of the derivatives of leucine and [1-^{13}C]leucine are shown in Fig. 2, with a fragmentation pattern based on the comparison of the spectra of the two isotopomers. The mass spectrum of norleucine (not shown), presents the same fragmentation pattern of leucine.

Mass Spectrometry Apparatus

Any conventional GC–MS instrument equipped with a capillary fused-silica column can be used. We currently use a Finnigan (ThermoQuest, Austin, TX) 4500 mass spectrometer. MS conditions are as follows: ion energy, 70 eV; emission current, 0.25 mA; conversion dynode, +3.0 kV; transfer line temperature, 280°; electromultiplier, 1.3 kV; preamplifier sensitivity, 10^{-8}; source temperature, 150°.

Gas Chromatography–SIM Analysis

Separation of amino acid and keto acid derivatives is performed on a Carlo Erba HR-GC with an SPB-20 fused-silica capillary column [30 m × 0.32 mm i.d.; film thickness, 0.25 μm (Supelco Inc., Bellefonte, PA)]. The column is coupled directly to the ion source of the mass spectrometer, which provides high inertness and separation efficiency, and helium is the carrier gas. The injector temperature is held at 280°. After 1 min at 100° the oven temperature is increased to 270° at 15°/min. Because accumulation of residues derived from the biological matrix in the capillary column would gradually alter the chromatographic separation, the glass sleeve preceding

[15] F. Magni, L. Arnoldi, G. Galati, and M. Galli Kienle, *Anal. Biochem.* **220**, 308 (1994).

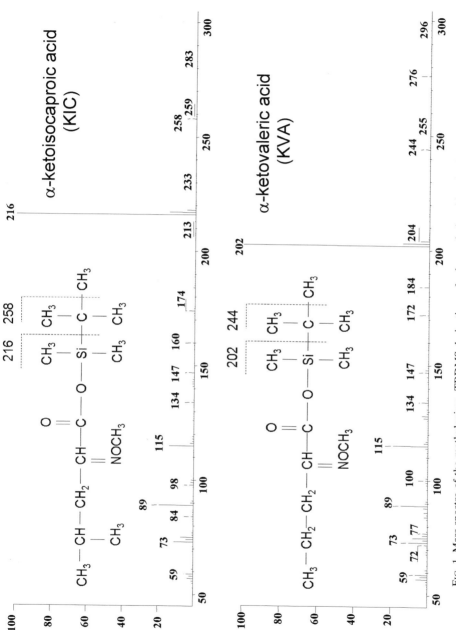

FIG. 1. Mass spectra of the methyloxime–TBDMS derivatives of α-ketovaleric acid and α-ketoisocaproic acid.

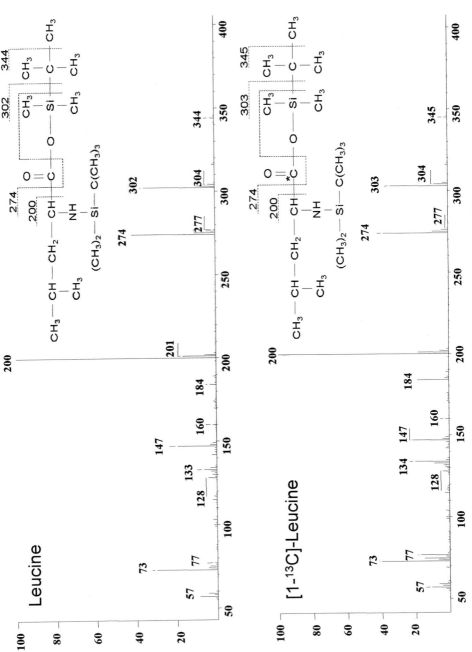

Fig. 2. Mass spectra of TBDMS derivatives of leucine and [1-¹³C]leucine.

the column is changed daily and the column is occasionally programmed to 270°. From time to time it is necessary to cut the first 10–15 cm of the silica-fused capillary column head connected with the injector to restore the original chromatographic behavior.

Under the GC conditions noted above, the analysis of leucine and KIC is not interfered with by amino acids and α-keto acids (α-ketoisovaleric acid, KIV; α-keto-β-methylvaleric acid, KMV) normally present in plasma (Fig. 3), or by matrix components. KVA and KIC give mainly the *anti* forms of the methyloxime derivatives (Fig. 3, peaks d and f in chromatograms of ions at m/z 202 and 216, respectively), whereas the *syn* forms (Fig. 3, peaks c and e) are less represented. For both KIV and KMV the two isoforms are obtained in comparable amounts (Fig. 3, peaks a, b and g, h in the chromatograms of ions at m/z 202 and m/z 216, respectively). Gas chromatographic separation of plasma amino acids and keto acids is optimized with control plasma and the identification of each peak is based on the comparison of the obtained mass spectrum with that of the pure standard compound. Although KMV could not be completely separated from KVA, SIM analysis allows differentiation of the two acids because no detectable signal is present in the KMV spectrum at m/z 202. Complete separation of valine, leucine, isoleucine, and norleucine is evident in Fig. 3. A typical analysis of plasma samples spiked with KVA is shown in Fig. 4, which presents chromatograms of ions selected for KIC (m/z 216) and KVA (m/z 202). Analysis of plasma samples spiked with norleucine is not shown

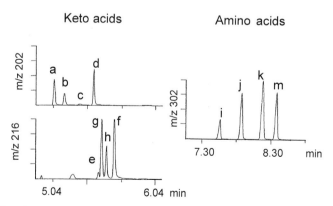

FIG. 3. Ion chromatograms of mixtures of methyloxime–TBDMS α-keto acids and TBDMS amino acids normally present in plasma. The two isoforms of α-ketoisocaproic acid (e and f), of α-ketovaleric acid (c and d), of α-ketoisovaleric acid (KIV; a and b), and of α-keto-β-methylvaleric acid (KMV; g and h) are indicated. The chromatogram at m/z 302 shows the separation of valine (i), leucine (j), isoleucine (h), and norleucine (m).

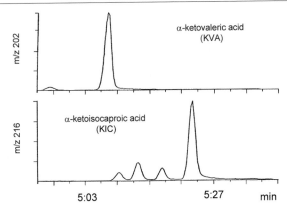

FIG. 4. Chromatograms of ions at m/z 202 and 216 obtained in analyzing the extract of a plasma sample spiked with α-ketovaleric acid (KVA). The main isoforms of KVA and endogenous α-ketoisocaproic acid (KIC) are clearly detectable.

because the ion chromatogram (m/z 302) is almost identical to that reported for amino acids in Fig. 3.

Leucine and KIC analysis is easily accomplished in 7–9 min and injections are performed every 15 min. Calculations of KIC concentration and enrichment are performed using only the peak area of the major form of the keto acid derivative.

Quantification

To evaluate plasma concentrations of leucine and KIC, SIM analysis is carried out by focusing ions at m/z 302 for leucine and norleucine and ions at m/z 216 and 202 for KIC and KVA, respectively. When enrichment with [1-^{13}C]leucine and [1-^{13}C]KIC is determined, ions at m/z 303 and 217 are also focused.

Plasma concentrations of leucine and KIC are calculated from area ratios of signals monitored by SIM at the retention times of the target compound and of the internal standard based on leucine/norleucine and KIC/KVA calibration curves.

Calibration Curves

A set of calibration curves is prepared by mixing aqueous solutions of standard norleucine (30 nmol) with leucine (0–30 nmol), and aqueous solutions of KVA (10 nmol) with KIC (0–10 nmol), to provide a series of samples with analyte-to-internal standard ratios 0–0.25–0.5–0.75–1. Calibration samples are then processed by the above-described analytical procedure. Ratios found for leucine/norleucine and KIC/KVA, obtained by

GC–SIM analysis, are plotted against the known molar ratios. Linear regression analysis is used to derive the equation of the line for each curve. It is worth noting that because of the absence of matrix interference (as determined by comparison of calibration curves obtained by spiking serum with standards with those prepared in the absence of the matrix),[15] the linearity of the method can be checked. The accuracy of the method is evaluated as percent bias between found and expected values using these curves. The percent bias ranges from −3.5 to 1.6% for leucine and from 0.7 to 4.5% for KIC at the various tested concentrations. Repeatability [percent coefficient of variation (% CV)] of leucine and KIC level evaluation is also determined by analyzing two or three aliquots of a plasma specimen (0.1 ml) on five different days; each aliquot is added with 75 nmol (50 μl) of norleucine and 20 nmol (50 μl) of KVA. The between-day imprecision (% CV) for the measurement of leucine and KIC plasma levels is 4.3 and 5.6%, respectively. The within-day imprecision, evaluated by triplicate injection of three aliquots of the same plasma, is 2.8 and 5.1% for leucine and KIC, respectively. The method has also been validated for the measurement of [13C]leucine–leucine and [13C]KIC–KIC ratios, when enrichment in plasma of subjects given [1-13C]leucine by intravenous injection is evaluated.[15] For this purpose calibration curves are prepared by mixing a fixed amount of unlabeled leucine and KIC with various amounts of standard solutions of [1-13C]leucine and [1-13C]KIC to obtain enriched samples with molar percent enrichment (moles of 13C per total moles) in the 0–10% range for both compounds. Curves are also prepared by spiking plasma aliquots (0.1 ml) with the same amounts of labeled and nonlabeled standards to obtain 13C-enriched plasma samples with molar percent enrichment in the 0–10% range. Calibration samples are then processed by the above-described analytical procedure. Leucine and KIC enrichments obtained by GC–SIM analysis are plotted against the known [13C]leucine–leucine and [13C]KIC–KIC molar ratios. Linear regression analysis is used to derive the equation of the line for each curve. Typical calibration curves of 13C enrichment of KIC and leucine have a linear relationship ($r \geq 0.997$ and $r \geq 0.998$, respectively) over the working range of 0–10% enrichment. No significant difference is found ($p > 0.05$) between slopes of curves for leucine and KIC prepared in plasma and in water as determined by comparison of the two calibration lines on three different days. Under the described experimental conditions the CV value for the raw isotopic ratios ($[M-57 + 1]^+/[M-57]^+$) is expected not to exceed 1% both in enriched and nonenriched samples (0–2 mol% enrichment) for both leucine and KIC. Details on the molar percent enrichment calculation were reported previously.[16]

---

[16] F. Magni, L. Monti, P. Brambilla, R. Poma, G. Pozza, and M. Galli Kienle, *J. Chromatogr.* **573**, 125 (1992).

Conclusions

Under the selected conditions silylation with MTBSTFA allows derivatization of both leucine and KIC and is efficient enough for the evaluation of plasma levels of the two compounds in the EI ionization mode. Compared with TMS derivatives,[17] TBDMS derivatives of amino acids offer the advantage of much higher stability to hydrolysis, improved separation by GC, and an intense molecular weight-indicative peak, $[M-57]^+$. Good linearity and high sensitivity are obtained. Optimal linearity obtained for $[1-^{13}C]$leucine–leucine calibration curves suggests that $[1-^{13}C]$leucine may be used instead of norleucine as the internal standard. Analysis of α-keto acids is more critical owing to the instability of these compounds, which easily decompose, dimerize, and decarboxylate.[10] Nevertheless, precision is rather good when using KIC and KVA working solutions prepared daily.

Direct derivatization of keto acids with MTBSTFA was not used because of the lack of linearity of the KIC–KVA calibration curve and low sensitivity of the analysis, likely attributable to the loss of the highly volatile compound from the samples.[18] The transformation of the acids into potassium salts before evaporation and silylation remarkably increased sensitivity, but linearity of the calibration curve still could not be obtained. Protection of the keto group as methyloxime before salt formation and evaporation to dryness under vacuum considerably decrease the instability of keto acids, so that under these conditions optimal linearity is reached ($r = 0.999$). Methoxymation and salification do not interfere in leucine analysis as demonstrated by the linearity of leucine–norleucine ($r = 1.000$) and $[^{13}C]$leucine–leucine ($r = 0.999$) calibration curves when MTBSTFA treatment is preceded by these steps. Compared with the other reported methods, quantification of both leucine and KIC and determination of their percent enrichment by the described procedure is rapid and simple because it requires only addition to plasma of internal standards (norleucine and KVA), deproteinization with CH_3CN in the presence of methoxyamine hydrochloride, centrifugation, evaporation to dryness, and treatment with MTBSTFA. Long purification steps such as extraction with solvents and cleaning up by ion-exchange resins described previously[3,14,19] are avoided. Moreover, previously published methods either require two different purification procedures for the leucine and KIC quantification[3,4] or need two distinct steps at the derivatization level.[14]

The imprecision of leucine determination is less than that of KIC deter-

[17] A. P. J. M. De Jong, J. Elma, and B. J. T. Van De Berg, *Biomed. Mass Spectrom.* **7,** 359 (1980).
[18] B. Schatowitz and G. Gercken, *J. Chromatogr.* **409,** 43 (1987).
[19] F. Rocchiccioli, J. P. Leroux, and P. Cartier, *Biomed. Mass Spectrom.* **8,** 160 (1981).

mination (2.8 and 5.1% CV, respectively), likely because of the α-keto acid instability, but the accuracy is on the same order of magnitude for both analytes (percent bias ranges from −3.5 to +1.6% for leucine and from +0.7 to +4.5% for KIC, respectively).

Between-day imprecision of plasma level evaluation is 4.5 and 5.5% for leucine and KIC, respectively, as determined by analyzing a plasma sample for five consecutive days.

In summary, the method presented here allows the evaluation of plasma levels of both leucine and KIC, using a single 0.1-ml plasma aliquot with norleucine and KVA as internal standards. The use of isotopomers of leucine and KIC as internal standards may improve the imprecision as reported by Kulik et al.[6] using deuterium-labeled standards. Compared to other methods,[20,21,22] precision, repeatability, and accuracy of the method, the simple and rapid preparation of the samples for GC–MS analysis, and the possibility of using an EI ion source render this method particularly useful for the quantification of amino acids and keto acids in studies with a large number of samples.

[20] D. E. Matthews, H. P. Schwarz, R. D. Yang, K. J. Motil, V. R. Young, and D. M. Bier, *Metabolism* **31,** 1105 (1982).
[21] N. K. Fukagawa, K. L. Minaker, J. W. Rowe, M. N. Goodman, D. E. Matthews, D. M. Bier, and V. R. Young, *J. Clin. Invest.* **76,** 2306 (1985).
[22] D. E. Matthews, K. J. Motil, D. K. Rohrbaugh, J. F. Burke, V. R. Young, and D. M. Bier, *Am. J. Physiol.* **238,** E473 (1980).

[8] Synthesis of Methacrylyl-CoA and (R)- and (S)-3-Hydroxyisobutyryl-CoA

By John W. Hawes and Edwin T. Harper

Introduction

In mammalian tissues valine is catabolized through a specific pathway distinct from that for other branched-chain amino acids. Unusual and requisite steps in this pathway include enzymatic hydration of methacrylyl-CoA, followed by cleavage of the resulting 3-hydroxyisobutyryl-CoA by a highly specific acyl-CoA hydrolase. This pathway differs from that for leucine and isoleucine in the production of a free branched-chain acid (3-hydroxyisobutyrate) and in utilizing enzymes (3-hydroxyisobutyryl-CoA hydrolase and 3-hydroxyisobutyrate dehydrogenase) dedicated to this specific function in valine degradation. It was proposed that this unusual pathway evolved to dispose irreversibly of methacrylyl-CoA, a toxic compound that can conjugate various cellular thiols.[1] The study of these enzymes, including crotonase and 3-hydroxyisobutyryl-CoA hydrolase, requires the availability of these branched-chain CoA-ester substrates. The present study presents methodology for the synthesis of methacrylyl-CoA and (R)- and (S)-3-hydroxyisobutyryl-CoA. Methods are described for reversed-phase high performance liquid chromatography (HPLC) purification of these CoA thioesters and their analysis by electrospray ionization (ESI) mass spectrometry and ^{1}H NMR.

Materials

Coenzyme A, Ellman reagent, crotonase, trifluoroacetic acid, and ethyl chloroformate are purchased from Sigma (St. Louis, MO). Methacrylic anhydride, methyl (R)-3-hydroxy-2-methylpropionate, and methyl (S)-3-hydroxy-2-methylpropionate are purchased from Aldrich (Milwaukee, WI). Tetrahydrofuran (THF) and acetonitrile are purchased from Fisher Scientific (Pittsburgh, PA).

Synthesis and Purification of Methacrylyl-CoA

Methacrylyl-CoA is synthesized by a method similar to that previously described by Stern *et al.*[2] Coenzyme A (150 mg, 195 μmol) is dissolved in 2 ml of 0.1 *M* sodium phosphate buffer, pH 8.5. To this solution, 0.5 ml of methacrylic anhydride (3 mmol) is slowly added, with stirring, at 0°. The solution is stirred on ice for 10 min, after which aliquots of 5 and 25 μl are removed and analyzed for free sulfhydryl groups by the Ellman assay.[3] Free sulfhydryl groups are completely absent, indicating that all of the CoA has reacted with methacrylic anhydride. The pH of this solution is lowered to pH 3.5 with HCl and the solution is extracted four times successively with 15 ml of water–saturated diethyl ether to remove excess methacrylic acid. The aqueous phase is then dried under a stream of nitrogen to remove residual ether. This product consists of crude methacrylyl-CoA and can be used as a substrate for crotonase or can be purified by reversed-phase HPLC. For purification, the entire sample is diluted to 9.5 ml with HPLC-grade water and loaded into a 10-ml injection loop on a Waters (Milford, MA) 600 HPLC system. The sample is injected onto a Vydac (Hesperia, CA) preparative C_{18} column (5 $\times$ 25 cm) and washed with 3 column volumes

[1] Y. Shimomura, T. Murakami, N. Fujitsuka, N. Nakai, Y. Sato, S. Sugiyama, N. Shimomura, J. Irwin, J. W. Hawes, and R. A. Harris, *J. Biol. Chem.* **269,** 14248 (1994).
[2] J. R. Stern, A. Campillo, and I. Raw, *J. Biol. Chem.* **218,** 971 (1956).
[3] G. L. Ellman, *Arch. Biochem. Biophys.* **82,** 70 (1959).

of 2% (v/v) solvent B in solvent A at a flow rate of 10 ml/min. Solvent A consists of 0.1% (v/v) trifluoroacetic acid (TFA) in water and solvent B consists of 90% acetonitrile, 0.1% (v/v) TFA. A solvent gradient is then applied to the column with a flow rate of 10 ml/min as shown in the tabulation below.

Time (min)	Percent A	Percent B
0	98	2
40	85	15
120	60	40
180	20	80
200	0	100

Several minor peaks elute throughout the gradient and one major peak, eluting at approximately 50% B, corresponds to methacrylyl-CoA. Elution is monitored by absorbance at 260 nm, using a Waters absorbance detector. The major peak is active as a substrate for crotonase, and is identified as methacrylyl-CoA by ESI mass spectrometry as described below. The fractions containing methacrylyl-CoA are combined and brought to neutral pH by addition of NaOH. This sample is lyophilized thee times, successively, to remove residual TFA and acetonitrile. The purified, lyophilized product is stored, desiccated, at −20°. Purified methacrylyl-CoA is quantitated by reading absorbance at 260 nm, using a molar extinction coefficient of 16,800. The purified product consists of 49 μmol of methacrylyl-CoA, corresponding to a yield of 30%.

Enzymatic Synthesis of (S)-3-Hydroxyisobutyryl-CoA

Methacrylyl-CoA is converted to S-3-hydroxyisobutyryl-CoA by incubation with crotonase. Twelve milligrams (14 μmol) of methacrylyl-CoA is dissolved in 750 μl of 0.1 M sodium phosphate buffer at pH 8.0. Five units of crotonase is added and this solution is incubated at 37° for 1 hr, producing an equilibrium mixture of methacrylyl-CoA and (S)-3-hydroxy-isobutyryl-CoA. This mixture is suitable for measurement of the activity of (S)-3-hydroxyisobutyryl-CoA hydrolase.[1,4] The methacrylyl-CoA and

[4] J. W. Hawes, J. Jaskiewicz, Y. Shimomura, B. Huang, J. Bunting, E. T. Harper, and R. A. Harris, *J. Biol. Chem.* **271,** 26430 (1996).

hydroxyisobutyryl-CoA are separated by reversed-phase HPLC (Fig. 1). In this case the sample is applied to a Vydac semipreparative C_{18} column (25 cm × 10 mm) and is washed and eluted with a gradient similar to that described above, at a flow rate of 2 ml/min, using a Perkin-Elmer (Norwalk, CT) 250 HPLC pump system. Methacrylyl-CoA elutes as a single peak at approximately 50% B and (S)-3-hydroxyisobutyryl-CoA elutes as a distinct peak at approximately 40% B (Fig. 1). Fractions corresponding to each peak are analyzed by ESI mass spectrometry, combined and neutralized by addition of NaOH. Each CoA ester is lyophilized three times repeatedly to remove excess TFA and acetonitrile.

For kinetic and mechanistic studies of S-3-hydroxyisobutyryl-CoA hydrolase, 2-deutero-3-hydroxyisobutyryl-CoA substrate is enzymatically synthesized by incorporating a stable deuterium at the C-2 position of the 3-hydroxyisobutyryl moiety. This is accomplished by performing the crotonase reaction as described above in 100% D_2O rather than H_2O. For

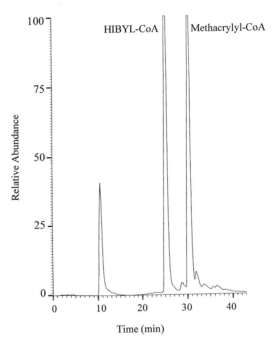

Fig. 1. Separation of methacrylyl-CoA and (S)-3-hydroxyisobutyryl-CoA. An equilibrium mixture of methacrylyl-CoA and (S)-3-hydroxyisobutyryl-CoA produced by the crotonase reaction was separated by reversed-phase HPLC, using a semipreparative C_{18} column (25 cm × 10 mm). Elution was monitored by absorbance at 260 nm and plotted with a scale of 0.5.

this reaction, lyophilized methacrylyl-CoA and crystallized crotonase are dissolved in D_2O. A 0.2 M sodium phosphate buffer is prepared by lyophilizing a 0.2 M sodium phosphate solution, pH 8.0, and resuspending it in an equal volume of D_2O. This lyophilization and resuspension procedure is repeated two times. An equal volume of the final buffer solution is then added to a D_2O solution containing 12 mg of methacrylyl-CoA and 5 units of crotonase. This solution is incubated at 37° for 1 hr. The resulting 2-deutero-3-hydroxyisobutyryl-CoA is purified by reversed-phase HPLC as described above, and analyzed by ESI mass spectrometry and ^1H NMR spectrometry.

Chemical Synthesis of (R)- and (S)-3-Hydroxyisobutyryl-CoA

To study the enantiomeric specificity of 3-hydroxyisobutyryl-CoA hydrolase, (R)- and (S)-3-hydroxyisobutyryl-CoA are synthesized, starting with (R)- and (S)-3-hydroxyisobutyrate methyl esters according to the method of Wieland and Rueff.[5] The methyl esters are hydrolyzed to form the corresponding free acids as previously described.[6] The hydroxyisobutyrates in free acid form are quantitated by a specific end-point assay using purified 3-hydroxyisobutyrate dehydrogenase.[6] Each isomer of 3-hydroxyisobutyrate (250 μmol) is lyophilized and resuspended in 2 ml of dry THF. This solution is placed in an ethanol–ice–salt bath (approximately −12°) and the atmosphere purged with nitrogen. To the chilled solution is added 300 μl of ethyl chloroformate to form a hydroxyisobutyrate–formate mixed anhydride. This mixture is stirred under nitrogen for 1 hr, purged of excess ethyl chloroformate under a stream of nitrogen, and resuspended in 5 ml of dry THF. The sample is then centrifuged at top speed in a clinical centrifuge to remove insoluble salts. Hydroxyisobutyrate–formate mixed anhydrides produced in this way are quantitated by the hydroxamate assay[7] using acetylphosphate as a standard. The mixed anhydrides are used to synthesize CoA esters immediately after synthesis and quantitation by the hydroxamate assay.

To produce (R)- and (S)-3-hydroxyisobutyryl-CoA from the corresponding mixed anhydrides, 50 μmol of coenzyme A in sodium phosphate buffer, pH 8.0, is added to a threefold excess of the anhydride while stirring on ice. This mixture is stirred on ice for 10 min, and then analyzed by the

[5] T. Wieland and L. Rueff, *Angew. Chem.* **65,** 186 (1953).

[6] P. M. Rougraff, R. Paxton, G. W. Goodwin, R. G. Gibson, and R. A. Harris, *Anal. Biochem.* **184,** 317 (1990).

[7] W. G. Robinson, R. Nagle, B. K. Bachhawat, F. P. Kupiecki, and M. J. Coon, *J. Biol. Chem.* **224,** 1 (1957).

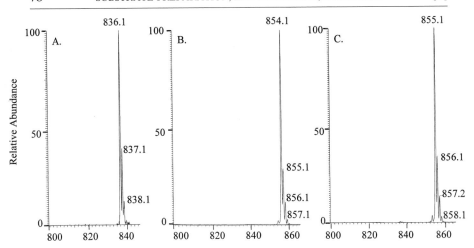

FIG. 2. Mass analysis of purified CoA esters. Electrospray ionization mass spectra were recorded with a Finnigan LCQ mass spectrometer as described in text. (A) Methacrylyl-CoA; (B) (S)-3-hydroxyisobutyryl-CoA; (C) (S)-2-deutero-3-hydroxyisobutyryl-CoA.

Ellman assay[3] as described above for methacrylyl-CoA. Complete conversion to the CoA ester is observed in each case. The resulting 3-hydroxyisobutyryl-CoA is purged of excess THF under a stream of nitrogen and brought to pH 3.5 by addition of HCl. This sample is extracted four times repeatedly with 15 ml of water-saturated ethyl ether. The (R)- and (S)-3-hydroxyisobutyryl-CoA are then lyophilized and stored desiccated at −20°. Aliquots of (R)- and (S)-3-hydroxyisobutyryl-CoA are also analyzed by the hydroxamate assay.[7] The hydroxamates produced are analyzed by paper chomatography in 2-butanol–formic acid–water (75/13/12, by volume), producing similar but distinctly different R_f values near 0.5. (R)- and (S)-3-hydroxyisobutyryl-CoA are also analyzed by thin-layer chomatography

TABLE I
RESONANCES ASSIGNED TO FUNCTIONAL GROUPS IN ACYL PORTION OF METHACRYLYL- AND 3-HYDROXYISOBUTYRYL-CoA

Functional group	Methacrylyl-CoA	2-[^{1}H]HIBYL-CoA[a]	2-[^{2}H]HIBYL-CoA
CH$_3$	1.80	1.11, 1.12 (d)	1.10 (s)
>CH−C=O		2.96	—
−CH$_2$OH		3.70 (m)	3.69 (m)
=CH	5.95		

[a] HIBYL, hydroxybutyryl.

in a solvent system consisting of ethanol–0.1 M sodium acetate, pH 4.5 (2 : 1, v/v), on 250-μm pore size Avicel plates (Analtech, Newark, DE).

Analysis of Methacrylyl-CoA and 3-Hydroxyisobutyryl-CoA by Electrospray Ionization Mass Spectrometry

For electrospray ionization mass spectrometry, methacrylyl-CoA and 3-hydroxyisobutyryl-CoA are diluted to a concentration of 0.1 μmol/ml in methanol containing 0.1% (w/v) acetic acid. Samples were analyzed on a Finnigan (ThermoQuest, Austin, TX) MAT LCQ by direct infusion with a flow rate of 5 μl/min. Samples are ionized with a spray voltage of 4.8 kV, a capillary voltage of 26 V, a capillary temperature of 200°, and a tube lens offset of 35 V. Nitrogen is used as a sheath gas, and no auxiliary gas is used. Spectra are scanned over an m/z range of 200 to 2000. Data are collected at 3 microscans per second and recorded as an average of 10 microscans. Purified methacrylyl-CoA is observed as a single charge ion (m$^+$ ion) at m/z 836.1 in good agreement with the calculated mass (Fig. 2A). Purified 3-hydroxyisobutyryl-CoA is also observed as an m$^+$ ion at m/z 854.1 (Fig. 2B). Deuterated 3-hydroxyisobutyryl-CoA is observed as an m$^+$ ion at m/z 855.1 (Fig. 2C).

^{1}H Nuclear Magnetic Resonance Analysis of 3-Hydroxyisobutyryl-CoA

The CoA esters are exhaustively lyophilized to remove traces of organic solvents, lyophilized several times from D$_2$O solution to remove exchangeable protons, and stored in a dry box prior to analysis. Samples for spectroscopy are prepared by dissolution of the ester in a buffer prepared by reconstitution of a lyophilized 50 mM potassium phosphate buffer, pH 7.5, in 100% D$_2$O containing sodium 3-trimethylsilyl 1-propionate as an internal chemical shift standard. Spectra are recorded on a Varian (Palo Alto, CA) 500-MHz spectrometer. Resonances due to the CoA portion of each ester are assigned by comparison of chemical shifts with those reported for other CoA esters.[8,9] Resonances assigned to the 3-hydroxyisobutyryl group are shown in Table I.

Acknowledgments

This work was supported by the Diabetes Research and Training Center of Indiana University School of Medicine (AM 20542).

[8] C. H. Lee and R. H. Sarma, *J. Am. Chem. Soc.* **97,** 1225 (1975).
[9] R. L. D'ordine, B. J. Bahnson, P. J. Tonge, and V. E. Anderson, *Biochemistry* **33,** 14733 (1994).

[9] Pathways of Leucine and Valine Catabolism in Yeast

By J. Richard Dickinson

Introduction

The budding yeast *Saccharomyces cerevisiae* is a model eukaryote for many researchers and is also an important industrial organism in both conventional and modern biotechnology industries. Despite this, catabolism of the branched-chain amino acids leucine, valine, and isoleucine has remained relatively little studied. The pathways used by most eukaryotes have been well understood for many years and yield, ultimately, acetyl-CoA and acetoacetate from leucine, succinyl-CoA from valine, and acetyl-CoA and propionyl-CoA from isoleucine.[1] All these metabolites can enter the tricarboxylic acid cycle. It has been known for many years that yeasts do not operate the same metabolic routes, because branched-chain amino acids can serve as sole source of nitrogen but not carbon. The predominant view has been that yeasts use the so-called Ehrlich pathway. This scheme (Fig. 1), first devised by Ehrlich in 1907,[2] and slightly modified later,[3] envisages that catabolism first involves transamination using α-ketoglutarate, followed by decarboxylation of the relevant keto acid to yield an aldehyde that is then reduced in an NADH-linked reaction producing the corresponding fusel alcohol.[4-6] Acceptance of this scheme has been problematic for at least four reasons. First, previous studies have never unambiguously established its existence. For example, merely showing that radioactively labeled amino acid was converted into isoamyl alcohol does not prove that the individual steps are those envisaged in the scheme. Other routes are conceivable. The enzyme(s) responsible for the key decarboxylation was never isolated. Some authors have stated that pyruvate decarboxylase is responsible,[6] but this had never been demonstrated. The second problem is that the Ehrlich pathway does not explain all the products that yeast makes: Some fusel alcohols do not correspond to any known amino acid. Third, if the pathway does exist as envisaged, explanations for the

---

[1] D. A. Bender, "Amino Acid Metabolism." John Wiley & Sons, New York, 1975.
[2] F. Ehrlich, *Ber. Dtsch. Chem. Ges.* **40,** 1027 (1907).
[3] O. Neubauer and K. Fromherz, *Hoppe-Seyler's Z. Physiol. Chem.* **70,** 326 (1911).
[4] T. G. Cooper, *in* "The Molecular Biology of the Yeast *Saccharomyces cerevisiae:* Metabolism and Gene Expression" (J. N. Strathern, E. W. Jones, and J. R. Broach, eds.), p. 39. Cold Spring Harbor Press, Cold Spring Harbor, New York, 1982.
[5] P. J. Large, *Yeast* **2,** 1 (1986).
[6] S. Derrick and P. J. Large, *J. Gen. Microbiol.* **139,** 2783 (1993).

0076-6879/00 $30.00

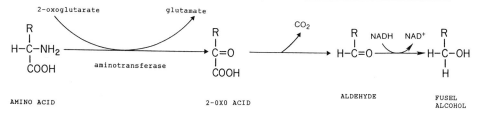

FIG. 1. The Ehrlich pathway.

production of fusel alcohols need to invoke a mixing of the synthetic and degradative reactions involving the branched-chain amino acids. This is something that cells normally avoid by coordinate regulation. Fourth, the Ehrlich pathway does not explain much of the known physiology, as Ehrlich himself was fully aware. Two of the most serious shortcomings are that the kinetics of amino acid utilization do not match fusel alcohol formation in complex media, and in media containing low levels of amino acids there is a poor correlation between the amino acid composition and the composition of the resulting fusel oil.[7] Academic curiosity is not the sole reason for seeking to understand the pathways of branched-chain amino acid catabolism in yeast. The fusel alcohol end products are significant flavor materials in all yeast-fermented products, contributing either desirable or undesirable flavor dependent on concentration. Knowledge of the pathways, enzymes, and genes involved is essential before these can be logically manipulated.

Our reexamination of the catabolism of the branched-chain amino acids in *S. cerevisiae* has confirmed that the first step is transamination.[8] Two distinct aminotransferases function both in amino acid biosynthesis and catabolism. One is mitochondrial (*TWT1* gene product) and one is cytosolic (*TWT2* gene product).[9,10] The genetic nomenclature here could confuse the unwary: *TWT1* (open reading frame *YHR208w*) and *TWT2* (*YJR148w*) have been referred to as *BAT1* and *BAT2* for branched-chain amino acid transaminase (respectively),[10] although the acronym "*BAT*" should be avoided because it has already been coined for another use. The two genes have also been referred to as *ECA39* and *ECA40* (respectively) because they are homologous to similarly named genes in several other eukaryotic systems.[9] It should be noted that *twt1Δ twt2Δ* double mutants still possess high levels of branched-chain amino acid aminotransferase activity in the cytosol,[10] indicating that other enzyme(s) exist with this activity. At present

[7] A. D. Webb and J. L. Ingraham, *Adv. Appl. Microbiol.* **5,** 317 (1963).
[8] J. R. Dickinson and V. Norte, *FEBS Lett.* **326,** 29 (1993).
[9] A. Eden, G. Simchen, and N. Benvenisty, *J. Biol. Chem.* **271,** 20242 (1996).
[10] G. Kispal, H. Steiner, D. A. Court, B. Rolinski, and R. Lill, *J. Biol. Chem.* **271,** 24458 (1996).

it is not clear what these enzymes are. Branched-chain α-keto acid dehydrogenase has also been demonstrated in *S. cerevisiae.* It has been purified and a number of its properties characterized.[11] However, branched-chain α-keto acid dehydrogenase is clearly not essential for yeast growth with leucine or valine as the sole source of nitrogen because *lpd1* mutants, which lack lipoamide dehydrogenase, the E3 component of branched-chain α-keto acid dehydrogenase, are still capable of catabolizing these branched-chain amino acids.[12,13] Hence, there is much that remains to be discovered about the pathways of branched-chain amino acid catabolism in yeast.

Media

YEPD medium for general yeast growth

Constituent	Per liter
Yeast extract (Difco, Detroit, MI)	10 g
Bacteriological peptone (Difco)	20 g
Glucose	20 g
Adenine	0.1 g
Uracil	0.1 g

YEPE medium for use when aerobic metabolism is required

Constituent	Per liter
Yeast extract (Difco)	10 g
Bacteriological peptone (Difco)	20 g
Ethanol	20 ml
Adenine	0.1 g
Uracil	0.1 g

Note: To prepare media containing ethanol all the constituents *except* the ethanol are added, mixed, aliquoted, and autoclaved. Sterile absolute ethanol (filtered through a 0.2 μm-grade filter) should be added immediately before inoculation with cells.

---

[11] J. R. Dickinson, *Methods Enzymol.* **324,** Chap. 36, 2000 (this volume).
[12] J. R. Dickinson, M. M. Lanterman, D. J. Danner, B. M. Pearson, P. Sanz, S. J. Harrison, and M. J. E. Hewlins, *J. Biol. Chem.* **272,** 26871 (1997).
[13] J. R. Dickinson, S. J. Harrison, and M. J. E. Hewlins, *J. Biol. Chem.* **273,** 25751 (1998).

Glucose minimal medium

Constituent	Per liter
Yeast nitrogen base (Difco)	1.67 g
Leucine or valine	20 g
Glucose	20 g

Ethanol minimal medium

Constituent	Per liter
Yeast nitrogen base (Difco)	1.67 g
Leucine or valine	20 g
Ethanol	20 ml

For minimal media auxotrophic requirements are added as needed to a final concentration of 20 μg/ml.

Culture Conditions and Handling of Yeast

The yeast strain should be raised in a starter culture of YEPD or YEPE, as appropriate, dependent on whether the subsequent experimentation calls for glucose or ethanol as the carbon source. The cells are grown in conical flasks filled to 40% of nominal capacity in a gyrorotary (orbital) shaker at 30°. The cells are grown to early stationary phase, which corresponds to an optical density of approximately 10 in YEPD medium, and then a small sample is removed and used to inoculate the minimal medium. When using minimal medium containing ^{13}C-labeled amino acid cost considerations mean that only a few milliliters of medium are prepared. The author often prepares a batch comprising only 2 ml in a 5-ml conical flask. This maintains a constant culture volume-to-flask size ratio in all experiments, thus ensuring that aeration conditions are similar whatever volume of culture is used. It can be difficult to incubate small flasks securely in most gyrorotary shakers. This problem can be simply overcome by placing the small (e.g., 5 or 10 ml) conical flask in a plastic 100-ml beaker and packing around the sides of the small flask with cotton (known as "cotton wool" in the UK). The plastic beaker will fit one of the regular sizes of retaining clip in the shaker.

Rationale of ^{13}C-Labeling Experiments and General Requirements

The choice of isotopically labeled substrate is crucial in any experiment. Although there are a number of suppliers of ^{13}C-labeled leucine and valine, the choice is restricted to amino acid labeled at C-1 or C-2, or uniformly labeled throughout the molecule. Leucine or valine labeled at C-1 would be almost useless; the α-keto acids produced by the initial transamination reaction (α-ketoisocaproic acid and α-ketoisovaleric acid, respectively), will be labeled at C-1 because, whatever pathway is used subsequently, there will be decarboxylation of the α-keto acid with consequent loss of the ^{13}C label. Hence all metabolites downstream of the keto acids will be unlabeled and the pathway will be "invisible" to ^{13}C nuclear magnetic resonance (NMR). Uniformly labeled leucine or valine is not preferred either, because each would have the disadvantage of resulting in an exceptionally complicated NMR spectrum that could take months, rather than days, to interpret. The spectrum would be much more complicated because instead of, e.g., leucine and all subsequent metabolites in the pathway having just one resonance each, there would be a resonance for every carbon atom in all the intermediates. In addition, most of the resonances would be complex multiplets due to carbon–carbon coupling because all adjacent carbon atoms would be labeled with ^{13}C. However, leucine or valine labeled at C-2 is ideal, as described below. It is advisable to purchase substrates with the highest isotopic enrichment available. The author uses [2 -^{13}C]leucine and [2-^{13}C]valine, with the C-2 carbons being 99.9% ^{13}C (from Cambridge Isotope Laboratories, Cambridge, MA).

All NMR experiments need to be performed under carefully controlled, standardized conditions. It could almost be said that the precise conditions do not matter as long as the conditions can be identically reproduced from one experiment to another. This is because the signals can vary in many ways under different conditions. Certain signals, particularly those of α-hydroxy acid intermediates, are exquisitely pH and concentration dependent. After obtaining a spectrum it is always best to identify the resonances in it. Only rarely is the identity of all the resonances known the first time a new analysis is done, and the simplest way to make some preliminary guesses is to inspect the literature or computer libraries of NMR data for resonances of possible candidate molecules. Hence, to compare and analyze results in this way, there is a compelling argument for acquiring spectra under conditions similar to those used previously by others. For many years the author has analyzed both "known" standard compounds and "unknowns" in a 50 mM potassium phosphate buffer, pH 6.0. The concentration of buffer chemicals is not insignificant: A requirement of ^{13}C NMR experiments is "proton decoupling" (subjecting the sample to broad-band radiation to eliminate the signals of all hydrogen atoms). Considerable amounts of energy are involved, which leads to the phenomenon of dielectric heating. This can be much reduced by lowering the buffer concentration to the minimum that is effective. An NMR spectrometer needs a "lock" signal; this is conveniently achieved by incorporating 15% (v/v) ^2H_2O in the sample.

Sample Preparation for Analysis by ^{13}C Nuclear Magnetic Resonance

For many studies of metabolism it has been necessary to prepare acid-soluble extracts from yeast. However, for studies of branched-chain amino acid catabolism it appears that the relevant metabolites are to be found in abundance in the growth medium. This is clearly a great convenience for the experimenter.

When the cells have been grown in ^{13}C-labeled medium for the required time two 1-ml aliquots should be removed to 1.5-ml microcentrifuge (Eppendorf) tubes. After centrifugation for 1 min the supernatants (i.e., growth medium) should be combined in a fresh tube. [If acid-soluble metabolites are required, 100 μl of ice-cold 15% (v/v) perchloric acid can be added to the pellet of cells in one of the tubes, followed by vortex mixing. The suspension is combined with the other pellet of cells. The last remaining traces of cells in the first tube can be transferred quantitatively to the second tube, using a further 100 μl of ice-cold perchloric acid. The whole sample should be vortex mixed and worked up as desired.] The growth medium should be made 50 mM with respect to phosphate buffer by the addition of 0.4 M potassium phosphate buffer, pH 6.0, and ^2H_2O added to a final concentration of 15% (v/v). This preparation will now occupy approximately 2.6 ml. About 350 μl may be transferred to a 5-mm NMR tube for analysis; the remainder should be stored frozen at $-20°$. It is worthwhile to prepare and store an excess of material in this way because at a later stage there will most likely be a need to confirm the identity of one or more resonances (see below).

Satisfactory spectra can be obtained from ^{13}C-labeling experiments with 10,000 accumulations. For an unlabeled pure reference compound as few as 500 accumulations will suffice, although 2000 would be better; and for a "spiking" experiment (explained in the next section) 15,000–20,000 would be sufficient. The spectra are recorded at 20°, using a Bruker (Coventry, UK) AMX360 NMR spectrometer operating at 90.5 MHz and using 32,000 data points over a spectral width of 22,000 Hz with Waltz-16 ^1H decoupling. The standard Bruker software has been used throughout. Any modern NMR spectrometer will suffice as long as ^1H decoupling is adequate and there is a sufficient delay (2–3 sec) between accumulations. All chemical shifts (δ ppm) are reported relative to external tetramethylsilane in C^2HCl_3; addition of sodium trimethylsilylpropane sulfonate gives a methyl signal at -2.6 ppm under the conditions described.

Identification of Signals in Nuclear Magnetic Resonance Spectrum

When the identity of a previously unidentified signal becomes evident the first thing to do is to record a spectrum of a *bona fide* sample of that chemical. This does not require the purchase of ^{13}C-labeled material. In nature approximately 1.1% of all carbon atoms are ^{13}C atoms, and thus a solution of unlabeled material (about 30 mg/ml in the same buffer) will be sufficient to obtain a good spectrum. Of course, this will show every atom present in that molecule. The identity of the resonance in the original biological specimen can then be confirmed by a "spiking" experiment, in which a sample of the prepared growth medium is mixed with a small portion of the solution containing the unlabeled suspect compound. Proof is obtained by seeing an increase in the intensity of the "unknown" resonance. The spectrum at this point needs careful inspection. A chemically similar compound, which alone appeared to have an identical position in the NMR spectrum, might have been suspected, but on mixing it with growth medium is seen to have a resonance that is merely very close. A spiking experiment can easily render the growth medium sample useless for further analysis because either the known compound had a resonance very close but not identical to that of an unidentified resonance, or one of the other signals in the unlabeled chemical is in an inconvenient position in relation to another unidentified resonance, or both. Hence the need (mentioned above) to store more prepared growth medium than is simply required to obtain a single NMR spectrum.

Analysis of Isoamyl and Isobutyl Alcohols by Gas
Chromatography–Mass Spectrometry

Aliquots (minimum volume, 2 ml) of culture are removed with a 5-ml disposable Luer-lock syringe fitted with a suitable length needle to reach into the culture flask. The needle is then removed and a cellulose acetate filter unit (0.45-μm pore size grade) is attached. The culture is filtered directly into a sterile autosampler bottle and sealed with a septum and screw cap. It may be stored at $-20°$ or analyzed immediately. If frozen, the samples should be agitated thoroughly after thawing to ensure homogeneity prior to analysis. The samples or standards (1 μl) should be injected onto a 30-m (0.32-mm i.d.) fused-silica capillary column with a 0.25-μm film of Supelcowax 10 (Supelco, Bellefonte, PA) in a Voyager gas chromatograph–mass spectrometer (GC–MS) (Finnigan, Manchester, UK). The injector temperature is 250° and the samples are chromatographed isothermally at 60°, using helium as the carrier gas, at a constant flow rate of 1 ml/min. Standard solutions of isoamyl alcohol or isobutyl alcohol give a linear calibration over a range of 1–500 μg/ml. The great advantage of GC–MS is that even when compounds chromatograph with similar retention times, they can still be independently quantitated because they almost invariably give rise to a unique ion. In the case of isoamyl alcohol (3-methyl-1-butanol) and active amyl alcohol (2-methyl-1-butanol) these may be discriminated because isoamyl alcohol produces a unique ion of mass 55 Da, which is not produced by active amyl alcohol.

Assay of α-Ketoisocaproate Reductase Activity in Yeast

Cells that have been grown in ethanol minimal medium with leucine as sole source of nitrogen are harvested by filtration on cellulose acetate membranes (0.45-μm pore size) or by centrifugation, washed with sterile distilled water, resuspended in ice-cold buffer A [50 mM potassium phosphate buffer (pH 7.4) containing 2 mM EDTA and 2 mM 2-mercaptoethanol] and disrupted with a Braun homogenizer as described in [36] in this

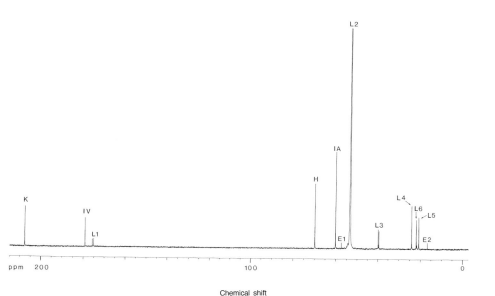

Chemical shift

Fig. 2. The ^{13}C NMR spectrum of a culture supernatant of a wild-type strain cultured in ethanol minimal medium with leucine as sole nitrogen source. The resonances marked are as follows: L1–L6, C-1 to C-6 (respectively) of leucine; K, C-2 of α-ketoisocaproate; IV, C-1 of isovalerate; H, C-2 of α-hydroxyisocaproate; IA, C-1 of isoamyl alcohol; E1, C-1 of ethanol; E2, C-2 of ethanol. [Reproduced, with permission, from J. R. Dickinson, M. M. Lanterman, D. J. Danner, B. M. Pearson, P. Sanz, S. J. Harrison, and M. J. E. Hewlins, *J. Biol. Chem.* **272,** 26871 (1997).]

volume.[11] Cell homogenate (30–50 μg) is used immediately as a source of enzyme.

α-Ketoisocaproate reductase is assayed at 25° in 1 ml of buffer A containing 0.2 mM NADH. The reaction is initiated by the addition of 30 μl of α-ketoisocaproate (30-mg/ml stock solution made up fresh each day and stored at 4°). NADH "disappearance" is monitored by monitoring the absorbance at 340 nm in a dual-beam recording spectrophotometer. The assay blank contains all components except α-ketoisocaproate.

Elucidation of Catabolic Pathways

Figure 2 shows the ^{13}C NMR spectrum of a culture supernatant of a wild-type strain cultured in a minimal medium in which ethanol was the

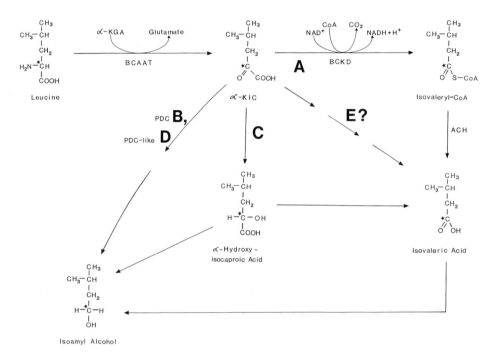

FIG. 3. Potential metabolic routes for the metabolism of leucine to isoamyl alcohol. The asterisks indicate carbon atoms in intermediates that were labeled with ^{13}C in the wild-type strain. Enzyme activities that are already known are abbreviated. BCAAT, branched-chain amino acid aminotransferase; BCKD, branched-chain α-keto acid dehydrogenase; ACH, acyl-CoA hydrolase; PDC, pyruvate decarboxylase. [Reproduced, with permission, from J. R. Dickinson, M. M. Lanterman, D. J. Danner, B. M. Pearson, P. Sanz, S. J. Harrison, and M. J. E. Hewlins, *J. Biol. Chem.* **272**, 26871 (1997).]

TABLE I

Isoamyl Alcohol Produced by Wild-Type and Mutant Strains Growing in Glucose Minimal Medium with Leucine as Sole Source of Nitrogen[a,b]

Strain	Relevant genotype	Isoamyl alcohol	
		μg/ml	μg/10^6 cells
IWD72	Wild type	318	3.4
MML22	*lpd1*	245	12.2
YSH5.127.-17C[c]	*pdc1 pdc5 pdc6*	1088	9.3
FY1679	Wild type	547	5.1
FY1679-YDL080c(a)	*ydl080c::Kan^r*	147	1.4
FY1679-YDL080c(α)	*ydl080c::Kan^r*	141	1.4
JRD815-1.2[c]	*pdc1 pdc5 pdc6 ydl080c::Kan^r*	63	1.6
JRD815-6.1[c]	*pdc1 pdc5 pdc6 ydl080c::Kan^r*	94	1.6

[a] Reproduced, with permission, from J. R. Dickinson, M. M. Lanterman, D. J. Danner, B. M. Pearson, P. Sanz, S. J. Harrison, and M. J. E. Hewlins, *J. Biol. Chem.* **272,** 26871 (1997).

[b] Isoamyl alcohol content was determined in early stationary phase. Values are the means of two separate experiments. Differences between samples were <4%.

[c] Strains YSH5.127.-17C, JRD815-1.2, and JRD815-6.1 lack pyruvate decarboxylase and hence cannot utilize glucose; these had ethanol (2%, v/v) as carbon source.

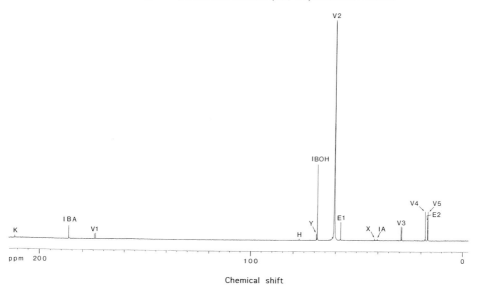

Chemical shift

Fig. 4. The ^{13}C NMR spectrum of a culture supernatant of a wild-type strain cultured in ethanol minimal medium with valine as sole nitrogen source. The resonances marked are as follows: V1–V5, C-1 to C-5 (respectively) of valine; K, C-2 of α-ketoisovalerate; IBA, C-1 of isobutyrate; H, C-2 of α-hydroxyisovalerate; IBOH, C-1 of isobutanol; E1 and E2, C-1 and C-2 of ethanol (respectively); IA, C-2 of isoamyl alcohol; X, impurity present in valine; Y, unidentified resonance. [Reproduced, with permission, from J. R. Dickinson, S. J. Harrison, and M. J. E. Hewlins, *J. Biol. Chem.* **273,** 25751 (1998).]

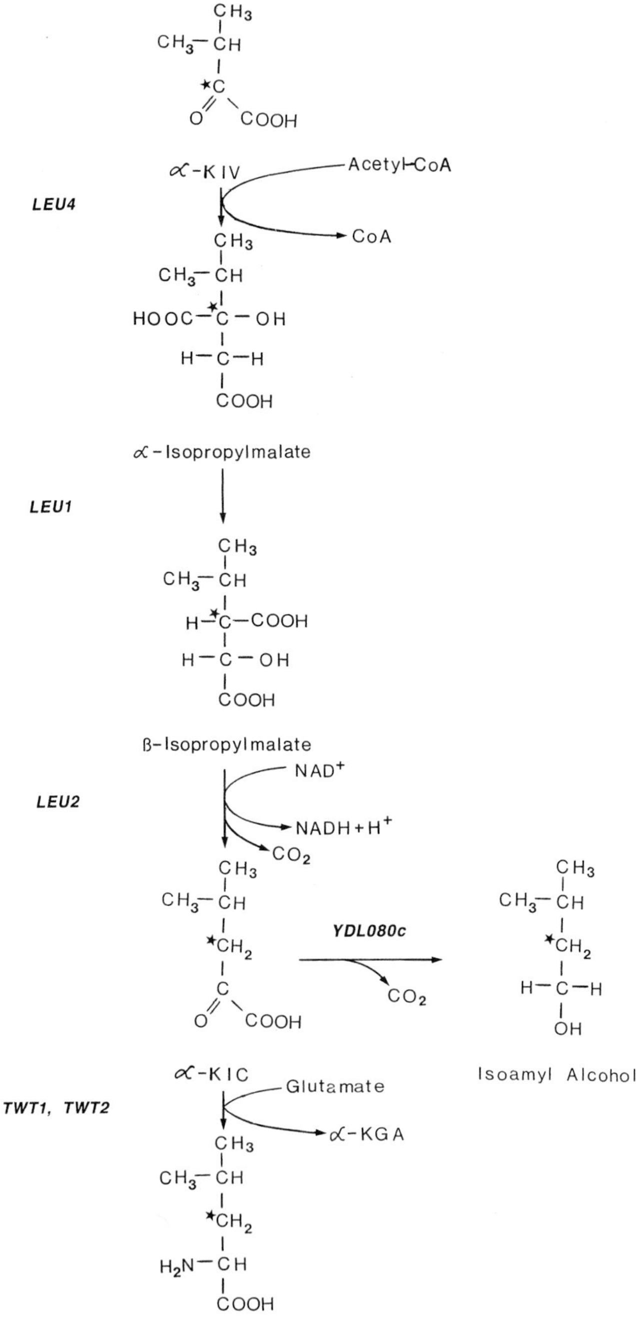

$$
\begin{array}{c}
CH_3 \\
| \\
CH_3-CH \\
| \\
{}^{*}C \\
O \diagup \quad \diagdown COOH
\end{array}
$$

LEU4

α-KIV ────── Acetyl-CoA

────→ CoA

$$
\begin{array}{c}
CH_3 \\
| \\
CH_3-CH \\
| \\
HOOC-{}^{*}C-OH \\
| \\
H-C-H \\
| \\
COOH
\end{array}
$$

α-Isopropylmalate

LEU1

$$
\begin{array}{c}
CH_3 \\
| \\
CH_3-CH \\
| \\
H-{}^{*}C-COOH \\
| \\
H-C-OH \\
| \\
COOH
\end{array}
$$

β-Isopropylmalate

LEU2

────── NAD⁺

────→ NADH+H⁺

────→ CO₂

$$
\begin{array}{c}
CH_3 \\
| \\
CH_3-CH \\
| \\
{}^{*}CH_2 \\
| \\
C \\
O \diagup \quad \diagdown COOH
\end{array}
$$

YDL080c ─────→

────→ CO₂

$$
\begin{array}{c}
CH_3 \\
| \\
CH_3-CH \\
| \\
{}^{*}CH_2 \\
| \\
H-C-H \\
| \\
OH
\end{array}
$$

Isoamyl Alcohol

α-KIC ────── Glutamate

TWT1, TWT2

────→ α-KGA

$$
\begin{array}{c}
CH_3 \\
| \\
CH_3-CH \\
| \\
{}^{*}CH_2 \\
| \\
H_2N-CH \\
| \\
COOH
\end{array}
$$

carbon source and [2-^{13}C]leucine the sole nitrogen source. From the metabolites identified and the positions labeled with ^{13}C, a number of routes between leucine and isoamyl alcohol are possible. All of these involve initial transamination to α-ketoisocaproate (Fig. 3). The first is via branched-chain α-keto acid dehydrogenase (route A in Fig. 3) to yield isovaleryl-CoA, which would then be converted to isovalerate by acyl-CoA hydrolase. The second possibility is via pyruvate decarboxylase (route B). Pathway C envisages α-ketoisocaproate reductase producing α-hydroxyisocaproate followed by decarboxylation to yield isoamyl alcohol. Route D proposes a pyruvate decarboxylase-like enzyme. Pathway E between α-ketoisocaproate and isovalerate is included as a theoretical possibility in Fig. 3, but cannot be explained using known or potential enzymes and is thus discarded.

Mutants may be used to deduce the actual *in vivo* pathway. Thus, route A is not used for the synthesis of isoamyl alcohol because abolition of branched-chain α-keto acid dehydrogenase in an *lpd1* disruption mutant does not prevent the formation of isoamyl alcohol (Table I). Route B (Fig. 3) via pyruvate decarboxylase is not required either, because the complete obliteration of pyruvate decarboxylase activity in a *pdc1 pdc5 pdc6* triple mutant has virtually no effect on the levels of isoamyl alcohol produced (Table I). The third possibility (route C, Fig. 3) is via α-ketoisocaproate reductase, a novel activity not previously known in *S. cerevisiae*. The true metabolic significance of this enzyme in yeast is not clear at this time, but it can have no role in the formation of isoamyl alcohol from α-hydroxyisocaproate because cell homogenates cannot convert α-hydroxyisocaproate to isoamyl alcohol. Route D (Fig. 3), a pyruvate decarboxylase-like enzyme encoded by *YDL080c*, appears to be the major route of decarboxylation of α-ketoisocaproate to isoamyl alcohol because strains with disruptions in this gene produce little isoamyl alcohol (Table I). The fact that *ydl080c* mutants produce some isoamyl alcohol suggests that another decarboxylase can substitute to a minor extent. The identity of this putative minor decarboxylase is still unknown.

Similar methods (^{13}C NMR spectroscopy using [2-^{13}C]valine as substrate combined with GC–MS and specific mutants) can be used to examine the

FIG. 5. The pathway of leucine biosynthesis. The asterisks indicate carbon atoms in intermediates that were labeled with ^{13}C. Enzymes are denoted by the structural genes that encode them. *LEU4*, α-isopropylmalate synthase; *LEU1*, β-isopropylmalate dehydratase; *LEU2*, β-isopropylmalate dehydrogenase; *TWT1* and *TWT2*, mitochondrial and cytoplasmic isozymes (respectively) of branched-chain amino acid transaminase; *YDL080c = KID1*, α-ketoisocaproate decarboxylase. [Reproduced, with permission, from J. R. Dickinson, S. J. Harrison, and M. J. E. Hewlins, *J. Biol. Chem.* **273**, 25751 (1998).]

catabolism of valine to isobutanol. This shows that the product of valine transamination, α-ketoisovalerate, is decarboxylated via pyruvate decarboxylase (i.e., the canonical interpretation of the Ehrlich pathway). This is confirmed by the observation that elimination of pyruvate decarboxylase activity in a *pdc1 pdc5 pdc6* triple mutant virtually abolishes isobutyl alcohol production. A single pyruvate decarboxylase isozyme is sufficient for isobutyl alcohol formation from valine.[13] Both the branched-chain α-keto acid dehydrogenase and the pyruvate decarboxylase-like enzyme encoded by *YDL080c* seem to be irrelevant because mutants defective in the respective activities produce wild-type levels of isobutanol.

^{13}C NMR spectroscopy reveals one further surprise: some ^{13}C from [2-^{13}C]valine appears at the C-2 position of isoamyl alcohol (see resonance marked "IA" in Fig. 4). The occurrence of isoamyl alcohol labeled at C-2 is readily explained by the pathway of leucine biosynthesis (Fig. 5). α-Ketoisovaleric acid labeled at C-2 is converted to α-isopropylmalate labeled at C-3, by α-isopropylmalate synthase. This is then converted by β-isopropylmalate dehydratase to β-isopropylmalate labeled at C-3 which in turn is converted to α-ketoisocaproic acid by β-isopropylmalate dehydrogenase. The next step in the leucine biosynthetic pathway is the formation of leucine, which would be labeled at C-3. No signal due to C-3 of leucine is observed; however, isoamyl alcohol labeled at C-2 was present (Fig. 4). This explanation is confirmed by the fact that a strain that lacks the *YDL080c*-encoded α-ketoisocaproate decarboxylase makes plenty of isobutanol but no isoamyl alcohol when valine is the sole nitrogen source (data not shown). Hence, the NMR study shows there is a mixing of valine catabolic and leucine biosynthetic pathways.

Concluding Remarks

It is now clear that neither the catabolism of leucine nor the catabolism of valine to the respective branched-chain alcohols requires branched-chain α-keto acid dehydrogenase. This makes the physiological role of this enzyme in yeast something of a mystery. The activity of branched-chain α-keto acid dehydrogenase is greater in complex medium when glycerol is the carbon source than in minimal media containing a branched-chain amino acid,[14] and so our present knowledge can be rationalized by concluding that branched-chain amino acid catabolism utilizes predominantly branched-chain α-keto acid dehydrogenase in complex media, but not at all in minimal media.

[14] J. R. Dickinson and I. W. Dawes, *J. Gen. Microbiol.* **138,** 2029 (1992).

Section II

Cloning, Expression, and Purification of Enzymes of
Branched-Chain Amino Acid Metabolism

[10] Isolation of Subunits of Acetohydroxy Acid Synthase Isozyme III and Reconstitution of Holoenzyme

By Maria Vyazmensky, Tsiona Elkayam, David M. Chipman, and Ze'ev Barak

Introduction

Bacterial acetohydroxy acid synthases (AHAS, EC 4.1.3.18, also known as acetolactate synthase) are composed of two kinds of subunit: large subunits with a molecular mass of about 62,000 Da, which apparently contain all the catalytic machinery of the enzyme,[1–3] and small subunits (9–17 kDa), which have a regulatory function. The small subunits are necessary for feedback inhibition by valine, one of the end products of the pathway in which AHAS is the first common enzyme. They are also necessary for the stabilization and activation of these enzymes, even in the case of AHAS isozyme II from *Escherichia coli* or *Salmonella typhimurium*,[2] which is unique in being insensitive to feedback inhibition. The bacterial AHASs appear to have a quaternary structure including two large subunits and two small subunits.[1,4,5]

We describe here methods for preparing separately the large and small subunits of *E. coli* AHAS III,[6] the products of genes *IlvI* and *IlvH*, respectively. The preparation of isolated subunits makes it possible to study the properties of each subunit alone, the nature of the interaction between them, and their effects on one another. AHAS III can be reconstituted from its subunits to form a holoenzyme apparently identical to the native enzyme. AHAS III shows a significant tendency to dissociate,[7] and the affinity of the small subunits for the large subunits can be studied by reconstitution titrations described below.

[1] L. Eoyang and P. M. Silverman, *J. Bacteriol.* **166**, 901 (1986).
[2] M. F. Lu and H. E. Umbarger, *J. Bacteriol.* **169**, 600 (1987).
[3] O. Weinstock, C. Sella, D. M. Chipman, and Z. Barak, *J. Bacteriol.* **174**, 5560 (1992).
[4] Z. Barak, J. M. Calvo, and J. V. Schloss, *Methods Enzymol.* **166**, 455 (1988).
[5] J. V. Schloss, D. E. VanDyk, J. F. Vasta, and R. M. Kutny, *Biochemistry* **24**, 4952 (1985).
[6] M. Vyazmensky, C. Sella, Z. Barak, and D. M. Chipman, *Biochemistry* **35**, 10339 (1996).
[7] C. Sella, O. Weinstock, Z. Barak, and D. M. Chipman, *J. Bacteriol.* **175**, 5339 (1993).

Isolation of Subunits of Acetohydroxy Acid Synthase Isozyme III

Principle

The isolation of the subunits of AHAS III is based on the separate subcloning of *IlvI* and *IlvH,* the genes encoding the two polypeptides. Because wild-type *E. coli* strains express two or three AHAS isozymes, care must be taken to obtain a bacterial extract in which the subunit of interest is present at a level much greater than that of any other AHAS polypeptide. One way of accomplishing this is to use a mutant host strain with no expressible genomic AHASs. We have obtained satisfactory yields of purified protein by using plasmids encoding one or the other separate subunits behind an *ilvIH* promoter, despite the shortcomings of the available AHAS⁻ strains.[6] Alternatively, the gene encoding the polypeptide may be inserted in an expression vector behind a strong inducible promoter, and the expression of genomic AHASs from the host strain (optimized for such a vector) repressed by growth in a medium rich in amino acids. The procedures described here are based on expression of *ilvI* and *ilvH* from an inducible expression vector. Because a much larger fraction of the protein in the crude extract is the desired AHAS subunit, isolation and purification of the polypeptide are easier than in the previously described procedure.[6]

Reagents

Sodium pyruvate, FAD, thiamine pyrophosphate (TPP), dithiothreitol (DTT), EDTA, creatine, isopropyl-β-D-thiogalactopyranoside (IPTG) (Sigma, St. Louis, MO)

Toyopearl Phenyl-650M (TosoHaas, Stuttgart, Germany)

Acrylamide, N',N'-methylenebisacrylamide, ammonium peroxydisulfate ammonium sulfate (Merck, Darmstadt, Germany)

1-Naphthol: Obtain from BDH (Poole, UK); material from some other sources yields dark precipitates that interfere with the assay

Solution A (used as the disruption buffer in the preparation of large subunit): 0.02 mM FAD, 10 mM EDTA, 1 mM DTT in 0.1 M Tris-HCl buffer at pH 7.8

Solution B (used as the disruption buffer in the preparation of small subunit): 10 mM EDTA, and 1 mM DTT in 0.1 M potassium phosphate buffer at pH 7.6

Solution C (used as dialysis buffer in preparation of small subunits): 1 mM EDTA and 1 mM DTT in 25 mM potassium phosphate buffer at pH 8.0

Bacterial Strains and Growth

The plasmid pUHE24S serves here as an IPTG-inducible expression vector. This construct is a derivative of pUHE24[8,9] with a unique *Nco*I site immediately downstream of the promoter (S. Leu, personal communication, 1997). Plasmid pUI is constructed by insertion of a *Nco*I–*Bgl*I fragment containing the entire *ilvI* gene[3] into the polylinker of pUHE24S, and expresses the large subunit of AHAS III without any modification. The *ilvH* coding region is amplified by polymerase chain reaction (PCR) from pCV88,[10] using a primer at the 5′ end that introduces several nucleotide changes: an *Nco*I site is created, the original start codon is converted to a glycine codon, and a new initial methionine codon is created immediately upstream. In addition, a unique *Pst*I site is inserted downstream of the stop codon. These changes allow an *Nco*I–*Pst*I fragment of the PCR product to be inserted into pUHE24S to produce the plasmid pUH, which expresses an altered small subunit of AHAS III, in which an additional glycine residue is inserted after the initial methionine (our unpublished results, 1998).

For production of the large subunit or small subunit of AHAS III, *E. coli* XL-Blue MRF′ host cells (Stratagene, La Jolla, Ca) are transformed with either pUI or pUH. The cells are grown in LB rich medium at 37° in shaker flasks (600 ml in 2-liter flasks at 200 rpm) to a turbidity of 0.5 optical density (OD) at 660 nm. IPTG (0.5 m*M*) is added to induce expression of the cloned gene and the cells are grown for 3 hr more and harvested by centrifugation. About 1.5 g of wet cells is obtained from 600 ml of culture.

Isolation of Large Subunit

Extraction. Harvested XL-Blue MRF′/pUI cells are washed twice with 0.1 *M* Tris buffer (pH 7.8), and stored at −20° if not immediately used. The bacteria are resupended in 10 ml of solution A for every gram of cell paste, cooled in an ice bath, and subjected to disruption in a model XL2015 ultrasonic liquid processor (Heat Systems, New York, NY) with a 1.3-cm flat-tip probe. Twelve cycles (20 sec each with 40-sec pauses) of sonication are used. The disrupted cells are centrifuged at 27,000g for 1 hr at 4° to remove cell debris.

[8] U. Deuschle, W. Kammerer, R. Gentz, and H. Bujard, *EMBO J.* **5,** 2987 (1986).
[9] E. Ben-Dov, S. Boussiba, and A. Zaritsky, *J. Bacteriol.* **177,** 2851 (1995).
[10] C. T. Lago, G. Sannia, G. Marino, C. H. Squires, J. M. Calvo, and M. DeFelice, *Biochim. Biophys. Acta* **824,** 74 (1985).

Purification of Large Subunit. The protein is purified in two steps, ammonium sulfate precipitation and hydrophobic column chromatography. The volume of the extract is measured and for every 100 ml of extract 25 g of ammonium sulfate is slowly added with stirring to the cell extract at room temperature. After additional stirring for 15 min, the precipitate is collected by centrifugation for 20 min at 27,000g and 4°. The pellet is redissolved in 5 ml of solution A for every gram of wet cells extracted. To this solution is slowly added one-half its volume of 1.4 M ammonium sulfate in solution A; this brings the final ammonium sulfate concentration to 0.7 M. The protein solution is then loaded on a 1.5 × 30 cm column of Toyopearl Phenyl-650M preequilibrated with 0.70 M ammonium sulfate in solution A. The column is washed at a flow rate of 1 ml min^{-1} with 600 ml of this buffer, and then eluted with a 100-ml linear gradient from the starting buffer to distilled water, followed by 100 ml more of water. The AHAS activity, measured as described below, peaks after about 20 ml of water. A typical yield is 120 mg of protein with a specific activity of 140 nmol mg^{-1} min^{-1}. The protein is concentrated by dialysis against 50% (v/v) glycerol containing 0.02 mM FAD, 10 mM EDTA, 1 mM DTT, and 0.1 M Tris-HCl at pH 7.8.

Isolation of Small Subunit

The harvested XL-Blue MRF′/pUH cells are washed twice with solution B and stored at −20° if not immediately used. The bacteria are resuspended in 5 ml of solution B for every gram of cell paste, cooled in an ice bath, and disrupted as described above. The disrupted cells are centrifuged at 27,000g for 1 hr at 4° to remove cell debris.

The volume of the extract is measured and for every 100 ml of extract 18 g of ammonium sulfate is slowly added. After 15 min of additional stirring at room temperature, the resulting precipitate is removed by centrifugation as described above. The volume of the supernatant is measured, and 9 g of additional ammonium sulfate for every 100 ml is slowly added with stirring. After 15 min, the precipitated protein is collected by centrifugation at 4° for 20 min at 27,000g. The precipitated protein is redissolved in 0.5 ml of solution C for every gram of the original cell paste.

The ammonium sulfate is removed by dialyzing the suspension against solution C overnight at 4°. The desalting leads to precipitation of the protein. The content of the dialysis sack is transferred to a centrifuge tube and the precipitate collected by centrifugation at 12,000g for 10 min at 4°. The precipitate is washed again in about 1 ml of solution C and then resuspended in a minimal volume of 0.1 M Tricine buffer, pH 8.5. To redissolve the protein, a saturated solution of $MgCl_2$ is added to the suspension, a drop at a time, with shaking until a clear solution results. This is diluted twofold with glycerol and stored at $-20°$.

The ammonium sulfate precipitate may alternatively be redissolved in 25 mM Tris-HCl buffer, pH 8.0, containing 1 mM EDTA and 1 mM DTT,[6] and desalted by dialyzing against the same buffer. For some mutant small subunits we find that this buffer results in more successful precipitation of the protein.

The activity of the small subunits can only be followed and defined with regard to their ability to reconstitute properties of the holoenzyme when combined with large subunits (see below).

Reconstitution of Acetohydroxy Acid Synthase III from Its Subunits

Principle

The activity of the large subunits of AHAS III alone is quite low (3–5% of the specific activity of the holoenzyme).[3,7] The reconstitution of the enzyme can thus be followed by measurement of enzyme activity under standard conditions after incubation of the large subunits together with varying amounts of the small subunits.[6] The large subunits of AHAS II show negligible activity by themselves and the reconstitution of this isozyme has been studied by using a similar method.[11] The one-point colorimetric assay for acetoin[12] (formed by decarboxylation of the product acetolactate) is more sensitive than the continuous spectrophotometric assay of pyruvate disappearance at 330 nm,[5] and thus more appropriate for assaying the activity of large subunits alone.

The standard assay mixture appropriate for AHAS III contains the substrate and cofactors at concentrations at which the activity of the holoenzyme is nearly saturated: 40 mM pyruvate, 0.1 mM TPP, 10 mM MgCl$_2$, and 0.05 mM FAD in a 0.1 M potassium phosphate buffer, pH 7.6. The final assay solution described below also contains 0.5 mM DTT and 5 mM EDTA, because they are present in the buffers used to stabilize the proteins involved. The assembly of the holoenzyme apparently requires neither TPP nor Mg^{2+}. Because it has been reported that the extended incubation of AHASs in the absence of substrate and the presence of these cofactors leads to inactivation of the enzyme[5] (possibly by formation of the thiazolone derivative of the cofactor; J. V. Schloss, personal communication, 1988), preincubation is carried out without TPP and in the presence of EDTA.

Another important characteristic of the reconstituted AHAS III holoen-

---

[11] C. M. Hill, S. S. Pang, and R. G. Duggleby, *Biochem. J.* **327,** 891 (1997).
[12] W. W. Westerfeld, *J. Biol. Chem.* **161,** 495 (1945).

zyme is its regained sensitivity to inhibition by valine. While the low but significant activity of AHAS III large subunits is totally insensitive to valine, the native or reconstituted holoenzyme is inhibited. In the presence of 40 mM pyruvate, inhibition is incomplete (~85%) at saturation with valine, with an apparent K_i of 0.02 mM.

Reagents

Solution D (used for reconstitution incubations): 0.1 mM FAD, 10 mM EDTA, and 1 mM DTT in a 0.1 M potassium phosphate buffer at pH 7.6

Solution E (concentrated substrate and cofactor mixture): 80 mM pyruvate, 0.2 mM TPP, and 20 mM MgCl$_2$ in a 0.1 M potassium phosphate buffer at pH 7.6

Enzyme Assay

The large subunits (3–20 μg) and small subunits (up to a fivefold molar excess) are added to 0.5 ml of solution D and preincubated for 15 min at 37°. The catalytic reaction is initiated by adding 0.5 ml of solution E. The reaction is stopped after 20 min by addition of 0.1 ml of 50% (v/v) H$_2$SO$_4$, which also converts acetolactate to acetoin. Acetoin is assayed by the Westerfeld method[12] as previously described.[13] If the range of the colorimetric assay is exceeded, the reaction time may be shortened or a smaller portion of the stopped reaction mixture taken for the acetoin assay. One unit (U) of activity is defined as that producing 1 μmol of acetolactate per minute under the above-described conditions.

The specific activity of the reconstituted holoenzyme can be defined as the specific activity of the large subunits at saturation with small subunits, corrected by multiplying by the weight fraction of large subunits in the holoenzyme, which is 0.78 [= 62,000/(62,000 + 17,000)] in the case of AHAS III. Saturation is assumed if the observed activity as a function of added small subunit reaches a plateau. The specific activity of the small subunit can be determined from its ability to activate the large subunit. Aliquots of the small subunit are preincubated with a fixed quantity of large subunit and the reaction carried out as described above. Ten to 20 μg of purified large subunit or a crude large subunit preparation containing about 0.002–0.005 U of AHAS activity may be used. The specific activity of the small subunit is then defined as the slope of a plot of observed AHAS activity as a function of the amount of small subunit protein added.

---

[13] S. Epelbaum, D. M. Chipman, and Z. Barak, *Anal. Biochem.* **191**, 96 (1990).

Analysis of Reconstitution

The process of association–dissociation between large and small sub-units can be analyzed by carrying out a complete activity titration of large subunits with small subunits. Quantitative analysis of the titration data requires assuming a model for the association process, although it does not appear that titration experiments are sufficient for testing such a model. The parameters obtained from the analysis are useful, however, for empirical description of the behavior of the system and, particularly, for comparison of the behavior of the wild-type and mutant polypeptides, homologous and heterologous reconstitution, and so on.

AHAS III large subunits appear to exist chiefly as monomers in the medium used for reconstitution, as judged by their behavior on a size-exclusion column, while the holoenzyme is composed of two large and two small subunits.[6] Although we do not know the pathway for assembly of the holoenzyme, we can use a simplified scheme to model the titration:

$$L + S \rightleftharpoons LS \rightleftharpoons L_2S_2$$

where L and S are the separate large and small subunits, respectively. LS is an intermediate complex and L_2S_2 is the active holoenzyme. The concentration of the active complex can be calculated numerically, given two dissociation constants:

$$K_d = [L] \times [S]/[LS]$$
$$K_2 = [LS]^2/[L_2S_2]$$

and taking into consideration the fact that the large subunits alone have measureable activity. Figure 1 shows examples of titrations of the same quantity of large subunits with either wild-type small subunits or small subunits encoded by the *ilvH* gene from a valine-resistant mutant.[6] The data can be modeled by assuming the holoenzyme has the same dissociation constant K_2 and the same activity whether it is reconstituted from wild-type or mutant small subunits. However, the apparent affinity of the mutant subunit for the large subunit is an order of magnitude lower than that of the wild type (Fig. 1).

Hill and co-workers[11] have reported a similar analysis in the case of AHAS II. However, the isolated large subunits of AHAS II appeared to exist primarily as dimers, so the data were fit by a different model:

$$L_2 \rightleftharpoons L_2S \rightleftharpoons L_2S_2$$

Fits to the data showed that the binding of the second small subunit is much stronger than the first.[11]

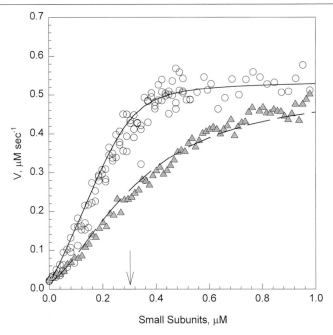

Fig. 1. Titration of AHAS III large subunits with small subunits. Each reaction mixture contained 18.7 μg of purified large subunit and varied amounts of wild-type ($\bigcirc$) or mutant ($\blacktriangle$) small subunit. The arrow indicates the concentration (0.30 μM) of the large subunit. After preincubation, the AHAS reaction was carried out under the standard conditions described and the amount of acetolactate formed in 20 min determined. The lines are calculated for the scheme with $K_d = 1.0 \times 10^{-7} M$ for the wild-type and $1.0 \times 10^{-6} M$ for the mutant, and $K_2 = 1.0 \times 10^{-8} M$ in both cases.

Discussion

The nature of the interactions among subunits remains unclear for many proteins, among them the AHASs. The cloning, expression, and purification of the large and small subunits of AHAS III make it possible to study the properties of isolated subunits and follow the reconstitution of the holoenzyme. The reconstitution of the enzyme from subunits with single mutations (derived by site-directed mutagenesis or isolated from spontaneous mutants) can be used to analyze the functional roles of various parts of the peptide chain within the quaternary structure. This procedure can be adapted for study of other enterobacterial AHAS isozymes or AHAS from other bacteria, as well as investigation of heterologous interactions between large and small subunits derived from different AHASs. Genes encoding putative small subunits from some eukaryotic organisms have now also been identified,[14–18] and we expect that reconstitution titrations with the catalytic subunits from the same organisms will become important tools in their biochemical characterization.

We should also point out the usefulness of isolated AHAS subunits in the study of the physiology of branched-chain amino acid biosynthesis. Because of the tendency of at least some of the AHAS isozymes to dissociate on dilution, the intracellular activities of these enzymes may be underestimated when they are assayed.[7,19] We have used the addition of small subunits to the assay mixture to correct this problem.[19] If, as now seems likely, eukaryotic AHASs also have a labile quaternary structure, the addition of the appropriate small subunits would also help clarify the real physiological activity of these AHASs.

Acknowledgment

This work was supported in part by Grant 243/98 from the Israel Science Foundation.

[14] N. Ohta, J. Plant Res. 110, 235 (1997).

[15] D. Chipman, Z. Barak, and J. V. Schloss, Biochim. Biophys. Acta 1385, 401 (1998).

[16] P. Bork, C. Ouzounis, C. Sander, M. Scharf, R. Schneider, and E. Sonnhammer, Protein Sci. 1, 1677 (1992).

[17] R. G. Duggleby, Gene 190, 245 (1997).

[18] M. E. Reith and J. Munholland, Plant Mol. Biol. Rep. 13, 333 (1995).

[19] S. Epelbaum, R. A. LaRossa, T. K. VanDyk, T. Elkayam, D. M. Chipman, and Z. Barak, J. Bacteriol. 180, 4056 (1998).

[11] Branched-Chain Amino-Acid Aminotransferase of Escherichia coli

By Hiroyuki Kagamiyama and Hideyuki Hayashi

Overview

In bacteria, branched-chain amino-acid aminotransferase (BCAT) catalyzes the last step of biosynthesis of branched-chain amino acids. Valine, leucine, and isoleucine are synthesized from the corresponding keto acids

METHODS IN ENZYMOLOGY, VOL. 324

by amino-group transfer from glutamate. In *Escherichia coli,* BCAT is encoded by the gene *ilvE,* which is a component of the *ilvGEDA* operon. Earlier studies of *E. coli* transaminases indicated that there are two kinds of transaminase activities. Transaminase A is active toward dicarboxylic amino acids, and transaminase B is active toward branched-chain amino acids.[1] Transaminase A was later resolved into two enzymes: one has narrower specificity for dicarboxylic amino acids, and the other has broader specificity and catalyzes the transamination of aromatic amino acids as well as dicarboxylic amino acids. The first, high-specificity enzyme was identified as the *E. coli* counterpart of aspartate aminotransferase (AspAT), which is prevalent throughout species and is the most abundant and best-studied aminotransferase.[2,3] The second enzyme was determined to be an aromatic amino acid aminotransferase, and has an important role in catalyzing transamination between glutamate and aromatic keto acids, the last step in the biosynthesis of phenylalanine and tyrosine.[2,3] Transaminase B was clearly distinguished from these aminotransferases by the absence of activity toward aspartate. Structural bases for this difference have been provided both in primary and in three-dimensional structures, and are summarized in this chapter.

Assay Method

Principle

The enzymatic transamination between branched-chain amino acids and 2-oxoglutarate can be followed by measuring the production of glutamate through coupling to the reduction of 3-(4,5-dimethyl-2-thiazolyl)-2,5-diphenyltetrazolium bromide, using glutamate dehydrogenase, NAD, and 1-methoxy-5-methylphenazinium methosulfate.[4] In the original procedure,[5] diaphorase and 2-(p-iodophenyl)-3-(p-nitropheny)-5-phenyltetrazolium chloride were used in place of 1-methoxy-5-methylphenazinium methosulfate and 3-(4,5-dimethyl-2-thiazolyl)-2,5-diphenyltetrazolium bromide.

Procedure

The assay mixture contains 50 mM HEPES–NaOH (pH 8.0), 0.1 M KCl, 10 mM isoleucine, 10 mM 2-oxoglutarate, 2.5 mM NAD (oxidized

[1] D. Rudman and A. Meister, *J. Biol. Chem.* **200,** 591 (1953).

[2] D. H. Gelfand and R. A. Steinberg, *J. Bacteriol.* **130,** 429 (1977).

[3] J. T. Powell and J. F. Morrison, *Eur. J. Biochem.* **87,** 391 (1978).

[4] K. Inoue, S. Kuramitsu, T. Ogawa, H. Ogawa, and H. Kagamiyama, *J. Biochem.* **104,** 777 (1988).

[5] R. Rej, *Anal. Biochem.* **119,** 205 (1982).

form), 30 μM 1-methoxy-5-methylphenazinium methosulfate, 0.2 mM 3-(4,5-dimethyl-2-thiazolyl)-2,5-diphenyltetrazolium bromide, and 0.2 mg of glutamate dehydrogenase, in a total volume of 1.0 ml. The reaction is started by the addition of enzyme, and the absorbance at 500 nm is continuously monitored. The molar extinction coefficient is determined by measuring the absorbance change on addition of known amounts of glutamate to the assay mixture. The value is generally about 2×10^4 M^{-1} cm^{-1}. Transamination between other amino acids and 2-oxoglutarate can be measured by substituting isoleucine with other amino acids.

Cloning of *ilvE* Gene

The chromosomal DNA of *E. coli* W3110 is prepared according to a standard method,[6] and digested partially with *Sau*3AI. The DNA fragments of about 1000 base pairs (b$_p$) are collected on a sucrose density gradient,[7] and ligated to the *Bam*HI site of pBR322. The plasmid is used to transform *E. coli* AB2227 (*ilvE⁻*). The transformants that suppress the branched-chain amino acid auxotroph are selected on minimum agar plates supplemented with ampicillin (50 μg ml^{-1}). The complete nucleotide sequence of the *ilvE* gene with its neighborhood and deduced amino acid sequence is shown in Fig. 1.

Purification Procedures

The methods for overproduction and purification of BCAT are modified from the previously published methods.[4]

Overproduction

The cloned *ilvE* gene is excised at the *Sal*I and *Xmn*I sites flanking the gene, and is ligated to the *Sal*I–*Sma*I site of pUC119. The resultant plasmid, named pUC119-*ilvE*, is used to transform *E. coli* JM103 cells. The transformed cells are grown in LB medium supplemented with ampicillin (50 μg ml^{-1}). After 18 hr of cultivation at 37°, the cells are harvested by centrifugation. It is usual to obtain 40 g of wet cells from 10 liters of culture medium.

Purification

Purification of BCAT is carried out at 0–4°. All buffers contain 2 mM 2-oxoglutarate, 0.2 mM EDTA, 5 mM 2-mercaptoethanol, and 0.1 mM

---

[6] L. Clarke and J. Carbon, *Methods Enzymol.* **68**, 396 (1979).
[7] H. Ogawa and J. Tomizawa, *J. Mol. Biol.* **23**, 265 (1967).

atgttggagaatttatcatgatgcaacatcaggtcacatgtatcggctcgcttcaatccag

```
         10        20        30        40        50        60        70        80        90       100       110       120
aaacctctagaacgtgtttacgtggtgcgctcatcgtggttccacgtctgctcaatgaactggcgcgccgccagctctgctcaatgaatggcgcgccgatgaccggttgccagcccac

        130       140       150       160       170       180       190       200       210       220       230       240
atgaccacgaagaagctgattacattggttcaatgggatgttcgctggaagaccggcaagtgcatgtgatgtcgcacgcgctgcactatggcacttcggttttgaaggcatc
 M  T  T  K  K  A  D  Y  I  W  F  N  G  E  M  V  R  W  E  D  A  K  V  H  V  M  S  H  A  L  H  Y  G  T  S  V  F  E  G  I

        250       260       270       280       290       300       310       320       330       340       350       360
cgttgcacgactcgcacaaaggaccggttgattccgccatcgtgacgatctcgcatatgcagcgtctgcatgactccgcaaaatctatcgctccggttcgcagagcattgatgagctgatg
 R  C  Y  D  S  H  K  G  P  P  V  V  F  R  H  R  E  H  M  Q  R  L  H  D  S  A  K  I  Y  R  F  P  P  V  S  Q  S  I  D  E  L  M

        370       380       390       400       410       420       430       440       450       460       470       480
gaagcttgctgtgacgtgatccgcaaaaacaatctcaccagcgcctatatccgtccgctatcgttttcgttcctctgaaggcgatgttctcctgaccgcgcagcagcaaacaccatcccgacggcgcaaaa
 E  A  C  R  D  V  I  R  K  N  N  L  T  S  A  Y  I  R  P  L  I  F  V  G  D  V  G  M  G  V  N  P  P  A  G  Y  S  T  D  V

        490       500       510       520       530       540       550       560       570       580       590       600
attatcgctttccgtgggggagcgtatctgggcgcagaagcgctggagcaggggatcgatgcgatggtttcctcgtggaatcgcgccgcacgttatatctctgaaggcgcaggcgaaaacctg
 I  I  A  A  F  P  W  G  A  Y  L  G  A  E  A  L  E  Q  G  I  D  A  M  V  S  S  W  N  R  A  A  P  N  T  I  P  T  A  A  K

        610       620       630       640       650       660       670       680       690       700       710       720
gcgggtaactacctctcttccctgctggtttccggtgctgtgtcaccccaccgttcacctcctcctcgcggtaccgctgatgccatcatcaaactggcgaaagagtctggaattgaagtacgtgagcag
 A  G  G  N  Y  L  S  S  L  L  V  G  S  E  A  R  R  H  G  Y  Q  E  G  I  A  L  D  V  N  G  Y  I  S  E  G  A  G  E  N  L

        730       740       750       760       770       780       790       800       810       820       830       840
tttgaagtgaaagatggtgttctgtttacgcccccgttcttcaccctcgtcaatcctgggtgttatgtccggatgagcggcacggccgcagaatcacgccagcgttcaggttgccgaaggccgttgtggcccg
 F  E  V  K  D  G  V  L  F  T  P  P  F  F  T  S  S  A  L  P  G  I  T  R  D  A  I  I  K  L  A  K  E  L  G  I  E  V  R  E  Q

        850       860       870       880       890       900       910       920       930       940       950       960
gttaccaaacgcattcagcagccttcttcggcctccttcactggcgaagataaatggggctggttagatcaagttaatcaataacaaaaatggacgacgcaccgtcc
 V  L  S  R  E  S  L  Y  L  A  D  E  V  F  M  S  G  T  A  A  E  I  T  P  V  R  S  V  D  G  I  Q  V  G  E  G  R  C  G  P

        970       980       990      1000      1010      1020      1030      1040      1050      1060      1070      1080
catttacgagacagcaactgggagtaaataaagtgcctaagtaccgttccgccaccaccatcatgttcgtaatatggcggtgctcgtgcgcgcaccggaatgaccga
 V  T  K  R  I  Q  Q  A  F  F  G  L  F  T  G  E  T  E  D  D  K  W  G  W  L  D  Q  V  N  Q  *

       1090      1100      1110      1120      1130      1140      1150      1160      1170      1180      1190      1200
cgccgattcggtaagccgattatccggttgtgactggtcgggtccagtccaccatttgtaccggtgactccgtcacgtcagatcctgcggaacaaattggtcgccgacaaattgaagcggctggcgg

       1210      1220      1230      1240      1250      1260      1270      1280      1290      1300
cgttgcaaagagttcaacaccattgggatgatgatggtgccacggggggatgctgttattcactgccattcgccatccgactgatc
```

pyridoxal 5'-phosphate (PLP). The cells (40 g) are resuspended in 100 ml of 0.1 M potassium phosphate buffer (KP$_i$), pH 7.0, and are disrupted sonically at 0° for 10 min with a Branson (Danbury, CT) sonifier model 350 set at output 6 and 50% duty. Cell debris is removed by centrifugation (10,000g, 40 min). The supernatant is applied to a DEAE-Toyopearl 650M column (2.5 × 25 cm) equilibrated with 20 mM KP$_i$, pH 7.0. The proteins are eluted with a 1-liter linear gradient of 0 to 0.6 M NaCl in 20 mM KP$_i$, and active fractions are combined. Solid ammonium sulfate is added to 20% saturation while maintaining the pH at 7.0 with 1 M KH$_2$PO$_4$. The enzyme solution is applied to a butyl-Toyopearl 650M column (2.5 × 25 cm) equilibrated with 20 mM KP$_i$, pH 7.0, containing 20% saturated ammonium sulfate. The proteins are eluted with a 1-liter gradient of 20 to 0% saturated ammonium sulfate in 20 mM KP$_i$, pH 7.0. The fractions containing BCAT are combined, and dialyzed against 4 liters of 20 mM KP$_i$, pH 7.0, for 6 hr, and then against 4 liters of 2 mM KP$_i$, pH 7.0, for 6 hr. The dialyzed solution is applied to a hydroxyapatite column (2.5 × 25 cm) equilibrated with 2 mM KP$_i$, pH 7.0. BCAT is eluted with a 1-liter linear gradient formed with 2 and 100 mM KP$_i$, pH 7.0. The high-purity fractions are combined and concentrated to about 10 ml with an Amicon (Danvers, MA) ultrafiltration cell, and the concentrated solution is applied to a Sephacryl S-200 column (3.8 × 90 cm) equilibrated with 10 mM KP$_i$, pH 7.0, containing 0.1 KCl. The enzyme is eluted with the same buffer. Generally, about 200 mg of BCAT is obtained from 40 g (wet weight) of *E. coli* JM103/pUC119-*ilvE* cells, with a yield of 40% and 20-fold purification. The final preparation of the enzyme has a specific activity of 0.4 kat kg^{-1}.

Crystallization and Crystallographic Analysis

Crystallization of BCAT is performed in the dark from polyethylene glycol 400 (PEG 400) by vapor diffusion of a hanging drop.[8] The reservoir solution contains 100 mM HEPES (pH 7.5), 200 mM MgCl$_2$ as the salt, 30% (w/v) PEG 400 as the precipitant, 1 mM NaN$_3$, 1mM EDTA, and 10 μM PLP. Two different crystals, one belonging to the monoclinic space

---

[8] K. Okada, K. Hirotsu, M. Sato, H. Hayashi, and H. Kagamiyama, *J. Biochem.* **121,** 637 (1997).

FIG. 1. Nucleotide sequence of the *ilvE* gene and its neighborhood [S. Kuramitsu, T. Ogawa, H. Ogawa, and H. Kagamiyama, *J. Biochem.* **97,** 993 (1985)]. The amino acid sequence deduced from the gene is shown in single-letter code. The DNA sequence is numbered from the beginning of the gene. A possible ribosome-binding site is underlined.

group $C2$ and the other to the orthorhombic space group $C222_1$, are obtained. An asymmetric unit cell contains three subunits of BCAT. Only the ethyl mercury thiosalicylate (EMTS) derivative of the monoclinic crystal is obtained as the heavy atom derivative. Therefore, selenomethionyl BCAT is prepared by expressing the BCAT protein in *E. coli* DL41 (*metA*⁻) cells harboring pUC119-*ilvE* in the presence of selenomethionine.[9] Both forms of the crystal are obtained for the selenomethionyl enzyme. Data collection is performed at 2.4- or 2.5-Å resolution. The monoclinic crystals of the native, the EMTS derivative, and the selenomethionyl enzymes are used for phase determination. Three mercury sites are located by using the isomorphous difference Patterson map calculated for the EMTS derivative. The positions of the 21 (7 × 3) selenium atoms are determined from difference Fourier maps phased with the mercury sites. It should be noted that the use of the selenomethionyl enzyme is not for applying the multiple anomalous dispersion method, as originally reported,[9] but for applying the conventional multiple isomorphous replacement (MIR) method. The initial MIR phase thus calculated is then improved by solvent flattening,[10] histogram matching,[11,12] and noncrystallographic symmetry averaging.[13] The improved electron density map is used for model building. Refinement is carried out with the X-PLOR program. The final R factor and free R factor are 18.8% for 37,577 reflections and 25.8% for 2044 reflections, respectively, with $F > 2$ [$\sigma(F)$] between 10.0- and 2.5-Å resolution, including 178 water molecules.

Properties

Covalent and Noncovalent Structures

The enzyme is composed of six identical subunits, each with M_r 33,960.[4,14,15] The primary structure of the BCAT subunit is 26% homologous to the subunit of the homodimeric enzyme D-amino-acid aminotransferase (DAAT) from *Bacillus* sp. YM-1 (Table I).[16] Each subunit of BCAT, like

[9] W. A. Hendrickson, J. R. Horton, and D. M. LeMaster, *EMBO J.* **9**, 1665 (1990).

[10] B.-C. Wang, *Methods Enzymol.* **115**, 90 (1985).

[11] K. Y. J. Zhang and P. Main, *Acta Crystallogr.* **A46**, 41 (1990).

[12] K. Y. J. Zhang and P. Main, *Acta Crystallogr.* **A46**, 377 (1990).

[13] G. Brigogne, *Acta Crystallogr.* **A32**, 832 (1976).

[14] F.-C. Lee-Peng, M. A. Hermodson, and G. B. Kohlhaw, *J. Bacteriol.* **139**, 339 (1979).

[15] S. Kuramitsu, T. Ogawa, H. Ogawa, and H. Kagamiyama, *J. Biochem.* **97**, 993 (1985).

[16] K. Tanizawa, S. Asano, Y. Masu, S. Kuramitsu, H. Kagamiyama, H. Tanaka, and K. Soda, *J. Biol. Chem.* **264**, 2440 (1989).

TABLE I
KINETIC PARAMETERS OF BCAT TOWARD VARIOUS AMINO ACIDS AND 2-OXOGLUTARATE[a]

Substrate	k_{cat}/sec^{-1}	K_m/mM		$\dfrac{k_{cat}K_m^{-1}}{M^{-1}\,sec^{-1}}$	
		Amino acid	2-Oxoglutarate	Amino acid	2-Oxoglutarate
Isoleucine	48	0.42	2.4	110,000	20,000
Leucine	78	2.2	6.6	22,000	7,300
Valine	19	2.7	1.7	7,000	11,000
Methionine	17	19	1.0	890	17,000
Phenylalanine	2.9	0.89	0.26	3,300	11,000
Tyrosine	2.2	7.0	0.24	310	9,200
Tryptophan	3.7	72	0.56	51	6,600

[a] The reaction was carried out in 50 mM HEPES–NaOH, pH 8.0, containing 0.1 M KCl at 25°. The overall ping–pong reactions were followed by measuring the glutamate formed from 2-oxoglutarate.[4] The $k_{cat}K_m^{-1}$ values toward 2-oxoglutarate are almost constant irrespective of the amino acid substrate, which is in accordance with theoretical considerations [H. Hayashi, K. Inoue, T. Nagata, S. Kuramitsu, and H. Kagamiyama, *Biochemistry* **32**, 12229 (1993)].

DAAT, contains one molecule of pyridoxal 5′-phosphate (PLP). PLP is covalently bound to Lys-159 via an azomethine linkage with the ε-amino group of the lysine side chain.[4] This and other active site residues, Arg-59 and Glu-193, later identified by X-ray crystallography,[8,17] are also conserved in DAAT at the corresponding position (Fig. 2). BCAT and DAAT have no apparent sequence homologies with other aminotransferases. Among the PLP-dependent enzymes whose sequences have been determined, 4-amino-4-deoxychorismate lyase,[18] encoded by *pabC* and catalyzing the last step of the biosynthesis of *p*-aminobenzoate (precursor of folate), is the only enzyme that is related in terms of primary structure to BCAT and DAAT (Fig. 2).

Spectroscopic Properties

Because Lys-159 of BCAT and PLP form a Schiff base (PLP–Lys-159), BCAT has a major absorption band at 410 nm and a minor absorption band at 330 nm, similar to DAAT.[4,19] Both enzymes show negative circular

[17] S. Sugio, G. A. Petsko, J. M. Manning, K. Soda, and D. Ringe, *Biochemistry* **34**, 9661 (1995).
[18] P. V. Tran and B. P. Nichols, *J. Bacteriol.* **173**, 3680 (1991).
[19] K. Tanizawa, Y. Masu, S. Asano, H. Tanaka, and K. Soda, *J. Biol. Chem.* **264**, 2445 (1989).

```
        1        10        20        30        40        50        60        70
BCAT    TTKKADYIWFNGEMVRWEDAKVHVMSHALHYGTSVFEGIRCYDSHKGPVVFRHREHMQRLHDSAKIYRFP
DAAT    -----GYTLWNDQIVKDEEVKIDKEDRGYQFGDGVYEVVKVYN----GEMFTVNEHIDRLYASAEKIRIT
         *    *    *    *        *   *   *      *    **   **  **  *
ADCM    -------FLINGHKQ----ESLAVSDRATQFGDGCFTTARVIDG----KVSLLSAHIQRLQDAC--QRLM
               *                       *                  *  **       *


                 80        90        100       110       120       130       140
BCAT    VSQSIDELMEACRDVIRKNNLTSAYIRPLIFVGDVGMGVNPPAGYSTDVIIAAFPWGAYLGAEALEQGID
DAAT    IPYTKDKFHQLLHELVEKNELNTGHIYFQVTRGTSPRAHQFPENTVKPVIIGYTKE-NPRPLENLEKGVK
          *        **  *     *       *          *      ***       * ** *
ADCM    ISCDFWPQLEQEMKTLAAE-QQNGVLKVVISRGSGGRGYSTLNSGPATRILSVTAYPAHYDRLRNE-GIT
                           *                  *            *         * *


                 150       160       170       180       190       200       210
BCAT    AMVSSWNRARPNTIPTAAKAGGNYLSSLLVGSEARRHGYQEGIALDVNGYISEGAGENLFEVKDGVLFTP
DAAT    ATFVE-D-IR-W-LRCDIKS-LNLLGAVLAKQEAHEKGCYEAILHRNN-TVTEGSSSNVFGIKDGILYTH
         *       * *   *  *  *  *  *   *    *   *    *  **  *  *  **  *  *** **
ADCM    LALSPVRLGR-NPHLAGIKH-LNRLEQVLIRSHLEQTNADEALVLDSEGWVTECCAANLFWRKGNVVYTP
          *        +    * *  *          *           *      *    * *   *    *


                 220       230       240       250       260       270       280
BCAT    PFTSSALPGITRDAIIKLAKELGIEVREQVLSRESLYLADEVFMSGTAAEITPVRSVDGIQVGEGRCGPV
DAAT    PANNMILKGITRDVVIACANEINMPVKEIPFTTHEALKMDELFVTSTTSEITPVIEIDGKLIRDGKVGEW
         *     * ***** *    * *    * *           ** *   *  ***** **     *   *
ADCM    RLDQAGVNGIMRQFCIRLLAQSSYQLVEVQASLEESLQADEMVICN---ALMPVMPVC--A-----CG--
          **  *   *          *              **         **          *


                 290       300
BCAT    TKRIQQAFFGLFTGETEDKWGWLDQVNQ
DAAT    TRKLQKQFETKIPKPLHI
         *    *    *
ADCM    DVSFSSATLYEYLAPLCERPN
```

FIG. 2. The amino acid sequence of *E. coli* BCAT [S. Kuramitsu, T. Ogawa, H. Ogawa, and H. Kagamiyama, *J. Biochem.* **97**, 993 (1985)], *Bacillus* sp. YM-1 DAAT [K. Tanizawa, S. Asano, Y. Masu, S. Kuramitsu, H. Kagamiyama, H. Tanaka, and K. Soda, *J. Biol. Chem.* **264**, 2440 (1989)], and *E. coli* 4-amino-4-deoxychorismate lyase [ADCM; P. V. Tran and B. P. Nichols, *J. Bacteriol.* **173**, 3680 (1991)]. The residues conserved between BCAT and DAAT are marked with asterisks below the sequence of DAAT. The residues conserved among the three enzymes are marked with asterisks below the sequence of ADCM. The PLP-binding lysine residues, identified in BCAT and DAAT, and proposed in ADCM, are marked with plus symbols.

dichroism (CD) bands at wavelengths corresponding to the absorption maxima.[4,19] This is in contrast to most PLP-dependent enzymes, which show positive CD at the absorption maxima,[20] indicating the similar, unique properties of the two enzymes.

[20] H. Hayashi, K. Inoue, T. Nagata, S. Kuramitsu, and H. Kagamiyama, *Biochemistry* **32**, 12229 (1993).

Catalytic Properties

BCAT catalyzes transamination of a variety of amino acids (Table I).[4] Besides branched-chain amino acids, BCAT is active toward phenylalanine and tyrosine, both of which have bulky side chains like branched-chain amino acids. Methionine is also a fairly good substrate. While catalylzing transamination of these hydrophobic amino acids, BCAT is also active toward glutamate and its corresponding keto acid 2-oxoglutarate. Thus, like many other aminotransferases, BCAT recognizes two structurally different sets of substrates, amino/keto acids with hydrophobic side chains, and amino/keto acids with carboxylic side chains.

Before the elucidation of the three-dimensional structures of BCAT and DAAT, an interesting stereochemical study had been carried out.[21] The key step in the catalytic reaction of aminotransferases is the 1,3-prototropic shift between the aldimine (Schiff base of an amino acid and PLP) and ketimine [Schiff base of a keto acid and pyridoxamine phosphate (PMP)] intermediates. For efficient removal of the α proton of the aldimine, the C_α–H bond should be perpendicular to the plane of the PLP pyridine ring. Therefore, there are two possible conformations of the PLP–substrate aldimine: in one conformation the α hydrogen resides in the *si* face of the aldimine, and in the other conformation in the *re* face. The base that assists the prototropic shift is expected to locate on one side of the aldimine. Therefore, if the aldimine takes the conformation in which the α hydrogen protrudes to the *si* (*re*) face, the *si* (*re*) face base would abstract (and exchange with solvent) both the α hydrogen of the aldimine and the *pro-S* (*pro-R*) hydrogen at C-4' of the ketimine. When apo-BCAT was reconstituted with (4' *R*)-[4'-³H]PMP and used as the catalyst for the transamination between 2-oxoglutarate and valine, the tritium was nearly completely released to the solvent. On the other hand, when (4' *S*)-[4'-³H]PMP was used for the reconstitution, essentially no radioactivity was detected in the solvent after the transamination reaction. Identical results were obtained for DAAT when it was used as a catalyst for the trasamination between D-alanine and 2-oxoglutarate. On the other hand, tritium was released only from (4' *S*)-[4'-³H]PMP in the catalytic reaction of AspAT (transamination between aspartate and 2-oxoglutarate), as expected from the crystallographic structure of AspAT.[22] Therefore, it was concluded that the base that catalyzes the 1,3-prototropic shift of the aldimine and ketimine is

[21] T. Yoshimura, T. Nishimura, J. Ito, N. Esaki, H. Kagamiyama, J. M. Manning, and K. Soda, *J. Am. Chem. Soc.* **115,** 3897 (1993).

[22] J. F. Kirsch G. Eichele, G. C. Ford, M. G. Vincent, J. N. Jansonius, H. Gehring, and P. Christen, *J. Mol. Biol.* **174,** 497 (1984).

located at the *re* face of the PLP–substrate aldimine in BCAT and DAAT. The crystallographic structures of BCAT[8] and DAAT[17] are consistent with this conclusion.

Crystallographic Structures

The hexameric structure of BCAT has D_3 symmetry and the overall shape of a triangular prism (Fig. 3, left). The interaction between the subunits indicates that the hexamer can be regarded as a trimer of dimers. Each subunit is composed of two domains (Fig. 3, right). The larger domain is composed of residues 137–302. Residues 127–136, corresponding to the loop connecting the two domains, and the N-terminal three residues are missing in the crystallographic structure, probably because of their high flexibility in the substrate-unbound form of the enzyme.

The orientation of the active site residues is similar to that of DAAT[17] (Fig. 4). Arg-59, which binds the phosphate group, and Glu-193, which interacts with the pyridine nitrogen of PLP, correspond to Arg-50 and Glu-177, respectively, of DAAT. The side chain of Lys-159 locates at the *re* face of the PLP–Lys-159 aldimine. In DAAT, the PLP–Lys-145 Schiff base has the same conformation. This is in accordance with the notion that the

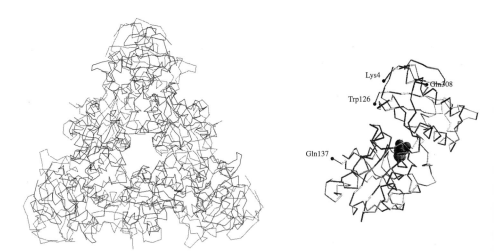

Fig. 3. The hexameric (*left*) and subunit (*right*) structures of BCAT (drawn using the coordinates of 1A3G). *Left:* Two sets of three subunits, related to each other by a noncrystallographic threefold axis, are separately drawn in gray (front) and black (back). *Right:* α-Carbon traces of one subunit. Residues 1–3 and 127–136 were not located on any electron density maps, and are not shown here. The PLP molecule is shown as a van der Waals surface model.

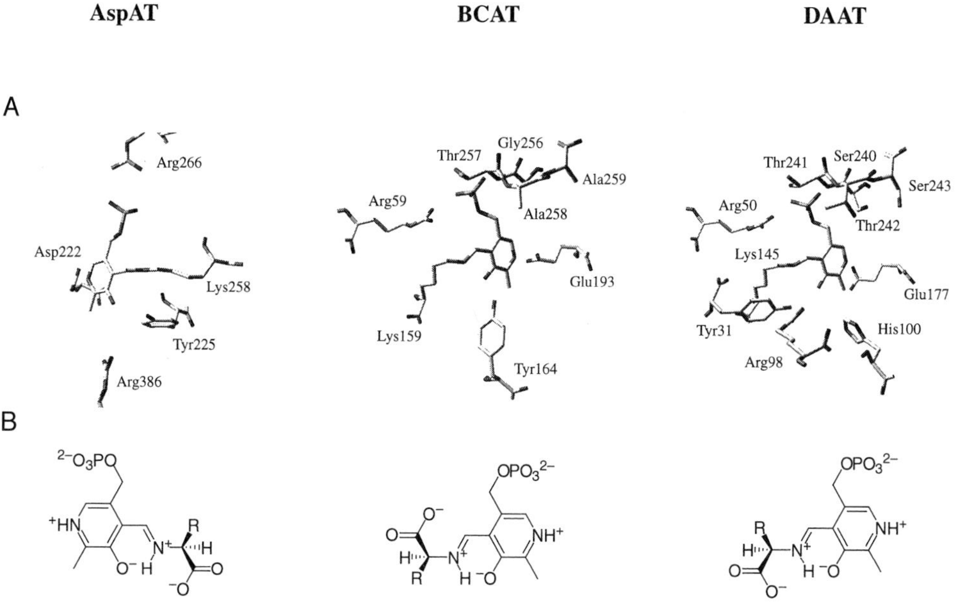

AspAT   **BCAT**   **DAAT**

A

B

Fig. 4. (A) The active site structures of *E. coli* AspAT (*left:* 1ARS), *E. coli* BCAT (*middle:* 1A3G), and *Bacillus* sp. YM-1 DAAT (*right:* 1DAA). (B) Schematic drawing of the coenzyme–substrate amino acid Schiff base (aldimine). The orientation of the coenzyme matches that in (A). In all structures, the α proton resides below the paper. The face below the paper is the *si* face of the aldimine in AspAT, and the *re* face of the aldimine in BCAT and DAAT.

ε-amino group of Lys-159 of BCAT and Lys-145 of DAAT act as the *re*-face base that catalyzes the 1,3-prototropic shift of the substrate–coenzyme Schiff base. The conformation of the substrate amino acid–PLP aldimine is schematically shown in Fig. 4B. In AspAT, Arg-386 is the residue that binds the α-carboxylate group of the substrate. The active site structure and the conformation of the amino acid–PLP aldimine of DAAT mirror those of AspAT, reflecting the enantiomeric specificity of these enzymes. Arg-98 and His-100 of DAAT are the proposed residues involved in the recognition of the substrate α-carboxylate group. While DAAT is active toward D-amino acids, BCAT recognizes L-amino acids. Therefore, compared with that in DAAT, the orientation of the substrate with respect to the coenzyme in BCAT is reversed. Thus, the α carboxylate is expected to point upward in Fig. 4. The amide nitrogens of Gly-256 and Thr-257 are the cadidates for the site that recognizes the α-carboxylate group of the substrate.

[12] Purification of Sodium-Coupled Branched-Chain Amino Acid Carrier of *Pseudomonas aeruginosa*

By YOSHIHIKO URATANI and TOSHIMITSU HOSHINO

The LIV-II carrier of *Pseudomonas aeruginosa* (see [13] in this volume[1]) is a sodium-coupled cotransporter for branched-chain amino acids L-leucine, L-isoleucine, and L-valine[1a,2] and can utilize lithium as well as sodium as coupling cation.[3] The LIV-II carrier genes (*braB*) of *P. aeruginosa* PAO and PML strains were cloned and sequenced,[4,5] showing that the only difference in primary structure between the carriers of PAO and PML involves an amino acid at position 292. The amino acid is threonine for PAO and alanine for PML. This substitution causes a difference in affinity of the carriers for Na$^+$, such that the leucine transport activity of the carriers is saturated at 0.1 mM NaCl for PAO and at 5.0 mM for PML.[5]

In this chapter, we describe a method for purification of the sodium-coupled branched-chain amino acid carrier protein (LIV-II carrier) of PML, i.e., the carrier with the lower affinity for sodium.[6] The LIV-II carrier is an integral membrane protein composed of 437 amino acids with a molecular mass of 45,249 Da and is detected as a protein band with an apparent molecular mass of 34 kDa on a sodium dodecyl sulfate (SDS)–polyacrylamide gel. The LIV-II carrier has been solubilized from *P. aeruginosa* membranes with detergent, *n*-octyl-β-D-glucopyranoside (octyl glucoside), and reconstituted into liposomes by a detergent–Sephadex gel-filtration procedure.[7] Prior to purification, the carrier protein should be overproduced with the cloned gene, because the content of the carrier in the *P. aeruginosa* cell membrane is low. The functional LIV-II carrier has been overproduced in *Escherichia coli* cells by using the T7 RNA polymerase/promoter system developed by Tabor and Richardson.[8] The carrier protein is solubilized from a membrane preparation of *E. coli* over-

[1] T. Hoshino and Y. Uratani, *Methods Enzymol.* **324,** Chap. 13, 2000 (this volume).
[1a] T. Hoshino, *J. Bacteriol.* **139,** 705 (1979).
[2] T. Hoshino and M. Kageyama, *J. Bacteriol.* **137,** 73 (1979).
[3] Y. Uratani, T. Tsuchiya, Y. Akamatsu, and T. Hoshino, *J. Membr. Biol.* **107,** 57 (1989).
[4] T. Hoshino, K. Kose, and Y. Uratani, *Mol. Gen. Genet.* **220,** 461 (1990).
[5] Y. Uratani and T. Hoshino, *J. Biol. Chem.* **264,** 18944 (1989).
[6] Y. Uratani, *J. Biol. Chem.* **267,** 5177 (1992).
[7] Y. Uratani, *J. Biol. Chem.* **260,** 10023 (1985).
[8] S. Tabor and C. C. Richardson, *Proc. Natl. Acad. Sci. U.S.A.* **82,** 1074 (1985).

producing the LIV-II carrier and purified to a homogeneous state by immu-noaffinity column chromatography. The purified protein is incorporated into liposomes for reconstitution and its transport activity is assayed.

Bacteria and Plasmids

Escherichia coli B7634 (*ileAhrbABCD*) lacking LIV transport activity is used as a recipient cell for the LIV-II carrier gene (*braB*) of PML and is grown in LB medium supplemented with isoleucine (50 μg/ml).[4,6] The T7 RNA polymerase/promoter plasmids pGP1-7 and pT7-6 are used for overproduction of the LIV-II carrier protein. Plasmid pUAH1 is a pUC18 recombinant plasmid with a 2.5-kilobase (kb) fragment containing the *braB* gene of PML.[5] The *braB* gene was transferred from pUAH1 downstream of the T7 promoter in the pT7-6 plasmid, generating pT7-6B.[6]

Overproduction of LIV-II Carrier

The *E. coli* B7634 cells containing both pGP1-7 and recombinant pT7-6B plasmids are grown at 30° in 1 liter of LB medium with kanamycin (30 μg/ml), carbenicillin (100 μg/ml), and isoleucine (50 μg/ml), using a rotary shaker. At an OD_{620} of 0.8, temperature is shifted to 42° and incubated for 25 min to induce rifampicin-insensitive T7 RNA polymerase. Rifampicin (20-mg/ml stock in methanol) is then added to give a final concentration of 100 μg/ml to inhibit endogenous protein synthesis in the *E. coli* cells. After a 5-min incubation with rifampicin, cells are supplemented with 20 amino acids (50 μg/ml each) and 20 mM glucose for overexpression of the carrier protein and incubated at 37° for 2 hr.

Preparation of Membranes

The following steps are performed at 4° or ice-cold temperature. Cells overproducing the LIV-II carrier are havested, washed twice with 250 ml of suspension buffer containing 10 mM HEPES–Tris buffer (pH 7.5), 0.1 M sucrose, and 5 mM MgCl$_2$ to avoid cell aggregation, and re-suspended in 15 ml of the suspension buffer containing in addition 1 mM dithiothreitol and deoxynuclease I (20 μg/ml). The cells are disrupted by a single passage through an Aminco (Silver Spring, MD) French pressure cell at 15,000 psi. After unbroken cells and cell debris are removed by centrifugation at 8600g for 10 min, the supernatant is centrifuged at 81,000g for 60 min to collect fragmented outer and cytoplasmic membranes. The membrane fragments are washed once with

250 ml of the suspension buffer, and resuspended in 2 ml of 50 mM potassium phosphate, pH 7.5. The protein concentration of the membrane preparation is approximately 15 mg/ml as determined by a modified method of Schaffner and Weissmann.[9,10] The membranes are frozen in an ethanol–dry ice bath and stored at $-70°$ until use.

Preparation of Immunoaffinity Column

This immunoaffinity purification method is adapted for purification of the LIV-II carrier, because it is one of the most powerful techniques for the isolation of proteins.[11] For this purpose, an antiserum directed against a synthetic 13-mer peptide, NH$_2$-Cys-Asp-Arg-Leu-Leu-Gly-Lys-Pro-Arg-Glu-Ala-Val-Ala-COOH (Ct13-mer peptide) is prepared. This peptide corresponds to the carboxyl-terminal portion of the LIV-II carrier.[4] The Ct13-mer peptide is conjugated to keyhole limpet hemocyanin, using m-maleimidobenzoyl-N-hydroxysuccinimide ester, a bifunctional linker reagent, and used as an antigen to immunize a rabbit for preparation of antiserum raised against the peptide. The anti-Ct13-mer antibody is purified from the rabbit antiserum by Ct13-mer peptide-affinity column chromatography. The peptide-affinity column is prepared with the Ct13-mer peptide (4.2 mg) coupled to EAH-Sepharose 4B beads (2.0 ml; Pharmacia, Piscataway, NJ), using m-maleimidobenzoyl-N-hydroxysuccinimide ester. The antibody is partially purified from the antiserum (8.0 ml) by ammonium sulfate precipitation between 25 and 50% saturation. After extensive dialysis against phosphate-buffered saline (PBS) consisting of 10 mM phosphate buffer (pH 7.2), 0.8% (w/v) NaCl, and 0.02% (w/v) KCl to remove residual ammonium sulfate, the partially purified antibody (28 mg of protein) is loaded onto the affinity column (bed volume of 2.0 ml), washed with PBS, and eluted stepwise with 1 ml of 100 mM glycine buffer, pH 2.5. Aliquots of the eluate containing anti-Ct13-mer antibody are immediately neutralized with 0.1 ml of 0.5 M sodium phosphate, pH 8.0. The purified antibody (1.0 mg of protein) is cross-linked with Affi-Gel–protein A beads (1.0 ml; Bio-Rad, Hercules, CA), using dimethylpimelimidate, transferred into a polypropylene Econo column (bed volume of 0.8 ml; Bio-Rad), and washed with PBS. The protein concentration of the anti-Ct13-mer peptide antibody is determined spectroscopically, using a value of 0.8 mg/ml per 1 OD at

[9] W. Schaffner and C. Weissmann, *Anal. Biochem.* **56,** 502 (1973).
[10] R. S. Kaplan and P. L. Pedersen, *Anal. Biochem.* **150,** 97 (1985).
[11] E. Harlow and D. Lane, "Antibodies: A Laboratory Manual." Cold Spring Harbor Laboratory Press, Cold Spring Harbor, New York, 1988.

280 nm. The yield of the purified antibody from partially purified antibody is approximately 3 to 4%.

Purification of LIV-II Carrier

All procedures for purification are carried out at 4° or on ice. *Escherichia coli* B7634 membranes (3.5 mg of protein in 250 μl) containing the overproduced carrier are diluted in 625 μl of 10 mM potassium phosphate buffer (pH 7.5) containing 0.1 M KCl, 20 mM NaNO$_3$, and 20 mM leucine (PSL). After addition of 200 μl of crude phosphatidylethanolamine (50 mg/ml) from *E. coli* (*E. coli* lipid) (type IX; Sigma, St. Louis, MO), partially purified with acetone[7] and sonicated in PSL with a Bransonic 12 bath sonicator (Branson, Danbury, CT), the membrane mixture is solubilized with 75 μl of 20% (w/v) n-dodecyl-β-D-maltopyranoside (dodecyl maltoside) for 10 min and centrifuged at 110,000g for 1 hr. The resultant supernatant is applied to an immunoaffinity column prewashed with 2 ml of PSL containing 1.5% dodecyl maltoside and *E. coli* lipid (1.0 mg/ml) and is circulated for 1 hr at a flow rate of 8 ml/hr, using a peristaltic pump. The column is then washed in the following order: (1) 4 ml of PSL containing 1.5% dodecyl maltoside and *E. coli* lipid (0.25 mg/ml), (2) 4 ml of PSL containing 1.5% n-octyl-β-D-glucopyranoside (octyl glucoside), *E. coli* lipid (0.25 mg/ml), and 0.5 M NaCl, and (3) 4 ml of 10 mM sodium phosphate, pH 6.0, containing 1.5% octyl glucoside and *E. coli* lipid (0.25 mg/ml). After washing, the carrier is eluted with 20 mM glycine buffer, pH 2.6, containing 1.5% octyl glucoside and *E. coli* lipid (0.25 mg/ml). Aliquots (400 μl) of the eluate are immediately mixed with 8 μl of 0.5 M sodium phosphate, pH 8.0, and 2 μl of 0.2 M dithiothreitol to adjust the pH to 5.9. Subsequently, the eluates containing LIV-II carrier are loaded onto an anion-exchange DE53 cellulose column (1.0 × 7.8 cm) equilibrated with 10 ml of a solution consisting of 1.5% octyl glucoside, *E. coli* lipid (0.25 mg/ml), 1 mM dithiothreitol, and 10 mM sodium phosphate, pH 6.0, and eluted with the same solution. The eluate is fractionated into 800-μl aliquots and immediately neutralized with 24 μl of 1 M Tris-HCl, pH 8.0, and 12 μl of *E. coli* lipid (50 mg/ml) to adjust the pH to 7.8. The carrier protein recovered in flowthrough fractions (total, 5 ml) is collected and concentrated fivefold with a Centricon-30 microconcentrator (Amicon, Danvers, MA). Purified LIV-II carrier is successively reconstituted into liposomes. The immunoaffinity column is regenerated by sequential washing with 4 ml of 0.1% dodecyl maltoside, 10 mM sodium phosphate (pH 8.0), 4 ml of 0.1% dodecyl maltoside, 100 mM tetraethylamine (pH 11.5), 16 ml of PBS, and 8 ml of PBS containing 0.01% (w/v) thimerosal, and stored at 4°. The column may be used at least 30 times.

Comments

The purity of the LIV-II carrier differs depending on the detergents used to solubilize the carrier from the membranes.[6] The membranes are solubilized with 1.5% (w/v) octyl glucoside (OG), dodecyl maltoside (DM), 3-[(3-chlolamidopropyl)diethylammonio]-1-propane sulfonate (CHAPS), or dodecyl/tetradecyl polyoxyethylene ether (Lubrol PX). The carrier is immunoaffinity purified from various detergent extracts and silver stained after SDS–polyacrylamide gel electrophoresis. DM is the most effective in terms of solubilization and gives a carrier preparation of more than 95% purity; OG is the second most effective. CHAPS and Lubrol PX are less effective, with both yield and purity of the carrier being low compared with OG and DM.

Replacement of DM with OG in the course of purification is crucial for subsequent reconstitution of the carrier. DM is the best detergent for solubilization but is not as effective for reconstitution, because its critical micelle concentration (CMC) (0.16 mM) is too low to allow its removal from the mixture of carrier, lipid, and the detergent. Thus, DM is replaced with OG (which has a high CMC, 25 mM) after the carrier in the DM extract is adsorbed by the immunoaffinity column.

A purification procedure for LIV-II carrier is shown in Table I.[6] The

TABLE I
PURIFICATION OF LIV-II CARRIER

Fraction	Protein		^{35}S-Labeled carrier[a]		Purification (fold)	Leucine counterflow activity[b] (nmol/mg protein/min)
	μg	%	10^3 cpm	%		
Membranes	3500	100	838	100	1.0	
Extract with dodecyl maltoside	1300	37	548	65	1.8	
Octyl glucoside	950	27	302	36	1.3 (1.0)	22 (1.0)
IA column peak[c]	28	0.8	199	24	30 (23)	452 (21)
DE53 column peak	7	0.2	61	7	35 (27)	546 (25)

[a] [^{35}S]Methionine-labeled LIV-II carrier is synthesized and the membrane preparation containing the ^{35}S-labeled carrier is mixed with a nonlabeled membrane preparation.
[b] Represents net Na$^+$-dependent leucine uptake for 1 min at 25° by subtracting leucine counterflow at 20 mM KNO$_3$ from that at 20 mM NaNO$_3$.
[c] IA, Immunoaffinity.

yield of carrier protein is examined in a membrane preparation containing [^{35}S]methionine-labeled carrier. The DE53 column peak fraction contains 7% of the ^{35}S-labeled carrier and 0.2% of the protein originally present in the membranes, resulting in a 35-fold purification of the ^{35}S-labeled carrier. This suggests that the content of the LIV-II carrier overproduced in the *E. coli* membranes is approximately 3% of total membrane proteins. A specific leucine counterflow activity assay shows that more than 90% of the purified carrier is functional.

Reconstitution of LIV-II Carrier

The purified LIV-II carrier is reconstituted into liposomes to measure Na^+ concentration gradient-driven transport activity and Na^+-dependent counterflow activity. For the Na^+ gradient-driven transport measurement, 1 ml of the frozen LIV-II carrier preparation is thawed quickly in a water bath at room temperature, kept on ice, and mixed with 350 μl of ice-cold *E. coli* lipid (50 mg/ml) sonicated in a solution of 0.1 M HEPES–Tris (pH 7.5), 10 mM KCl, 1 mM dithiothreitol (HTPD). The carrier mixture is solubilized completely with 20–30 μl of 20% octyl glucoside to become transparent. The cleared mixture is applied to a Sephadex G-50 column (1.5 × 40 cm) equilibrated with HTPD and is eluted with HTPD at a flow rate of 12 ml/hr at 18°. Proteoliposomes in turbid fractions are collected and centrifuged at 81,000g for 1 hr. The pellet is suspended in 800 μl of HTPD, frozen in an ethanol–dry ice bath, and stored at $-70°$. For the leucine counterflow measurement, proteoliposomes are prepared by the same procedure, but replacing HTPD with a solution of 0.1 M KCl, 20 mM NaNO$_3$, 20 mM potassium phosphate (pH 7.5), 1 mM dithiothreitol, and 20 mM leucine.

Transport Assays

Na$^+$ Concentration Gradient-Driven Transport Assay

Frozen proteoliposomes are thawed rapidly in a room temperature water bath and sonicated at room temperature for 15 sec. After dilution with 4 ml of HTPD, the sonicated proteoliposomes are centrifuged at 81,000g for 1 hr at 4°. The pellet is suspended in 25 μl of HTPD and kept on ice. For the transport assay, proteoliposomes are prewarmed at 25° for 3 min and diluted 100-fold in a buffer solution of 0.1 M HEPES–Tris, pH 7.5, containing 10 mM NaCl and a ^{14}C-labeled amino acid (1.0 μCi/ml) to initiate the reaction. The concentrations used for transport of leucine, isoleucine, and valine are 3.5, 3.1, and 3.8 μM, respectively. The reaction mixture is stirred during the assay. Aliquots (100 μl) are removed at time points of 20 sec and 1, 2, 3, 4, and 5 min; diluted with 4 ml of 0.15 M LiCl; filtered through an HA-type filter (0.45-μm pure size; Millipore, Bedford, MA) presoaked in water; and washed once with 4 ml of 0.15 M LiCl. At time zero, 1 μl of proteoliposome preparation is added to 100 μl of the reaction mixture and rapidly diluted with 4 ml of 0.15 M LiCl. After drying the filters under an infrared lamp, the radioactivity of the filters is counted with a liquid scintillation counter.

Na$^+$-Dependent Leucine Counterflow Assay

After the concentrated proteoliposome preparation for counterflow measurement is prewarmed at 25°, leucine counterflow is initiated by diluting the proteoliposomes 100-fold into a solution of 0.1 M KCl, 20 mM NaNO$_3$ or KNO$_3$, 20 mM potassium phosphate (pH 7.5), and 24 μM [^{14}C]leucine (0.9 μCi/ml). Aliquots (100 μl) are removed at appropriate intervals, diluted, filtered, and washed in the same manner as described above. After drying the filters, the radioactivity is counted.

Comments

The Na$^+$ gradient-driven cotransport reaction is one in which Na$^+$ and a substrate equilibrate across the membrane barrier such that the translocation of Na$^+$ on a cotransport protein (carrier protein) is coupled to the translocation of the substrate in the same direction. The electrochemical gradient of Na$^+$ that impels the downhill movement of Na$^+$ drives an accumulation of the substrate against its concentration gradient.[12,13] The Na$^+$-dependent counterflow reaction is one in which in the presence of no concentration gradient of Na$^+$, translocation of a substrate across the membrane barrier is caused by the translocation of the same substrate in the opposite direction as a result of the reverse reaction of transport on the carrier. The chemical gradient of substrate that impels the downhill substrate movement can drive an uphill movement of the (labeled) substrate against its concentration gradient. Both Na$^+$ gradient-driven leucine transport and Na$^+$-dependent leucine counterflow by purified LIV-II carrier increase with time and gradually decrease after maximal

[12] F. M. Harold, *Bacteriol. Rev.* **36**, 172 (1972).
[13] G. J. Kaczorowski and H. R. Kaback, *Biochemistry* **18**, 3691 (1979).

TABLE II

NET SODIUM GRADIENT-DRIVEN TRANSPORT AND SODIUM-DEPENDENT
COUNTERFLOW ACTIVITIES OF PURIFIED LIV-II CARRIER[a]

Substrate	Transport activity[b] (nmol/mg protein/min)	Counterflow activity[c] (nmol/mg protein/min)
Leucine	11.5	533
Isoleucine	8.0	—
Valine	3.0	—

[a] At 25°.
[b] Represents net Na[+] gradient-driven uptake for 1 min by subtracting substrate transport at 10 mM KCl from that at 10 mM NaCl.
[c] Defined in Table I.

uptake around 5 to 6 min. The two transport activities of the carrier are shown in Table II.

Concluding Remarks

Purification of several cotransporters (carriers and permeases) of *E. coli* by various methods has been reported, following the success in overproduction of the cotransporter proteins with high efficiency (more than 10% of total membrane proteins). Those proteins include lactose carrier (permease),[14–16] melibiose permease,[17] and proline carrier.[18]

Among the methods reported, the immunoaffinity purification method as described here for the purification of the LIV-II carrier of *P. aeruginosa* is one of the best methods for a small-scale purification. In addition, one-step purification is possible, depending on the purity of the anti-Ct13-mer antibody. For a large-scale purification, it is reasonable to begin by scaling up all steps of the procedure. Another way to succeed would be to overproduce the LIV-II carrier (i.e., more than 10% of total membrane proteins); however, such overproduction has not yet been attained.

[14] M. J. Newman, D. L. Foster, T. H. Wilson, and H. R. Kaback, *J. Biol. Chem.* **256**, 11804 (1981).
[15] M. G. P. Page, J. P. Rosenbusch, and I. Yamato, *J. Biol. Chem.* **263**, 15897 (1988).
[16] T. G. Consler, B. L. Persson, H. Jung, K. H. Zen, K. Jung, G. G. Prive, G. E. Verner, and H. R. Kaback, *Proc. Natl. Acad. Sci. U.S.A.* **90**, 6934 (1993).
[17] T. Pourcher, S. Leclercq, G. Brandolin, and G. Leblanc, *Biochemistry* **34**, 4412 (1995).
[18] K. Hanada, I. Yamato, and Y. Anraku, *J. Biol. Chem.* **263**, 7181 (1988).

[13] Reconstitution of *Pseudomonas aeruginosa* High-Affinity Branched-Chain Amino Acid Transport System

By Toshimitsu Hoshino *and* Yoshihiko Uratani

Introduction

Active transport of branched-chain amino acids L-leucine, L-isoleucine, and L-valine across the *Pseudomonas aeruginosa* cytoplasmic membrane is mediated by three distinct systems: LIV-I, LIV-II, and LIV-III. The high-affinity LIV-I system is a multicomponent system, whereas the low-affinity LIV-II and LIV-III systems are unicomponent systems, mediated by Na^+- and H^+-coupled carrier proteins, respectively. All the genes required for these transport systems (*bra* genes) have been located on the *P. aeruginosa* PAO chromosome, cloned, and sequenced.[1–4]

The LIV-I transport system is an ABC (ATP-binding cassette) transporter[5] and consists of five *bra* gene products: BraC, a periplasmic binding protein for branched-chain amino acids; BraD and BraE, integral membrane proteins; and BraF and BraG, putative ATP-binding proteins.[1] By using a T7 RNA polymerase/promoter system, the BraD, BraE, BraF, and BraG proteins (BraDEFG proteins), membrane components of LIV-I, can be overproduced as a complex in *Escherichia coli* cytoplasmic membrane. The membrane components are solubilized from the *E. coli* membrane with octyl glucoside and then incorporated into liposomes to reconstitute the LIV-I transport system. In this reconstituted system, branched-chain amino acid transport depends solely on the presence of all five Bra proteins, including BraC added externally and ATP loaded internally to the proteoliposomes.[6] The *in vitro* reconstitution studies of a few periplasmic binding protein-dependent transport systems, including LIV-I, have clarified that the transport systems are ATPases that couple ATP hydrolysis with substrate translocation across the membrane in a binding protein-dependent

[1] T. Hoshino and K. Kose, *J. Bacteriol.* **172,** 5531 (1990).
[2] T. Hoshino and K. Kose, *J. Bacteriol.* **172,** 5540 (1990).
[3] T. Hoshino, K. Kose, and Y. Uratani, *Mol. Gen. Genet.* **220,** 461 (1990).
[4] T. Hoshino, K. Kose, and Y. Uratani, *J. Bacteriol.* **173,** 1855 (1991).
[5] C. F. Higgins, *Annu. Rev. Cell Biol.* **8,** 67 (1992).
[6] T. Hoshino, K. Kose, and K. Sato, *J. Biol. Chem.* **267,** 21313 (1992).

manner,[6-8] solving a long-standing controversy over the energy-coupling mechanism of the transport systems in this category.

In this chapter, several methods are presented for the reconstitution study of the *P. aeruginosa* LIV-I transport system with Bra proteins over-produced in *E. coli*. The method of constructing the *E. coli* strain that is used to overexpress the BraDEFG proteins is also described, as the genetic background of the strain used for overproduction of membrane components seems crucial for successful reconstitution, as described below.

Methods for Preparation of BraC

Overexpression of BraC in Escherichia coli Cells

The BraC protein can be purified either from cold shock fluid[9] of *P. aeruginosa* cells or from osmotic shock fluid[10,11] of *E. coli* cells with a plasmid harboring the *braC* gene. A much higher yield of BraC is attained from *E. coli* cells than from *P. aeruginosa* cells if an appropriate expression plasmid is used. For this purpose, we have placed a 1.5-kilobase pair (kbp) DNA fragment harboring the *braC* gene downstream of the *lac* promoter of pUC18,[12] generating pUTH83,[13] and deposited this plasmid in *E. coli* JM109 [*recA1 endA1 gyrA1 hsdR17* Δ(*lac-proAB*) (F' *traD36 proAB lacI*q ZΔM15) λ−].

The *E. coli* JM109 cells carrying pUTH83 are grown at 37° to the midexponential phase in LB medium[14] containing carbenicillin (100 μg/ ml) with vigorous aeration (2 liters in a 5-liter flask shaken at 130 rpm on a rotating shaker platform). The cells are then supplemented with isopropyl-β-D-thiogalactopyranoside (IPTG, 1 mg/ml) to induce BraC expression and grown at 37° for an additional 3 to 4 hr. All the procedures described below are carried out at 4° unless otherwise indicated. The cells are harvested by centrifugation and subjected to an osmotic shock procedure exactly as described by Amanuma and Anraku.[11] The resultant shock fluid typically

[7] L. Bishop, R. Agbayani, Jr., S. V. Ambudkar, P. C. Maloney, and G. F.-L. Ames, *Proc. Natl. Acad. Sci. U.S.A.* **86**, 6953 (1989).
[8] A. L. Davidson and H. Nikaido, *J. Biol. Chem.* **265**, 4254 (1990).
[9] T. Hoshino, *J. Bacteriol.* **139**, 705 (1979).
[10] H. C. Neu and L. A. Heppel, *J. Biol. Chem.* **240**, 3685 (1965).
[11] H. Amanuma and Y. Anraku, *J. Biochem.* (*Tokyo*) **76**, 1165 (1974).
[12] C. Yanisch-Perron, J. Vieira, and J. Messing, *Gene* **33**, 103 (1985).
[13] T. Hoshino and K. Kose, *J. Bacteriol.* **171**, 6300 (1989).
[14] J. Sambrook, E. F. Fritsch, and T. Maniatis, "Molecular Cloning: A Laboratory Manual," 2nd Ed. Cold Spring Harbor Laboratory Press, Cold Spring Harbor, New York, 1989.

shows more than 90% purity of the BraC protein on a sodium dodecyl sulfate (SDS)–polyacrylamide gel stained with Coomassie Brilliant Blue. The osmotic shock fluid (total, 1000 ml from a 20-liter culture) is concentrated approximately 10-fold by ultrafiltration with a Diaflo YM10 filter (Amicon, Danvers, MA) and stored at $-20°$ until used for purification of BraC. The concentrated shock fluid typically contains 3 to 5 mg of protein per milliliter.

Purification of BraC

Frozen shock fluid is thawed and subjected to ammonium sulfate precipitation. Proteins precipitated between 60 and 95% saturation are collected by centrifugation at 12,000g for 20 min, suspended in TMC buffer [10 mM Tris-HCl (pH 7.3), 1 mM MgCl$_2$, and 1 mM CaCl$_2$] containing 125 mM NaCl, and dialyzed overnight against TMC buffer–125 mM NaCl. The dialyzed ammonium sulfate fraction is applied to a 50-ml DEAE-Sephadex A-50 column equilibrated with TMC buffer–125 mM NaCl. After the column is washed with the equilibration buffer, BraC is eluted with a 400-ml gradient of 125 to 250 mM NaCl in TMC buffer. Fractions are inspected for BraC by SDS–polyacrylamide gel electrophoresis. The fractions containing BraC, which is eluted at about 180 mM NaCl, are pooled, concentrated by ultrafiltration with a Diaflo YM10 filter, and dialyzed overnight against T buffer (10 mM Tris-HCl, pH 7.3). The slight precipitate that has formed during dialysis is removed by centrifugation at 10,000g for 10 min. The resultant supernatant is saved and stored at $-20°$ as a purified BraC preparation. This procedure typically yields at least 250 mg of BraC from a 20-liter culture. The purified BraC is more than 99% homogeneous on an SDS–polyacrylamide gel stained with Coomassie Brilliant Blue.

Preparation of Ligand-Free BraC

The BraC protein has high affinities for its substrates, giving K_d values of 0.3 to 0.5 μM for branched-chain amino acids[15] and of 3 to 5 μM for alanine and threonine.[6] Therefore, it is likely that the purified BraC preparation still retains substrates that have not been dissociated during purification. Those substrates bound to BraC may disturb the accurate determination of transport activities of reconstituted proteoliposomes when using a ^{14}C-labeled amino acid as a transport substrate. Thus, BraC protein devoid of bound substrates is prepared by the sequential dialysis method of Amanuma et al.[16] Briefly, 60 mg of BraC in 2 ml of T buffer is mixed

[15] T. Hoshino and M. Kageyama, J. Bacteriol. 141, 1055 (1980).
[16] H. Amanuma, J. Itoh, and Y. Anraku, J. Biochem. (Tokyo) 79, 1167 (1976).

with 6 ml of 8 M urea dissolved in T buffer and dialyzed for 2 days against 6 M urea in T buffer to eliminate bound substrates by changing the dialysis buffer three times (1.2 liters each time). The ligand-free BraC is then renatured by extensive dialysis against T buffer for 3 days by changing the buffer at least five times (2 liters each time). More than 80% of BraC is recovered to the final dialysate, which is divided into small aliquots and stored at $-20°$ until use.

Methods for Preparation of Proteoliposomes

Construction of Plasmids and Strains

The BraDEFG proteins are present at a low content in *P. aeruginosa* cells, which limits the biochemical analysis of the LIV-I transport system. Because the overproduction of the BraDEFG proteins is toxic to both *P. aeruginosa* and *E. coli* cells, an expression system that can be tightly regulated is required for overproduction of these proteins. We have used the T7 RNA polymerase/promoter system developed by Tabor and Richardson[17] to overproduce the Bra proteins in *E. coli*. For this purpose, the 3.9-kbp DNA fragment containing the *braD*, *braE*, *braF*, and *braG* gene cluster has been cloned downstream of the T7 promoter of plasmid pT7-5 (carbenicillin resistant, Cb^r), generating pTDG50.[1]

The LIV-I transport system requires ATP as an energy source. When the LIV-I system is reconstituted into proteoliposomes, using a crude extract of *E. coli* membranes, the proteoliposomes are contaminated by *E. coli* ATPases, which may dissipate ATP loaded internally to the proteoliposomes and also disturb the measurement of any ATPase activity of the reconstituted LIV-I system. In addition, the presence of proteins of the *E. coli* branched-chain amino acid transport systems may further disturb the measurement of the transport activity of the reconstituted system. Thus, an *E. coli* strain lacking both F_0F_1-ATPase, most powerful ATPase in the *E. coli* membrane, and the entire branched-chain amino acid transport activity has been constructed.[6] A deletion mutation of the ATPase of *E. coli* DK8 (*unc*)[18] is introduced into *E. coli* B7634,[19] which is defective in the branched-chain amino acid transport systems (*hrb*), by P1 cotransduction[20] with the closely linked *ilv*Q::Tn*10*, generating strain TH1. Strain TH2

[17] S. Tabor and C. C. Richardson, *Proc. Natl. Acad. Sci. U.S.A.* **82,** 1074 (1985).
[18] D. J. Klionsky, W. S. A. Brusilow, and R. D. Simoni, *J. Bacteriol.* **160,** 1055 (1984).
[19] I. Yamato and Y. Anraku, *J. Bacteriol.* **144,** 36 (1980).
[20] T. J. Silhavy, M. L. Berman, and L. W. Enquist, "Experiments with Gene Fusions." Cold Spring Harbor Laboratory Press, Cold Spring Harbor, New York, 1984.

is then constructed by transforming TH1 with plasmid pGP1-2 (kanamycin resistant, Km^r)[17] carrying the T7 RNA polymerase gene under the control of the bacteriophage λ promoter P_L and the temperature-sensitive repressor cI857.

Preparation of Membranes

The strain TH2 cells carrying plasmid pTDG50 are grown aerobically at 30° to the midexponential phase in 2× YT medium[14] (1.5 liters of culture medium in a 5-liter flask) supplemented with 0.5% (w/v) glucose, leucine (100 μg/ml), isoleucine and valine (200 μg/ml each), carbenicillin (100 μg/ml), kanamycin (30 μg/ml), and 0.2 mM isopropyl-β-D-thiogalactoside. The cells are then incubated at 42° for 15 min with rifampin (100 μg/ml) and 20 amino acids (50 μg/ml each) to induce the expression of the T7 RNA polymerase. The cells are grown at 37° for another 2.5 hr, harvested by centrifugation, and washed once in chilled MMK buffer consisting of 25 mM potassium morpholine propanesulfonic acid (MOPS, pH 7.0), 2.5 mM MgCl$_2$, and 50 mM KCl. The cells are resuspended in MMK buffer to give a cell density of 0.25 to 0.3 g (wet weight)/ml and then disrupted by two passages through a French pressure cell at 10,000 psi (68.9 MPa). After the disrupted cell suspension is incubated with DNase I (50 μg/ml) on an ice bath for 10 min, unbroken cells and cell debris are removed by three successive centrifugations at 2400g for 10 min. The membranes are collected by centrifugation at 100,000g for 30 min, and washed twice with MMK buffer by centrifugation at 100,000g for 30 min. The washed membranes are suspended in MMK buffer containing 50% (v/v) glycerol at a protein concentration of 10 to 20 mg of protein per milliliter, divided into aliquots, and stored at −70° until use. Typically, this method yields a membrane vesicle preparation containing 10 mg of protein from 1 g (wet weight) of cells.

Solubilization of BraDEFG Proteins and Reconstitution into Proteoliposomes

Solubilization and reconstitution are carried out by a modification of the method of Ambudkar and Maloney.[21] In this protocol, crude phosphatidylethanolamine extracted from E. coli (type IX; Sigma, St. Louis, MO) is used as E. coli phospholipids after partial purification with acetone.[22,23] To solubilize membrane proteins, frozen membranes (4 mg of protein) are

---

[21] S. V. Ambudkar and P. C. Maloney, J. Biol. Chem. **261,** 10079 (1986).
[22] G. Rouser, A. N. Siakotos, and S. Fleischer, Lipids **1,** 85 (1966).
[23] Y. Uratani, J. Biol. Chem. **260,** 10023 (1985).

thawed on ice, diluted to 2 mg of protein per milliliter of MMKD buffer (MMK buffer with 1 mM dithiothreitol) containing 20% (v/v) glycerol, 0.37% (w/v) *E. coli* phospholipids, and 1.2% (w/v) octyl glucoside and incubated on ice for 30 min. Solubilized membrane proteins are obtained as a supernatant fraction after centrifugation at 100,000g for 1 hr. Two milliliters of the supernatant fraction is mixed with 250 μl of bath-sonicated *E. coli* lipids (40 mg/ml of MMKD buffer) and 13 μl of 20% (w/v) octyl glucoside to give a final concentration of 1.2%. The mixture is incubated on ice for 30 min. Two milliliters of the mixture is then dispersed into 50 ml of MMKD buffer containing 10 to 20 mM nucleotide such as ATP. The dilution buffer containing nucleotides should be prepared and adjusted to pH 7 with potassium hydroxide just before use. The diluted mixture is left for 15 min at room temperature to form proteoliposomes containing the nucleotide internally. The proteoliposomes thus formed are collected by centrifugation at 100,000g for 1 hr, washed once in 10 ml of MMKD by centrifugation at 100,000g for 20 min, and resuspended in 1 ml of MMKD buffer. The proteoliposome suspension is kept on ice and immediately used for transport assay. Typically, 30% of total membrane proteins are solubilized with 1.2% (w/v) octyl glucoside and 20 to 30% of the solubilized proteins are incorporated into liposomes, giving a proteoliposome suspension (1 ml) containing 0.2 to 0.4 mg of protein as determined by the method of Peterson[24] and 5 to 8 mg of phospholipids as determined by the method of Gerlach and Deuticke[25] after extraction by the method of Bligh and Dyer.[26] Liposomes without proteins can also be prepared as described above by replacing the membrane preparation with MMKD buffer containing 50% (v/v) glycerol.

Transport Assay

A reaction cocktail for the transport assay consists of 200 μl of proteoliposome suspension (0.2 to 0.4 mg of protein per milliliter of MMKD buffer) and 50 μl of a mixture of ligand-free BraC and [^{14}C]leucine (50 μCi/μmol) (50 μM each in T buffer). Typically, reactions are carried out at 30°. The proteoliposome suspension (200 μl) is preincubated for 1 min. To this proteoliposome suspension, a 50-μl mixture of BraC and [^{14}C]leucine is added to initiate transport reaction. In most assays, BraC and [^{14}C]leucine are mixed 2 to 3 hr before the assay to form liganded BraC, although the dissociation–association reaction of BraC and [^{14}C]leucine seems to reach

[24] G. L. Peterson, *Anal. Biochem.* **83**, 346 (1977).
[25] E. Gerlach and B. Deuticke, *Biochem. Z.* **337**, 477 (1963).
[26] E. G. Bligh and W. J. Dyer, *Can. J. Biochem. Physiol.* **37**, 911 (1959).

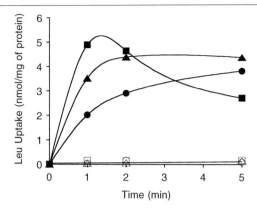

Fig. 1. Leucine transport by LIV-I reconstituted into proteoliposomes. Leucine uptake by proteoliposomes loaded internally with ATP (solid symbols) or ADP (open symbols) was assayed at 25° (circles), 30° (triangles), and 37° (squares) as described in text.

equilibrium within 1 hr. At the final concentrations of BraC and [^{14}C]leucine added (10 μM each), the concentration of liganded BraC is estimated to be 8 to 9 μM, using the K_d value of 0.3 μM. At the specified times 45-μl aliquots are withdrawn, diluted into 2 ml of MMKD buffer, and filtered through 0.45-μm HAWP Millipore (Bedford, MA) filters presoaked in H$_2$O. The filters are then washed once with 2 ml of MMKD buffer and dried. Radioactivity retained on the filters is measured by liquid scintillation counting. By this filtration procedure more than 90% of the proteoliposomes in the reaction mixture are trapped on the filter as determined by the retention of ATP loaded inside the proteoliposomes. Radioactivity of a reaction mixture devoid of proteoliposomes is also measured to determine the nonspecific binding of unincorporated [^{14}C]leucine to the filters. An example of the transport assay by this method is shown in Fig. 1. Under the conditions used, omission of any one of the constituents results in no transport, showing that the transport activity of the proteoliposomes is specifically due to the *P. aeruginosa* LIV-I transport system. The data obtained are reproducible although the specific activity of proteoliposomes varies considerably with batches of membrane preparations used for solubilizing the BraDEFG proteins.

Conclusion

No reliable *in vitro* assay system has yet been established for periplasmic binding protein-dependent ABC transporters, including *P. aeruginosa* LIV-I, using either shocked cells or membrane vesicles. Thus, the *in vitro* system with proteoliposomes is the best available at present to approach biochemically the molecular mechanisms of the ABC transporters.

The methods presented here are applicable to many other bacterial periplasmic binding protein-dependent ABC transporters, particularly the high-affinity branched-chain amino acid transport systems in *E. coli*[27] and *Salmonella typhimurium*,[28] because these systems are basically identical, in terms of genetic and molecular organization, to the *P. aeruginosa* LIV-I transport system. As suggested from the uptake at 37° in Fig. 1, leucine accumulated in proteoliposomes leaks out to a considerable extent, which hinders the quantitative analysis of reconstituted LIV-I. The use of a substrate that is more hydrophilic than leucine (e.g., threonine), however, may improve the quantifiability of the assay.

[27] M. D. Adams, L. M. Wagner, T. J. Graddis, R. Landick, T. K. Antonucci, A. L. Gibson, and D. L. Oxender, *J. Biol. Chem.* **265**, 11436 (1990).
[28] K. Matsubara, K. Ohnishi, and K. Kiritani, *J. Biochem.* (*Tokyo*) **112**, 93 (1992).

[14] Purification of *Pseudomonas putida* Branched-Chain Keto Acid Dehydrogenase E1 Component

By Kathryn L. Hester, Jinhe Luo, and John R. Sokatch

Introduction

Pseudomonas putida is a soil and water organism that uses a wide variety of organic compounds as carbon and energy sources; as such, the organism plays important roles in the study of catabolic pathways and in bioremediation. Branched-chain keto acid dehydrogenase (BCKAD) of *P. putida* is the second enzyme in the common pathway for the metabolism of valine, leucine, and isoleucine. *Pseudomonas putida* BCKAD is a multienzyme complex that is induced by growth on branched-chain amino or keto acid[1] and repressed by growth in the presence of glucose or ammonium ion.[2] The four structural genes encoding the three components of *P. putida* BCKAD have been cloned and sequenced; the *bkd* operon (Fig. 1[2]) is expressed as a single, polycistronic message. The positive transcriptional

[1] V. P. Marshall and J. R. Sokatch, *J. Bacteriol.* **110**, 1073 (1972).
[2] P. Sykes, G. Burns, J. Menard, K. Hatter, and J. R. Sokatch, *J. Bacteriol.* **169**, 1619 (1987).

0076-6879/00 $30.00

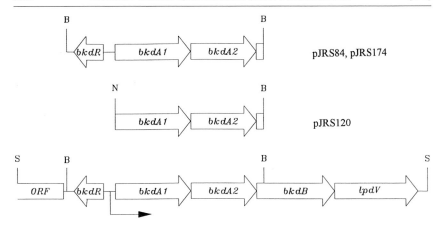

FIG. 1. Constructs used in this study. The structural genes of the *bkd* operon of *P. putida* are as follows: *bkdR* (BkdR, the positive transcriptional regulator of the *bkd* operon), *bkdA1* (E1α subunit), *bkdA2* (E1β subunit), *bkdB* (E2 component), and *lpdV* (Lpd-Val, the specific E3 component of branched-chain keto acid dehydrogenase of *P. putida*). The cloning vector for pJRS84 and pJRS174 is pKT230.[2a] The difference between the two plasmid constructs is that the first 13 codons of *bkdA2* have been deleted from pJRS174, so that only the 37-kDa form of E1β is produced (see Fig. 6).

regulator of the operon, BkdR, has been identified[3] and its DNA-binding properties characterized.[4]

The following procedure for the purification of E1$\alpha_2\beta_2$ of *P. putida* BCKAD has produced active protein that has been used for molecular mass determinations to confirm its heterotetrameric nature,[5] and also for crystallization studies leading to elucidation of the structure of the E1 component.[6] Along with previous data on the structures of the E2 and E3 components, this allows for visualization of the structure of the entire multienzyme complex and the possible modes of interactions between the three components.

Properties of E1 Component of Branched-Chain Keto Acid
    Dehydrogenase from *Pseudomonas putida*

The E1 component of *P. putida* BCKAD expressed in *P. putida* is a heterotetramer, E1$\alpha_2\beta_2$, with a calculated molecular mass of 167 kDa. The

[2a] M. Bagdasarian, R. Lurz, B. Rückert, F. C. H. Franklin, M. M. Bagdasarian, J. Frey, and K. N. Timmis, *Gene* **16**, 237 (1981).

[3] K. T. Madhusudhan, D. Lorenz, and J. R. Sokatch, *J. Bacteriol.* **175**, 3934 (1993).

[4] K. T. Madhusudhan, G. Huang, and J. R. Sokatch, *J. Bacteriol.* **177**, 636 (1995).

[5] K. Hester, J. Luo, G. Burns, E. H. Brasswell, and J. R. Sokatch, *Eur. J. Biochem.* **233**, 828 (1995).

[6] Ævarsson, K. Seger, S. Turley, J. R. Sokatch, and W. G. J. Hol, *Nature Structural Biology* **6**, 785 (1999).

```
CTGGGGGTTTGAGATGAACGACCACAACAACAGCATCAACCCGGAAACCGCCATGGCCACCACTACCATG
L  G  V  ***  L  N  D  H  N  N  S  I  N  P  E  T  A  M  A  T  T  T  M     39 kDa E1β
                                                  A  T  T  T  M     37 kDa E1β
```

FIG. 2. Deduced and observed amino acid sequences of the 37- and 39-kDa forms of E1β. LGV, The last three amino acids of E1α; ***, the translational terminator for *bkdA1*. The two initiating methionine codons of *bkdA2* are underlined. Trans, Translation of the nucleotide sequence of *bkdA2*; 39 kDa and 37 kDa, the experimentally obtained N-terminal sequences.

molecular mass obtained by sedimentation equilibrium agrees with this value.[5] There are two translational start sites for E1β, which are 13 codons apart (see Fig. 2), and two forms of E1β are produced in wild-type *P. putida* PpG2 with molecular masses of 37 and 39 kDa. The first codon of the 39-kDa form is leucine, even though the initiating codon is ATG.[5] In the crystal structure the cofactor thiamine pyrophosphate can be seen residing in the active site of the heterotetramer, making contacts with both α and β subunits.[6]

When the *P. putida* BCKAD E1 component is expressed in *Escherichia coli*, only one translational start site for E1β is recognized, and only the 39-kDa product is produced. In addition, the N-terminal amino acid of the 39-kDa E1β is methionine instead of leucine.[5]

Materials and Methods

Bacterial Strains

Pseudomonas putida PpG2: The wild-type strain; originally obtained from I. C. Gunsalus, University of Illinois

JS388: A PpG2 *recA* mutant that has been described previously[5]

JS390: A PpG2 mutant in which the *bkd* operon has been deleted between the *Nco*I sites in *bkdR* and *lpdV* and replaced with a tetracycline resistance cassette; construction of this mutant has been described previously[5]

JS330: A methionine auxotroph[7]

Escherichia coli DH5α: Obtained from GIBCO-BRL (Gaithersburg, MD)

Plasmids

PMC1871: A fusion vector from Pharmacia Biotech (Piscataway, NJ); GenBank accession number L08936

pJRS84 (Fig. 1): Contains a 3.19-kb *Bsp*HI fragment containing *bkdR*, *bkdA1,* and *bkdA2* cloned into pKT230[5]

---

[7] P. Sykes, J. Menard, V. McCully, and J. R. Sokatch, *J. Bacteriol.* **162,** 203 (1985).

pJRS174 (Fig. 1): Similar to pJRS84, with the exception that the first 13 codons of *bkdA2* have been deleted from pJRS174, so that only the 37-kDa form of E1β is produced[6]

pJRS120 (Fig. 1): Contains *bkdA1* and *bkdA2* cloned into pCYTEXP1, and *E. coli* expression plasmid,[8] with *Nde*I sites engineered at the 3' end of the *atpE* translation-initiation region and at the 5' end of *bkdA1*.[5] pCYTEXP1 utilizes the heat-sensitive *c*I857 repressor to control expression, the *atpE* translation-initiation region to express cloned genes, and the λ $P_R P_L$ promoters[8]

pJRS3: Contains *bkdB* and *lpdV* genes in pUC19[2]

Media and Buffers

YT broth (2×): 16 g of Tryptone, 10 g of yeast extract, and 5 g of NaCl per liter of water

Valine–isoleucine medium: 3 g of L-valine, 1 g of L-isoleucine, 100 ml of nitrogen-free Basal G (see below), and 890 ml of water. The medium is sterilized, and then 10 ml of Salts S (see below) is added. Twenty milliliters $1M$ glucose can be added to give valine–isoleucine–glucose medium when growing *bkd* operon mutants

The compositions of the salt solutions are as follows:

Nitrogen-free Basal G: 43.5 g of K_2HPO_4, 17 g of KH_2PO_4, and water to 1 liter

Salts S: 39.44 g of $MgSO_4 \cdot 7H_2O$, 5.58 g of $MnSO_4 \cdot 2H_2O$, 1.11 g of $FeSO_4 \cdot 7H_2O$, 0.33 g of $CaCl_2$, 0.12 g of NaCl, 1.0 g of ascorbic acid, and water to 1 liter

BCKAD buffer: 50 mM potassium phosphate (pH 7.0), 1 mM EDTA, 1 mM L-valine, 0.5 mM dithiothreitol, 0.5 mM thiamine pyrophosphate α-Ketoisovalerate (0.1M)

BCKAD Enzyme Assay of Purified E1α$_2$β$_2$

Principle of Assay. The overall reaction catalyzed by branched-chain keto acid dehydrogenase is shown in Fig. 3. R–CO–COO⁻ can be α-ketoisovalerate, α-ketoisocaproate, or α-keto-β-methylvalerate. Pyruvate is usually inactive or at best a poor substrate. The exception is the pyruvate/branched-chain keto acid dehydrogenase of *Bacillus subtilis*, which not only oxidizes pyruvate for energy but also provides branched-chain fatty acids for membrane synthesis.[9]

The E1 component catalyzes the reductive acylation of the lipoyl residue of E2, with oxidative decarboxylation of α-ketoisovalerate, α-ketoisocapro-

[8] T. N. Belev, M. Singh, and J. E. G. McCarthy, *Plasmid* **26**, 147 (1992).

[9] R. N. Perham and P. N. Lowe, *Methods Enzymol.* **166**, 330 (1983).

(1) $R\text{-CO-COO}^- + E1\text{-TPP} \longrightarrow E1\text{-TPP}=C(OH)\text{-R} + CO_2$

(2) $E2\text{-LipS}_2 + E1\text{-TPP}=C(OH)\text{-R} \rightleftharpoons E2\text{-lip} \begin{smallmatrix} S\text{-CO-R} \\ \\ SH \end{smallmatrix} + E1\text{-TPP}$

(3) $E2\text{-lip} \begin{smallmatrix} S\text{-CO-R} \\ \\ SH \end{smallmatrix} + CoASH \rightleftharpoons E2\text{-lip(SH)}_2 + CoAS\text{-CO-R}$

(4) $E2\text{-lip(SH)}_2 + E3\text{-FAD(S)}_2 \rightleftharpoons E2\text{-lip-S}_2 + E3\text{-FAD(SH)}_2$

(5) $E3\text{-FAD(SH)}_2 + NAD^+ \rightleftharpoons E3\text{-FAD(S}_2) + NADH + H^+$

Net: $R\text{-CO-COO}^- + NAD^+ + CoASH \longrightarrow CoAS\text{-CO-R} + CO_2 + NADH + H^+$

FIG. 3. Reactions catalyzed by the three components of branched-chain keto acid dehydrogenase. The reaction catalyzed by $E1\alpha_2\beta_2$ is shown in reactions (1) and (2) and results in the acylation of the E2 component. The net reaction is shown on the bottom line and is the reaciton used to follow the purification of $E1\alpha_2\beta_2$ in the presence of excess E2 and E3 (Lpd-Val) components.

ate, and α-keto-2-methylvalerate (Fig. 3). There is no specific assay for the reaction catalyzed by the E1 component alone. Dye reduction assays with 2,6-dichlorophenolindophenol or ferricyanide have been used,[10] sometimes in combination with $1\text{-}^{14}C$-labeled substrate. 2,6-Dichlorophenolindophenol can be used with the complex from *P. putida,* but ferricyanide is inactive.

The assay described in this chapter provides excess E2 and E3 component, so that measuring the reduction of NAD^+ follows the overall reaction. E2 and E3 are provided with extracts of *E. coli* containing pJRS3, a pUC-derived plasmid that carries the genes expressing the E2 and E3 components of branched-chain keto acid dehydrogenase from *P. putida.*[2]

Reagents

Potassium phosphate (pH 7.0), 0.1 M
EDTA, 0.1 M
L-Valine, 0.1 M
NAD^+, 0.1 M
Coenzyme A, 0.01 M in 0.02 M dithiothreitol
Thiamine pyrophosphate, 0.021 M
Magnesium chloride, 0.1 M
2-Ketoisovalerate, 0.1 M
Cell-free extract of *E. coli* TB1 (pJRS3), 3–5 units/ml

[10] D. T. Chuang, *Methods Enzymol.* **166,** 146 (1988).

Procedure. The reaction mixture contains 100 μl of phosphate buffer (pH 7.0), 20 μl of NAD$^+$, 10 μl of coenzyme A plus dithiothreitol (DTT), 10 μl of thiamine pyrophosphate, 10 μl of magnesium chloride, 50 μl of L-valine, 0.06 units of BCKAD E2 and E3 components from TB1(pJRS3) extract, and 0.004–0.04 unit of BCKAD E1 component. Enough water is added to make the final volume 0.96 ml, and the absorbance is read at 340 nm for 1 min to correct for any endogenous reaction. The reaction is started with 40 μl of 0.1 M 2-ketoisovalerate and the absorbance is read at 340 nm for 1 min. The assay is linear with absorbance changes of up to 0.2 per minute. One unit is the amount of enzyme that produces 1 μmol of NADH per minute.

The amount of E2 and E3 components to be added to the assay is determined by titration with excess E1 component. Increasing amounts of TB1 (pJRS3) cell-free extract are assayed as described above with 0.05–0.1 unit of E1 component until the absorbance change plateaus, usually between 0.3 and 0.4 per minute.

Purification of E1 Component

Purification from Pseudomonas putida Carrying pJRS84 or pJRS174. A typical purification procedure is shown in Table I. The preparation from *P. putida* hosts begins with a 5-ml 2× YT culture inoculated from a single colony that is grown for 8 hr at 30°; *P. putida* JS388(pJRS84) is grown in medium supplemented with tetracycline (200 μg/ml) and kanamycin (90 μg/ml). This procedure works well with any of the hosts

TABLE I

PURIFICATION OF E1$\alpha\beta$ FROM JS388(pJRS84) AND *Escherichia coli* DH5α(pJRS120)[a]

Purification step	pJRS84		pJRS120	
	Total protein (mg)	Specific activity[b] (U/mg)	Total protein (mg)	Specific activity (U/mg)
Cell-free extract[c]	447	5.2	382	5.2
DEAE-Sepharose CL-6B pool	93	15	46	23
FPLC pool	17	26	12	38

[a] Reprinted by permission from K. Hester, J. Luo, G. Burns, E. H. Brasswell, and J. R. Sokatch, *Eur. J. Biochem.* **233,** 828 (1995).

[b] Specific activity is micromoles of NADH produced per minute per milligram of protein at 30°.

[c] Values from typical purifications of 2 liters of culture.

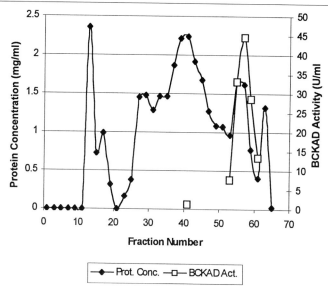

FIG. 4. Elution profile of BCKAD E1αβ produced by DEAE-Sepharose CL-6B chromatography. A cell-free extract of JS388(pJRS84) was applied to the column, and BCKAD E1αβ was eluted as described in text. Fractions were assayed for protein concentration by the Bradford method, and BCKAD assays were performed as described in text.

derived from *P. putida*. Flasks (500 ml) of valine–isoleucine–glucose medium are inoculated with the 8-hr culture and grown overnight at 30° with aeration until the OD_{660} reaches 0.8, at which point the cells are harvested and washed once with 0.15 M sodium chloride. Cells are resuspended in BCKAD buffer at a buffer-to-cell paste mass ratio of 6:5, and cell extracts are prepared by sonic oscillation. In the examples given, a Heat Systems (Farmingdale, NY) oscillator (model W-225R set at 3) was used. Each tube contained a total volume of 3–5 ml and received four 1-min bursts with rest in between each burst.

The broken cell suspension is centrifuged at 90,000g for 1 hr in an ultracentrifuge at 4°. This step greatly reduces membrane-bound enzymes, which interfere with the enzyme assay by oxidizing NADH in the absence of keto acid substrates. The cell extract is loaded onto a DEAE-Sepharose CL-6B column (Sigma, St. Louis, MO) with a 30-ml bed volume. The column is washed with 50 ml of BCKAD buffer, and the complex is eluted with a linear gradient; the receiving flask contains 200 ml of BCKAD buffer, and the reservoir flask contains 200 ml of BCKAD buffer plus 0.55 M sodium chloride. Fractions of 5 ml are collected at a flow rate of 1 ml/min.

The fractions are assayed for E1$\alpha_2\beta_2$, and active fractions are pooled and dialyzed against 10 volumes of BCKAD buffer. A typical chromatogram is shown in Fig. 4. The pooled fractions are applied to a Resource Q or Mono Q HR 5/5 (Pharmacia) column and eluted with a gradient of 0 to 0.8 M sodium chloride in BCKAD buffer. Again, the fractions are assayed for E1$\alpha_2\beta_2$, and the active fractions are pooled and dialyzed against 10 volumes of BCKAD buffer. A typical chromatogram is shown in Fig. 5.

Purification from Escherichia coli DH5α(pJRS120). To purify *P. putida* E1$\alpha_2\beta_2$ from *E. coli*, pJRS120 is used. pJRS120 contains *bkdA1* and *bkdA2* cloned into the *Nde*I sites of pCYTEXP1 (Fig. 1). The preparation of E1$\alpha_2\beta_2$ from *E. coli* hosts begins with a 2-ml 2× YT culture inoculated from a single colony and grown for 8 hr at 30°. *Escherichia coli* DH5α(pJRS120) is grown in medium supplemented with ampicillin (200 μg/ml). A 50-ml volume of 2× YT medium is inoculated with the 2-ml 8-hr culture and grown overnight at 30°, and then used to inoculate 500-ml flasks of 2× YT medium. The cultures are grown at 30° with shaking to an OD$_{600}$ of 0.8, and then

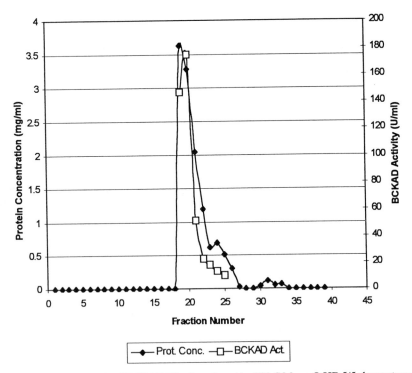

FIG. 5. Elution profile of BCKAD E1$\alpha\beta$ produced by FPLC Mono Q HR 5/5 chromatography. Pooled fractions 54–60 from Fig. 4 were applied to the Mono Q column, and assays were performed as in Fig. 4.

shifted to 42° for 3 hr to induce the culture. The cells are harvested by centrifugation and washed once with 0.15 M sodium chloride. The purification of $E1\alpha_2\beta_2$ then proceeds as described above.

Comments on Purification Procedure

The results from one purification of $E1\alpha_2\beta_2$ from JS388(pJRS84) are summarized in Table I and Fig. 6, a sodium dodecyl sulfate (SDS)–polyacrylamide gel in which the E1 α and β subunits are visible at each purification step. Three major proteins are visible: the 45-kDa protein is E1α, and the 37- and 39-kDa proteins have been shown to be two separate translational products of *bkdA2* (Fig. 2),[5] which are also produced in wild-type PpG2.[5] The N-terminal amino acid of the 39-kDa protein is leucine instead of methionine, and we have attempted to determine the molecular mechanism of this unique translational event: the mRNA sequence of the intergenic region between *bkdA1* and *bkdA2* corresponded to the DNA sequence[5]; insertion of guanine residues preceding the initiating ATG of *bkdA2* did not affect the production of two forms of the β subunit, or the

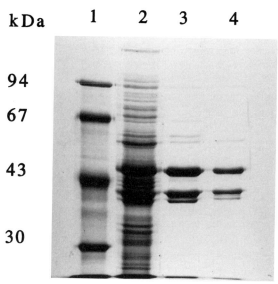

FIG. 6. Purification of $E1\alpha_2\beta_2$ from *P. putida* JS388(pJRS84). SDS–polyacrylamide gel of fractions obtained in the purification of E1$\alpha\beta$. Lane 1, molecular mass standards; lane 2, soluble fraction after centrifugation at 90,000g, 50 μg of protein; lane 3, pooled fractions obtained by DEAE-Sepharose CL-6B chromatography, 10 μg of protein; lane 4, pooled fractions obtained by FPLC Mono Q chromatography, 1 μg of protein. [Reprinted by permission from K. Hester, J. Luo, G. Burns, E. H. Brasswell, and J. R. Sokatch, *Eur. J. Biochem.* **233**, 828 (1995).]

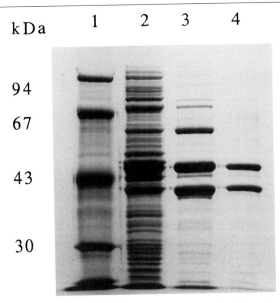

FIG. 7. Purification of E1$\alpha_2\beta_2$ from *E. coli* DH5α(pJRS120). SDS–polyacrylamide gel of fractions obtained in the purification procedure. Lane 1, molecular mass standards; lane 2, soluble fraction after 90,000g centrifugation, 10 μg; lane 3, pooled fractions obtained by DEAE-Sepharose CL-6B chromatography, 5 μg; lane 4, pooled fractions obtained by FPLC Resource Q chromatography, 1 μg. [Reprinted by permission from K. Hester, J. Luo, G. Burns, E. H. Brasswell, and J. R. Sokatch, *Eur. J. Biochem.* **233**, 828 (1995).]

N-terminal leucine of the 39-kDa form[5]; a plasmid with *bkdA2* as the only structural gene of the *bkd* operon also produced N-terminal leucine in the 39-kDa form[11]; and a plasmid in which *bkdA2* was replaced with *lacZ* expressed β-galactosidase with the N-terminal amino acid removed.[12] From these results, we can conclude that the mechanism does not involve RNA editing or ribosome jumping, and that the "signal sequence" is at the N-terminal end of E1β or at the 5' end of *bkdA2*.

To determine if the translation of *bkdA2* was unique to *P. putida*, pJRS120 was constructed as described above. Good expression was seen in *E. coli* DH5α, and the purification results are shown in Table I and Fig. 7. Two major proteins are present in the purified preparations: the 45-kDa protein in E1α, and the 39-kDa protein is E1β.[5] In *E. coli*, translation begins at the first AUG of *bkdA2* mRNA, and the N-terminal amino acid of E1β is methionine.

[11] K. L. Hester, J. Luo, and J. R. Sokatch, "Biochemistry and Physiology of Thiamin-Diphosphate Enzymes" (H. Bisswanger and A. Schellenberger, eds.), p. 374. A. u. C. Intemann, Wissenschaftlicher Verlag, Prien, 1996.

[15] *Pseudomonas mevalonii* 3-Hydroxy-3-methylglutaryl-CoA Lyase

By HENRY M. MIZIORKO and CHAKRAVARTHY NARASIMHAN

Introduction

$$\text{3-Hydroxy-3-methylglutaryl-CoA} \xrightarrow{M^{2+}} \text{acetoacetate} + \text{acetyl-CoA}$$

3-Hydroxy-3-methylglutaryl-CoA (HMG-CoA) cleavage is a divalent cation-dependent irreversible reaction.[1] It has been well established to occur in the ketogenic tissues (e.g., liver, kidney) of animals, providing a freely soluble lipid-derived metabolic fuel.[2] Several bacterial DNA sequences encoding HMG-CoA lyase have been reported.[3,4] The requirement for this enzymatic reaction in prokaryotes has been most convincingly explained[5] in the case of the enzyme from *Pseudomonas mevalonii*. In this bacterium, which utilizes mevalonic acid as a carbon source, the production of acetyl-CoA by HMG-CoA lyase supports a variety of biosynthetic pathways. The isolated enzyme has demonstrated utility in preparation of (*R*)-HMG-CoA, a potent inhibitor of HMG-CoA reductase, because of its ability to quantitatively deplete the physiologically active metabolite, (*S*)-HMG-CoA, from the mixed isomers produced by chemical synthesis of HMG-CoA.[6] In addition, the recombinant form of *P. mevalonii* HMG-CoA lyase, as expressed and isolated from *Escherichia coli,* represents the protein first used in studies that identified components of the enzyme active site.[7] For mechanistic or protein chemistry work that must be performed in the absence of high levels of reducing compounds [e.g., dithiothreitol (DTT), mercaptoethanol], the preferred enzyme would be bacterial HMG-CoA lyase.

The procedure described here for preparation of recombinant *P. mevalonii* HMG-CoA lyase represents straightforward methodology that has been routinely employed for preparation of both wild-type[8] and mutant

[1] L. Stegink and M. J. Coon, *J. Biol. Chem.* **243,** 5272 (1968).
[2] A. M. Robinson and D. H. Williams, *Physiol. Rev.* **60,** 143 (1980).
[3] D. H. Anderson and V. W. Rodwell, *J. Bacteriol.* **171,** 6468 (1989).
[4] M. Baltscheffsky, M. Brosche, T. Hultman, L. Lundvik, P. Nyren, Y. Sakai-Nore, A. Severin, and A. Strid, *Biochim. Biophys. Acta* **1337,** 113 (1997).
[5] D. S. Scher and V. W. Rodwell, *Biochim. Biophys. Acta* **1003,** 321 (1989).
[6] K. M. Bischoff and V. W. Rodwell, *Biochem. Med. Metab. Biol.* **48,** 149 (1992).
[7] P. W. Hruz, C. Narasimhan, and H. M. Miziorko, *Biochemistry* **31,** 6842 (1992).
[8] C. Narasimhan and H. M. Miziorko, *Biochemistry* **31,** 11224 (1992).

forms[9] of the enzyme. This strategy supports isolation of significant amounts of high specific activity enzyme with good homogeneity, assuming that robust expression of soluble, active enzyme has been achieved during bacterial propagation and that enzyme stability is maintained during protein isolation. Preparation of the recombinant protein from *E. coli* in which expression has been suboptimal produces enzyme of reasonably high purity and specific activity that is useful for many experimental applications; yield, however, will be substantially lower. Experimental details that are important to optimizing expression are included in the protocol described below.

Assay Methods

Principle

HMG-CoA cleavage may be followed spectrophotometrically by using coupled enzyme assays that measure production of either of the two reaction products, acetyl-CoA or acetoacetate. Alternatively, the reaction may be followed with [14C]HMG-CoA, estimating substrate cleavage by measuring conversion of the original acid-stable radioactivity to a volatile radiolabeled product. Different experimental situations have demanded use of each of these assays.

Citrate Synthase-Coupled Assay of HMG-CoA Lyase Activity

Reagents

Tris-HCl (pH 8.2), 1.0 M
$MgCl_2$, 1.0 M
DTT, 100 mM
NADH, 5 mM
NAD, 30 mM
L-Malate, 50 mM (neutralized as potassium salt)
HMG-CoA (pH 4.5), 6 mM [synthetic HMG-CoA is a mixture of (R,S)-isomers]
Citrate synthase (pig heart), 215 U/mg
Malate dehydrogenase (pig heart), 900 U/mg

Procedure. This spectrophotometric assay represents a modification of the method of Stegink and Coon[1] in which acetyl-CoA, produced along with acetoacetate on HMG-CoA cleavage, is coupled to a citrate synthase assay. Reaction of acetyl-CoA to produce citrate requires oxaloacetate, generated in the cuvette by malate dehydrogenase-catalyzed oxidation of

[9] C. Narasimhan, J. R. Roberts, and H. M. Miziorko, *Biochemistry* **34,** 9930 (1995).

malate with the reduction of NAD^+ and the resulting increase in $A_{340\,nm}$. The rate of NAD^+ reduction is proportional to the amount of HMG-CoA lyase added to a limiting value of 1.5 ΔA/min and linear until a total absorbance increase of >0.4 has occurred. The complete reaction mixture (1.0 ml) contains 200 μmol of Tris-HCl (pH 8.2), 10 μmol of $MgCl_2$, 1.5 μmol of NAD^+, 0.05 μmol of NADH, 5.0 μmol of DTT, 2.5 μmol of L-malate, malate dehydrogenase (9 U), citrate synthase (4 U), and the sample containing HMG-CoA lyase. After the mixture has been incubated at 30° and a stable baseline recorded, HMG-CoA (0.12 μmol) is added and the rate of increase in $A_{340\,nm}$ is measured. An extinction coefficient of 6.2 × $10^3 M^{-1}$ is used to quantitate product acetyl-CoA formation. This assay is used routinely for measuring HMG-CoA lyase activity at stages of enzyme isolation where NADH oxidase contaminants do not require significant correction of the measured rate of absorbance increase. The small volumes of *E. coli* extracts needed to measure activity of highly overexpressed HMG-CoA lyase do not cause serious interference. This assay may be unsuitable for inhibition studies that involve acyl-CoA analogs that also significantly inhibit citrate synthase.

β-Hydroxybutyrate Dehydrogenase-Coupled Assay of HMG-CoA Lyase Activity

Reagents

Tris-HCl (pH 8.2), 1.0 M
$MgCl_2$, 1.0 M
DTT, 100 mM
NADH, 5 mM
HMG-CoA (pH 4.5), 6 mM
β-Hydroxybutyrate dehydrogenase (*Rhodopseudomonas sphaeroides*), 36 U/ml

Procedure. In this spectrophotometric assay,[10] the acetoacetate produced along with acetyl-CoA on HMG-CoA cleavage is coupled to the hydroxybutyrate dehydrogenase assay.[11] Tris-HCl (pH 8.2, 100 μmol), NADH (0.1 μmol), $MgCl_2$ (5.0 μmol),* DTT (2.0 μmol), β-hydroxybutyrate dehydrogenase (0.5 IU), and the sample containing HMG-CoA lyase are added to a 500-μl assay mixture and allowed to incubate at 30° for 10 min, during which time any baseline drift in $A_{340\,nm}$ may be observed. HMG-CoA (0.06 μmol) is added and the rate of decrease in $A_{340\,nm}$ is measured.

[10] P. W. Hruz, V. E. Anderson, and H. M. Miziorko, *Biochim. Biophys. Acta* **1162,** 149 (1993).
[11] M. N. Berry, *Biochim. Biophys. Acta* **92,** 156 (1964).
* $MnCl_2$ (0.4–0.6 μmol) can also be used as the activator divalent cation in this assay.[10]

An extinction coefficient of $6.2 \times 10^3 \, M^{-1}$ is used to quantitate the production of acetoacetate. This assay is used to measure activity of alternative substrates that produce products that do not efficiently couple to citrate synthase or when measuring HMG-CoA lyase activity in the presence of acyl-CoA analogs that inhibit citrate synthase.

Radioactive Assay of HMG-CoA Lyase Activity

Reagents

Tris-HCl (pH 8.2), 1.0 *M*
MgCl$_2$, 1.0 *M*
DTT, 100 m*M*
[^{14}C]HMG-CoA (pH 4.5), 2.4 m*M* (~3 mCi/mmol)

Procedure. Improved sensitivity, required when sample activity is limited (e.g., in the assay of a catalytically impaired mutant HMG-CoA lyase), is afforded by measuring the cleavage of [^{14}C]HMG-CoA by the method of Clinkenbeard *et al.*[12] Although either commercially available, chemically synthesized [3-^{14}C]HMG-CoA or enzymatically synthesized [5-^{14}C]HMG-CoA is suitable for the assay, the chemically synthesized substrate is a mixture of (R,S)-isomers and only 50% of the total radioactivity will be volatilized on exhaustion of this substrate. The reaction mixture (0.2 ml) contains Tris-HCl (pH 8.2, 40 μmol), MgCl$_2$ (2 μmol), DTT (2 μmol), and the sample containing HMG-CoA lyase. The reaction, performed at 30°, is initiated by addition of [^{14}C]HMG-CoA (24 nmol; 6000 dpm/nmol). At various time points, aliquots (40 μl) are removed, pipetted into glass shell vials (15 × 45 mm; flat bottomed), and acidified with 6 *N* HCl (0.1 ml) prior to heating to dryness at 95°, using an aluminum heating block [Lab-Line (Melrose Park, IL) 2073 or equivalent] in a fume hood. The acid-stable radioactivity attributable to [^{14}C]HMG from unreacted substrate is determined by liquid scintillation counting. Depletion of acid-stable radioactivity is a measure of enzymatic cleavage of substrate to form a volatile radiolabeled product ([3-^{14}C]acetoacetate from [3-^{14}C]HMG-CoA; [1-^{14}C]acetoacetate from [5-^{14}C]HMC-CoA).

Units. A unit of enzymatic activity is defined as the amount of enzyme necessary to convert 1 μmol of HMG-CoA to products acetyl-CoA and acetoacetate in 1 min under the conditions described. Specific activity is expressed in units per milligram of protein. In the published reports from which these methods have been compiled, protein concentration has been

---

[12] K. D. Clinkenbeard, W. D. Reed, R. A. Mooney, and M. D. Lane, *J. Biol. Chem.* **250**, 3108 (1975).

determined by the Bradford method,[13] with bovine serum albumin used as a calibration standard.

Expression of Active HMG-CoA Lyase

Escherichia coli BL21 (DE3) transformed with the expression vector pT7-2600 containing the *P. mevalonii* HMG-CoA lyase gene (*mvaB*) is a generous gift of V. Rodwell (Purdue University, Lafayette, IN). The fraction of total *E. coli* BL21 (DE3) protein represented by the *mvaB* gene product is a function of bacterial growth conditions (namely, temperature).[8] For example, at 37° more than half of the total protein expressed is represented by HMG-CoA lyase. At 30 and 22°, the fraction of total expressed protein that is represented by lyase decreases somewhat, although it remains the major component of the samples. Even more dependent on these expression conditions is the level of HMG-CoA lyase activity. Although the expression levels of lyase protein are high at both 37 and 30°, the specific activities of enzyme in those fractions are comparatively low. On comparison of soluble extracts with the corresponding crude homogenates, it becomes apparent that lyase protein is largely insoluble when expressed under temperature conditions typical for bacterial propagation. Growth at 22°, however, markedly improves the expression of catalytically active enzyme. Most of the enzyme that has been expressed remains soluble after high-speed centrifugation of the crude extract. Therefore, the 22° growth condition is now routinely used to overexpress active *P. mevalonii* HMG-CoA lyase.

Purification of HMG-CoA Lyase

Bacterial Growth and Induction

Bacteria are grown in LB medium [10 g of Bacto tryptone (Difco, Detroit, MI), 5 g of Bacto yeast extract, and 5 g of NaCl per liter] at 22° in flasks shaken at 250 rpm. Growth is monitored by periodically measuring the optical density (OD) at 600 nm. Starter cultures (15 ml) are grown overnight in LB medium containing ampicillin (200 μg/ml). This overnight culture is diluted to 1.5 liters with LB medium containing ampicillin (200 μg/ml) and allowed to grow at 22° to an OD of 0.6. At this point, expression of HMG-CoA lyase is induced by the addition of isopropyl-β-D-thiogalactopyranoside (IPTG) to a final concentration of 1 mM and growth continued until late-log phase (OD ~3.0). The cells are harvested by low-speed centrif-

---

[13] M. M. Bradford, *Anal. Biochem.* **72**, 248 (1976).

ugation (3000g, 10 min) at 4° and the pellets may be stored at −20° prior to cell lysis.

Bacteria are suspended in 50 ml of cold buffer that consists of 20 mM phosphate (pH 7.2), 1 mM EDTA, DNase I (10 μg/ml), RNase A (10 μg/ml), and 100 μM phenylmethylsulfonyl fluoride (PMSF). The cells are lysed with a French pressure cell at 16,000 psi. This crude extract is centrifuged at 181,000g for 60 min and the resultant high-speed supernatant containing the active HMG-CoA lyase decanted and kept on ice. This high-speed supernatant is promptly brought to 40% $(NH_4)_2SO_4$ saturation by addition of the solid salt with slow stirring. The mixture is stirred slowly at 4° for an additional 4.5 hr. The precipitated protein is centrifuged at 12,100g for 20 min at 4°. After removing the supernatant from the pellet, the pellet is dissolved in a minimum volume of 10 mM phosphate buffer, pH 7.2, containing 20% (v/v) glycerol. At this point, the inclusion of glycerol in buffer is crucial to retention of enzyme activity. The dissolved ammonium sulfate fraction is desalted by rapidly passing the sample, divided into appropriately sized aliquots, over Sephadex G-50 centrifugal desalting columns[14] equilibrated with 10 mM phosphate buffer containing 20% (v/v) glycerol. The desalted sample is immediately applied to a Q-Sepharose Fast Flow anion-exchange column (1.5 × 45 cm) equilibrated with the desalting buffer. The bound material is eluted with a 10–100 mM phosphate gradient (1.0 liter), pH 7.2, containing 20% (v/v) glycerol. The protein peak containing the enzyme activity is pooled and concentrated with an Amicon (Danvers, MA) concentrator equipped with a PM30 membrane.

The pooled active fractions are essentially homogeneous by sodium dodecyl sulfate–polyacrylamide gel electrophoresis (SDS–PAGE) (Fig. 1). The purification steps are summarized in Table I. The enzyme is recovered after the final step in about 41% yield. The preparation represents an approximate 10-fold purification from the crude extract to the homogeneous enzyme, reflecting the high level of initial overexpression.

Characterization of *Pseudomonas Mevalonii* HMG-CoA Lyase

Enzyme Stability

The recombinant enzyme recovered from the Q-Sepharose anion-exchange column retains full activity for several months if stored at −80° in the phosphate buffer containing 20% (v/v) glycerol, pH 7.2. The recombinant enzyme exhibits, after early stages in the purification, a marked requirement for stabilizing agents such as glycerol or substrate. This situation

[14] H. S. Penefsky, *J. Biol. Chem.* **252**, 2891 (1977).

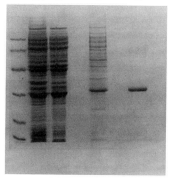

FIG. 1. SDS–PAGE of *P. mevalonii* HMG-CoA lyase at various stages of purification. Lane 1, molecular weight standards (14.4, 21.5, 31.0, 45.0, 66.2, and 97.4 kDa); lane 2, total bacterial cell extract; lane 3, high-speed supernatant obtained by centrifuging the extract at 181,000g for 60 min; lane 4, sample from ammonium sulfate precipitation (0–40% saturation); lane 5, HMG-CoA lyase after Q-Sepharose column chromatography. [Reprinted, with permission, from C. Narasimhan and H. M. Miziorko, *Biochemistry* **31**, 11224 (1992).]

contrasts sharply with the stability observed with the avian liver enzyme and can, in some cases, complicate experimental design. The bacterial enzyme is labile at ambient or higher temperatures. For example, even in the presence of 20% (v/v) glycerol, the enzyme loses 63% of its initial activity when incubated at ambient temperature over a period of 5 hr. Addition of 10 mM MgCl$_2$ to this sample does not improve the stability of

TABLE I

PURIFICATION OF RECOMBINANT *Pseudomonas mevalonii* HMG-CoA LYASE[a]

Purification step	Total units[b]	Total protein[c] (mg)	Specific activity (units/mg)	Yield (%)
Crude extract	3416	490	7.0	100
Soluble extract	3122	446	7.0	91
0–40% (NH$_4$)$_2$SO$_4$ fractionation	1990	95	21.0	58
Q-Sepharose chromatography	1384	19.8	70.0	41

[a] Reprinted, with permission, from C. Narasimhan and H. M. Miziorko, *Biochemistry* **31**, 11224 (1992).
[b] Activity measured by the citrate synthase-coupled spectrophotometric assay.
[c] Protein measured by the Bradford dye-binding assay, employing bovine serum albumin as the calibration standard.

the enzyme at ambient temperature. The presence of 0.2 mM HMG-CoA helps, however, to maintain the activity (>90% after 8 hr) of the enzyme at 22°.

The recombinant *P. mevalonii* enzyme is attractive in that it does not exhibit the stringent requirement for exogenous thiol that characterizes the avian enzyme.[15,16] When isolated in buffers that contain no exogenous thiol, enzyme shows only about a twofold enhancement in activity when 5 mM DTT is added. Conversely, after separating DTT-preincubated enzyme from reducing agents by centrifugal gel filtration, about 50% of the optimal activity is retained. This property promises to facilitate physical and protein chemistry studies that would otherwise be hindered by the requirement to either maintain an elevated level of exogenous thiol or work under anaerobic conditions.

Molecular Properties

The native molecular weight and quaternary structure of recombinant *P. mevalonii* HMG-CoA lyase remain to be resolved. The enzyme has been subjected to gel filtration on a Superose 12 column (12 mm × 30 cm; Pharmacia, Uppsala, Sweden) equilibrated with 50 mM Tris, pH 8.00, containing 0.1 M NaCl, 10% (v/v) glycerol, and calibrated with appropriate molecular weight markers. Under these conditions, the native recombinant lyase, which elutes slightly before a carbonic anhydrase marker, exhibits a molecular weight approximating that reported for the HMG-CoA lyase protomer (Table II; 27 kDa,[15] 32 kDa[5]) and measured for the recombinant enzyme under denaturing SDS–PAGE conditions (Fig. 1). However, in subsequent spin-labeling studies,[9,17] a nitroxide-containing substrate analog has been demonstrated to bind the protein, exhibiting a rotational correlation time of 20 nsec. If enzyme is assumed to be spherical, a Stokes radius of 27 Å would be predicted from this rotational correlation time. This value agrees well with the estimate of 28.2 Å reported by Scher and Rodwell.[5] Using a partial specific volume of 0.75 ml/g, a protein with the predicted Strokes radius of 27 Å would have a molecular mass of 64 kDa, in good agreement with the value anticipated for a dimeric enzyme. On the basis of this estimate, the bacterial lyase would resemble the enzyme isolated from avian liver and the recombinant human enzyme, which have clearly been demonstrated by cross-linking studies to exist as dimers.

[15] P. R. Kramer and H. M. Miziorko, *J. Biol. Chem.* **255,** 11023 (1980).

[16] P. W. Hruz and H. M. Miziorko, *Protein Sci.* **1,** 1144 (1992).

[17] C. Narasimhan, W. E. Antholine, and H. M. Miziorko, *Arch. Biochem. Biophys.* **312,** 467 (1994).

TABLE II
PROPERTIES OF HMG-CoA LYASES[a]

| | *P. mevalonii* lyase[b] | | |
Property	Homologously expressed	Recombinant	Avian lyase[c]
Reduced thiol requirement in assay	Marked dependency	Twofold increase	Marked dependency
Cation requirement	Marked dependency	Marked dependency	Marked dependency
Specific activity (units/mg)	22.1	70	351
K_m (HMG-CoA, μM)	100	20	8
K_a (Mn^{2+}, μM)	—	2	6[d]
K_a (Mg^{2+}, μM)	—	6,900	56[d]
Subunit	32,000[e]	32,000	27,000
Molecular mass (Da)	31,600[f]		

[a] Reprinted, with permission, from C. Narasimhan and H. M. Miziorko, *Biochemistry* **31,** 11224 (1992).

[b] For the homologously expressed *P. mevalonii* lyase, the data are from Scher and Rodwell[5]; data for the recombinant lyase are from Narasimhan and Miziorko.[8]

[c] Data for the avian lyase are from Kramer and Miziorko.[15]

[d] Apparent K_m values, estimated at saturating HMG-CoA concentration, are from Hruz et al.[10]

[e] Molecular weight determined by SDS–PAGE.

[f] From the cDNA sequence (Anderson and Rodwell[3]).

Catalytic Properties

The specific activity observed for the homogeneous recombinant protein is 70 U/mg, a value severalfold higher than reported for the homologously expressed *P. mevalonii* enzyme.[5] The presence of multiple protein components in that preparation may partially account for this discrepancy. In contrast, the avian[15] and human[18] enzymes exhibit specific activities approximately fivefold higher than measured with the prokaryotic enzyme. While such a difference may reflect actual higher catalytic efficiency in the eukaryotic enzymes, the lability of the recombinant enzyme, despite precautions taken during isolation, may partially account for the observed differences.

In comparing the substrate-binding properties of the various HMG-CoA lyases (Table II), variations comparable in magnitude to those observed for specific activities are apparent. Apparent K_m values for the (S)-isomer of HMG-CoA are in the 10^{-5}–10^{-4} M range. HMG-CoA lyase utilizes only

[18] J. R. Roberts, C. Narasimhan, P. W. Hruz, G. A. Mitchell, and H. M. Miziorko, *J. Biol. Chem.* **269,** 17841 (1994).

one isomer from an (R,S)-mixture of chemically prepared HMG-CoA.[1] As discussed by Higgins *et al.*,[19] lyase uses the same isomer that is metabolized by HMG-CoA reductase to form (R)-mevalonate. Because the reductase reaction does not affect the stereochemistry at C-3 and because the stereochemistry at C-3 of (R)-mevalonate is equivalent to that at C-3 of (S)-HMG-CoA, it follows that (S)-HMG-CoA is the substrate for reductase as well as lyase. This analysis has more recently been experimentally validated by Rodwell and colleagues.[5]

Isolated Enzyme Containing Tightly Bound Cations

Inhibition by Metal Chelators. Recombinant *P. mevalonii* HMG-CoA lyase is sensitive to metal chelators. On a 4-hr incubation with *o*-phenanthroline (4 mM), enzyme activity gradually decreases to 43% of its initial activity. That inhibition is attributable to chelation of a tightly bound cation is confirmed by control experiments that demonstrate the nonchelating isomer, *m*-phenanthroline, to be without effect. Attempts to restore enzyme activity by resupplementing the enzyme with exogenous cation after *o*-phenanthroline treatment have not been successful. Incubation of recombinant HMG-CoA lyase with 0.2 mM EDTA for up to 4 hr does not result in any significant inhibition of enzyme activity.

Metal Analyses. Enzyme purified after expression in *E. coli* grown in standard LB broth has been analyzed by atomic absorption and electron paramagnetic resonance (EPR) methodologies. Copper is the predominant tightly bound cation identified, although zinc and iron are also present at lower levels. The concentrations of cations associated with the isolated recombinant lyase do not approach the level expected for stoichiometric binding to enzyme protomer. A survey of 20 metals by inductively coupled plasma emission spectroscopy failed to detect significant levels of other cations. However, when growth medium is supplemented with 1 mM divalent copper (CuSO$_4$) prior to induction of HMG-CoA lyase synthesis, the enzyme subsequently isolated shows an order of magnitude increase in metal content, with bound cation approaching stoichiometric levels with respect to enzyme subunits. At this elevated level of tightly bound cation, there is good agreement between atomic absorption and EPR estimates, suggesting that all the bound copper is the divalent, paramagnetic species. Furthermore, the measured EPR parameters ($A_{\parallel} = 152$ G; $g_{\parallel} = 2.28$) suggest that enzyme liganding creates a type II copper center[20] that may involve protein nitrogen ligand(s). These observations make the possibility of adventitious cation binding appear unlikely.

[19] M. J. P. Higgins, J. A. Kornblatt, and H. Rudney, *in* "The Enzymes" (P. D. Boyer, ed.), 3rd Ed., Vol. VII, pp. 432–434. Academic Press, New York, 1972.

[20] J. Peisach and W. Blumberg, *Arch. Biochem. Biophys.* **165,** 691 (1974).

On the basis of our observation that specific activity of freshly isolated HMG-CoA lyase does not strongly correlate with differences in copper content, assignment of a structural, rather than catalytic, role to tightly bound cation seems reasonable. This hypothesis is supported by the observation that enzyme low in copper content is more unstable to thermal denaturation than enzyme that has been copper enriched. Clearly, an expanded investigation of the function of bound cation in prokaryotic lyase, as well as a survey of the levels of bound cation in eukaryotic lyases, will be required before any precise role can be assigned.

Enzyme Activation by Dissociable Divalent Cations

While added Zn^{2+} or Cu^{2+} shows no stimulatory effects on activity of the isolated enzyme, micromolar concentrations of Mn^{2+} and millimolar concentrations of Mg^{2+} markedly stimulate enzyme activity. In the case of HMG-CoA lyase, Mn^{2+} supports enzyme activity at only about 40% of the optimal Mg^{2+}-supported level. Concentration dependence studies[21] for each of these cations have been performed[8] to determine activator constants (K_a values) of 2 μM and 6.8 mM, respectively. It has also been observed that Mn^{2+} at concentrations higher than 100 μM inhibits the enzyme activity while Mg^{2+} at concentrations lower than 300 μM has virtually no stimulatory effect on the enzyme activity. The 10^3-fold difference between affinity constants for Mg^{2+} and Mn^{2+} does not depend on content of tightly bound copper, as comparable differences have been measured with enzyme produced by induction in copper-supplemented LB broth.

The three orders of magnitude difference between the activator constants, which is much larger than the difference usually observed between binding constants of Mg^{2+} and Mn^{2+} to small molecules, may have structural implications. Similar differences between Mn^{2+} and Mg^{2+} binding to other enzyme systems have been reported.[22–24] Sussman and Weinstein[25] have investigated the marked discrimination that proteins may exhibit in binding cations of different ionic radii. Their study suggests that Mn^{2+} may preferentially be stabilized in a hydrophobic binding pocket within the HMG-CoA lyase active site. Mg^{2+}, which exhibits a strong aqueous solvation preference, would be expected to bind much less tightly in such a model.

Acknowledgment

Pseudomonas mevalonii HMG-CoA lyase studies performed in the author's laboratory have been supported by NIH DK-21491.

[21] A. S. Mildvan and M. Cohn, *J. Biol. Chem.* **240,** 238 (1965).
[22] B. H. Lee and T. Nowak, *Biochemistry* **31,** 2165 (1992).
[23] M. E. Lee and T. Nowak, *Biochemistry* **31,** 2172 (1992).
[24] M. D. Denton and A. Ginsburg, *Biochemistry* **8,** 1714 (1969).
[25] F. Sussman and H. Weinstein, *Proc. Natl. Acad. Sci. U.S.A.* **86,** 7880 (1989).

[16] Human 3-Hydroxy-3-methylglutaryl-CoA Lyase

By Henry M. Miziorko, Chakravarthy Narasimhan, and
Jacqueline R. Roberts

Introduction

3-Hydroxy-3-methylglutaryl-CoA lyase (EC 4.1.3.4) catalyzes the cleavage of HMG-CoA to form acetyl-CoA and acetoacetate [Eq. (1)].

$$\text{HOOC}-CH_2-\overset{\overset{\displaystyle OH}{|}}{\underset{\underset{\displaystyle CH_3}{|}}{C}}-CH_2-\overset{\overset{\displaystyle O}{\|}}{C}-\text{SCoA} \xrightarrow{M^{2+}}$$

$$CH_3-\overset{\overset{\displaystyle O}{\|}}{C}-CH_2-\text{COOH} + CH_3-\overset{\overset{\displaystyle O}{\|}}{C}-\text{SCoA} \quad (1)$$

This reaction is a key step in ketogenesis and also represents the last step in the leucine catabolic pathway. Deficiencies in enzyme activity account for the human inherited metabolic disease, hydroxymethylglutaric aciduria.[1] Although enzyme preparations from mammalian liver have been reported and characterized,[2,3] the first homogeneous, high specific activity preparation from eukaryotic tissue was reported for the avian enzyme.[4] This protein, as well as a homogeneous preparation of the recombinant *Pseudomonas mevalonii* enzyme,[5] were used in affinity labeling studies that identified elements of the catalytic apparatus.[6] Work with the avian enzyme also suggested that an intersubunit thiol/disulfide exchange represents a potential mechanism for regulating activity of the eukaryotic enzyme.[7]

Investigation of catalytic and potential regulatory mechanisms, as well as the modeling of mutations that result in human hydroxymethylglutaric aciduria, required the development of a convenient recombinant source of stable, homogeneous human enzyme. For this reason, human lyase-encod-

[1] K. M. Gibson, J. Breuer, and W. L. Nyhan, *Eur. J. Pediatr.* **148,** 180 (1988).
[2] B. K. Bachawat, W. G. Robinson, and M. J. Coon, *J. Biol. Chem.* **216,** 727 (1955).
[3] L. D. Stegink and M. J. Coon, *J. Biol. Chem.* **243,** 5272 (1968).
[4] P. R. Kramer and H. M. Miziorko, *J. Biol. Chem.* **255,** 11023 (1980).
[5] C. Narasimhan and H. M. Miziorko, *Biochemistry* **31,** 11224 (1992).
[6] P. W. Hruz, C. Narasimhan, and H. M. Miziorko, *Biochemistry* **31,** 6842 (1992).
[7] P. W. Hruz and H. M. Miziorko, *Protein Sci.* **1,** 1144 (1992).

ing cDNA[8] was used to develop a bacterial expression system for the mature mitochondrial form of human HMG-CoA lyase.[9] The methodology described in this account is used for routine expression and isolation of the active recombinant mitochondrial isoform of the enzyme. Subsequently, analogous approaches were used to produce a protein containing the leader sequence that is proteolysed concomitant with import of HMG-CoA lyase into the mitochondrion; this full-length protein appears to represent a peroxisomal isoform of the enzyme.[10] Isolation of this isozyme has been reported; the protocol[11] involves only slight modifications of the procedure outlined for the mitochondrial isoform.

Assay Methods

Principle

HMG-CoA cleavage may be followed spectrophotometrically by using coupled enzyme assays that measure production of either of the two reaction products, acetyl-CoA (citrate synthase-coupled assay) or acetoacetate (β-hydroxybutyrate assay). In addition, the reaction may be followed with increased sensitivity using [^{14}C]HMG-CoA, estimating substrate cleavage by measuring conversion of the acid-stable radioactivity of the substrate to a volatile radiolabeled product. Each of these assays has been required in different experimental situations; procedures for each assay are documented in this volume[12] in [15], on *P. mevalonii* HMG-CoA lyase. To support the isolation of recombinant human mitochondrial HMG-CoA lyase, the citrate synthase-coupled spectrophotometric assay has been routinely used; this assay is described below.

Citrate Synthase-Coupled Assay of HMG-CoA Lyase Activity

Reagents

Tris-HCl (pH 8.2), 1.0 M
MgCl$_2$, 1.0 M

---

[8] G. A. Mitchell, M. F. Robert, P. Hruz, S. Wang, G. Fontaine, C. Behnke, L. Mende-Mueller, K. Schappert, C. Lee, K. M. Gibson, and H. Miziorko, *J. Biol. Chem.* **268,** 4376 (1993).

[9] J. R. Roberts, C. Narasimhan, P. W. Hruz, G. A. Mitchell, and H. M. Miziorko, *J. Biol. Chem.* **269,** 17841 (1994).

[10] L. Ashmarina, N. Rusnak, H. M. Miziorko, and G. A. Mitchell, *J. Biol. Chem.* **269,** 31929 (1994).

[11] L. Ashmarina, M. F. Robert, M. A. Elsliger, and G. A. Mitchell, *Biochem. J.* **315,** 71 (1996).

[12] H. M. Miziorko and C. Narasimhan, *Methods Enzymol.* **324,** Chap. 15, 2000 (this volume).

Dithiothreitol (DTT), 100 mM
NADH, 5 mM
NAD, 30 mM
L-Malate, 50 mM (neutralized as potassium salt)
HMG-CoA, 6 mM, pH 4.5
Citrate synthase (pig heart), 215 U/mg
Malate dehydrogenase (pig heart), 900 U/mg

Procedure. This spectrophotometric assay represents a modification[4] of the method of Stegink and Coon,[3] in which acetyl-CoA, produced along with acetoacetate on HMG-CoA cleavage, is coupled to a citrate synthase assay. Reaction of acetyl-CoA to produce citrate requires oxaloacetate, generated in the cuvette by malate dehydrogenase-catalyzed oxidation of malate with the reduction of NAD^+ and the resulting increase in $A_{340 nm}$. The rate of NAD^+ reduction is proportional to the amount of HMG-CoA lyase added to a limiting value of 1.5 ΔA/min and linear until a total absorbance increase of >0.4 has occurred. The complete reaction mixture (1.0 ml) contains 200 μmol of Tris-HCl (pH 8.2), 10 μmol of $MgCl_2$, 1.5 μmol of NAD^+, 0.05 μmol of NADH, 5.0 μmol of DTT, 2.5 μmol of L-malate, malate dehydrogenase (9 U), citrate synthase (4 U), and the sample containing HMG-CoA lyase. After the mixture has been incubated at 30° and a stable baseline recorded, HMG-CoA (0.12 μmol) is added and the rate of increase in $A_{340 nm}$ is measured. An extinction coefficient of 6.2 $\times$ $10^3 M^{-1}$ is used to quantitate product acetyl-CoA formation. This assay is used routinely for measuring HMG-CoA lyase activity at stages of enzyme isolation where NADH oxidase contaminants do not require significant correction of the measured rate of absorbance increase. The small volumes of *Escherichia coli* extracts needed to measure activity of recombinant HMG-CoA lyase do not cause serious interference. This assay may be unsuitable for inhibition studies that involve acyl-CoA analogs that also significantly inhibit citrate synthase or for evaluation of alternative substrates that produce an acyl-CoA product that does not couple with citrate synthase.

Units. A unit of enzymatic activity is defined as the amount of enzyme necessary to convert 1 μmol of HMG-CoA to products acetyl-CoA and acetoacetate in 1 min under the conditions described. Specific activity is expressed in units per milligram of protein. In the published reports from which these methods have been compiled, protein concentration has been determined by the Bradford method[13] with bovine serum albumin used as a calibration standard.

[13] M. M. Bradford, *Anal. Biochem.* **72,** 248 (1976).

Design of Expression Plasmid for Human Mitochondrial HMG-CoA Lyase

In designing an expression system, it is necessary to decide whether to insert into the expression vector the complete cDNA, encoding the transit peptide-containing precursor to the mature mitochondrial matrix protein, or a shortened sequence encoding a protein closer in sequence to the mature HMG-CoA lyase. Catalytically active precursor forms of mitochondrial matrix enzymes have been produced[14,15] by recombinant DNA methodology. However, our laboratory has directly determined the N terminus of the mature avian HMG-CoA lyase.[9] Alignment of these data with the deduced sequences of the human[8] and mouse[16] precursor proteins has indicated sufficient homology to allow the assignment of T28 as the N terminus of the mature human enzyme. The M26-G27 sequence of the precursor protein suggests the mutation of the CTATGG sequence in the cDNA (bases 74–79) to a CCATGG sequence, which encodes an *Nco*I cleavage site that overlaps a start codon. This change is implemented by polymerase chain reaction (PCR) mutagenesis techniques.[17] The modified cDNA is thus predicted to encode a protein with a two-amino acid (Met-Gly) extension in comparison with the mature mitochondrial lyase. Because N-terminal methionines are typically cleaved from proteins expressed in *E. coli,* a single glycine residue extension is anticipated for the purified recombinant lyase. In view of the fact that slight heterogeneity at the extreme N terminus is apparent on comparison of deduced sequences for bacteria,[18] cricket,[8] mouse,[16] and human enzymes,[8] this modest structural perturbation does not raise serious concerns over consequences to the structural integrity and catalytic activity of the recombinant protein. The last issue addressed in design of the expression system involves changing the GAAGCC sequence (bases 994–999) to GGATCC by PCR mutagenesis. This produces a *Bam*HI site downstream from the stop codon (bases 976–978) in the cDNA.

[14] F. Altieri, J. R. Mattingly, Jr., F. J. Rodriguez-Berrocal, J. Youssef, A. Iriarte, T. Wu, and M. Martinez-Carrion, *J. Biol. Chem.* **264,** 4782 (1989).
[15] J. Jeng and H. Weiner, *Arch Biochem. Biophys.* **289,** 221 (1991).
[16] S. Wang, J. H. Nadeau, A. Duncan, M. F. Robert, G. Fontaine, K. Schappert, K. R. Johnson, E. Zietkiewicz, P. Hruz, H. Miziorko, and G. A. Mitchell, *Mammalian Genome* **4,** 382 (1993).
[17] S. N. Ho, H. D. Hunt, R. M. Horton, J. K. Pullen, and L. R. Pease, *Gene* **77,** 51 (1989).
[18] D. H. Anderson and V. W. Rodwell, *J. Bacteriol.* **171,** 6468 (1989).

Expression of Mitochondrial HMG-CoA Lyase

Bacterial Growth

Escherichia coli BL21(DE3) is used when transformation involves a pET-3d-derived expression plasmid. Strain JM105 is used for transformation involving the pTrc99A-derived plasmid. Bacteria are grown at the indicated temperatures in LB broth [10 g of Bacto tryptone (Difco, Detroit, MI), 5 g of Bacto yeast extract, and 5 g of NaCl per liter] supplemented with ampicillin (50 μg/ml). Growth is monitored by measurement of optical density (OD) at 600 nm. After cultures reach an OD of ~0.6, expression of HMG-CoA lyase is induced by addition of isopropyl-β-D-thiogalactopyranoside (IPTG) to a final concentration of 1 mM. Bacteria are harvested at late log phase; induction at 22° typically requires overnight growth before harvesting. Cells are harvested by centrifugation (4°) at 3000g for 45 min. Pellets are stored at −20° prior to cell lysis with a French pressure cell.

Construction and Evaluation of Human HMG-CoA Lyase
* Expression Plasmids*

Expression of the cDNA for human HMG-CoA lyase, modified as described above to generate *Nco*I and *Bam*HI restriction sites, has been attempted with two different vectors. Into the *Nco*I/*Bam*HI-digested T7 expression plasmid, pET-3d, is ligated the appropriately digested human lyase cDNA, generating pETHL-1. Similarly, the doubly digested human lyase cDNA is ligated into the *Nco*I/*Bam*HI-digested plasmid pTrc99A (Pharmacia, Piscataway, NJ) to generate the expression plasmid pTrcHL-1 (Fig. 1). In these expression plasmids, the transcription of the lyase gene is under the control of the *lac* and *trc* promoters, respectively, and thus inducible by IPTG.

On the basis of our earlier success in producing active recombinant bacterial HMG-CoA lyase[5] using expression that relied on T7 polymerase, the expression vector pET-3d, which contains the appropriate *Nco*I and *Bam*HI sites, was initially selected for insertion of the modified lyase-encoding cDNA insert (Fig. 1). The resulting expression plasmid (pETHL-1) is transformed into *E. coli* BL21. As observed earlier in our expression of recombinant bacterial lyase, growth and expression of the transformed bacteria at 37° result in recovery of the target protein as a large fraction of total *E. coli* protein, but most of the expressed target is insoluble and inactive. In contrast to the substantial improvement in recovery of active, soluble lyase that we enjoyed when the bacterial lyase was expressed at lower temperatures, only modest improvement is observed when a similar strategy is applied to production of recombinant human HMG-CoA lyase

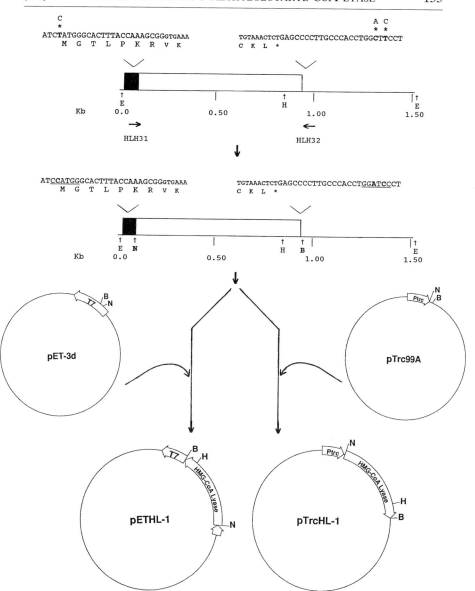

FIG. 1. Construction of plasmids for expression of human HMG-CoA lyase in *E. coli.*

TABLE I
EXPRESSION OF RECOMBINANT HUMAN HMG-CoA LYASE UNDER
VARIOUS GROWTH CONDITIONS[a]

Expression plasmid	Temperature (growth/induction)	Activity (units/ml culture)
pETHL-1	37°/37°	0.2
	22°/22°	0.7
	22°/15°	0.5
	22°/12°	0.6
	15°/15°	0.6
pTrcHL-1	37°/37°	0.8
	37°/30°	1.2
	37°/22°	2.3

[a] Reprinted, with permission, from J. R. Roberts, C. Narasimhan, P. W. Hruz, G. A. Mitchell, and H. M. Miziorko, *J. Biol. Chem.* **269,** 17841 (1994).

in this expression system (Table I). There is precedent[19] suggesting that such problems may be circumvented by using a different expression vector; different expression rates and levels of target may correlate with recovery of a higher percentage of total expressed target protein as active, soluble enzyme. The expression vector pTrc99A, which conveniently contains unique and appropriately positioned *Nco*I and *Bam*HI sites, affords a straightforward way to test this possibility. The same coding insert is, therefore, ligated into appropriately restricted vector and the resulting expression plasmid (pTrcHL-1; Fig. 1) is transformed into *E. coli* JM105. Although IPTG induction of the transformed bacteria does not result in overexpression of lyase at the exaggerated levels that are supported by the pET system, a significant amount of lyase protein is produced. Regardless of induction temperature (Table I), most of the expressed lyase is soluble and active, accounting for a considerable improvement in recovery of enzyme units in comparison with the pETHL-1 expression system. This observation suggests that the pTrcHL-1 system might be exploited for production of high specific activity human HMG-CoA lyase.

Construction of a C323S mutant form of human HMG-CoA lyase has been described.[9] This protein lacks a cysteine close to the C terminus

[19] Z. Zhang, G. Bai, S. Deans-Zirattu, M. F. Browner, and E. Y. C. Lee, *J. Biol. Chem.* **267,** 1484 (1992).

of the protein; this residue has been implicated in a thiol/disulfide exchange that markedly affects *in vitro* activity.[7] The coding sequence has also been inserted into pTrc99A to produce expression plasmid pTrcHL-C233S.

Purification of Human HMG-CoA Lyase

Preliminary work (not shown) has led to a purification protocol that yields less than 1 mg of highly purified human HMG-CoA lyase from 1 liter of bacteria transformed with pETHL-1. The yield of high specific activity enzyme from pTrcHL-1-transformed cells (Table I) represents about a fivefold improvement over our results with pETHL-1-based expression. Thus, the protocol outlined below has been optimized for enzyme produced by expression using PTrcHL-1.

HMG-CoA lyase is purified from the pellet of a pTrcHL-1-transformed *E. coli* JM105 culture (1 liter) harvested by low-speed centrifugation ($3000g$, 45 min, 4°) at late log phase after induction with IPTG at 22°. The pellet is resuspended and homogenized in 50 ml of cold (0°) lysis buffer [10 mM potassium phosphate (pH 7.8), 5 mM EDTA, 0.1 mM phenylmethylsulfonyl fluoride (PMSF), DNase (10 μg/ml), and RNase (10 μg/ml)]. The resuspended cells are lysed in a French pressure cell at 16,000 psi. The crude extract is centrifuged at $100,000g$ for 1 hr at 4°. The high-speed supernatant is loaded onto a Q-Sepharose anion-exchange column (14 × 1.5 cm) equilibrated in 10 mM potassium phosphate, pH 7.8, containing 1 mM DTT (buffer A). At pH 7.8, most of the *E. coli* protein will bind to the anion-exchange column while human HMG-CoA lyase is recovered in the unbound fraction. Protein in this recovered material is diluted to ~0.30–0.40 mg/ml with buffer A. The diluted HMG-CoA lyase from the anion-exchange eluate is brought to 40% $(NH_4)_2SO_4$ saturation by slow addition of the salt with constant stirring at 4°. After 2 hr of stirring, the 40% $(NH_4)_2SO_4$ fraction is centrifuged at $12,100g$ for 20 min at 4°. The supernatant (containing the lyase activity) is immediately brought to 65% $(NH_4)_2SO_4$ by addition of the solid salt with constant stirring at 4° for 16–17 hr. The precipitated HMG-CoA lyase is then centrifuged at $12,100g$ for 30 min at 4°. *Note:* Omission of DTT from the purification buffers used through this stage of the purification does not markedly affect yield or specific activity of wild-type enzyme, but DTT is absolutely necessary in the remaining purification steps. The pellet from the ammonium sulfate precipitation procedure is resuspended in 1.0 M $(NH_4)_2SO_4$ in buffer B [buffer A supplemented with 20% (v/v) glycerol]. The resulting solution is loaded onto a phenyl-agarose column (18 × 1.0 cm) that has been equilibrated in

1.0 M (NH$_4$)$_2$SO$_4$ in buffer B. The column is washed with 10 ml of the equilibrating buffer, followed by 10 ml of 0.7 M (NH$_4$)$_2$SO$_4$ in buffer B. HMG-CoA lyase is eluted from the hydrophobic resin with a reverse gradient of 0.7–0.0 M (NH$_4$)$_2$SO$_4$ in buffer B (60 ml); 1.5-ml fractions are collected. Because the human lyase elutes at the low ionic strength end of the gradient, an additional 20 ml of buffer B is washed through the column to ensure complete recovery of activity. The fractions are analyzed spectrophotometrically for activity, and by sodium dodecyl sulfate–polyacrylamide gel electrophoresis (SDS–PAGE) for purity. Those fractions containing only HMG-CoA lyase and one high molecular mass contaminant (~60-kDa subunit) are pooled for further purification. The pooled fractions are concentrated to ~5 ml in an Amicon (Danvers, MA) stirred cell, and further concentrated (<2 ml) in an Amicon minicell with a YM10 membrane. The concentrated sample is loaded onto a Superose 12 (preparative grade) column (50 × 2 cm) equilibrated in 0.1 M NaCl in buffer B. HMG-CoA lyase is eluted at a flow rate of 0.22 ml/min, and fractions of 1.2 ml are collected. The high molecular weight contaminant (>240,000 native molecular weight) that copurifies through all the prior steps elutes first from the molecular sieve column, cleanly separating from HMG-CoA lyase, which SDS–PAGE indicates to be homogeneous at this stage. The isolated human HMG-CoA lyase is stored at −80°. Enzyme recovery and specific activity at different steps in the purification are summarized in Table II. Figure 2 shows SDS–PAGE data for enzyme at various stages in the purification.

The C323S mutant lyase is purified as described for wild-type recombinant enzyme. In contrast to wild-type enzyme, specific activity of C323S lyase is not substantially affected by omission of 1 mM DTT from the

TABLE II
PURIFICATION OF RECOMBINANT HUMAN HMG-CoA LYASE[a]

Purification step	Total units	Total protein (mg)	Specific activity (units/mg)	Yield (%)
Crude extract	3625	851	4.26	100
Soluble extract	3795	748	5.10	104
Q-Sepharose	3467	212	16.35	96
40–65% (NH$_4$)$_2$SO$_4$ fractionation	2355	82.5	28.5	65
Phenyl-agarose	1000	10.5	95.2	28
Superose 12	932	5.85	159	26

[a] Reprinted, with permission, from J. R. Roberts, C. Narasimhan, P. W. Hruz, G. A. Mitchell, and H. M. Miziorko, *J. Biol. Chem.* **269**, 17841 (1994).

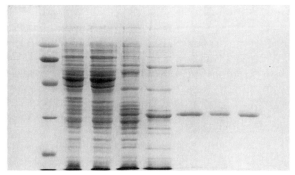

FIG. 2. SDS–PAGE of recombinant human HMG-CoA lyases at various stages of purification. Lane 1, molecular weight standards; lane 2, bacterial cell extract; lane 3, high-speed supernatant from centrifugation at 100,000g for 1 hr; lane 4, sample after Q-Sepharose; lane 5, 40–65% ammonium sulfate fraction; lane 6, phenyl-agarose eluate; lane 7, Superose 12-purified wild-type human lyase; lane 8, Superose 12-purified C323S lyase. [Reprinted, with permission, from J. R. Roberts, C. Narasimhan, P. W. Hruz, G. A. Mitchell, and H. M. Miziorko, *J. Biol. Chem.* **269,** 17841 (1994).]

buffers used for the two final purification steps. As in the case of wild-type enzyme, homogeneous enzyme (Fig. 2) is recovered from the purification protocol. Isolated C323S lyase is stable for months when stored at −80°. Table III summarizes recovery and specific activity of C323S HMG-CoA lyase at different steps in the purification.

TABLE III
PURIFICATION OF RECOMBINANT C323S HUMAN HMG-CoA LYASE[a]

Purification step	Total units	Total protein (mg)	Specific activity (units/mg)	Yield (%)
Crude extract	4258	982	4.33	100
Soluble extract	4376	768	5.70	103
Q-Sepharose	4114	161	25.6	97
40–65% $(NH_4)_2SO_4$ fractionation	2335	60.5	38.6	55
Phenyl-agarose	1971	7.9	249	46
Superose 12	1756	5.05	348	41

[a] Reprinted, with permission, from J. R. Roberts, C. Narasimhan, P. W. Hruz, G. A. Mitchell, and H. M. Miziorko, *J. Biol. Chem.* **269,** 17841 (1994).

Characterization of Human HMG-CoA Lyase

Edman degradation of isolated wild-type enzyme indicates that the N-terminal methionine has been cleaved from approximately 80% of the protein, confirming the prediction presented above in the outline of expression strategy. While yield of the isolated human protein (per liter of cultured bacteria) is about threefold lower than experienced with recombinant bacterial lyase,[5] recovery of the higher specific activity recombinant human lyase is adequate to support most mechanistic, protein chemistry, and even structural investigations. Such work will also be facilitated by the improved stability exhibited by the recombinant eukaryotic protein in comparison with the prokaryotic enzyme.

While HMG-CoA lyase has not been isolated from human tissue, a comparison of the properties of the recombinant wild-type and C323S enzymes (Table IV) with those of the protein isolated from avian liver suggests that these enzymes represent appropriate models. Human C323S

TABLE IV
PROPERTIES OF HMG-CoA LYASES[a]

| Property | Recombinant human | | Avian[b] |
	Wild-type	C323S	
Specific activity (units/mg)	159	348	350
K_m (HMG-CoA, μM)	24	45	8
K_m (Mn^{2+}, μM)	0.34	0.37	10
K_m (Mg^{2+}, μM)	233	322	50
Stoichiometry of butynoyl-CoA modification[c]	1.0	1.0	0.9
DTT stimulation of activity	10-fold	2-fold	100-fold
Subunit molecular mass (kDa)	31.6[d]	31.6[d]	31.4[d]
	34[e]	34[e]	27[e]
Native molecular mass (kDa)	48.6	48.6	49.0

[a] Reprinted, with permission, from J. R. Roberts, C. Narasimhan, P. W. Hruz, G. A. Mitchell, and H. M. Miziorko, *J. Biol. Chem.* **269,** 17841 (1994).

[b] Properties of avian HMG-CoA lyase have been described by Kramer and Miziorko[4] and Hruz *et al.*[6]

[c] Stoichiometry is calculated as $mol_{label}/mol_{subunit}$.

[d] Molecular mass calculated from deduced amino acid sequence, assuming that mature enzyme has been processed to produce an N terminus comparable to that measured for avian HMG-CoA lyase (T28).

[e] Apparent subunit molecular mass determined on the basis of mobility on SDS–PAGE.

lyase matches avian enzyme in specific activity. Wild-type human lyase, isolated by the current protocol, shows a twofold diminution in specific activity despite precautions to maintain enzyme in a reduced state. This value exceeds any other reports for lyase preparations from mammalian tissues.[2,3] The lower specific activity of wild-type lyase cannot be attributed to copurification, along with catalytically active enzyme, of protein that is inactive due to gross perturbations in structure. On parallel treatment of both wild-type and C323S lyases with 2-butynoyl-CoA, which we previously documented to function as an affinity label,[6] both enzymes are inactivated. Moreover, comparable stoichiometries of modification (1.0 per site; Table IV) are observed for both enzymes. These values are in good agreement with the estimate reported for the avian liver enzyme. Thus, despite a diminished specific activity, wild-type HMG-CoA lyase appears to possess a full complement of intact substrate-binding sites.

Activity of both wild-type and C323S lyases is markedly stimulated by divalent cations. Catalytic activity afforded by Mn^{2+} is comparable to that supported by the physiological cation, Mg^{2+}. Wild-type and C323S human lyases exhibit K_m values for Mg^{2+} (Table II) that are four- to sixfold higher than reported for avian enzyme. K_m values for Mn^{2+} are approximately three orders of magnitude lower. The observation that a cation activator such as Mn^{2+} binds with such high affinity has proven useful in the design of studies[20] that identified amino acid H235 as a ligand for this cation.

Recombinant human lyases exhibit, on SDS–PAGE, a subunit molecular mass that is in reasonable aggreement with the cDNA-deduced value of 31.6 kDa. Molecular sieve chromatography suggests that the native enzyme is a dimer, as is the avian enzyme.[4] The validity of this assignment is also suggested by the results of protein cross-linking experiments.[9] Oxidation of C323, or modification of this residue with the bifunctional reagent o-phenylenedimaleimide, results in conversion of wild-type lyase to a species that exhibits, on SDS–PAGE, a 65-kDa dimer instead of the 32-kDa monomer observed with untreated, reduced enzyme. Similar treatment of C323S lyase does not produce a 65-kDa protein species.

Acknowledgments

Human HMG-CoA lyase studies performed in the authors' laboratory have been supported by NIH DK-21491. Collaborative interaction with the laboratory of Dr. Grant A. Mitchell (Hopital Ste.-Justine, University of Montreal) led to the isolation of cDNA encoding human HMG-CoA lyase; the contributions of these colleagues are gratefully acknowledged.

[20] J. R. Roberts and H. M. Miziorko, *Biochemistry* **36**, 7594 (1997).

[17] Branched-Chain α-Keto Acid Dehydrogenase Kinase

By Kirill M. Popov, Yoshiharu Shimomura, John W. Hawes, and Robert A. Harris

Introduction

Branched-chain α-keto acid dehydrogenase (BCKDH) kinase catalyzes phosphorylation and inactivation of the BCKDH complex, an intramitochondrial multienzyme complex responsible for regulating branched-chain amino acid catabolism.[1,2] BCKDH kinase exists bound to the complex and is subject to both short-term and long-term regulation of its activity.[2,3] Before isolation of the kinase had been accomplished, only its enzyme activity had been identified as a reaction catalyzed by the purified BCKDH complex.[4–6] Thereafter, it was recognized that dithiothreitol (DTT), which was routinely used in buffers during purification steps, is critically important for binding of the kinase to the complex.[7,8] Purification of the kinase was further facilitated by the development of a method for isolation of the BCKDH complex based on phenyl-Sepharose chromatography.[9] BCKDH kinase remains tightly associated with the complex during chromatography on this matrix.

Principle for Purification of Branched-Chain α-Keto Acid Dehydrogenase Kinase

The initial step for purification of BCKDH kinase relies on isolation of BCKDH complex with high kinase activity. The crucial point for preservation of high kinase activity during purification of BCKDH complex is

[1] A. E. Harper, R. H. Miller, and K. P. Block, *Annu. Rev. Nutr.* **4**, 409 (1984).

[2] R. A. Harris, B. Zhang, G. W. Goodwin, M. J. Kuntz, Y. Shimomura, P. Rougraff, P. Dexter, Y. Zhao, R. Gibson, and D. W. Crabb, *Adv. Enzyme Regul.* **30**, 245 (1990).

[3] R. Kobayashi, Y. Shimomura, T. Murakami, N. Nakai, N. Fujitsuka, M. Otsuka, N. Arakawa, K. M. Popov, and R. A. Harris, *Biochem. J.* **327**, 449 (1997).

[4] H. R. Fatania, K. S. Lau, and P. J. Randle, *FEBS Lett.* **132**, 285 (1981).

[5] R. Paxton and R. A. Harris, *J. Biol. Chem.* **257**, 14433 (1982).

[6] R. Lawson, K. G. Cook, and S. J. Yeaman, *FEBS Lett.* **157**, 54 (1983).

[7] Y. Shimomura, N. Nanaumi, M. Suzuki, K. M. Popov, and R. A. Harris, *Arch. Biochem. Biophys.* **283**, 293 (1990).

[8] N. Nanaumi, Y. Shimomura, and M. Suzuki, *Biochem. Int.* **25**, 137 (1991).

[9] Y. Shimomura, R. Paxton, T. Ozawa, and R. A. Harris, *Anal. Biochem.* **163**, 74 (1987).

the presence of fresh dithiothreitol throughout the procedure.[7] The kinase can then be released from the complex by oxidation of the dithiothreitol with potassium ferricyanide followed by purification by high-speed centrifugation, immunoadsorption chromatography, and anion-exchange chromatography.[7] The kinase can also be dissociated from the complex by chromatography on Q-Sepharose followed by additional purification by chromatography on Mono Q HR.[10] Another effective method for purification of BCKDH kinase from the BCKDH complex isolated from bovine kidney mitochondria has been described by Lee et al.[11]

Assay Methods for Branched-Chain α-Keto Acid Dehydrogenase Complex

The activity of the BCKDH complex is assayed spectrophotometrically at 30°.[12] One unit of BCKDH complex catalyzes the formation of 1 μmol of NADH per minute.

Assay Methods for Branched-Chain α-Keto Acid Dehydrogenase Kinase

Reagents

Assay buffer (5×): 100 mM N-2-hydroxyethylpiperazine-N'-2-ethane-sulfonic acid (HEPES), 7.5 mM MgCl$_2$, and 10 mM dithiothreitol (pH 7.35 at 30° with NaOH)

K$_2$HPO$_4$, 250 mM (pH 7.35 at 30° with HCl; final concentration in assay, 50 mM)

Glycerol (80%, v/v; final concentration in assay, 20%)

ATP, 5 mM (neutralized with NaOH; final concentration in assay, 0.5 mM)

Bovine serum albumin, 16 mg/ml (final concentration in assay, 0.4 mg/ml)

Kinase-depleted BCKDH complex, 4.5 mg of protein/ml (at least 5 units/mg of protein). Methods for preparation are described below.

[10] K. M. Popov, Y. Shimomura, and R. A. Harris, *Protein Express. Purif.* **2,** 278 (1991).
[11] H.Y. Lee, T. B. Hall, S. M. Kee, H. Y. L. Tung, and L. J. Reed, *BioFactors* **3,** 109 (1991).
[12] N. Nakai, R. Kobayashi, K. M. Popov, R. A. Harris, and Y. Shimomura, *Methods Enzymol.* **324,** Chap. 6, 2000 (this volume).

Procedure

BCKDH kinase activity can be assayed either by measuring the rate of BCKDH complex inactivation in the presence of ATP (method 1) or by the incorporation of ^{32}P into the BCKDH complex from $[\gamma\text{-}^{32}P]ATP$ (method 2). Kinase activity is best measured by BCKDH inactivation during the course of purification of BCKDH kinase complex whereas incorporation of ^{32}P into the complex is the method of choice for the purified complex.

Method 1: BCKDH Complex Inactivation. Mix 40 μl of 5× assay buffer, 40 μl of 250 mM K_2HPO_4, 50 μl of 80% (v/v) glycerol, 0.1–0.2 unit of kinase-depleted BCKDH complex associated with the kinase (2–4 molecules per molecule of complex), and the right amount of water to bring the volume to 180 μl. Oligomycin (25 μg/ml) is added to the assay when kinase activity is being measured in the crude tissue extract. After prewarming the assay cocktail at 30° for 2 min, the kinase reaction is initiated by addition of 20 μl of 5 mM ATP. The mixture is incubated at 30°, and a portion (20 μl) of the mixture is taken into the assay cocktail for the BCKDH complex at appropriate time points (at least four) and the BCKDH activity is immediately measured. Apparent first-order rate constants for BCKDH inactivation are calculated by least-squares linear regression analysis.

Method 2: ^{32}P Incorporation into Complex. Mix 40 μl of 5× assay buffer, 40 μl of 250 mM K_2HPO_4, 50 μl of 80% (v/v) glycerol, 5 μl of bovine serum albumin (16 mg/ml), 20 μl of kinase-depleted BCKDH complex, an appropriate amount of the purified kinase (2–4 molecules per molecule of complex), and the amount of water necessary to bring the volume to 180 μl. After prewarming at 30° for 2 min, the kinase reaction is initiated by addition of 20 μl of 5 mM $[\gamma\text{-}^{32}P]ATP$ (specific activity, 330–400 cpm/ pmol). The mixture is incubated at 30°, and a portion (20 μl) of the mixture is taken for measuring protein-bound ^{32}P as described previously.[5] One unit of BCKDH kinase corresponds to 1 μmol of phosphate incorporated into BCKDH complex per minute.

Addition of K_2HPO_4 to the kinase assay at the indicated concentration is required for maximum kinase activity and optimal inhibitor sensitivity.[13] The addition of glycerol to the assay improves the accuracy at high kinase activity-to-dehydrogenase activity ratios. Glycerol at the concentration used in the assay (20%, v/v) inhibits kinase activity by approximately 40%. Adding glycerol to the assay is not necessary provided the kinase-to-dehydrogenase activity ratios can and are adjusted to establish optimal assay conditions for both activities.

[13] Y. Shimomura, M. J. Kuntz, M. Suzuki, T. Ozawa, and R. A. Harris, *Arch. Biochem. Biophys.* **266**, 210 (1988).

Purification of Rat Liver Branched-Chain α-Keto Acid
Dehydrogenase–Kinase Complex

Buffers and Reagents

Frozen rat livers are obtained from Pel-Freez Biologicals (Rogers, AR)

Buffer A: 30 mM HEPES, 1 mM EDTA, 0.15 M KCl, 0.5% (v/v) Triton
X-100, 2% (v/v) rat (or bovine) serum, 0.1 mM phenylmethylsulfonyl
fluoride (PMSF), 0.01 mM N-tosyl-L-phenylalanine chloromethyl
ketone (TPCK), trypsin inhibitor (10 μg/ml), 1 μM leupeptin, 0.1
mM thiamine pyrophosphate, 3 mM dithiothreitol (pH 7.5 at 4° with
KOH). Triton X-100 and rat serum are added to the buffer before
addition of PMSF and TPCK from 100 and 10 mM stock solutions
in ethanol, respectively

Buffer B: 30 mM HEPES, 0.1 mM EDTA, 0.1 M KCl, 10 mM MgSO$_4$,
0.1% (v/v) Triton X-100, trypsin inhibitor (10 μg/ml), 1 μM leupeptin, 5 mM dithiothreitol (pH 7.5 at 4° with KOH)

Buffer C: 50 mM potassium phosphate, 0.1 mM EDTA, trypsin inhibitor (10 μg/ml), 0.1 μM leupeptin, 3 mM dithiothreitol (pH 7.5 at 4°
with KOH); 2× buffer C without addition of dithiothreitol is prepared as a stock solution

Buffer D: 30 mM potassium phosphate (pH 7.5), 0.1 mM EDTA, 1
mM MgSO$_4$, 0.05% (v/v) Triton X-100, 3 mM dithiothreitol

Special note: Dithiothreitol is weighed and added directly to the buffers
just prior to use

Phenyl-Sepharose CL-4B (Pharmacia LKB Biotechnology, Piscataway, NJ)

Hydroxylapatite, Bio-Gel HTP (Bio-Rad, Hercules, CA), suspended
in water for at least 1 day before use. Small particles of hydroxylapatite are removed by decantation

Polyethylene glycol 6000 (PEG), 50% (w/v)

Procedure

The original method using phenyl-Sepharose for purification of BCKDH
complex[9] has been modified for purification of the complex with high kinase
activity[7] (Table I). All procedures are performed at 0–4° unless specified
otherwise. Frozen ($-80°$) rat livers (400 g) are homogenized without being
thawed in 2 volumes of buffer A with a Waring blender at full speed for
4 min. The homogenate is centrifuged at 9000g for 20 min. The supernatant
is filtered through four layers of cheesecloth. The pellet is rehomogenized
for 1 min with 1.5 volumes of buffer A and the homogenate is centrifuged
as described above. The supernatants are combined and made 2% (w/v)

TABLE I

PURIFICATION OF RAT LIVER BRANCHED-CHAIN α-KETO ACID DEHYDROGENASE COMPLEX
WITH HIGH KINASE ACTIVITY[a]

Fraction	Total protein (mg)	BCKDH complex activity			Kinase activity[b] (min^{-1})
		Specific activity (units/mg protein)	Total activity (units)	Yield (%)	
Mg-PEG	77,320	0.06	4,573	100	ND
First Phenyl-Sepharose and PEG	3,060	1.02	3,125	68	0.10
Second Phenyl-Sepharose and PEG	1,242	2.36	2,926	64	1.03
Third Phenyl-Sepharose and centrifugation	447	4.87	2,176	48	1.13
Hydroxylapatite and centrifugation	192	8.21	1,576	34	1.01

[a] Adapted with permission from Y. Shimomura, N. Nanaumi, M. Suzuki, K. M. Popov, and R. A. Harris, *Arch. Biochem. Biophys.* **283,** 293 (1990).
[b] BCKDH kinase activity was assayed by measuring the rate of loss of BCKDH complex activity at 0.5, 1, 2, and 3 min after addition of ATP. Kinase activity is expressed as a first-order rate constant for inactivation of the dehydrogenase activity of the BCKDH complex.

in PEG by slow addition with constant mixing. Gentle stirring of the solution for 20 min is followed by centrifugation of the solution at 9000g for 45 min. The supernatant is collected, made 4.5% (w/v) in PEG, gently stirred for 20 min, and centrifuged at 9000g for 20 min.

The pellet is suspended in 400 ml of buffer B (400 g of liver) and incubated at 37° for 40–50 min with gentle stirring to fully activate BCKDH complex by dephosphorylation. The suspension is cooled to approximately 5° by stirring on ice, made to 600 ml with buffer B, and then brought to 3% (w/v) in PEG. After 20 min of gentle stirring, the solution is centrifuged at 9000g for 40 min. The pellet obtained is suspended in 400 ml of buffer C containing 10% (v/v) glycerol and 0.15 M KCl.

Phenyl-Sepharose gel (400 ml) equilibrated with buffer C containing 10% (v/v) glycerol and 0.15 M KCl is added to the suspension and the mixture is kept overnight on ice. The mixture is then poured on top of a layer of 100 ml of phenyl-Sepharose gel equilibrated with the same buffer on a sintered-glass filter (9-cm diameter) funnel. The gel is washed with 2 liters of the same buffer and then with 1 liter of buffer C containing 10% (v/v) glycerol. The BCKDH complex is eluted from the gel with 2 liters of buffer C containing 10% (v/v) glycerol and 3% (w/v) Tween 20. The eluate is made 50 mM in ammonium sulfate by addition of the solid salt followed by slow addition of PEG to 6.5% (w/v). Glycerol inhibits precipitation of BCKDH complex by PEG, but this effect is reversed by the addition of ammonium sulfate. After 20 min of gentle mixing, the suspension is centrifuged at 9000g for 20 min. The pellet obtained is suspended in 50 ml of buffer C containing 10% (v/v) glycerol and stockpiled at $-80°$.

The procedure described above is repeated several times until approximately 2.4 kg of rat livers has been processed as shown in Table I.

The preparations are thawed, combined, and made to approximately 400 ml with buffer C containing 10% (v/v) glycerol and 0.15 M KCl. After further addition of 2 mM dithiothreitol as the solid, the suspension is applied to a phenyl-Sepharose column (5 × 12 cm) equilibrated with buffer C containing 10% (v/v) glycerol and 0.15 M KCl. The column is washed with 400 ml of the same buffer at a flow rate of approximately 100 ml/hr and further washed with 300 ml of buffer C containing 10% (v/v) glycerol. BCKDH complex is then eluted from the column with buffer C containing 10% (v/v) glycerol and 2% (w/v) Tween 20. Fractions containing BCKDH activity are combined, supplemented with ammonium sulfate, and the complex precipitated with PEG as described above. The pellet obtained is suspended in approximately 70 ml of buffer C containing 10% (v/v) glycerol and 0.15 M KCl.

The suspension is applied to a phenyl-Sepharose column (2.5 × 20 cm) equilibrated with the same buffer. The column is washed with 100 ml of the same buffer at a flow rate of approximately 70 ml/hr and further washed with 100 ml of buffer C containing 10% (v/v) glycerol. The complex is eluted from the column with buffer C containing 10% (v/v) glycerol and 2% (w/v) Tween 20. The fractions containing the BCKDH activity are combined and centrifuged at 230,000g for 120 min. The pellet is suspended in 30 ml of buffer C containing 0.1% (w/v) Tween 20.

The suspension is applied to a hydroxylapatite column (3.1 × 7.2 cm) equilibrated with buffer C containing 0.1% (w/v) Tween 20. The column is washed with 150 ml of the same buffer at a flow rate of 26 ml/hr and further washed with 250 ml of buffer C containing 0.1% (w/v) Tween 20 and 50 mM potassium phosphate (total concentration of potassium phosphate, 100 mM). BCKDH complex is eluted from the column with buffer C containing 0.1% (w/v) Tween 20 and 200 mM potassium phosphate (total concentration of potassium phosphate, 250 mM). After further addition of 1 mM dithiothreitol, the eluate is centrifuged at 230,000g for 80 min. The pellet obtained is used as BCKDH–kinase complex for further purification of the kinase as described below.

The overall purification is summarized in Table I.

Purification of Rat Heart Branched-Chain α-Keto Acid
Dehydrogenase–Kinase Complex

Rat heart has considerably more BCKDH kinase activity than rat liver,[14] and therefore is a better starting material for purification of the kinase. The BCKDH–kinase complex can be purified from frozen rat hearts obtained from Pel-Freez Biologicals by a modification of the method used for purification of the liver complex. The hearts are homogenized in 2.5 volumes of buffer A instead of 2 volumes, and the first PEG fractionation of the heart extract is made to 1% (w/v) PEG rather than 2% (w/v). Approximately 30 mg of BCKDH–kinase complex with a specific activity of at least 5 U/mg of protein for BCKDH can be purified from 550 g of frozen rat hearts.

Procedure for Concurrent Purification of Branched-Chain α-Keto
Acid Dehydrogenase– and Pyruvate Dehydrogenase–Kinase
Complexes from Rat Heart

In this procedure[10] frozen ($-70°$) rat hearts (200–$300g$) are homogenized in a Waring blender at full speed for 4 min in 3 volumes of buffer A. The homogenate is centrifuged at $10,000g$ for 20 min. The pellet is homogenized again with 5 volumes of buffer A and centrifuged as described above. The combined supernatants are filtered through four layers of cheesecloth. The supernatant is made to 1% (w/v) in PEG from a 50% (w/v) solution and is stirred for 20 min before centrifugation at $10,000g$ for 40 min. The supernatant is made to 4.5% (w/v) in PEG and treated as described above. The pellet is suspended in 1.5 volumes of buffer B with a Potter homogenizer and incubated at 37° for 50 min with gentle stirring. After incubation, the suspension is cooled on ice, made to 3% (w/v) in PEG, and stirred for 30 min. The precipitate is collected by centrifugation at $10,000g$ for 40 min. The pellet is resuspended in a Potter homogenizer with 1 volume of buffer C with 10% (v/v) glycerol and 0.15 M KCl and applied to a phenyl-Sepharose column (2.5×70 cm) equilibrated with the same buffer at a flow rate of 50 ml/hr. The column is washed with starting buffer until the absorbance of the eluate at 280 nm decreases to near zero and then further washed with 2 volumes of buffer C with 10% (v/v) glycerol. Protein is eluted by buffer C containing 10% (v/v) glycerol and 2% (w/v) Tween 20. Fractions of 10–12 ml are collected. The BCKDH and pyruvate dehydrogenase (PDH) complexes are eluted from the column prior to the major peak of absorbance. Fractions containing enzyme activity are pooled and applied at

[14] R. Paxton, M. Kuntz, and R. A. Harris, *Arch. Biochem. Biophys.* **244,** 187 (1986).

a flow rate of 50 ml/hr to a hydroxyapatite column (2.5 × 30 cm) equilibrated with 5 volumes of buffer C with 0.1% (w/v) Triton X-100. The column is washed with starting buffer until the absorbance at 280 nm of the eluate decreases to near zero and then developed stepwise with 100 and 200 mM potassium phosphate prepared in buffer C with 0.1% (w/v) Triton X-100. The PDH complex-containing fractions, eluted at 100 mM phosphate, are pooled and the enzyme precipitated by centrifugation at 200,000g for 4 hr. The pellet is resuspended in a Potter homogenizer with buffer C and 10% (v/v) glycerol to a protein concentration of 3 mg/ml and stored at −70°. Fractions containing the BCKDH complex, eluted with 200 mM phosphate, are pooled and made to 6.5% (w/v) in PEG and stirred for 40 min before centrifugation at 10,000g for 40 min. Precipitated enzyme is suspended in a Potter homogenizer with a minimum volume of buffer D and applied at a flow rate of 70 ml/hr to a Q-Sepharose column (1.5 × 20 cm) equilibrated with buffer D. The column is washed extensively until the absorbance of the eluate decreases to zero and developed with a stepwise KCl gradient (300 and 600 mM). Fractions of 5–6 ml are collected. The BCKDH complex elutes at a KCl concentration of 600 mM. Fractions containing the BCKDH complex are pooled and centrifuged at 200,000g for 4 hr. The pellet is resuspended in buffer C with 10% (v/v) glycerol at a protein concentration of 2 mg/ml and stored at −70°. Both complexes possess high kinase activity (first-order rate constants for ATP-dependent inactivation of 3.5 min^{-1} for PDH kinase and 1.4 min^{-1} for BCKDH kinase).[10]

Purification of Branched-Chain α-Keto Acid Dehydrogenase Kinase

Reagents

Potassium ferricyanide, 100 mM (prepare fresh daily)
Buffer E: 30 mM potassium phosphate, 0.1 mM EDTA (pH 7.5 at 4°);
2× buffer E is prepared as a stock solution
DEAE-Sephacel CL-4B (Pharmacia LKB Biotechnology)
Protein A-Sepharose CL-4B (Pharmacia LKB Biotechnology)/
Anti-BCKDH complex antibody-bound protein A-Sepharose: The E1 component of rat liver BCKDH complex prepared according to the method of Cook and Yeaman[15] is used to prepare antibody against the BCKDH complex.[7] The antiserum prepared with the E1 preparation contains both anti-E1 and anti-E2 antibodies, because the E1 preparation is contaminated with a trace of E2 and E2 is very antigenic; however, the antiserum is devoid of antibodies against

[15] G. Cook and S. J. Yeaman, *Methods Enzymol.* **166**, 303 (1988).

BCKDH kinase. Anti-BCKDH complex antibody-bound protein A-Sepharose is prepared by mixing, at room temperature for 20 min, 2 ml of the antiserum with 0.8 ml of protein A-Sepharose gel equilibrated with buffer D containing 0.1% (w/v) Tween 20 and 0.15 M KCl and then washing with the same buffer

Liver Branched-Chain α-Keto Acid Dehydrogenase Kinase

All procedures are performed at 0–4° unless specified otherwise. The liver BCKDH–kinase complex (approximately 190 mg; see Table II) is suspended in 40 ml of 250 mM potassium phosphate, 0.1 mM EDTA, trypsin inhibitor (10 μg/ml), 0.1 μM leupeptin, 0.1% (w/v) Tween 20, and 1 mM dithiothreitol (pH 7.5 at 4°). The suspension is made 3 mM in potassium ferricyanide from the 100 mM solution to completely oxidize the dithiothreitol present in the solution. The mixture should be yellow in color because the amount of potassium ferricyanide exceeds that of dithiothreitol in the mixture. After mixing for 20 min, the solution is centrifuged at 230,000g for 100 min. The pellet contains the BCKDH complex, which can be further processed as described below to prepare the kinase-depleted complex. The supernatant containing the kinase is made 5 mM

TABLE II

PURIFICATION OF LIVER AND HEART BRANCHED-CHAIN α-KETO ACID
DEHYDROGENASE KINASE[a]

	Liver			Heart		
Fraction	Protein (mg)	Specific activity[b] (mU/mg protein)	Yield (%)	Protein (mg)	Specific activity (mU/mg protein)	Yield (%)
BCKDH–kinase complex	192	2.39	100	27.0	5.2	100
Ferricyanide treatment–PEG	5.1	8.04	9.1	1.03	21.5	22
Immunoadsorption chromatography	4.3	9.30	8.7	0.69	25.7	13
DEAE-Sephacel chromatography	1.6	21.3	7.6	—	—	—
Protein A-Sepharose chromatography	0.92	26.8	5.3	0.49	35.7	13

[a] Adapted with permission from Y. Shimomura, N. Nanaumi, M. Suzuki, K. M. Popov, and R. A. Harris, *Arch. Biochem. Biophys.* **283**, 293 (1990).

[b] BCKDH kinase activity was assayed in the presence of glycerol by measuring the incorporation of ^{32}P into the enzyme from [γ-^{32}P]ATP at 30° for 1 min. A milliunit (mU) of enzyme activity is defined as 1 nmol of ^{32}P incorporated per minute.

in dithiothreitol and then 6.5% (w/v) in PEG. After gentle stirring for 20 min the mixture is centrifuged at 17,000g for 15 min. The pellet is suspended in 1.5 ml of buffer D containing 10% (v/v) glycerol and 3 mM dithiothreitol. The suspension may still be contaminated with a minor amount of BCKDH complex, which can be removed by passage through an anti-BCKDH complex antibody-bound protein A-Sepharose column (0.5 × 4 cm) equilibrated with buffer E containing 0.1% (w/v) Tween 20 and 0.15 M KCl. After application of the suspension, the column is washed with the same buffer at a flow rate of approximately 20 ml/hr. Fractions containing protein are immediately made 3 mM in dithiothreitol and combined. After addition of an equal volume of buffer E containing 3 mM dithiothreitol to decrease the salt concentration, the eluate is concentrated to 3–4 ml by ultrafiltration through a YM10 membrane (Amicon, Danvers, MA). The concentrate is applied to a DEAE-Sephacel column (0.7 × 4 cm) equilibrated with buffer E containing 3 mM dithiothreitol and the column is washed with the same buffer at a flow rate of 18 ml/hr. The majority of the kinase activity is not bound to the column and is recovered in the eluate. The kinase fraction is contaminated with a minor amount of immunoglobulins, which can be removed by passing the preparation through a protein A-Sepharose column (0.7 × 0.6 cm) equilibrated with buffer E containing 0.05% (w/v) Tween 20 and 3 mM dithiothreitol at a flow rate of 20 ml/hr. The eluate contains almost homogeneous kinase, which can be stocked at −80°. The overall purification of liver BCKDH kinase by this procedure is summarized in Table II.[7]

Heart Branched-Chain α-Keto Acid Dehydrogenase Kinase

The heart kinase can be purified from heart BCKDH–kinase complex by the same procedure described for the liver kinase, except that the DEAE-Sephacel column chromatography is omitted (Table II).[7] The heart kinase preparation is nearly homogeneous as determined by sodium dodecyl sulfate–polyacrylamide gel electrophoresis (SDS–PAGE) (Fig. 1).[7]

Heart Branched-Chain α-Keto Acid Dehydrogenase Kinase by Q-Sepharose and Mono Q HR Chromatography

A second strategy based on the tendency of the BCKDH kinase to dissociate from the complex during chromatography on Q-Sepharose has also been developed.[10] The BCKDH–kinase complex (15–20 mg protein) is desalted on a PD10 column equilibrated with buffer D containing 0.2 mM oxidized glutathione. Desalted protein is applied at a flow rate of 2 ml/min at room temperature to a Q-Sepharose column (1.6 × 10 cm) equilibrated with the same buffer (Fig. 2). The column is washed with 3 volumes of starting buffer. BCKDH kinase is eluted during washing of the

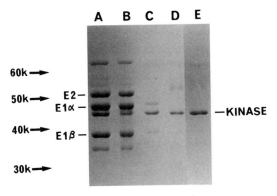

FIG. 1. Sodium dodecyl sulfate–polyacrylamide gel electrophoresis patterns obtained during purification of rat heart BCKDH kinase. Lane A, BCKDH–kinase complex (7 μg); lane B, BCKDH complex obtained after ferricyanide treatment (7 μg); lane C, the kinase fraction obtained after separation from the BCKDH complex (3 μg); lane D, the proteins obtained by immunoadsorption chromatography (2 μg); lane E, the purified kinase (2 μg). Proteins were stained with Coomassie blue. [Adapted with permission from Y. Shimomura, N. Nanaumi, M. Suzuki, and K. M. Popov, and R. A. Harris, *Arch. Biochem. Biophys.* **283**, 293 (1990).]

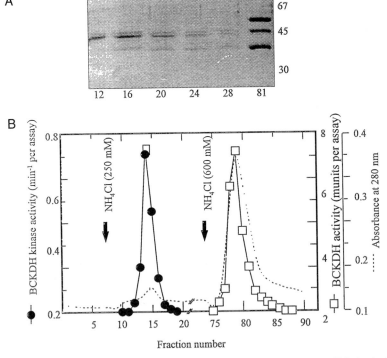

FIG. 2. Purification of BCKDH kinase by Q-Sepharose chromatography. (A) Analysis of the proteins contained in fractions by SDS–PAGE and silver staining. (B) Elution profiles of protein and kinase activity. [Adapted with permission from K. M. Popov, Y. Shimomura, and R. A. Harris, *Protein Express. Purif.* **2**, 278 (1991).]

column with 10 column volumes of buffer D containing 0.2 mM oxidized
glutathione and 0.25 M NH$_4$Cl. The first fractions of the 0.25 M NH$_4$Cl
wash contain essentially pure BCKDH kinase. Further elution results in
some release of the E1 component of BCKDH complex bound to the
column. Extensive washing of the column at this step is essential for complete separation of BCKDH kinase from BCKDH and is used, in spite of
some loss of the E1 component from the complex, to prepare kinase-
depleted BCKDH (see below). Pooled kinase fractions are supplemented
with 5 mM dithiothreitol (final concentration) and dialyzed against buffer
D with 5 mM dithiothreitol. The dialyzed preparation is applied to a Mono
Q HR 5/5 column (flow rate, 1 ml/min) equilibrated with buffer D plus 5
mM dithiothreitol (Fig. 3). The column is washed with 3 column volumes

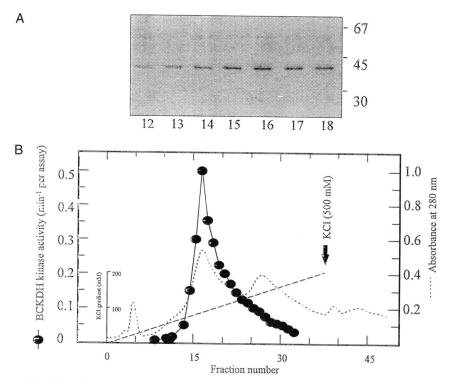

FIG. 3. Purification of BCKDH kinase by Mono Q HR 5.5 chromatography. (A) Analysis
of proteins contained in factions by SDS–PAGE and silver staining. (B) Elution profiles of
protein and kinase activity. [Adapted with permission from K. M. Popov, Y. Shimomura, and
R. A. Harris, *Protein Express. Purif.* **2**, 278 (1991).]

TABLE III
PURIFICATION OF BRANCHED-CHAIN α-KETO ACID DEHYDROGENASE KINASE
FROM RAT HEART[a]

| | BCKDH kinase | | |
Purification stage	Protein (mg)	Specific activity[b] (mU/mg protein)	Yield (%)
BCKDH–kinase complex	20	6.4	100
Q-Sepharose chromatography	1.8	54	76
Mono Q chromatography	0.64	90	45

[a] Adapted with permission from K. M. Popov, Y. Shimomura, and R. A. Harris, *Protein Express. Purif.* **2,** 278 (1991).

[b] BCKDH kinase activity was assayed in the absence of glycerol by measuring the incorporation of ^{32}P into the enzyme from [γ-^{32}P]ATP at 30° for 1 min. A milliunit (mU) of enzyme activity corresponds to 1 nmol of ^{32}P incorporated per minute.

of starting buffer before the kinase is eluted with a linear KCl gradient from 0 to 200 mM prepared in buffer D with 5 mM dithiothreitol. BCKDH kinase is eluted at a KCl concentration near 100 mM and is essentially free of the dehydrogenase component (Fig. 3). Pooled kinase fractions are concentrated by vacuum dialysis and dialyzed against 10 mM imidazole–asparagine buffer (pH 7.4) containing 0.1 mM EDTA and 5 mM dithiothreitol. Although best stored at 4° in this solution, the kinase activity is not stable and the protein has a tendency to aggregate. Purification of rat heart BCKDH kinase by this procedure is summarized in Table III.[10]

Preparation of Kinase-Depleted Branched-Chain α-Keto Acid Dehydrogenase Complex

Reagents

Potassium ferricyanide, 100 mM (prepare fresh daily)
NH$_4$Cl, 2 M

Kinase-Depleted Liver Branched-Chain α-Keto Acid Dehydrogenase Complex

The pellet of liver BCKDH complex obtained after ferricyanide treatment in the kinase purification is suspended in 20 ml of buffer D containing 10% (v/v) glycerol and 3 mM dithiothreitol. Because this BCKDH complex preparation still contains a significant amount of the kinase, the preparation is further treated to prepare kinase-depleted BCKDH complex as follows. The suspension is made 1 M in NH_4Cl from a 2 M solution and 4 mM in potassium ferricyanide from a 100 mM solution. The mixture is then centrifuged at 230,000g for 80 min. The pellet (kinase-depleted BCKDH complex) is suspended in buffer D containing 10% (v/v) glycerol and 3 mM dithiothreitol. The kinase activity of the preparation is routinely less than 10% of the activity of the original BCKDH–kinase complex. The specific activity of BCKDH of the preparation is approximately 5 U/mg of protein.[7] It can be stored at $-80°$.

Kinase-Depleted Heart Branched-Chain α-Keto Acid Dehydrogenase Complex

Extensive washing of the Q-Sepharose column used to prepare the heart BCKDH kinase results in removal of essentially BCKDH kinase from the BCKDH complex bound to the column. Kinase-depleted BCKDH is then eluted from the column with buffer D containing 0.2 mM oxidized glutathione and 0.6 M NH_4Cl. Fractions containing kinase-depleted BCKDH are pooled, precipitated with PEG as described above, resuspended in a minimum volume of buffer C, and dialyzed overnight against buffer C with 10% (v/v) glycerol.[10]

Branched-Chain α-Keto Acid Dehydrogenase Kinase cDNA

Partial Amino Acid Sequence of Branched-Chain α-Keto Acid Dehydrogenase Kinase and Generation of Oligonucleotide Probe for Branched-Chain α-Keto Acid Dehydrogenase Kinase

In this procedure[16] amino-terminal protein sequencing is performed with an Applied Biosystems (Foster City, CA) model 477A Pulse Liquid protein sequencer, and the phenylthiohydantoin derivatives are analyzed by reversed-phase light-performance liquid chromatography with a model 120A analyzer. Internal amino acid sequence analysis is obtained by *in situ* trypsin digestion on nitrocellulose[17] and separation of the peptide fragments by narrow-bore reversed-phase high-performance liquid chromatography. Degenerate, inosine-containing oligonucleotides synthesized according to the sequences of the amino terminus and an internal peptide of BCKDH

[16] K. M. Popov, Y. Zhao, Y. Shimomura, M. J. Kuntz, and R. A. Harris, *J. Biol. Chem.* **267,** 13127 (1992).

[17] R. H. Aebersold, J. Leavitt, R. A. Saavedra, L. E. Hood, and S. B. H. Kent, *Proc. Natl. Acad. Sci. U.S.A.* **84,** 6970 (1987).

kinase are used to amplify rat heart mRNA by reverse transcriptase-polymerase chain reaction (RT-PCR). The first strand of cDNA is synthesized with Moloney murine leukemia virus reverse transcriptase, using random hexamers according to the instructions of the manufacturer (Perkin-Elmer Cetus, Norwalk, CT). Forty cycles of amplification are performed with primers   GCIACIGA(T/C)ACICA(C/T)CA(C/T)GTIGA(A/G)CTICC and GG(C/T)TT(T/C)GCIACIAC(A/G)TC, corresponding to residues 4–13 and residues 30–36, respectively, of the NH_2 terminus of the protein as determined by direct protein sequencing. The amplification product is subcloned into M13 mp18 (Bethesda Research Laboratories, Gaithersburg, MD) and sequenced. A perfectly matching 24-mer oligonucleotide (GAAC-GCTCCAAGACTGTTACCTCC), synthesized according to the sequence of the PCR product, is used for library screening.

Library Screening

The oligonucleotide probe is labeled to a specific activity of 10^9 cpm/ μg with $[\gamma-^{32}P]$ATP by the reaction catalyzed by T_4 kinase. This probe is used to screen 5×10^5 clones from rat heart library. Positive clones are purified through three more cycles of screening. The inserts from the phage clones are recovered as M13 mp18 phagemids and both strands sequenced. The cDNA has an open reading frame that encodes a 30-amino acid leader sequence and a 382-amino acid mature protein with a predicted molecular mass of 43,280 Da.[16]

Expression of Recombinant Branched-Chain α-Keto Acid Dehydrogenase Kinase in *Escherichia coli*

Plasmids

The pGroESL plasmid is kindly supplied by A. Gatenby (Central Research and Development, Du Pont, Wilmington, DE). A prokaryotic expression vector for rat BCKDH kinase[18] is constructed as follows. The original rat BCKDH kinase cDNA is ligated to the *Eco*RI sites of pBluescript II SK⁻. Single-stranded DNA is produced by helper phage rescue procedures according to the manufacturer protocol (Stratagene, La Jolla, CA). To remove the DNA encoding the mitochondrial leader sequence,

[18] J. W. Hawes, R. J. Schnepf, A. E. Jenkins, Y. Shimomura, K. M. Popov, and R. A. Harris, *J. Biol. Chem.* **270**, 31071 (1995).

site-directed mutagenesis is utilized to add an *Sst*I restriction site just prior to the first codon of the mature protein-reading frame. Mutant plasmids are analyzed by restriction analysis, and the cDNA containing the mature protein reading frame is isolated by treatment with *Sst*I and *Not*I. This 2-kilobase pair (kbp) cDNA fragment is ligated to the *Sst*I and *Not*I sites of pET28a (Novagen, Madison, WI) to produce an amino-terminal fusion with a histidine tag (His-Tag) encoded by the vector. The reading frame of this construct is verified by double-stranded DNA sequencing.

Recombinant Enzyme Expression and Purification

HMS174(DE3) or BL21(DE3) *Escherichia coli* cells are double transfected with the pET-BK plasmid (kanamycin resistant) and the pGroESL plasmid (chloramphenicol resistant). Double transfectants are selected by growth at 37° in TY medium containing a 70-μg/ml concentration of both kanamycin and chloramphenicol. BL21(DE3) cells provide a significantly higher yield of soluble kinase protein (8- to 10-fold higher) than HMS174(DE3) cells. When the cell cultures obtain an optical density of 0.8, isopropyl-1-thio-β-D-galactopyranoside is added to a final concentration of 0.5 mM, and growth of the cultures is continued for 18–20 hr at 37°. The induced bacterial cultures are pelleted, resuspended in 10 volumes of buffer F [20 mM Tris-HCl (pH 8.0), 0.5 M NaCl, 10 mM 2-mercaptoethanol, 5 mM imidazole] supplemented with 0.5% (v/v) Triton X-100, benzamidine (100 μg/ml), and PMSF. The cells are lysed by sonication and the resulting homogenates are centrifuged at 20,000g for 30 min at 4°. The supernatants are applied to 10 ml of Ni^{2+}-NTA-agarose (Qiagen, Chatsworth, CA) previously equilibrated in buffer F. After sample loading, the columns are washed successively with 3 volumes of buffer F containing 20 mM imidazole, 40 mM imidazole, and 60 mM imidazole, followed by elution of the kinase with buffer F containing 200 mM imidazole. Eluted protein is collected as 250-μl fractions and the most concentrated fractions are pooled. Glycerol is added to the preparation to give a final concentration of 50% (v/v) to prevent aggregation. The final preparation is stored in small aliquots at $-80°$. Prior to being used for experiments, the enzyme is diluted 100-fold to minimize inhibition of kinase activity by glycerol. Alternatively, glycerol and salts can be removed quickly and easily by passage through a 10-ml Sephadex G-50 desalting column. Triton X-100 (0.1%, v/v) is added to the solution prior to passage through Sephadex desalting columns to prevent retarded flow of the protein through the column as a result of hydrophobic interactions.

Properties of Native Branched-Chain α-Keto Acid Dehydrogenase
Kinase, cDNA Encoding Branched-Chain α-Keto Acid
Dehydrogenase Kinase, and Wild-Type Recombinant Branched-
Chain α-Keto Acid Dehydrogenase Kinase

The liver and heart kinase preparations are nearly homogeneous, with a single polypeptide of molecular mass of 44–45 kDa as determined by SDS–PAGE (Figs. 1–3).[7,10,11] The purified kinase is readily reconstituted with the kinase-depleted BCKDH complex simply by mixing each of them in the kinase assay buffer. The specific activities of liver and heart kinase preparations are 27 and 36 mU/mg of protein, respectively, as measured by ^{32}P incorporation into the kinase-depleted BCKDH complex in the presence of glycerol (Table II).[7] Rat heart kinase prepared by the Q-Sepharose method has a specific activity of 90 mU/mg of protein when measured in the absence of glycerol (Table III).[10] BCKDH kinase is specific to the BCKDH complex and does not inactivate the PDH complex.[10] BCKDH kinase requires the E2 component of the BCKDH complex to express the maximal activity for inactivation (phosphorylation) of the E1 component.[19] Autophosphorylation of the purified kinase is not detectable.[7] The purified BCKDH kinase is sensitive to inhibition by α-chloroisocaproate and dichloroacetate, with 50% inhibition occurring at 14 μM and 1.8 mM, respectively.[7] BCKDH kinase will phosphorylate histones, but the rate of phosphorylation is low relative to that of the E1 component of the BCKDH complex.[7,10]

The recombinant enzyme has a specific activity of about 50 mU/mg of protein (assayed in the absence of glycerol), appears nearly pure by SDS–PAGE with Coomassie blue staining, and exhibits properties quite similar to those of the native enzyme in terms of kinetics and inhibitor sensitivity.

Acknowledgments

This work was supported by grants from the U.S. Public Health Services (NIH DK19259 and DK40441 to R.A.H., GM51262 to K.M.P.), the Diabetes Research and Training Center of the Indiana University School of Medicine (AM 20542), the Grace M. Showalter Trust (to R.A.H.), the Uehara Memorial Foundation (to Y.S.), and the University of Tsukuba Project Research Fund (to Y.S.).

[19] R. A. Harris, K. M. Popov, Y. Shimomura, Y. Zhao, J. Jaskiewicz, N. Nanaumi, and M. Suzuki, *Adv. Enzyme Regul.* **32,** 267 (1992).

[18] Expression of E1 Component of Human Branched-Chain α-Keto Acid Dehydrogenase Complex in *Escherichia coli* by Cotransformation with Chaperonins GroEL and GroES

By R. Max Wynn, James R. Davie, Jiu-Li Song, Jacinta L. Chuang, and David T. Chuang

Introduction

The decarboxylase (E1) component of the mammalian branched-chain α-keto acid dehydrogenase (BCKD) complex is a thiamine pyrophosphate (TPP)-dependent enzyme that catalyzes decarboxylation of α-keto acids derived from leucine, isoleucine, and valine. The E1 component is a heterotetramer consisting of two α (monomers, 45.5-kDa) and two β (monomers, 37.8-kDa) subunits. The crystal structures of the E1 components of *Pseudomonas putida*[1] and human[2] BCKD complexes revealed that residues from both E1α and E1β subunits form the TPP-binding pocket. A pentameric coordination involving an Mg^{2+} ion in the TPP pocket forms the active site of mammalian E1. Human mutations in both E1α (type IA) and E1β (type IB) subunits of E1 have been described in patients with heritable maple syrup urine disease (MSUD).[3] These mutations have been shown to perturb normal functions of E1, resulting in inactivation of the BCKD complex.

To elucidate structural–functional properties of mammalian E1 and how MSUD mutations disrupt E1 function, we have undertaken to express recombinant mammalian E1 in *Escherichia coli*. This bacterial host offers an advantage in that it does not contain a BCKD complex. In earlier studies, we individually expressed the E1α and E1β subunits in *E. coli*. However, the recombinant E1β subunit was largely insoluble,[4] and *in vitro* mixing of E1α and the soluble fraction of E1β in bacterial lysates did not result in reconstitution of active E1 heterotetramers. This is in contrast to studies with pyruvate dehydrogenase E1 of *Bacillus stearothermophilus,* where a

---

[1] A. Ævarsson, K. Seger, S. Turley, J. R. Sokatch, and W. G. J. Hol, *Nature Struct. Biol.* **6,** 785 (1999).

[2] A. Ævarsson, J. L. Chuang, R. M. Wynn, S. Turley, D. T. Chuang, and W. G. J. Hol, *Structure* **8,** 277 (2000).

[3] D. T. Chuang and V. E. Shih, "The Metabolic and Molecular Basis of Inherited Disease," 7th Ed, p. 1239. McGraw-Hill, New York, 1995.

[4] R. M. Wynn, J. L. Chuang, J. R. Davie. C. W. Fisher, M. A. Hale, R. P. Cox, and D. T. Chuang, *J. Biol. Chem.* **267,** 1881 (1992).

functional E1 heterotetramer was reconstitued *in vitro* by using the individually expressed and purified E1α and E1β proteins.[5] To express E1 of the mammalian BCKD complex, our laboratory later developed a strategy of coexpression of the mature sequences for both E1α and E1β subunits from the same plasmid in *E. coli.*[6] This resulted in expression and assembly of functional E1 heterotetramers. However, the yield was marginal, with a large fraction of misfolded recombinant proteins in the aggregated form.

Molecular chaperones have been shown to promote proper folding and assembly of many proteins.[7] Inside mitochondria, heat shock protein 60 (Hsp60) and Hsp10 have been shown to be essential for biogenesis of matrix F1 ATPase.[8] Because mammalian E1 is a mitochondrial protein, we developed a strategy to express E1 heterotetramers by cotransformation of *E. coli* with bacterial chaperonins GroEL/GroES, which are homologs of mitochondrial Hsp60/Hsp10, respectively. This resulted in a markedly improved expression and assembly of E1 heterotetramers.[8a] Here, we describe the methods for chaperonin-assisted expression of mammalian E1, and the utility of this efficient expression system to study alterations in the E1 assembly state caused by MSUD mutations.

Induction of Mammalian Maltose-Binding Protein–E1 Fusion by Cotransformation with Chaperonins GroEL/GroES

Plasmids and Cell Host

The pH1 plasmid, which coexpresses maltose-binding protein (MBP)–E1α (human) and unfused E1β (bovine) mature sequences, is constructed as described previously.[6] The vector contains a factor X_a protease cleavage site (IEGR ↓) that links the C terminus of the MBP sequence to the N terminus of the E1α subunit. The pH1 plasmid contains pTac and pTrc promoters that drive MBP–E1α and E1β expression, respectively. The pGroESL plasmid, which contains the *E. coli* GroES and GroEL operon, is obtained from A. Gatenby (Du Pont Experimental Station, Wilmington, DE). pGroESL contains a pTac promoter and a downstream heat shock-sensitive element. *Escherichia coli* CG-712 (ES^{ts}) and CG-714 (EL^{ts}) are also supplied by Du Pont.

[5] I. A. Lessard and R. N. Preham, *J. Biol. Chem.* **269**, 10378 (1994).

[6] J. R. Davie, R. M. Wynn, R. P. Cox, and D. T. Chuang, *J. Biol. Chem.* **267**, 16601 (1992).

[7] F. U. Hartl, *Nature (London)* **381**, 5712 (1996).

[8] M. Y. Cheng, F. U. Hartl, J. Martin, R. A. Pollock, F. Kalousek, W. Neupert, E. M. Hallberg, R. L. Hallberg, and A. L. Horwich, *Nature (London)* **337**, 620 (1989).

[8a] R. M. Wynn, J. R. Davie, R. P. Cox, and D. T. Chuang, *J. Biol. Chem.* **267**, 12400 (1992).

Cell Culture Medium and Buffers

YTGK medium: 16 g of Yeast extract, 10 g of Bacteriological peptone, 5 g of NaCl, 10 ml of glycerol, 0.75 g of KCl, autoclaved, followed by addition of 100 mg of ampicillin, and 50 mg of chloramphenicol in 1 liter at 23°

Cell wash buffer: 50 mM Tris-HCl (pH 7.5), 100 mM NaCl, 5 mM EGTA, and 5 mM EDTA

Cell lysis buffer I: 50 mM potassium phosphate (KP$_i$) (pH 7.0), 100 mM NaCl, 2 mM MgCl$_2$, 0.2 mM TPP, 0.1% (v/v) Triton X-100, 0.01% (v/v) NaN$_3$, and 0.1 mM EDTA. The solution is filtered with a 0.22-μm pore size filter. To the filtrate, 0.25 mM dithiothreitol (DTT), lysozyme (1 mg/ml), and protease inhibitors [1 mM phenyl-methylsulfonyl fluoride (PMSF) and 1 mM benzamidine] are freshly added

Column buffer A: 50 mM KP$_i$ (pH 7.0), 2 mM MgCl$_2$, 0.2 mM TPP, 0.1% (v/v) Triton X-100, 0.75 mM NaN$_3$, 0.1 mM EDTA, 0.25 mM DTT, and 500 mM NaCl

Amylose resin: Purchase from New England BioLabs (Beverly, MA) or synthesize as described previously[9]

Enzyme storage buffer: 50 mM KP$_i$ (pH 7.5), 1.0 mM DTT, 10% (v/v) glycerol, and 250 mM KCl

Cotransformation of Bacterial Host and Maltose-Binding Protein–E1 Expression

The pH1 plasmid (containing a pBR-322-based ColE1 replicon) and the pGroESL plasmid (carrying a pACYC-based p15A replicon) are suitable for cotransformation based on plasmid compatibility. The pH1 plasmid is first transformed into CG-712 or CG-714 competent cells according to a standard procedure.[10] Cell colonies are selected for ampicillin resistance on LB ampicillin plates. Colonies are isolated and grown in liquid culture with ampicillin and converted to competent cells for transformation with the pGroESL plasmid. Cell colonies are then doubly selected for ampicillin–chloramphenicol resistance on LB ampicillin–chloramphenicol plates.

Cotransformed CG-712 or CG-714 cells are grown as an overnight culture at 37° in YTKG medium with antibiotics to maintain the expression plasmid (Amp^R) and the F' plasmid (Tet^R) that carries the *lacI* transcriptional repressor gene. The overnight cultures are diluted 3:1000 into 1-liter aliquots of the sterile YTGK medium prewarmed to 37°. Cultures are

[9] O. K. Kellerman and T. Ferenci, *Methods Enzymol.* **90,** 459 (1982).
[10] J. Sambrook, E. F. Fritsch, and T. Maniatis, "Molecular Cloning: A Laboratory Manual." Cold Spring Harbor Laboratory Press, Cold Spring Harbor, New York, 1989.

grown under heat shock conditions at 42° to induce expression of GroEL and GroES. When the cell density reaches an $A_{550\,nm}$ of 0.6, growth temperature is lowered to 37°. Cells are subsequently treated with 1 mM isopropyl-β-D-thiogalactopyranoside (IPTG) to induce MBP–E1 and chaperonin expression. Cultures are grown continuously in wide-mouth, unbaffled 2-liter Erlenmeyer flasks at 37° with moderate shaker agitation (220 rpm) for 15–18 hr.

Extraction of Maltose-Binding Protein–E1 and Measurement of Enzyme Activity

Harvested cells from 8-liter cultures are suspended in 300 ml of lysis buffer I containing protease inhibitors. Cell suspensions are sonicated three times with a Branson (Danbury, CT) 450 sonifier at a power setting of 6 for 4–5 min (with 5- to -10-min cooling cycles), using a normal probe tip. Cell lysates are centrifuged at 12,000g for 15 min at 4°. Clarified lysates are assayed for E1 activity by a radiochemical assay with α-keto[1-^{14}C]isovalerate as a substrate and 2,6-dichlorophenol-indophenol as an artificial electron acceptor.[6]

Figure 1A shows the levels of E1 activity in cells with and without cotransformation with chaperonins GroEL/GroES, which are referred to as double and single transformants, respectively. With CG-712 (ES^{ts}) cells, single transformants carrying the pH1 plasmid alone exhibit low residual E1 activity. Cotransformation with pH1 and pGroESL plasmids results in a greater than 500-fold increase in E1-specific activity in the ES^{ts} host, when one compares the double transformants with single transformants. A smaller, 30-fold increase in E1-specific activity is also observed in double transformants of the EL^{ts} host, compared with single transformants. Sodium dodecyl sulfate–polyacrylamiade gel electrophoresis (SDS–PAGE) analysis indicates that the marked induction of E1α and E1β in doubly transformed ES^{ts} and EL^{ts} hosts, relative to singly transformed hosts, which coincides with overexpression of chaperonins GroEL/GroES (Fig. 1B). It is noted that no soluble E1β subunit is detected in single transformants (Fig. 1B, lanes 1 and 3), supporting the dependence of this subunit on chaperonins for proper folding. The results establish that chaperonins GroEL/GroES are essential for efficient expression and assembly of mammalian E1 in *E. coli.* This forms the basis for the development of a hexahistidine (His$_6$)-tagged expression system to investigate the assembly of wild-type and mutant mammalian E1, as described below.

Affinity Purification of Maltose-Binding Protein–E1 with Amylose Resin

Mammalian MBP–E1 overexpressed in CG-712 cells is efficiently purified by amylose resin affinity chromatography. All lysis steps are performed

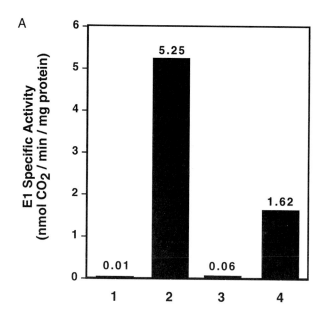

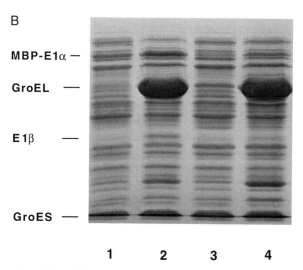

FIG. 1. Induction of E1 activity and subunit concentrations by overexpression of chaperonins GroEL and GroES in *E. coli*. CG-712(ES^{ts}) and CG-714 (EL^{ts}) mutant strains are singly transformed with the pH1 plasmid encoding the MBP–E1α fusion protein and E1β or doubly transformed with pH1 and the GroESL plasmid, which overexpresses GroEL and GroES. Cell lysates are assayed for E1 activity (A) and analyzed by SDS–PAGE followed by Coomassie blue staining (B). Lane 1, ES^{ts} + pH1 (single transformants); lane 2, ES^{ts} + pH1 + pGroESL (double transformants); lane 3, EL^{ts} + pH1; lane 4, EL^{ts} + pH1 + pGroESL. [From R. M. Wynn, J. R. Davie, R. P. Cox, and D. T. Chuang, *J. Biol. Chem.* **267**, 12400 (1992), with permission.]

at 4°. The NaCl concentration in MBP–E1 lysate is increased from 100 to 500 mM by the addition of concentrated 5 M NaCl stock solution. An equal volume of column buffer A (containing 500 mM NaCl) and a one-third volume of settled amylose resin equilibrated in buffer A are added to the lysate. The relatively high salt concentration is required to maintain the solubility of MBP–E1. After batch mixing for 2–3 hr, the amylose resin slurry is loaded into a 2.5-cm-diameter column and the excess lysate is allowed to drain off. The amylose resin is washed with 6 bed volumes of column buffer A. MBP–E1 bound to amylose resin is eluted with 2 bed volumes of column buffer A with added 10 mM maltose. Fractions containing MBP–E1 are identified by a spectrophotometric assay as described previously,[11] followed by SDS–PAGE analysis. Additional purification steps include ion-exchange chromatography on Q Sepharose HR resin and gel-filtration chromatography on Sephacryl S-300 HR. The highly purified MBP–E1 is apparently homogeneous as judged by SDS–PAGE, with a yield of approximately 50 mg/liter of cell culture. The purified MBP–E1 is stored at $-80°$ in the enzyme storage buffer at a concentration of 50 mg/ml.

Efficient Expression and Purification of Hexahistidine-Tagged Human E1

The MBP–E1 fusion protein is relatively soluble and can be expressed in large quantities. However, to facilitate crystal structure studies and characterization of human mutant E1, we have developed a simplified His$_6$ tag expression system for human E1. This procedure takes advantage of efficient expression mediated by chaperonin proteins GroEL/GroES, similar to expression of MBP–E1. The His$_6$-tagged human E1 heterotetramers are readily purified by Ni^{2+}-NTA extraction. The relatively high affinity of the His$_6$ tag for Ni^{2+} ions (K_D of 10^{-10} M) facilitates the development of a pulse–chase method to measure the rate of assembly for wild-type and mutant E1 subunits.[12]

Expression Vectors and Bacterial Strains

The pHisT-E1 expression vector, which coexpresses human His$_6$-tagged E1α and untagged E1β mature sequences, is constructed as described previously.[12] The vector contains a tobacco etch virus (TEV) protease cleavage

[11] K. S. Lau, A. J. L. Cooper, and D. T. Chuang, *Biochim. Biophys. Acta* **1038,** 360 (1990).

[12] R. M. Wynn, J. R. Davie, J. L. Chuang, C. D. Cote, and D. T. Chuang, *J. Biol. Chem.* **273,** 13110 (1998).

site (LDNLYFQ ↓ S) that links the His$_6$ tag to the N terminus of human E1α. Cleavage by the TEV protease is significantly more specific than that by factor X$_a$ protease, which is present in the pH1 vector for expression of MBP–E1. The expression of His$_6$–E1α and untagged E1β is mediated by the pTrc promoter. The pGroESL plasmid and CG-712 cells are described above.

Cell Growth and Expression of Hexahistidine-Tagged E1

The phHisT-E1 vector carries the ColE1 replicon, which is compatible with the p15A replicon in the pGroESL plasmid with respect to cotransformation. The procedures for stepwise cotransformation of the two plasmids into the CG-712 cells, cell growth, and induction for human His$_6$–E1 expression are similar to those described above for expression of MBP–E1.

Purification of Human Hexahistidine-Tagged E1 Heterotetramers

Buffers

Cell lysis buffer II: 50 mM KP$_i$ (pH 7.0), 500 mM NaCl, 2 mM MgCl$_2$, 0.2 mM TPP, 0.1% (v/v) Triton X-100, 0.01% (w/v) NaN$_3$, and 0.1 mM EDTA. The solution is filtered with a 0.22-μm pore size filter. To the filtrate, 2 mM 2-mercaptoethanol, lysozyme (1 mg/ml), and protease inhibitors (1 mM PMSF and 1 mM benzamidine) are freshly added

Column buffer B: 100 mM KP$_i$ (pH 7.5), 500 mM KCl, 0.2% (w/v) Tween 20, 2 mM 2-mercaptoethanol, 0.1 mM MgCl$_2$, and 0.1 mM TPP

E1 dialysis buffer: 20 mM KP$_i$ (pH 7.5), 0.2 mM EDTA, 0.5 mM 2-mercaptoethanol, and 250 mM KCl

Column buffer C: 50 mM KP$_i$ (pH 7.5), 1.0 mM DTT, 5% (v/v) glycerol, and 100 mM KCl

Column buffer D: 50 mM KP$_i$ (pH 7.5), 1.0 mM DTT, 5% (v/v) glycerol, and 350 mM KCl

Ni^{2+}-NTA Affinity Chromatography. Five milliliters of settled Ni^{2+}-NTA resin (Qiagen, Chatsworth, CA) is washed with a 10-fold excess of H$_2$O by centrifugation to remove ethanol from the storage solution. The washed resin is suspended with 5 ml of column buffer B. Cell lysates are prepared as described under Extraction of Maltose-Binding Protein–E1, except for the use of cell lysis buffer II in place of lysis buffer I. DTT in cell lysis buffer I cannot be used because this sulfhydryl reagent inactivates Ni^{2+}-NTA resin. Ten milliliters of washed Ni^{2+}-NTA resin is mixed with 300 ml of cell lysate (from 8 liters of culture) for 2 hr on a rocking platform. After being packed into a column, the resin is washed with 10 bed volumes of column buffer B with added 15 mM imidazole. His$_6$–E1 is eluted with 15–250 mM imidazole in the same buffer. Column fractions are analyzed by the E1 spectrophotometric assay as described previously[11] and by SDS–PAGE analysis.

Figure 2 shows the elution profile of E1 activity (Fig. 2, bottom) with the imidazole gradient, and the Coomassie blue-stained SDS–PAGE profile of column fractions (Fig. 2, top). Human His$_6$–E1 in the peak fractions is approximately 95% pure. The protein elution profile coincides with the

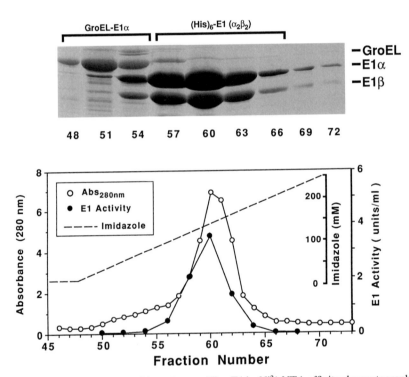

FIG. 2. Purification of recombinant human His$_6$–E1 by Ni^{2+}-NTA affinity chromatography. *Escherichia coli* strain CG-712 (ES^{ts}) is cotransformed with the pHisT-E1 plasmid (coexpressing human His$_6$–E1α and untagged E1β subunits) and the pGroESL plasmid (overexpressing GroEL and GroES). Cell lysates are extracted with Ni^{2+}-NTA resin, and His$_6$–E1 is eluted with a 15–250 mM imidazole gradient. Column fractions (2.0 ml each) are analyzed by SDS–PAGE (*top*). E1 activity (*bottom*) in each fraction is assayed by a spectrophotometric method,[11] in which the reduction in $A_{600\ nm}$ of 2,6-dichlorophenol-indophenol is monitored. One unit is defined as 1 μmol of the dye reduced per minute. [From R. M. Wynn, J. R. Davie, J. L. Chuang, C. D. Cote, and D. T. Chuang, *J. Biol. Chem.* **273,** 13110 (1998), with permission.]

presence of E1 activity, indicating that E1α and E1β subunits assemble into native $\alpha_2\beta_2$ heterotetramers. The latter point is confirmed by the correct molecular mass of 171 kDa as determined by a calibrated gel-filtration column (data not shown). Earlier fractions corresponding to GroEL–E1α–E1β complex represent folding intermediates of His$_6$–E1.[13]

Further Purification by Fast Protein Liquid Chromatography. To further purify human His$_6$-tagged E1 by fast protein liquid chromatography (FPLC), enzyme fractions from the Ni^{2+}-NTA column are pooled and dialyzed against E1 dialysis buffer (see above). This removes imidazole from the Ni^{2+}-NTA column fractions; imidazole has reversible inhibitory effects on E1 activity. The dialyzed enzyme solution is loaded onto a 0.5 × 5 cm FPLC Mono Q ion-exchange column (Amersham Pharmacia Biotech, Piscataway, NJ) equilibrated with column buffer C. His$_6$–E1 is eluted with a 100–400 mM KCl gradient in the same buffer. The enzyme is eluted at 6.1 mS ionic strength. Peak fractions are concentrated by centrifugation, using Millipore (Bedford, MA) Ultrafree-15 filter devices with a 30-kDa cutoff Biomax membrane. The concentrated enzyme is applied to an FPLC HiLoad Superdex 200 gel-filtration column equilibrated with column buffer D. Human His$_6$–E1 eluted from the HiLoad Superdex 200 is apparently homogeneous. The enzyme is stored at −80°, at a concentration of 3–20 mg/ml in enzyme storage buffer (see above).

Isolation and Separation of an $\alpha\beta$ Heterodimeric
Assembly Intermediate

We have shown previously that the assembly of human E1 $\alpha_2\beta_2$ hetero-tetramers proceeds through an inactive $\alpha\beta$ heterodimeric intermediate both *in vivo*[12] and *in vitro*.[13] When transformed CG-712 cells are grown for 16 hr at 37° in YTGK medium only the predominant His$_6$-tagged heterotetra-mers are isolated. However, when cells are grown in a minimal medium, the heterodimeric intermediate accumulates during the early phase of culture. This section describes cell culture conditions for expression of the heterodimeric intermediate and its isolation. The availability of purified heterodimers facilitates studies of their interactions with chaperonins GroEL/GroES, which is essential for conversion into active E1 heterotet-ramers.[13]

[13] J. L. Chuang, R. M. Wynn, J.-L. Song, and D. T. Chuang, *J. Biol. Chem.* **274,** 10395 (1999).

Medium

C2 broth minimal medium: 2 g of NH_4Cl, 6 g of $Na_2HPO_4 \cdot 7H_2O$, 3 g of KH_2PO_4, 3 g of NaCl, 6 g of yeast extract, autoclaved, followed by addition of 40 μM TPP, 50 mg of carbenicillin, and 50 mg of chloramphenicol in 1 liter as modified from the C broth medium.[14]

Method

CG-712 cells cotransformed with pHisT-E1 and pGroESL plasmids are grown at 37° in C2 minimal medium with low sulfur and amino acid content. When cell density reaches an $A_{550\ nm}$ of 0.6, 0.5 mM IPTG is added to the culture to induce the expression of E1 subunits for 3 hr. Lysates prepared from harvested cells are extracted with Ni^{2+}-NTA resin. The column is washed and bound proteins are eluted with an imidazole gradient as described above. Because the E1β subunit is untagged, its extraction with Ni^{2+}-NTA indicates assembly with the His_6–E1α subunit. Fractions containing $\alpha\beta$ heterodimers and $\alpha_2\beta_2$ heterotetramers are concentrated in Millipore Ultrafree-15 filter devices with a 30-kDa cutoff Biomax membrane. Concentrated protein species in 1 ml are separated on 10 ml of a 10–25% (w/v) sucrose density gradient spun at 210,000g for 18 hr at 4°. Fractions (0.7 ml) are analyzed by SDS–PAGE and assayed radiochemically for E1 activity (see [19] in this volume[15]). Assembled His_6–E1α and E1β subunits that sediment near the top of the gradient and do not exhibit E1 activity are $\alpha\beta$ heterodimers. Fractions 3–5, which contain $\alpha\beta$ heterodimers, are combined and proteins further purified on a preparative HiLoad Superdex 200 column (2.6 × 60 cm) with samples collected in 5-ml fractions. Figure 3A shows that the heterodimeric species appears as the major peak

[14] L. M. Guzman-Verduzco and Y. M. Kupersztoch, *J. Bacteriol.* **169**, 5201 (1987).

[15] J. L. Chuang, J. R. Davie, R. M. Wynn, and D. T. Chuang, *Methods Enzymol.* **324**, Chap. 19, 2000 (this volume).

FIG. 3. FPLC gel-filtration profiles of $\alpha\beta$ heterodimers and $\alpha_2\beta_2$ heterotetramers. (A) His_6-tagged $\alpha\beta$ heterodimers are purified by Ni^{2+}-NTA extraction and sucrose density gradient centrifugation, followed by FPLC separation on a HiLoad Superdex 200 column (2.6 × 60 cm). The peak fraction of His_6-tagged heterodimers is fraction 42, which corresponds to a molecular mass of 85.5 kDa. (B) His_6-tagged $\alpha_2\beta_2$ heterotetramers are purified similarly and eluted at fraction 39, with a molecular mass of 171 kDa. The molecular mass markers used for calibration (in kDa) are as follows: chymotrypsinogen A (23), ovalbumin (44), bovine serum albumin (67), human E3 (110), aldolase (158), and GroEL (840). [From J. L. Chuang, R. M. Wynn, J.-L. Song, and D. T. Chuang, *J. Biol. Chem.* **274**, 10395 (1999), with permission.]

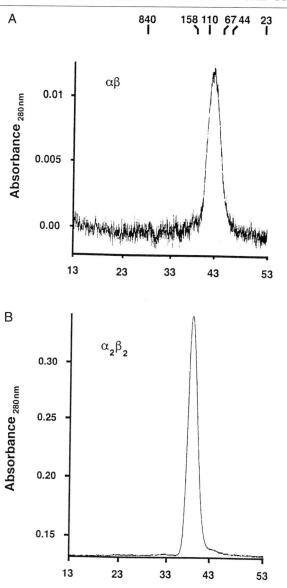

at fraction 42, with a size of 85.5 kDa. The sucrose density gradient profile and the absence of E1 activity confirm the identity of $\alpha\beta$ heterodimers. The native heterotetramer is expressed in CG-712 cells grown in YTGK medium and purified as described above. The heterotetramer is eluted at fraction 39 of the HiLoad Superdex 200 FPLC column as a 171-kDa species (Fig. 3B). Purified heterodimers are stored at $-80°$ in enzyme storage buffer at a concentration less than 1 mg/ml. The purified heterodimer forms a complex with GroEL, which can be used in the refolding studies of human E1.[13]

Expression of Mutant Human E1 in Aberrant Assembly State

The chaperonin-augmented expression system has been employed to investigate the effect of MSUD mutations on folding and assembly of human E1. Type IA MSUD mutations (Y393N-α, T265R-α, and Y368C-α) that occur in the E1α subunit[3] are engineered into the pHisT-E1 vector by subcloning of the reverse transcriptase-polymerase chain reaction (RT-PCR) product derived from homozygous MSUD patients.[13] The pHisT-E1 plasmid carrying one of these type IA mutations is cotransformed with pGroESL into CG-712 cells as described above, and transformed cells are grown in YTGK medium. For wild-type, Y393N-α, and Y368C-α mutant E1, expression is induced with IPTG and cell growth maintained at 37° for 16 hr. The T265R-α mutant E1 is expressed similarly, except that the culture temperature is lowered to 28° to prevent aggregation of the mutant protein. Cell lysates containing wild-type and mutant E1 proteins are subjected to Ni^{2+}-NTA extraction. Isolated protein species are separated on a 10–25% sucrose density gradient to analyze the subunit assembly state.

Figure 4 depicts the sedimentation profiles of the wild-type E1 protein and mutant E1 proteins carrying the Y393N-α, T265R-α, or Y368C-α mutation. The wild-type E1 protein induced at 37° migrated as a heterotetrameric species of 171 kDa. In contrast, the Y393N-α mutant E1 expressed at 37° migrated as a heterodimeric species with a molecular mass of 85.5 kDa. The mutant E1 containing the F364C-α mutation was also present entirely as heterodimers in sucrose density gradient centrifugation. Interestingly, the mutant T265R-α, when expressed at 28° but not at 37°, was able to remain soluble and assemble with the normal E1β subunit. The sedimentation profile indicates that the T265R-α mutant E1 migrated predominantly as heterotetramers, although lesser amounts of lower molecular weight species are present and sediment in early gradient fractions. Similarly, the mutant E1 bearing the Y368C-α mutation

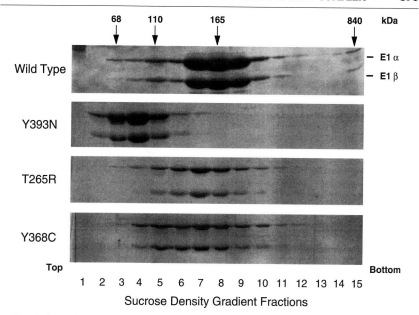

Fig. 4. Assembly state of wild-type and mutant E1 proteins as analyzed by sucrose density gradient centrifugation. Wild-type and mutant E1 carrying different mutations in the E1 subunits are expressed in cotransformed CG-712 cells (ES[ts]) and purified by Ni^{2+}-NTA extraction, followed by gel filtration on a Sephacryl S-100 HR column. Purified wild-type and mutant E1 proteins are fractionated on a 10–25% sucrose density gradient. Fractions (0.7 ml each) are collected from top to bottom and analyzed by SDS–PAGE and Coomassie blue staining. The molecular mass markers used are as follows (in kDa): serum albumin (68), human E3 (110), E1 heterotetramers (165), and GroEL (840). [From R. M. Wynn, J. R. Davie, J. L. Chuang, C. D. Cote, and D. T. Chuang, *J. Biol. Chem.* **273**, 13110 (1998), with permission.]

sediments as both heterotetramers and lower molecular weight species when expressed at 37°.

The results show that the C-terminal aromatic amino acids in the E1α subunit are important for proper assembly of E1 heterotetramers. The Y393N-α mutation occurs at high prevalence in homozygous Mennonite MSUD patients.[3] It is striking that this mutation completely prevents the conversion of heterodimers to native heterotetramers, which exist in the normal assembly pathway. It is also of interest that overexpression of chaperonins GroEL/GroES promotes the expression of mutant E1 proteins to some degree, but their assembly state is apparently determined by the nature of the mutations.

[19] Production of Recombinant Mammalian Holo-E2 and E3 and Reconstitution of Functional Branched-Chain α-Keto Acid Dehydrogenase Complex with Recombinant E1

By Jacinta L. Chuang, James R. Davie, R. Max Wynn, and David T. Chuang

The dihydrolipoyl transacylase (E2) component of the branched-chain α-keto acid dehydrogenase (BCKD) complex is a lipoic acid-bearing enzyme. E2 consists of 24 identical subunits forming a cubic structure, and is the structural core of the BCKD complex.[1] For structure–function studies, we previously expressed bovine E2 in *Escherichia coli* and showed that the expressed E2 is devoid of lipoic acid.[2] Despite the absence of the prosthetic group, the recombinant apo-E2 was able to assemble into a native 24-meric cubic structure as observed by electron microscopy. The apo-E2 possesses full transacylase activity with the active site located in the inner-core domain. However, the apoenzyme was unable to reconstitute with recombinant E1 and E3 into an active BCKD complex. The cloning of the bacterial lipolyating enzyme LplA by Cronan's group[3] prompted us to investigate whether mammalian branched-chain apo-E2 can be lipoylated *in vitro* by this enzyme. Here, we describe efficient *in vitro* lipoylation of apo-E2 with bacterial LplA. The availability of recombinant holo-E2 and E3 components allowed us to reconstitute a functional mammalian BCKD complex with recombinant E1. Using an improved radiochemical assay for the overall reaction, activities of individual E1 and E2 components can be measured with a high degree of sensitivity. The development of this sensitive reconstituted assay enabled us to monitor the *in vitro* refolding of $\alpha_2\beta_2$ tetramers of E1 in the presence of bacterial chaperonins GroEL and GroES.[4,5]

The following sections describe the expression and purification of recombinant enzymes required for the reconstitution of the functional BCKD complex. The expression of recombinant E1 is described in [18] in this

[1] F. H. Pettit, S. J. Yeaman, and L. J. Reed, *Proc. Natl. Acad. Sci. U.S.A.* **75,** 4881 (1978).
[2] T. A. Griffin, K. S. Lau, and D. T. Chuang, *J. Biol. Chem.* **263,** 14008 (1988).
[3] T. W. Morris, K. E. Reed, and J. E. Cronan, Jr., *J. Biol. Chem.* **269,** 16091 (1994).
[4] J. L. Chuang, R. M. Wynn, J.-L. Song, and D. T. Chuang, *J. Biol. Chem.* **274,** 10395 (1999).
[5] Y. S. Huang and D. T. Chuang, *J. Biol. Chem.* **274,** 10405 (1999).

volume.[5a] The improved radiochemical assay for the BCKD complex in 24-well tissue culture plates is also reported.

Production of Functional E2 and E3 Componants

Expression and Purification of Bovine Apodihydrolipoyl Transacylase (E2)

The pKK 233-2 expression vector (Amersham Pharmacia Biotech, Piscataway, NJ) harboring cDNA encoding the mature bovine E2 protein[2] is transformed into XL1-Blue cells (Stratagene, La Jolla, CA). Single colonies are selected from LB–carbenicillin (100 μg/ml) plates and used to inoculate 50-ml starter liquid cultures. After overnight growth at 37° with shaking, 10 ml of the starter culture is used to inoculate 1 liter of medium containing LB–carbenicillin. After 3 hr at 37°, cells are exposed to 0.75 mM isopropyl-β-D-thiogalactopyranoside (IPTG) and grown overnight (37°, rotation speed of 220 rpm). The cells are harvested by centrifugation at 4000 rpm for 20 min at 4°. The cell pellet from each liter of culture is resuspended in 50 ml of ice-cold E2 buffer [50 mM potassium phosphate buffer (pH 7.5), 350 mM NaCl, 5% (v/v) glycerol with 1 mM dithiothreitol (DTT), 1 mM benzamidine, and 1 mM phenylmethylsulfonyl fluoride (PMSF)], and lysed in a French pressure cell at 15,000–18,000 psi. The lysate is centrifuged at 25,000 rpm in a 60 Ti rotor for 30 min at 4°. E2 protein is precipitated from the supernatant by addition of an equal volume of 4 M ammonium sulfate. The precipitated E2 protein is further purified on a Sephacryl S-400 column (Amersham-Pharmacia Biotech) (2.6 × 100 cm) equilibrated in E2 buffer. Fractions are collected, analyzed by sodium dodecyl sulfate–polyacrylamide gel electrophoresis (SDS–PAGE), and stained with Coomassie blue. E2-containing fractions are pooled and precipitated with an equal volume of 4 M ammonium sulfate. Purified E2 is dissolved in E2 buffer and protein concentrations are determined with the Coomassie Plus protein assay reagent (Pierce, Rockford, IL), using bovine serum albumin (BSA) as a standard. The resulting recombinant E2 protein is approximately 95% pure as judged by SDS–PAGE and Coomassie blue staining. However, the recombinant E2 is not lipoylated *in vivo* even with the supplementation of 0.2 mM DL-6,8-thioctic acid in the culture medium.[6] The apo-E2 has approximately the same specific transacylase activity as

---

[5a] R. M. Wynn, J. R. Davie, J.- L. Song, J. L. Chuang, and D. T. Chuang, *Methods Enzymol.* **324,** chap. 18, 2000 (this volume).

[6] G. A. Radke, K. Ono, S. Ravindran, and T. E. Roche, *Biochem. Biophys. Res. Commun.* **190,** 982 (1993).

the native E2 from purified bovine BCKD complex (185 versus 173 nmol/min/mg).[2]

Expression and Purification of Escherichia coli Lipoate-Protein Ligase A

For in vitro lipoylation studies, E. coli TM202[BL21(λDE3)/pTM70-1] carrying the gene encoding the bacterial lipoylating enzyme LplA[3] was kindly provided by J. Cronan (University of Illinois, Urbana-Champaign, IL). One liter of LB–kanamycin (100 μg/ml) medium is inoculated with 10 ml of an overnight culture of TM202. After 3 hr of growth at 37°, IPTG (final concentration, 0.75 mM) is added and the culture is agitated at 220 rpm at 37° overnight. The cell pellet is collected by centrifugation at 4000 rpm for 20 min at 4° and resuspended in 50 ml of ice-cold E2 buffer. The cell suspension is lysed in a French pressure cell at 15,000–18,000 psi. The lysate is centrifuged at 25,000 rpm in a 60 Ti rotor for 30 min at 4°. The LplA protein is precipitated from the clarified lysate by the addition of 4 M ammonium sulfate to a final concentration of 1 M. The precipitated protein is pelleted by centrifugation at 12,000 rpm for 30 min and redissolved in E2 buffer. The protein solution is dialyzed against the E2 buffer, followed by centrifugation at 25,000 rpm in a 60 Ti rotor for 30 min at 4°. The clarified supernatant contains the 38-kDa monomeric LplA protein at greater than 98% purity. The protein concentration is determined as described above.

In Vitro Lipoylation of Recombinant Bovine Apo-E2 Protein

To produce holo-E2, a 20 μM concentration (based on 45-kDa monomers) of purified recombinant bovine apo-E2 is lipoylated by incubation, at 37° for 2 hr to overnight, in 50 mM potassium phosphate (pH 7.5), 350 mM NaCl, 1.5 mM MgCl$_2$, 1 mM DTT, 1.5 mM DL-6,8-thioctic acid, 0.5 mM MgATP, and 50 μM purified LplA protein. Lipoylated E2 is separated from LplA by fast protein liquid chromatography (FPLC) on a Superdex 200 column equilibrated in E2 buffer. Holo-E2 is eluted in the void volume, and the fractions are combined and concentrated on an Ultrafree-15 centrifugal filter device (Millipore, Bedford, MA).

Expression and Purification of Human Dihydrolipoyl Dehydrogenase (E3)

To construct an expression vector for human E3, the first-strand cDNA is synthesized from total RNA from human lymphoblasts, using a primer based on the published cDNA sequence.[7] The entire coding sequence for human E3 is amplified from the first-strand cDNA and ligated into the pTrcHisB vector (InVitrogen, Carlsbad, CA). To XL1-Blue cells harboring the pTrcHis-hE3 plasmid, 1 mM IPTG is added to induce E3 expression, and cells are grown in LB–carbenicillin (100 μg/ml) overnight at 30°. The cell pellet from 1 liter of culture is lysed by sonication in 100 ml of lysis buffer [100 mM potassium phosphate (pH 7.5), 500 mM NaCl, 0.2 mM EDTA, 0.5% (v/v) Triton X-100, 0.5% (w/v) Tween 20, 1 mM PMSF, 1 mM benzamidine, and lysozyme (1 mg/ml)]. The lysate is clarified by centrifugation at 17,300g for 30 min at 4°. Ten milliliters of Ni^{2+}-NTA resin suspension in water (1:1, v/v) (Qiagen, Valencia, CA) is mixed with the collected supernatant at 4° for 1 hr. The suspension is packed into a column, and the resin is washed with 100 ml of buffer A [100 mM potassium phosphate (pH 7.5), 500 mM NaCl, 0.1 mM EDTA, 0.1 mM EGTA, 10 mM 2-mercaptoethanol, and 10% (v/v) glycerol] containing 15 mM imidazole. E3 protein is eluted batchwise from Ni^{2+}-NTA resin with buffer A containing 250 mM imidazole. To the eluate, 4 M ammonium sulfate is added to a final concentration of 1.33 M. After centrifugation at 17,300g for 30 min at 4°, the floating pellet containing mostly contaminating proteins is removed by filtering through a nylon mesh. E3 protein is precipitated by addition of 4 M ammonium sulfate to final concentration of 2.5 M. After centrifugation, the E3 pellet is dissolved in E3 buffer [50 mM potassium phosphate (pH 6.5), 0.15% (v/v) Triton X-100, and 1 mM DTT] and dialyzed against three 1-liter volumes of same buffer at 4°. Recombinant E3 is bright yellow in color and can be visually monitored throughout the purification. Protein concentrations are determined as described above. The purified protein is assayed spectrophotometrically for E3 activity on the basis of the reduction of DL-lipoamide.[8] The recombinant E3 has a specific activity of 585 μmol/min/mg of protein.

Radiochemical Assays for E1 or E2 Based on Overall Reaction of a
   Reconstituted Recombinant Branched-Chain α-Keto Acid
   Dehydrogenase Complex

The improved-sensitivity radiochemical assays for E1 and E2 are performed according to reaction (1), catalyzed by the BCKD complex, using α-keto[1-^{14}C]isovalerate ([1-^{14}C] KIV) as substrate. ^{14}CO$_2$ evolved is absorbed in 2 N NaOH and counted for radioactivity.

[7] G. Otulakowski and B. H. Robinson, *J. Biol. Chem.* **262**, 17313 (1987).
[8] D. A. Stumpf and J. K. Parks, *Ann. Neurol.* **4**, 366 (1978).

$$\text{RCO–}^{14}\text{COOH} + \text{CoASH} + \text{NAD}^+ \xrightarrow{\text{TPP, Mg}^{2+}} \text{RCO–SCoA}$$
$$+ \text{NADH} + \text{H}^+ + {}^{14}\text{CO}_2 \uparrow \quad (1)$$

Reagents

Potassium phosphate buffer (pH 7.5), 1 M

HEPES buffer (pH 7.5), 1 M

Tris-HCl buffer (pH 8.8), 1.5 M

NAD^+ (Sigma, St. Louis, MO), 50 mM; store in aliquots at $-20°$

NaCl, 5 M

MgCl_2, 1 M

Coenzyme A (CoA; Sigma), lithium salt, 13 mM; store in aliquots at $-20°$

DL-6,8-Thioctic acid (lipoic acid; Sigma), 400 mM in 95% (v/v) ethanol; store at $-20°$

DL-Dithiothreitol (DTT; Sigma), 100 mM; store in aliquots at $-20°$

MgATP (Sigma), 200 mM [551 mg of ATP dissolved in 2 ml of 1.5 M Tris (pH 8.8), 1 ml of 1 M MgCl_2, 2 ml of H_2O]; store in aliquots at $-20°$

Mg-TPP (thiamine pyrophosphate; Sigma), 200 mM [92.16 mg of TPP dissolved in 200 μl of 1 M MgCl_2, 800 μl of 1 M HEPES (pH 7.9)]; store in aliquots at $-20°$

NaOH, 2 N

Triton X-100, 20% (v/v)

Trichloroacetic acid (TCA), 15% (w/v)

α-Ketoisovalerate (KIV; Sigma), 100 mM; store in aliquots at $-80°$

α-Keto[1-^{14}C]isovalerate ([1-^{14}C]KIV), 6.4 mM in 15 mM HCl (specific activity, 1800 cpm/nmol). The preparation of the labeled substrate is as described previously,[9] except that [1-^{14}C]valine is purchased from American Radiolabeled Chemicals (St. Louis, MO)

Holo-E2 in E2 buffer (see above), 1 mg/ml or 0.9 μM based on 1116-kDa 24-mers

E1[5a] in E1 buffer [50 mM potassium phosphate (pH 7.5), 250 mM KCl, 1 mM DTT, 5% (v/v) glycerol], 2 mg/ml or 12 μM based on 168-kDa heterotetramers

E3 in E3 buffer (see above), 4 mg/ml or 36 μM based on 110-kDa homodimers

Definition of Unit and Specific Activity. One unit is the amount of E1 that catalyzes the evolution of 1 μmol of $^{14}\text{CO}_2$ per minute under the defined assay conditions. Specific activity is expressed as units per milligram

[9] D. T. Chuang, *Methods Enzymol.* **166**, 146 (1988).

FIG. 1. The setup of a 24-well plate for radiochemical decarboxylation assays. Each well holds the reaction mixture and a center cup that houses an NaOH-impregnated paper wick. Wells are sealed with a self-adhesive film during the reaction. See the text for the assay procedure.

of protein. Protein is determined by Coomassie Plus protein assay reagent as described above.

Equipment and Supplies

 Tissue culture plates, 24-well (any source)
 Clear adhesive film for microtiter plates (USA/Scientific, Ocala, FL)
 Plastic center cups with stem removed (Kontes, Vineland, NJ)
 Filter paper wicks (0.3 × 0.7 cm)
 Motorized microliter pipette (Rainin, Woburn, MA)
 Disposable syringe (1 ml)
 Needle, 21 gauge
 Scotch tape

Improved Radiochemical Assay for E1

 The radiochemical decarboxylation assay is carried out on 24-well tissue culture plates as shown in Fig. 1. This is a significantly simplified version of the microtiter plate assay developed previously for directly measuring the level of decarboxylation in leukocytes and cultured fibroblasts.[10] The

---

[10] U. Wendel, W. Wöhler, H. W. Goedde, U. Langenbeck, E. Passarge, and H. W. Rüdiger, *Clin. Chim. Acta* **45**, 433 (1973).

use of tissue culture plates rather than small pear-shaped reaction flasks described previously[9] greatly improves the speed of assays. To assay for E1 based on reconstituted BCKD activity, the assay mixture in 300 μl contains 100 mM potassium phosphate (pH 7.5), 2.5 mM NAD$^+$, 2 mM MgCl$_2$, 100 mM NaCl, 0.26 mM CoA, 2 mM DTT, 200 μM Mg-TPP, 0.1% (v/v) Triton X-100, 7 nM holo-E2 24-mers, and 318 nM E3 homodimers. The amounts of holo-E2 and E3 are in excess in the reaction mixture based on the stoichiometry of each E2 homo 24-mer capable of binding 12 E1 heterotetramers.[1] The assay mixture is pipetted into each well of a 24-well tissue culture plate, which is kept on ice. One center cup housing a filter paper wick is placed into each well that contains assay mixture. Twenty-five microliters of 2 N NaOH is pipetted directly onto each paper wick. The E1 enzyme solution (up to 3 μg in 5–50 μl) to be assayed is added to the assay mixture in each well. A motorized microliter pipette is used to quickly add 10 μl of [1-^{14}C]KIV substrate (final concentration, ~0.2 mM) to the assay mixture in each well. The entire plate is sealed with a clear adhesive film, and transferred to a 37° shaking water bath. At the end of a 20-min incubation, the plate is replaced on ice. Using a 1-ml disposable syringe fitted with a 21-gauge needle, 0.5 ml of 15% (w/v) TCA is injected directly into the reaction mixture in each well. Needle holes are sealed with strips of Scotch tape and the plate is transferred back to the 37° water bath. The plate is further incubated for 45 min to recover residual released ^{14}CO$_2$ from the reaction mixture. At the completion of incubation, the filter paper wick is transferred to 4 ml of scintillation cocktail[9] to determine radioactivity in a scintillation spectrometer.

The sensitivity of the preceding E1 radiochemical assay, based on the overall reaction of a reconstituted BCKD complex, is compared with that of a previously described E1 radiochemical assay.[9] The latter assay measures the level of decarboxylation by E1 alone in the presence of an artificial electron acceptor. Figure 2 shows that the decarboxylation rate by E1 alone, with 2,6-dichlorophenol-indophenol as the electron acceptor and [1-^{14}C]KIV as substrate, is linear for up to 59 nM or 3 μg of E1 protein, with a k_{cat} of 19.69 min^{-1}. When decarboxylation by E1 is assayed in a reconstituted BCKD system in the presence of excess holo-E2 and E3, the rate is linear up to 29 nM or 1.5 μg of E1 heterotetramers. More significantly, the k_{cat} of 364 min^{-1} obtained in the reconstituted system is 18-fold higher than that measured with E1 alone. The results show the high sensitivity of the reconstituted E1 assay, which can be used in functional studies of this enzyme component.

Radiochemical Assay for E2

The reconstituted recombinant BCKD complex system can also be used to assay for E2 activity. The assay mixture is essentially identical to that

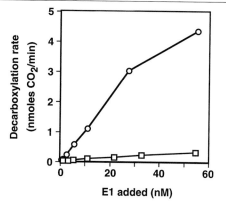

FIG. 2. Decarboxylation of [1-^{14}C]KIV by recombinant E1 or the reconstituted recombinant BCKD complex. Decarboxylation of [1-^{14}C]KIV by the E1 componant alone (□) was assayed in the presence of an artificial electron acceptor, 2,6-dichlorophenol-indophenol, as described previously.[9] The rate of decarboxylation by E1 in the reconstituted BCKD complex system in the presence of excess holo-E2 and E3 (○) was determined as described in text.

for the preceding E1 activity assay except that a constant amount of recombinant E1, e.g., a 12.5 nM concentration, is present while the E2 concentration is varied. The radiochemical E2 assay is employed to study reconstituted BCKD activity with apo-E2 lipoylated to different degrees. The partially lipoylated E2 is produced by incubation of apo-E2 with different

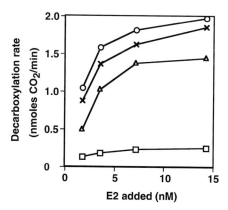

FIG. 3. Decarboxylation of [1-^{14}C]KIV by a reconstituted BCKD complex with recombinant E2 lipoylated to various degrees. The apo-E2 protein was lipoylated *in vitro* with different molar ratios of LplA monomers to E2 monomers. The reaction mixture contained 12.5 nM recombinant E1 tetramers, 318 nM E3 dimers, and varying concentrations of E2 24-mers. The decarboxylation by the reconstituted BCKD complex was assayed as described in text. The molar ratios of LplA to E2 were as follows: (□) 0; (△) 0.83; (×) 2.5; (○), 3.3.

amounts of LplA enzyme. As shown in Fig. 3, reconstitution of the BCKD complex with apo-E2 without *in vitro* lipoylation results in minimal decarboxylation of [1-^{14}C]KIV. Maximal decarboxylation is achieved when apo-E2 is lipoylated with threefold molar excess of LplA protein. The results demonstrate the utility of the radiochemical E2 assay in assessing the efficiency of *in vitro* lipoylation with apo-E2.

[20] Production of Recombinant E1 Component of Branched-Chain α-Keto Acid Dehydrogenase Complex

By JOHN W. HAWES, YU ZHAO, KIRILL M. POPOV, YOSHIHARU SHIMOMURA, and ROBERT A. HARRIS

Introduction

The activity state of mitochondrial branched-chain α-keto acid dehydrogenase (BCKDH) is highly regulated according to the physiological requirements for protein synthesis or degradation of excess branched-chain amino acids.[1] This regulation is primarily exerted through reversible phosphorylation of the catalytic subunit of BCKDH by a specific protein kinase and a specific phosphoprotein phosphatase. Two phosphorylation sites exist on the E1α subunit although regulation of BCKDH activity results exclusively from phosphorylation of site 1, Ser-293 in rat E1α. Phosphorylation of site 2, Ser-303 in rat E1α, occurs at a significantly slower rate[2] and is silent with respect to regulation of BCKDH activity.[3] Thiamine pyrophosphate (TPP) and branched-chain α-keto acids (cofactor and substrates for the BCKDH-catalyzed reaction) inhibit the phosphorylation by BCKDH kinase, and it has been suggested that BCKDH kinase inactivates BCKDH E1 by placing a covalently linked phosphate directly into the active site of the dehydrogenase.[4] Studies of the mechanism of inactivation of BCKDH by phosphorylation require a ready source of purified subunits of this multienzyme complex that can be reconstituted into an active BCKDH complex under various

[1] R. A. Harris, K. M. Popov, Y. Zhao, and Y. Shimomura, *J. Nutr.* **124,** 1499S (1994).

[2] R. Paxton, M. J. Kuntz, and R. A. Harris, *Arch. Biochem. Biophys.* **244,** 187 (1986).

[3] K. G. Cook, A. P. Bradford, S. J. Yeaman, A. Aitken, I. M. Fearnley, and J. E. Walker, *Eur. J. Biochem.* **145,** 587 (1984).

[4] R. A. Harris, J. W. Hawes, K. M. Popov, Y. Zhao, Y. Shimomura, J. Sato, J. Jaskiewicz, and T. D. Hurley, *Adv. Enzyme Regul.* **37,** 271 (1997).

conditions. To this end we have expressed the BCKDH E1 component as a recombinant enzyme[5] in a form that can be easily purified and reconstituted with the purified transacylase core (E2) of the BCKDH complex. The purified recombinant E1 subunit is enzymatically active when combined with E2, can be inactivated by treatment with the purified recombinant BCKDH kinase and ATP, and can be manipulated by site-directed mutagenesis.

Materials

R408 and VCS M13 are purchased from Stratagene (La Jolla, CA). Vectors pET15 and pET21 are purchased from Novagen (Madison, WI). Nickel-NTA-agarose is purchased from Qiagen (Valencia, CA). Restriction enzymes, restriction enzyme buffers, and other DNA-modifying enzymes are purchased from BRL (Gaithersburg, MD). All chemical reagents for enzyme assay are purchased from Sigma (St. Louis, MO). The site-directed mutagenesis system is purchased from Amersham (Arlington Heights, IL). Native rat liver BCKDH E2 subunit is prepared by a modification of the method of Cook *et al.*[6] according to Shimomura *et al.*[7] The pGroESL plasmid was a kind gift of A. Gatenby (Central Research and Development, Du Pont, Wilmington, DE).

Construction of Prokaryotic Expression Vector for Branched-Chain
   α-Keto Acid Dehydrogenase E1

A prokaryotic expression vector[5] containing rat BCKDH E1α and E1β cDNAs is constructed as follows. A *Bam*HI site is introduced into the 5′ end of rat E1β cDNA by site-directed mutagenesis. The 1220-base pair (bp) *Bam*HI fragment, containing nucleotides 62–1281 of rat E1β, is ligated to the *Bam*HI site of pET-21a (Novagen) to produce plasmid pET-E1β. This plasmid carries the E1α-coding sequence in frame for expression from the T7 promoter. A 2474-bp *Mae*I fragment, containing nucleotides 87–1683 of rat E1α cDNA and nucleotides 450–1301 of pUC19, is isolated from the rat E1α cDNA and ligated to the *Nde*I site of pET-15b (Novagen) to produce the plasmid pET-E1α. This plasmid carries the E1α cDNA with an in-frame fusion of a hexahistidine tag and a T7 promoter. A fragment carrying the T7 promoter, the E1α-coding region, and the T7 terminator is isolated from the pET-E1α plasmid by *Cla*I digestion, followed by "blunt-

[5] Y. Zhao, J. W. Hawes, K. M. Popov, J. Jaskiewicz, Y. Shimomura, D. W. Crabb, and R. A. Harris, *J. Biol. Chem.* **269,** 18583 (1994).
[6] K. G. Cook, A. P. Bradford, and S. J. Yeaman, *Biochem. J.* **225,** 731 (1985).
[7] Y. Shimomura, R. Paxton, T. Ozawa, and R. A. Harris, *Anal. Biochem.* **163,** 74 (1987).

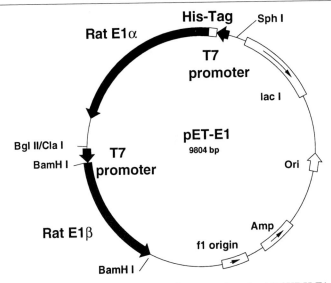

FIG. 1. Prokaryotic vector construct for expression of rat BCKDH E1.

ing" with T4 DNA polymerase, and *Sph*I digestion. The pET-E1β plasmid is then digested with *Bgl*II followed by "blunting" with T4 DNA polymerase and digestion with *Sph*I. The E1α blunt-ended *Sph*I fragment is then ligated to the blunt-ended *Sph*I-cut pET E1β plasmid. This ligation produced an expression vector with two separate "cassettes" with a T7 promoter and T7 terminator for the E1α and E1β cDNAs. The final expression vector (Fig. 1) was prepared by performing site-directed mutagenesis to remove 57 extra bases (encoding vector sequence and a portion of the mitochondrial leader sequence) from the 5′ end of the E1β cDNA. E1α in this construct has 11 extra amino acids at the N terminus corresponding to the leader sequence, and 19 extra amino acids from the vector that include the histidine tag and a thrombin cleavage site. This vector is designated pET-E1.

Site-Directed Mutagenesis of Branched-Chain α-Keto Acid
  Dehydrogenase E1α

Single-stranded pET-E1 DNA is isolated by single-strand rescue, using either R408 or VCS M13 helper phage. Single-strand rescue and purification of the single-stranded DNA are performed exactly according to the manufacturer protocols (Stratagene). Site-directed mutagenesis[5,8] is performed

[8] J. W. Hawes, R. J. Schnepf, A. E. Jenkins, Y. Shimnomura, K. M. Popov, and R. A. Harris, *J. Biol. Chem.* **270,** 31071 (1995).

with the Sculptor system (Amersham), following the manufacturer protocol. Mutants are identified by dideoxynucleotide sequencing methods, using the double-stranded plasmid DNA as template.

Expression and Purification of Recombinant Branched-Chain α-Keto Acid Dehydrogenase E1

In this procedure[5,8] BL21(DE3) *Escherichia coli* cells are double transfected with the pET-E1 plasmid (ampicillin resistant) and the pGroESL plasmid (chloramphenicol resistant). Double transfectants are selected by growth at 37° in TY medium containing both kanamycin and chloramphenicol (70 μg/ml each). When the cell cultures obtain an optical density at 600 nm (OD$_{600\,nm}$) of approximately 0.8, isopropyl-β-D-thiogalactopyranoside (IPTG) is added to a final concentration of 0.5 mM and growth is continued for 18–20 hr at 37°. The induced bacterial cultures are then pelleted and resuspended in 10 volumes of buffer A [20 mM Tris-HCl (pH 8.0), 0.5 M NaCl, 10 mM 2-mercaptoethanol, 5 mM imidazole] containing 0.5% (v/v) Triton X-100, and a 100-μg/ml concentration of both benzamidine and phenylmethylsulfonyl fluoride (PMSF). Cells are lysed by sonication and the resulting homogenates are centrifuged at 20,000g for 30 min at 4°. The soluble fraction is applied to a 10-ml nickel-NTA (Qiagen) column previously equilibrated in the same buffer. The column is washed with 5 volumes of buffer A containing 20 mM imidazole, 5 volumes of buffer A containing 40 mM imidazole, and 5 volumes of buffer A containing 60 mM imidazole. The histidine-tagged protein is eluted with buffer A containing 200 mM imidazole. Elution fractions containing E1 protein are pooled, brought to 20% (v/v) glycerol, and stored at −70°. Purified enzyme is analyzed by sodium dodecyl sulfate–polyacrylamide gel electrophoresis (SDS–PAGE) and appears as two bands (E1α and E1β proteins) in equal relative quantities. The coexpression of GroEL and GroES increases the yield of soluble E1 by approximately threefold in this system.

Assay of Branched-Chain α-Keto Acid Dehydrogenase Activity

Enzyme activity is measured spectrophotometrically at 340 nm by monitoring the generation of NADH produced at 30° by the combined action of the BCKDH E1, E2, and E3 subunits. Each assay consists of 1 ml of cocktail containing 30 mM potassium phosphate (pH 7.5), 0.1% (v/v) Triton X-100, 1 mM NAD$^+$, 2 mM dithiothreitol (DTT), 0.4 mM coenzyme A (CoA), 0.4 mM TPP, 2 mM MgCl$_2$, 7.5 units of porcine heart dihydrolipoamide reductase (E3), 1 μg of E1, and 0.75 μg of E2. A baseline absorbance is measured for several minutes and the reaction is then initiated by the addition of 0.5 mM α-ketoisovalerate (KIV). In this assay, a nonlinear phase, or "lag phase," is observed for the activity of the reconstituted complex. This nonlinear phase extends for approximately 5 min, followed by a linear rate of NADH production. The cause of this initial nonlinear phase has not been determined. For kinetic studies, the KIV solutions are standardized by an end-point assay using purified BCKDH complex.[9] The concentrations of TPP solutions are measured by absorbance at 267 nm ($E = 8520\ M^{-1}\ cm^{-1}$).

Reconstitution of Branched-Chain α-Keto Acid Dehydrogenase E1/E2 Complex

BCKDH E1/E2 subcomplex is reconstituted[5,8] with purified recombinant E1s and purified native E2. Increasing the ratio of E1 to E2 protein in the reconstituted complex increases BCKDH activity until the E1/E2 ratio is approximately 1.[10] Activity reaches a plateau at E1/E2 ratios above 1, suggesting that E1 is rate limiting below a ratio of 1 and that E2 is rate-limiting above a ratio of 1 (Fig. 2).

Assay of Branched-Chain α-Keto Acid Dehydrogenase E2 Binding by Mutant Branched-Chain α-Keto Acid Dehydrogenase E1 Enzymes

A number of site-directed mutants have been produced with this system. Mutations of the phosphorylation site 1, Ser-293, affect dehydrogenase activity and E1 kinetic parameters, as do other mutations surrounding this site. It is important to determine if such mutations block E1 activity by affecting the E1 active site, or by affecting the interaction of E1 and E2 during reconstitution of the complex. Interaction of inactive mutant E1 proteins with the BCKDH E2 subunit is measured by a method similar to that of Cook et al.[6] This assay is based on the observation that inactive mutant E1 proteins can inhibit the reconstitution of enzymatically active complex with wild-type E1 and E2. A constant amount of E2 protein is combined with a constant amount of E1 protein, using various ratios of wild-type and mutant E1s. The amount of E1 is in excess of the amount of E2 to assure that E2 is rate limiting for BCKDH activity and that E1-binding sites on E2 are saturated. Reconstituted complexes with different ratios of mutant and wild-type E1 are assayed for BCKDH activity as

---

[9] G. W. Goodwin, M. J. Kuntz, R. Paxton, and R. A. Harris, *Anal. Biochem.* **162,** 536 (1987).

[10] Y. Zhao, K. M. Popov, Y. Shimomura, N. Y. Kedishvili, J. Jaskiewicz, M. J. Kuntz, J. Kain, B. Zhang, and R. A. Harris, *Arch. Biochem. Biophys.* **308,** 446 (1994).

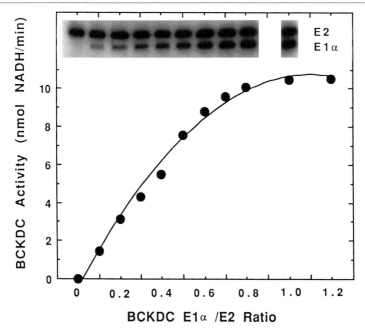

FIG. 2. BCKDH activity of reconstituted complexes with different E1/E2 ratios. A constant amount of purified native E2 subunit (25 pmol) was reconstituted with different amounts of purified recombinant BCKDH E1 (0–30 pmol). Excess bovine E3 subunit was present in the assay mixutres. *Inset:* Western analysis with E2- and E1α-specific antisera. [Reproduced with permission from Y. Zhao, K. M. Popov, Y. Shimomura, N. Y. Kedishvili, J. Jaskiewicz, M. J. Kuntz, J. Kain, B. Zhang, and R. A. Harris, *Arch. Biochem. Biophys.* **308**, 446 (1994).]

described above. Mutant E1 proteins that appear to reconstitute with E2 produce approximately 50% inhibition when present at an equimolar ratio with wild-type E1 (Fig. 3).[8]

Characteristics of Wild-Type and Mutant Branched-Chain α-Keto Acid Dehydrogenase E1s

BCKDH subcomplex reconstituted with wild-type E1 and native E2 displays enzyme activity and kinetic parameters similar to those of native BCKDH complex (Table I).[5] Mutations of the two known phosphorylation sites on the E1α subunit (S293 and S303) produce differing effects on activity and kinetic parameters. S303A and S303E mutant E1s display a specific activity and kinetic parameters unchanged from that of wild-type E1 (Table I). S293A E1 displays a 12-fold increase in K_m for KIV without a significant change in V_{max} and S293E E1 is enzymatically inactive.[5] These

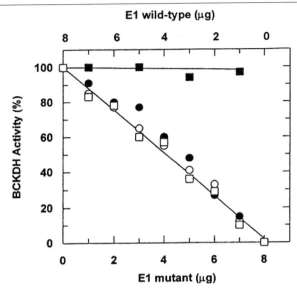

FIG. 3. BCKDH activity of reconstituted complexes with different ratios of wild-type and mutant E1. A constant amount of purified native E2 (0.75 μg) was used for reconstitution with a constant total amount of E1 (8 μg) but with various ratios of mutant and wild-type E1. (■) Amount of wild-type E1 varied as indicated without addition of mutant E1; (○) amount of total E1 kept constant (8 μg) by varying R288A and wild-type E1; (●) by varying H292A and wild-type E1; (□) by varying D296A and wild-type E1. [Reproduced with permission from J. W. Hawes, R. J. Schnepf, A. E. Jenkins, Y. Shimomura, K. M. Popov, and R. A. Harris. *J. Biol. Chem.* **270**, 31071 (1995).]

TABLE I
KINETIC PARAMETERS OF WILD-TYPE AND MUTANT E1 PROTEINS[a]

Recombinant E1	K_m (α-ketisovalerate) (μM)	V_{max} (U/mg protein)[b]
Wild-type	45 ± 13[c]	5.4 ± 0.7
S293E	—	—
S293A	532 ± 98	5.5 ± 0.6
S303E	56 ± 9	5.2 ± 0.2
S303A	58 ± 15	5.8 ± 0.7

[a] Reproduced in part with permission from Y. Zhao, J. W. Hawes, K. M. Popov, J. Jaskiewicz, Y. Shimomura, D. W. Crabb, and R. A. Harris, *J. Biol. Chem.* **269**, 18583 (1994).
[b] Refers to units (μmoles of NADH produced per minute) per milligram of recombinant E1 protein.
[c] Mean ± SE; data obtained with four concentrations of α-ketoisovalerate spanning the K_m value.

data suggest a role for this residue in α-keto acid substrate binding, and it has been shown that this residue is completely conserved in the amino acid sequences of all known BCKDH E1α proteins as well as those of pyruvate dehydrogenase and α-ketoglutarate dehydrogenase.[8] Alanine substitutions of three other residues surrounding S293 also abolish the enzymatic activity of E1.[8] These include R288A, H292A, and D296A. These three mutant E1s are devoid of enzyme activity at all pH values tested and all substrate and cofactor concentrations tested; however, they appear to bind to the E2 subunit as efficiently as the wild-type E1 (Fig. 3). Evidence has been obtained for the involvement of His-292 in thiamine pyrophosphate binding.[8]

Acknowledgments

This work was supported by grants from the U.S. Public Health Services (NIH DK19259 to R.A.H.), the Diabetes Research and Training Center of Indiana University School of Medicine (AM 20542), the Grace M. Showalter Trust (to R.A.H.), the Uehara Memorial Foundation (to Y.S), and the University of Tsukuba Project Research Fund (to Y.S.).

[21] Mammalian Methylmalonate-Semialdehyde Dehydrogenase

By NATALIA Y. KEDISHVILI, GARY W. GOODWIN,
KIRILL M. POPOV, and ROBERT A. HARRIS

Introduction

Methylmalonate-semialdehyde dehydrogenase (MMSDH), located in the mitochondrial matrix space, catalyzes the irreversible oxidative decarboxylation of malonate semialdehyde and methylmalonate semialdehyde to acetyl-CoA and propionyl-CoA, respectively. These reactions are in the distal portions of the valine and pyrimidine catabolic pathways.[1,2] MMSDH belongs to the aldehyde dehydrogenase superfamily,[3,4] but is unique among

[1] J. R. Sokatch, L. E. Sanders, and V. P. Marshall, *J. Biol. Chem.* **243**, 2500 (1968).
[2] G. W. Goodwin, P. M. Rougraff, E. J. Davis, and R. A. Harris, *J. Biol. Chem.* **264**, 14965 (1989).
[3] N. Y. Kedishvili, K. M. Popov, P. M. Rougraff, Y. Zhao, D. W. Crabb, and R. A. Harris, *J. Biol. Chem.* **267**, 19724 (1992).
[4] J. Perozich, H. Nicholas, B.-C. Wang, R. Lindahl, and J. Hempel, *Protein Sci.* **8**, 137 (1999).

the members of this family because coenzyme A is required for the reaction and a CoA ester is produced.[1,2] MMSDH, like other aldehyde dehydrogenases, has esterase activity.[5] An active site S-acyl enzyme is common to both the esterase reaction and the aldehyde dehydrogenase/CoA ester synthetic reaction.[5] An active site cysteine residue (Cys-285 in rat MMSDH[3]) is acetylated during hydrolysis of p-nitrophenyl acetate by MMSDH. Acetyl-CoA is produced instead of acetate from p-nitrophenyl acetate when the reaction is conducted in the presence of coenzyme A.[5] Long-chain fatty acyl-CoA esters also inactivate MMSDH by acylation of its active site cysteine residue,[6] presumably by reversal of the above process. Active site acylation has been proposed as a mechanism for the regulation of MMSDH activity *in vivo* by long-chain fatty acids.[7]

Purification of Native Methylmalonate-semialdehyde Dehydrogenase from Rat Liver

Reagents

Buffer A: 20 mM Ammonium acetate (pH 7.5 at 4°), 0.1 mM EDTA, 2 mM dithiothreitol (DTT), 0.5 mM NAD$^+$ (grade AA-1)
Buffer B: 25 mM potassium phosphate (pH 7.5), 0.1 mM EDTA, 2 mM DTT, 0.5 mM NAD$^+$ (grade AA-1), 10% (v/v) glycerol
Buffer C: 10 mM Tris-HCl (pH 8.0 at 4°), 0.1 mM EDTA, 2 mM DTT, 0.5 mM NAD$^+$ (grade AA-1)

Purification Procedure

In this procedure,[2,8] the enzyme is purified from 300 g of frozen rat liver, previously stored at −70°. Livers are allowed to partially thaw at 4°, and then are homogenized in a Waring blender at the high setting for 1 min in 4 volumes of buffer A supplemented with protease inhibitors (see below, in the section on expression of MMSDH in *E. coli*). The suspension

[5] K. M. Popov, N. Y. Kedishvili, and R. A. Harris, *Biochim. Biophys. Acta* **1119,** 69 (1992).
[6] L. Berthiaume, I. Deichaite, S. Peseckis, and M. D. Resh, *J. Biol. Chem.* **269,** 6498 (1994).
[7] I. Deichaite, L. Berthiaume, S. M. Peseckis, W. F. Patton, and M. D. Resh. *J. Biol. Chem.* **268,** 1338 (1993).
[8] N. Y. Kedishvili, K. M. Popov, and R. A. Harris, *Arch. Biochem. Biophys.* **290,** 21 (1991).

is further homogenized in five portions with a Polytron PT 10 homogenizer (Brinkmann, Westbury, NY) at a setting of 4 for 1 min. The pH is adjusted to 7.5 at 4°, and insoluble material is removed by centrifugation at 100,000g for 60 min at 4°. The clear supernatant is carefully decanted, the pH is adjusted to 6.5 at 4° with acetic acid, and the extract mixed with 600 ml of CM-Sepharose equilibrated with buffer A, pH 6.5, at 4°. The slurry is stirred gently for 30 min and then unbound material (containing the MMSDH) is removed by filtration. The CM-Sepharose is washed twice with 1 volume of buffer A. Filtrates are combined, adjusted to pH 7.0 at 4° with NH_4OH, and mixed with 800 ml of DEAE-Sephacel equilibrated with buffer A, pH 7.0 at 4°. The slurry is mixed for 30 min and unbound material containing MMSDH is removed by filtration. The DEAE-Sephacel is washed three times with buffer A. All washes are combined, and the pH adjusted to 7.5 at 4° with NH_4OH. This extract is applied at a flow rate of 60–80 ml/hr to a hydroxylapatite column (2.5 × 20 cm) equilibrated with buffer B. MMSDH is eluted with a linear gradient of potassium phosphate (total volume, 500 ml) from 100 to 300 mM prepared in buffer B. The enzyme solution is concentrated to a volume of 10–20 ml under N_2 pressure with a YM10 membrane (Amicon, Danvers, MA) and applied at a flow rate of 50 ml/hr to a Sephacryl S-300 column (2.5 × 95 cm) equilibrated with buffer A (pH 7.5 at 4°). Fractions containing MMSDH are pooled, the pH is adjusted to 6.0 at 4° with acetic acid, and the extract is applied to an S-Sepharose Fast Flow column (1.5 × 10 cm) equilibrated with buffer A (pH 6.0) with 10% (v/v) glycerol. In the presence of NAD^+, MMSDH does not bind to S-Sepharose and elutes in the void volume, whereas most other proteins remain bound. The purified enzyme is concentrated on a phenyl-Sepharose column dialyzed against buffer C, divided into small aliquots, and stored at $-70°$. Ten milligrams of the enzyme protein can be purified from 100 g of rat liver with a specific activity of 7–9 units/mg of protein measured with malonate semialdehyde as substrate. One unit is 1 μmol/min at 30°.

Apoenzyme Preparation

To prepare enzyme[5] depleted of NAD^+, MMSDH at a concentration of 1–2 mg/ml in buffer C is supplemented with $(NH_4)_2SO_4$ to 2 M, incubated for 15 min at room temperature, and applied at a flow rate of 1 ml/hr to a Sephadex G-25 Fast Flow column equilibrated with 15 mM potassium phosphate (pH 7.8), 0.1 mM EDTA, and 1 mM DTT. To measure the residual amount of NAD^+ in MMSDH, 1 mg of enzyme is precipitated with 6% (w/v, final concentration) perchloric acid, the extract is neutralized with potassium hydroxide, and NAD^+ is measured by an enzymatic end-point assay.[4] Usually, less than 0.05 mol of NAD^+ per mole of enzyme is detected.

Activity Assay

Preparation of Malonate Semialdehyde and Methylmalonate Semialdehyde

The ethyl ester diethyl acetal of methylmalonate semialdehyde is synthesized as described by Kupiecki and Coon.[9] Hydrolysis is carried out at 50° for 4 hr with H_2SO_4. The product is then cautiously neutralized on ice with 6 N KOH, brought to pH 6.4 with 1 M KH_2CO_3, cold-filtered through Whatman (Clifton, NJ) No. 1 filter paper, and stored in small aliquots at −70°. Racemic ethylmalonate semialdehyde is prepared by an identical procedure, starting from the corresponding ethyl ester diethyl acetal (ethylhydroacrylic acid). The ethyl ester diethyl acetal of malonate semialdehyde (ethyl 3,3-diethyloxypropionate; Aldrich, Milwaukee, WI) is hydrolyzed in a similar manner except that saponification is completed at room temperature for 2 hr. The neutralized, filtered product is used immediately.

Procedure

Enzyme activity is routinely measured by following the reduction of NAD^+ at 340 nm with a cocktail consisting of 30 mM sodium pyrophosphate, pH 8.0, adjusted with HCl at room temperature, 2 mM DTT, 2 mM NAD^+, 0.5 mM CoA, and 0.5 mM malonate semialdehyde or methylmalonate semialdehyde. Reactions are initiated with enzyme. Enzyme activity can also be measured by a coupled assay based on the generation of methylmalonate semialdehyde from L-3-hydroxyisobutyrate by 3-hydroxyisobutyrate dehydrogenase.[2] MMSDH also hydrolyzes p-nitrophenyl acetate. Esterase activity of the enzyme is determined in 50 mM potassium phosphate (pH 7.8) and 0.1 mM EDTA at 30°.[5] p-Nitrophenyl acetate, prepared in acetone to minimize spontaneous hydrolysis, is added to a final concentration of 0.25 mM to initiate the reaction. Acetone does not affect enzyme activity provided its concentration is less than 2% in the assay solution. Esterase activity is followed by p-nitrophenol production at 400 nm ($E = 16 \times 10^3$ M^{-1} cm^{-1}).

[9] F. P. Kupiecki and M. J. Coon, *Biochem. Prep.* **7,** 69 (1960).

Cloning Strategy Used to Obtain Methylmalonate-semialdehyde Dehydrogenase cDNA

Partial Amino Acid Sequence of Methylmalonate-semialdehyde Dehydrogenase

In this procedure,[3] N-terminal protein sequencing is performed with an Applied Biosystems (Foster City, CA) model 477A Pulse Liquid protein sequencer, and the phenylthiohydantoin derivatives are analyzed by reversed-phase high-performance liquid chromatography with a model 120A analyzer. To obtain additional sequence information, MMSDH at a final concentration of 50 μg/ml is digested with lysyl endopeptidase C in 10 mM potassium phosphate (pH 7.5), 0.1 mM EDTA, and 0.1 mM DTT for 60 min at 30° (MMSDH-to-protease ratio, 300:1). A 50-kDa proteolytic fragment produced by this treatment is separated by sodium dodecyl sulfate–polyacrylamide gel electrophoresis (SDS–PAGE), blotted onto a polyvinylidene membrane, and subjected to amino acid sequence analysis.

Oligonucleotide Probes

The peptide sequences available (SSSSVPTVKLFINGKFVQ and NMNLYSYRLPLGVCAGIATFNFPAG for the N terminus of MMSDH and the 50-kDa proteolytic fragment, respectively) did not allow the design of an oligonucleotide probe with low degeneracy and high melting temperature.[3] Degeneracy is reduced, therefore, by incorporating inosine residues in the design of two oligonucleotides based on the peptide sequences underlined above.

Cloning

Degenerate, inosine-containing oligonucleotides are synthesized on the basis of peptide sequence and used to amplify rat liver cDNA by polymerase chain reaction (PCR).[3] The first strand of cDNA is synthesized with Maloney murine leukemia virus (Mo-MuLV) reverse transcriptase, using random hexamers according to the manufacturer instructions (Perkin-Elmer Cetus, Norwalk, CT). Thirty-five cycles of PCR are performed with primers A(A/G)CTITT(T/C)ATIGA(T/ C)GGIAA(A/G)TT(C/T)GTIGA and GCNGG(G/A)AA(G/A)TT(G/ A)AAIGGIGCIAT, using 1 min at 42° for annealing, 3 min at 72° for extension, and 1 min at 94° for denaturation. The PCR product is gel purified, subcloned into M13mpl8 (Bethesda Research Laboratories, Gaithersburg, MD) and sequenced with Sequenase version 2.0 according to suggested procedures (U.S. Biochemical, Cleveland, OH). Nondegenerate oligonucleotides GGACCTTTATTCCTACC-

GCCTGCCTCTGGGGGTG and GGAATCCAAAAGTGACAAATG-GATTGACATCCAC are synthesized on the basis of the sequence of PCR product and used for the library screening. Oligonucleotide probes are labeled with [γ-^{32}P]ATP, using T4 kinase. These probes are used to screen 1.5×10^6 individual plaque-forming units (PFU) from rat liver λgt11 cDNA library (Clontech Laboratories, Palo Alto, CA) essentially as described by Sambrook et al.[10] Hybridization conditions are as follows: 6× SSC (150 mM sodium chloride, 15 mM sodium citrate, pH 7.5), 5× Denhardt's solution [0.1% (w/v) bovine serum albumin, 0.1% (w/v) polyvinylpyrrolidone, 0.1% (w/v) Ficoll 400], 0.1% (w/v) SDS, heat-denatured salmon sperm DNA (0.1 mg/ml), and the radiolabeled probe (2×10^6 cpm/ml) at 55° for 17 hr. The filters are washed with 6× SSC–0.1% (w/v) SDS four times at room temperature, and once with 2× SSC–0.1% (w/v) SDS at 55° for 5 min and autoradiographed. Positive plaques are purified through four cycles of screening.

Procedure Used to Obtain 5' Coding Region of Methylmalonate-semialdehyde Dehydrogenase cDNA

In this procedure,[3] a specific primer is designed to hybridize to bases 145–177 of the coding strand of the rat cDNA: oligo 1 (GAAAGAAGAT-GCTGGATACCATGTGGAGTTTAC). External primers (one for each insert orientation) are synthesized to correspond to bases 4266–4288 (GGTGGCGACGACTCCGGAGCCCG) and bases 4323–4352 (TTGA-CACCAGACCAACTGGTAATGGTAGCG) of the Escherichia coli lactose operon flanking the EcoRI insertion site in λgt11 (primers L and R, respectively). Phage DNA from 1 ml of amplified cDNA library (Clontech Laboratories) stock (titer, 10^{10} PFU/ml) is purified by conventional techniques[10] and used as a template for PCR. Each reaction mixture consists of specific primer, one external primer, and 300–500 ng of purified phage DNA along with deoxynucleoside triphosphates, buffer, and Taq polymerase, as per the manufacturer instructions (GeneAmp; Perkin-Elmer Cetus). Template is denatured at 94° for 1 min, primers are annealed at 55° for 1 min, and chains are polymerized at 72° for 2 min for 35 cycles with a 7-min extension at 72° added to the final cycle. A PCR product is found to hybridize specifically to a 27-bp internal oligonucleotide corresponding to bases 118–144 (TTTAGAAGAAACCTGCAGGATCCGGGC) of the rat cDNA. The band is subcloned and sequenced.

[10] J. Sambrook, E. F. Fritsch, and T. Maniatis, "Molecular Cloning: A Laboratory Manual," pp. 1–90. Cold Spring Harbor Laboratory Press, Cold Spring Harbor, New York, 1985.

Expression of Wild-Type Methylmalonate-semialdehyde
Dehydrogenase in *Escherichia coli*

A full-length cDNA encoding MMSDH is constructed from two partial cDNA clones. The 700-bp cDNA and the 1400-bp cDNA corresponding to the 5' and 3' ends of the mRNA, respectively, are digested with *Kpn*I and *Aat*II restriction endonucleases. The fragments are purified by agarose gel electrophoresis and ligated using T4 DNA ligase. The resulting construct is 2 kb, which includes coding sequence for 10 amino acids of the 32-residue leader peptide. The leader peptide and the 3'-noncoding sequence are subsequently removed by PCR amplification with primers flanking the mature coding region. The cDNA encoding mature polypeptide is ligated into the pGEX-2T expression vector and cotransfected into TG-1 *Escherichia coli* cells with the expression plasmid pGroESL (obtained as a kind gift from A. Gatenby, Du Pont, Wilmington, DE). To facilitate purification, the expression vector encodes an MMSDH–glutathione *S*-transferase fusion protein (MMSDH–GST) that is cleavable with thrombin. pGroESL carries the cDNAs for chaperonins GroES and GroEL. Although MMSDH is detected in *E. coli* in the absence of chaperonins by Western blot analysis, the resulting protein is insoluble (Fig. 1B, lane 2), and was inactive. Coexpression with chaperonins has a dramatic effect on the recovery of soluble, active MMSDH activity (Fig. 1A, lane 4). Thus, coexpression of GroEL and GroES is now routinely used in the preparation of recombinant MMSDH. TG-1 cells transfected with pGEX-2T-MMSDH and pGroESL are grown in TY medium supplemented with ampicillin (200 μg/ml) and chloramphenicol (100 μg/ml) to an OD$_{600}$ of 0.5–0.7. Induction is initiated by adding isopropyl-β-D-thiogalactopyranoside (IPTG) to 0.2 mM. Cells are harvested by centrifugation and suspended in phosphate-buffered saline (PBS) containing 0.1% (v/v) 2-mercaptoethanol and protease inhibitors phenylmethylsulfonyl fluoride (50 μg/ml), leupeptin (50 μg/ml), and benzamidine (5 mM). Cells are sonicated and centrifuged to remove the insoluble fraction. The supernatant is applied to glutathione-agarose under gravity flow. The agarose column is washed with PBS until protein cannot be detected in the flowthrough (<1 μg/ml) by the Bradford assay. The fusion protein is then eluted with 10 mM glutathione in 50 mM Tris-HCl (pH 8.0)–1 mM DTT. Free glutathione is removed by dialysis against thrombin cleavage buffer [50 mM Tris-HCl (pH 8.0), 150 mM NaCl, 2.5 mM CaCl$_2$, 0.1% (v/v) 2-mercaptoethanol]. Optimal conditions required to fully cleave the fusion protein with thrombin are determined experimentally (Fig. 2). Four milligrams of MMSDH–GST is incubated with 6 μg of thrombin at room temperature. Aliquots of the reaction mix are taken at the indicated times and the reaction is stopped by boiling in SDS–PAGE loading buffer.

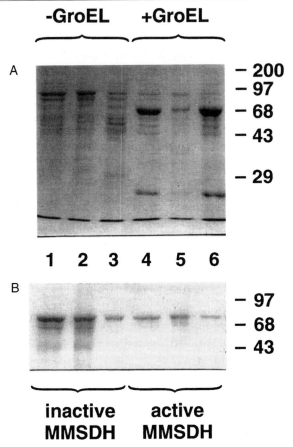

FIG. 1. Expression of MMSDH–GST fusion protein in the presence and absence of GroEL and GroES "chaperonin" proteins. Lanes 1–3, expression of MMSDH in the absence of GroEL and GroES; lanes 4–6, expression of MMSDH in the presence of GroEL and GroES; lanes 1 and 4, homogenates after sonication; lanes 2 and 5, pellet of insoluble proteins after centrifugation; lanes 3 and 6, soluble fraction. (A) SDS–PAGE analysis; (B) corresponding Western blot analysis of MMSDH expression.

Complete cleavage is achieved after 15 min of incubation with thrombin (Fig. 2). MMSDH is separated from glutathione transferase by a Mono Q column (Pharmacia, Piscataway, NJ) with a linear salt gradient [0–500 mM NaCl in 10 mM Tris-HCl (pH 8.5), 1 mM DTT]. Fractions containing active MMSDH are combined and concentrated with Amicon concentrators (molecular mass cutoff, 30 kDa). Approximately 1.3 mg of active recombinant protein with a specific activity of 0.14 U/mg as measured by a coupled assay is obtained from 1 liter of culture (Fig. 3).

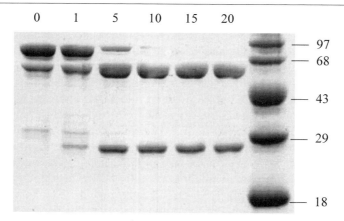

FIG. 2. SDS–PAGE analysis of MMSDH–GST cleavage with thrombin. Samples of 30 μg were taken at the indicated time points (0, 1, 5, 10, 15, and 20 min) from a mixture of 4 mg of MMSDH–GST and 6 μg of human thrombin (Sigma) incubated at room temperature, and the reaction was stopped by boiling in SDS–PAGE gel loading buffer.

Preparation and Expression of Methylmalonate-semialdehyde Dehydrogenase Mutants

Cys-285 in MMSDH[3] corresponds to Cys-243 in rat cytosolic class 3 aldehyde dehydrogenase,[4] which is conserved among all aldehyde dehydrogenases[4] and is part of the active site. To test the role of the corresponding cysteine residue in MMSDH, Cys-285 has been replaced in MMSDH with

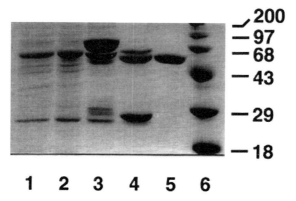

FIG. 3. Purification of recombinant MMSDH. Lane 1, cell homogenate; lane 2, soluble protein fraction; lane 3, eluate from affinity glutathione–agarose column; lane 4, cleavage of fusion protein with thrombin; lane 5, purified recombinant MMSDH after separation from GST on a Mono Q column; lane 6, molecular weight standards.

alanine by site-directed mutagenesis. Glu-268, conserved among all aldehyde dehydrogenases[4] except MMSDH, has been suggested to act as a base
to deprotonate the active site cysteine residue.[11] Asn-251 in MMSDH,
which aligns with Glu-209 in rat cytosolic class 3 aldehyde dehydrogenase,[4]
has been replaced with glutamate by site-directed mutagenesis. Mutagenesis
is performed on MMSDH cDNA cloned into single-stranded M13 vector
with a Sculptor Mutagenesis kit according to the manufacturer instructions
(Amersham, Arlington Heights, IL). The mutations are confirmed by sequencing and the resulting mutant cDNAs are cloned into pGEX-2T expression vector. Expression vectors for wild-type MMSDH, cysteine-to-alanine,
and asparagine-to-glutamate mutants are cotransfected with pGroESL into
TG-1 cells. Only wild-type MMSDH is active. Both mutant proteins are
expressed in large amounts in the soluble fraction but both are inactive.
Thus, Cys-285 is essential for MMSDH function, and substitution of asparagine in position 251 with negatively charged glutamate (as in aldehyde
dehydrogenase) also inactivates the enzyme, probably by preventing the
binding of a negatively charged substrate. Whereas Glu-268 in aldehyde
dehydrogenase may participate in the catalytic hydrolysis of the intermediate acyl enzyme,[11] in the case of MMSDH this step involves CoA and the
formation of propionyl-CoA without hydrolysis.

Characteristics of Native Methylmalonate-semialdehyde
    Dehydrogenase cDNA Encoding Rat Methylmalonate-semialdehyde
    Dehydrogenase, and Wild-Type Recombinant Methylmalonate-
    semialdehyde Dehydrogenase

The monomer molecular mass of the native enzyme, determined by
SDS–PAGE, is 58 kDa.[2] The rat cDNA encodes a mature protein of 503
amino acids, to give a molecular mass of 55,330 Da.[3] The native molecular
mass, estimated by gel filtration, is 250 kDa, suggesting a tetrameric structure.[2] The cDNA also encodes a 32-amino acid leader with characteristics
expected of a mitochondrial targeting sequence.[3] Kinetic constants for the
various substrates indicate that both malonate and methylmalonate semialdehydes are physiological substrates.[2] The purified rat liver enzyme exhibits
K_m and V_{max} values of 4.5 μM and 9.4 units/mg of protein with malonate
semialdehyde and 5.3 μM and 2.5 units/mg protein with methylmalonate
semialdehyde. The pH optimum with methylmalonate semialdehyde is
about pH 8, which is appropriate for the mitochondrial matrix space.[2]
MMSDH appears to use both stereoisomers of methylmalonate semialdehyde, but the substrate may racemize spontaneously. Nevertheless, a single

[11] X.-P. Wang and H. Weiner, *Biochemistry* **34,** 237 (1995).

semialdehyde dehydrogenase contributes to the oxidation of valine [(S)-isomer], thymine [(R)-isomer], and several compounds catabolized via β-alanine (transaminates to malonate semialdehyde).[2] Either no activity or low activity is found with the following compounds: acetaldehyde, butyraldehyde, isobutyraldehyde, α-methylbutyraldehyde, and malondialdehyde.[2] Low but nevertheless detectable activity is found with succinate semialdehyde. It is likewise found that ethylmalonate semialdehyde, an intermediate in alloisoleucine metabolism, is a poor substrate, giving rates less than 10 milliunits/mg protein.[2] That the enzyme uses this compound poorly is surprising, because an organic academia has been described[12,13] that is consistent with a defect in a single semialdehyde dehydrogenase responsible for the oxidation of malonate, methylmalonate, and ethylmalonate semialdehydes. On the other hand, evidence has been presented for an isolated deficiency in MMSDH in a child with developmental delay, mild methylmalonic aciduria without any increase in malonate, ethylmalonate, or β-alanine.[14] MMSDH is quite sensitive to inhibition by NADH (K_i of 3.1 μM, competitive with NAD$^+$).[2] Thus, an increase in the mitochondrial NADH-to-NAD$^+$ ratio in response to fatty acid oxidation may inhibit MMSDH and perhaps divert methylmalonate semialdehyde to β-aminoisobutyrate by transamination.

MMSDH possesses the catalytic ability to hydrolyze p-nitrophenyl acetate with concomitant formation of an S-acyl enzyme.[5] Deacylation of the S-acyl enzyme intermediate is the rate-limiting step in the esterase reaction, and the activity is regulated by NAD$^+$, NADH, and CoA, suggesting that coenzyme binding in the active site may induce conformational changes in the enzyme that affects the accessibility of an enzyme–thioester intermediate for deacylation.[5] Limited proteolysis studies carried out with lysyl endopeptidase C, chymotrypsin, and trypsin demonstrated NAD$^+$ protection against proteolysis in both the N-terminal and C-terminal parts of the intact enzyme.[8] Indeed, our initial efforts to purify the enzyme from rat liver were frustrated until we discovered the protection afforded by NAD$^+$.[2] On the other hand, S-acylation of the enzyme with p-nitrophenyl acetate prevents the stabilizing conformational change induced by NAD$^+$.[8] These findings may have physiological significance because they suggest that acylation may make the enzyme more susceptible to proteolysis. MMSDH is also subject to S-acylation at its active site cysteine residue by long-chain fatty

[12] R. J. Pollit, A. Green, and R. Smith, *J. Inher. Metab. Dis.* **8**, 75 (1985).

[13] K. M. Gibson, C. F. Lee, M. J. Bennett, B. Holmes, and W. L. Nyhan, *J. Inher. Metab. Dis.* **16**, 563 (1993).

[14] C. R. Roe, E. Struys, R. M. Kok, D. S. Roe, R. A. Harris, and C. Jakobs, *Mol. Gen. Metab.* **65**, 35 (1998).

acyl-CoA esters.[6] Acylation by long-chain fatty acids has been proposed to be a physiologically important mechanism for the regulation of MMSDH activity.[6,7]

Site-directed mutagenesis studies have revealed that Cys-285 is essential for MMSDH function, and that substitution of asparagine in position 251 with negatively charged glutamate (corresponds to Glu-268 in aldehyde dehydrogenase) inactivates the enzyme, probably by preventing the binding of a negatively charged substrate. Whereas Glu-268 in aldehyde dehydrogenase may participate in the catalytic hydrolysis of an intermediate S-acyl enzyme, in the case of MMSDH this step involves CoA and the formation of propionyl-CoA without hydrolysis.

Acknowledgments

This work was supported in part by National Institutes of Health Grant DK 40441 (to R.A.H.), the Diabetes Research and Training Center of Indiana University School of Medicine (AM 20542), and the Grace M Showalter Trust (to R.A.H.).

[22] Mammalian 3-Hydroxyisobutyrate Dehydrogenase

By John W. Hawes, David W. Crabb, Rebecca J. Chan, Paul M. Rougraff, Ralph Paxton, and Robert A. Harris

Introduction

The catabolism of valine differs from that of the other branched-chain amino acids (leucine and isoleucine) in that a free branched-chain acid, 3-hydroxyisobutyrate (HIBA), is formed in the pathway. This is unique in the catabolic pathways for branched-chain amino acids because it is not esterified to coenzyme A. HIBA is produced by a hydrolytic reaction catalyzed by a highly specific acyl-CoA thioesterase, 3-hydroxyisobutyryl-CoA hydrolase.[1] HIBA produced in this pathway is reversibly oxidized by another highly specific enzyme, 3-hydroxyisobutyrate dehydrogenase (HIBADH). This dehydrogenase is a member of a previously unrecognized family of enzymes, the 3-hydroxyacid dehydrogenases.[2] This family includes

[1] J. W. Hawes, J. Jaskiewicz, Y. Shimomura, B. Huang, J. Bunting, E. T. Harper, and R. A. Harris, *J. Biol. Chem.* **271,** 26430 (1996).

[2] J. W. Hawes, E. T. Harper, D. W. Crabb, and R. A. Harris, *FEBS Lett.* **389,** 263 (1996).

6-phosphogluconate dehydrogenase, D-phenylserine dehydrogenase, D-threonine dehydrogenase, and a number of hypothetical 3-hydroxyacid dehydrogenases encoded by unidentified microbial open reading frames.[3] Rabbit liver HIBADH has been purified and partially characterized.[4] Rat liver HIBADH and *Pseudomonas* HIBADH have been structurally characterized at the level of cDNA sequences[5,6] and expressed sequence tags corresponding to mouse, human, and microbial HIBADHs appear in various genetic databases. Structural and mechanistic studies of HIBADH require a ready source of recombinant enzyme that can be easily purified and manipulated by site-directed mutagenesis. Purified HIBADH is also a useful reagent for a specific assay for quantitation of 3-hydroxyisobutyrate in tissues.[7] This chapter presents methods for purification of the native rabbit liver enzyme, cloning of rat and human cDNAs, the construction of prokaryotic expression vectors to produce rat HIBADH in *Escherichia coli* as a glutathione transferase-tagged protein or a six histidine-tagged protein, and the expression of recombinant rat enzyme in *E. coli*.

Assay of 3-Hydroxyisobutyrate Dehydrogenase Activity

Materials

3-Hydroxyisobutyrate is synthesized as previously described.[4] Briefly, methyl 3-hydroxyisobutyrate (Sigma, St. Louis, MO) is added to a slurry of 12 ml of water and 6 g of Na_2CO_3 and incubated at 45° in a shaking water bath until the organic phase disappears (8–12 hr). The slurry is cooled and ice-cold 10 N H_2SO_4 (27 ml) is added dropwise. $(NH_4)_2SO_4$ is added to saturation and the 3-hydroxyisobutyrate is extracted six times with diethyl ether. The ether extract is dried with solid Na_2SO_4 and evaporated under vacuum. The residue of 3-hydroxyisobutyrate (free acid form) is resuspended in water and titrated to neutrality with DL-α-phenylethylamine. The phenylethylamine salt is dried by lyophilization and crystallized from tetrahydrofuran.

[3] J. W. Hawes, E. T. Harper, D. W. Crabb, and R. A. Harris, *Enzymol. Mol. Biol. Carbonyl Metab.* **6,** 395 (1996).

[4] P. M. Rougraff, R. Paxton, M. J. Kuntz, D. W. Crabb, and R. A. Harris, *J. Biol. Chem.* **263,** 327 (1988).

[5] P. M. Rougraff, B. Zhang, M. J. Kuntz, R. A. Harris, and D. W. Crabb, *J. Biol. Chem.* **264,** 5899 (1989).

[6] M. I. Steele, D Lorenz, K. Hatter, A. Park, and J. R. Sokatch, *J. Biol. Chem.* **267,** 13585 (1992).

[7] P. M. Rougraff, R. Paxton, G. W. Goodwin, R. G. Gibson, and R. A. Harris, *Anal. Biochem.* **184,** 317 (1990).

All other materials required for the spectrophotometric assay are obtained from Sigma.

Spectrophotometric Assay of 3-Hydroxyisobutyrate Dehydrogenase Activity

HIBADH is assayed spectrophotometrically at 30° in a solution containing 50 mM glycine-K (pH 10), 1 mM NAD$^+$, 1 mM dithiothreitol (DTT), 1 mM EDTA-K, and 2 mM (S)-3-hydroxyisobutyrate.[4] The rate of NADH production is followed at 340 nm after initiation of the reaction with (S)-3-hydroxyisobutyrate. For kinetic studies, the hydroxyisobutyrate substrate is quantitated by an enzymatic end-point assay.[7] To measure the end point of HIBA oxidation the preceding assay conditions are used with the inclusion of 130 mM hydrazine sulfate. Absorbance at 340 nm is measured until no further NADH production occurs.

Native Rabbit Liver 3-Hydroxyisobutyrate Dehydrogenase

Materials

Frozen rabbit livers are purchased from Pel-Freez (Rogers, AR). CM-Sepharose is purchased from Pharmacia (Piscataway, NJ). Affi-Gel blue is purchased from Bio-Rad Laboratories (Hercules, CA). Ultrogel AcA-34 is purchased from LKB (Bromma, Sweden). All other materials are purchased from Sigma.

Procedure for Purification of Rabbit Liver 3-Hydroxyisobutyrate Dehydrogenase

In this procedure,[4] frozen ($-70°$) rabbit liver ($\sim$1.2 kg) is homogenized until a smooth mixture is obtained in a Warning blender with 2 volumes (w/v) of ice-cold 50 mM Tris-HCl, pH 8.0, containing 0.1% (v/v) Triton X-100, 2 mM EDTA-K, 2 mM DTT, 2% (v/v) bovine serum, 0.5 μM leupeptin, 0.5 μM pepstatin A, aprotonin (1 μg/ml), trypsin inhibitor (25 μg/ml), and phenylmethylsulfonyl fluoride (PMSF, 25 μg/ml). This mixture is homogenized a second time with a Polytron PT-10 homogenizer for 1 min at a setting of 4 in 0.3-liter portions. All subsequent procedures are done between 4 and 8°. The homogenates are combined and centrifuged for 30 min at 9000 rpm in a Sorvall (Newtown, CT) GS-3 rotor. The supernatant is made 6.5% (w/v) in polyethylene glycol (PEG 8000) by slow addition of an ice-cold 50% (w/v) solution and slowly mixed for 30 min before centrifugation of the mixture as described above. The supernatant is made to pH 8.0 with 2 M Tris and applied to a column of DEAE-Sephacel (4 $\times$ 36 cm) equilibrated with buffer A [30 mM Tris-HCl (pH 8.0), 1 mM dithiothreitol]. Nonadsorbed proteins are eluted with 0.5 liter of buffer A before the enzyme is eluted with a 1.5-liter linear KCl gradient from 0 to 0.5 M in buffer A. Fractions with HIBADH activity are pooled and made 35% saturated with ammonium sulfate, and centrifuged for 20 min at 13,000 rpm in a Sorvall GSA rotor. The resulting supernatant is made 65% saturated with ammonium sulfate and centrifuged as described above. The precipitate is gently resuspended in 30 ml of buffer A to produce an ammonium sulfate concentration of 1 M (measured by conductivity), and the pH is readjusted to pH 8.0. This sample, clarified by centrifugation at 10,000 rpm for 15 min in a Sorvall SS-34 rotor, is applied to a column of phenyl-Sepharose (5.4 × 14 cm) previously equilibrated in buffer A containing 1 M ammonium sulfate. The column is washed with successive solutions of 0.6 liter each of buffer A containing 1 M ammonium sulfate, buffer A containing 0.5 M ammonium sulfate, and finally with buffer A alone. HIBADH activity elutes with the last solution. Pooled fractions are made to 30 mM MES [2-(N-morpholino)ethane sulfonate] by addition of 1 M MES and adjusted to pH 5.5 by slow addition of 10% (w/v) acetic acid. This sample is applied to a carboxymethyl-Sepharose column (4 × 15 cm) previously equilibrated in buffer B [30 mM MES (pH 5.5), 1 mM dithiothreitol]. The column is washed with buffer B and the enzyme eluted with a 1-liter linear KCl gradient from 0 to 275 mM in buffer B. The fractions with HIBADH activity are pooled and made to 30 mM Tris-HCl, pH 8.0, and applied to a column of Affi-Gel blue (1 × 27 cm) previously equilibrated in buffer A. The column is washed with successive solutions of buffer A alone, buffer A containing 0.3 M KCl, and buffer A containing 0.3 M KCl and 0.2 mM NADH. Fractions containing HIBADH activity are pooled and concentrated by ammonium sulfate precipitation as described above. The ammonium sulfate precipitate is resuspended in 2 ml of buffer A containing 2% (v/v) glycerol and applied to a column of Ultrogel AcA-34 (1.5 × 100 cm) previously equilibrated in the same buffer. Fractions containing HIBADH activity are pooled, and applied to a small (1.5 × 1.5 cm) Affi-Gel blue column and eluted with buffer A made to 0.3 M KCl and 1 mM NADH. The concentrated sample is dialyzed twice against 1 liter of a solution of 50 mM HEPES-K (pH 7.5), 40% (v/v) glycerol, 5 mM dithiothreitol, and 0.5 mM EDTA. HIBADH activity is stable for at least 12 months when stored at $-70°$ in this buffer. The procedure provides an 1800-fold purification with a yield of about 20% of the enzyme present in rabbit liver. The enzyme corresponds to a single protein band on sodium dodecyl sulfate–polyacrylamide gel electrophoresis (SDS–PAGE) (Coomassie blue staining) with an apparent molecular weight of 34,000.

Recombinant Rat 3-Hydroxyisobutyrate Dehydrogenase

Materials

A λ gt11 rat liver cDNA library is purchased from Clontech Laboratories (Palo Alto, CA). pBluescript KS[+] and VCS M13 are purchased from Stratagene (La Jolla, CA). pET28a is purchased from Novagen (Madison, WI). pGEX-KG is a kind gift from J. Dixon (University of Michigan, Ann Arbor, MI). Nickel-NTA-agarose is purchased from Qiagen (Valencia, CA). Restriction enzymes and restriction enzyme buffers are purchased from BRL (Gaithersburg, MD). EST T67941 is purchased from the American Type Culture Collection (ATCC, Rockville, MD).

Rat and Human 3-Hydroxyisobutyrate Dehydrogenase cDNAs

A 1.7-kb cDNA clone encoding rat HIBADH is obtained by screening a rat liver λ gt11 library with a 17-base oligonucleotide probe corresponding to a portion of the N-terminal amino acid sequence of the rabbit liver enzyme.[5] The cDNA contains an open reading frame of 1038 base pairs (bp) that encodes a 300-amino acid mature protein but lacks a full-length mitochondrial leader sequence. Several human cDNAs corresponding to HIBADH appear in the GenBank databases. One of these (accession number T67941) can be purchased from the ATCC. Clone T67941 contains a full-length mitochondrial leader sequence and an open reading frame encoding the entire 300-amino acid mature protein. The deduced amino acid sequence for human HIBADH is 91% identical to that of the rat.

Construction of Vectors for Prokaryotic Expression of Glutathione S-Transferase-Tagged and Hexahistidine-Tagged 3-Hydroxyisobutyrate Dehydrogenase

Digestion with XbaI and partial digestion with DdeI of the rat liver HIBADH cDNA yields a 962-bp DNA fragment that contains the entire coding region of the mature HIBADH minus the first three residues of the N terminus (Ala-Ser-Lys). These missing amino acids as well as a BamHI site can be added to the construct with two complementary oligonucleotides (5'-GATCCGGTGGTGGTGGTGGTCATATGGCTTC-3' and 5'-TTA-GAAGCCATATGACCACCACCACCACCG-3'), which are phosphorylated with T4 polynucleotide kinase, annealed, and ligated to the DdeI–XbaI fragment. The resulting full-length cDNA is gel purified and ligated to the BamHI and XbaI sites of vector pGEX-KG,[8] placing the HIBADH sequence in frame with glutathione S-transferase (GST), and separating them

---

[8] K. L. Guan and J. E. Dixon, Anal. Biochem. 192 (1991).

by the "kinker" peptide Gly-Gly-Gly-Gly-Met and a thrombin cleavage site. The fidelity of this construct is verified by dideoxynucleotide DNA sequencing.

For expression of rat HIBADH with an N-terminal hexahistidine tag, the cDNA from pGEX-HIBADH is isolated by digestion with *Nde*I and *Hind*III. This cDNA fragment is purified by agarose gel electrophoresis and ligated to the *Nde*I and *Hind*III sites of vector pET28a, placing an in-frame six-histidine repeat and a thrombin cleavage site directly before the codon for the N-terminal alanine residue of HIBADH. The sequence of this construct is verified by DNA sequencing.

Expression and Purification of Glutathione S-Transferase-Tagged 3-Hydroxyisobutyrate Dehydrogenase

Wild-type and mutant forms of HIBADH are expressed in *E. coli* and purified by glutathione affinity chromatography.[9] The plasmid pGEX-HIBADH is transformed into TG1 *E. coli* cells that are cultured in TY medium containing ampicillin (100 μg/ml) Three liters of culture is used to purify each form (wild-type and mutants) of HIBADH. For expression of the GST–HIBADH fusion protein, 3 liters of the recombinant TG1 *E. coli* culture is grown at 37° until the optical density of the culture at 600 nm reaches about 1.0. Isopropyl-β-D-thiogalactopyranoside (IPTG) is added to a final concentration of 0.5 mM and the cells cultured for another 15 to 18 hr. The cells are pelleted by centrifugation and resuspended in 10 volumes of buffer A [138 mM NaCl, 2.7 mM KCl, 1.2 mM KH_2PO_4, 8.1 mM Na_2HPO_4, 0.5% (v/v) Triton X-100, 0.2 mM NAD^+, 10 mM 2-mercaptoethanol, and 100 μg/ml each of benzamidine and phenylmethylsulfonyl fluoride]. Cells are lysed by sonification with a sonifier cell disrupter W185 (Branson Sonic Power, Plainview, NY). The resulting homogenates are centrifuged at 20,000g for 30 min at 4° and the extracts are immediately used to purify the recombinant fusion proteins by immobilized glutathione affinity chromatography according to the method of Guan and Dixon.[8] The purified GST–HIBADH fusion proteins are cleaved with thrombin at a ratio of 1 μg of thrombin per 2 mg of fusion protein,[8] and filtered through fresh glutathione-agarose. The second glutathione-agarose fractions (flow-through) are applied to 1 ml of Reactive Red 120-agarose previously equilibrated in thrombin buffer [50 mM Tris-HCl (pH 8.0), 150 mM NaCl, 2.5 mM $CaCl_2$, and 10 mM 2-mercaptoethanol]. The column is washed with 25 volumes of thrombin buffer followed by elution with buffer containing 0.5 M NaCl and 10 mM NAD^+. The fractions containing protein are pooled,

[9] J. W. Hawes, D. W. Crabb, R. M. Chan, P. Rougraff, and R. A. Harris, *Biochemistry* **34,** 4231 (1995).

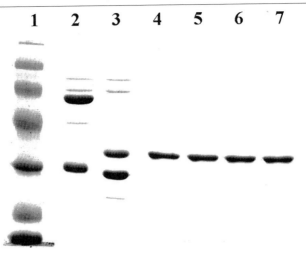

Fig. 1. SDS–PAGE analysis of purified wild-type and mutant HIBADHs. Lane 1, molecular weight standards (i.e., myosin, phosphorylase *b*, BSA, ovalbumin, carbonic anhydrase, β-lactoglobulin, and lysozyme); lane 2, 5 μg of wild-type GST–HIBADH; lane 3, 5 μg of wild-type GST–HIBADH after thrombin cleavage; lanes 4–7, 2 μg of purified HIBADH (wild-type, Y162F, Y162A, Y32F). Proteins were reduced, denatured, and electrophoresed through a standard 10% (w/v) polyacrylamide gel, and then stained with Coomassie blue. [Reprinted with permission from J. W. Hawes, D. W. Crabb, R. M. Chan, P. Rougraff, and R. A. Harris, *Biochemistry* **34,** 4231 (1995). Copyright © 1995 American Chemical Society.]

concentrated by ultrafiltration, dialyzed against 4 liters of buffer A, and stored at 4°. The purified enzyme is stable for more than 2 months. Figure 1 shows an SDS–PAGE analysis of purified GST–HIBADH, thrombin cleaved GST–HIBADH, and Reactive Red-agarose-purified HIBADH (wild-type and mutant forms).

Expression and Purification of Hexahistidine-Tagged 3-Hydroxyisobutyrate Dehydrogenase

HIBADH is also readily expressed as a hexahistidine-tagged enzyme from the pET28a construct described above. This protein is expressed in BL21(DE3) *E. coli* cells and purified in one step from the recombinant *E. coli* extracts by nickel chelation chromatography using nickel-NTA agarose.[2] Three liters of the recombinant BL21(DE3) *E. coli* culture is grown at 37° until the optical density of the culture at 600 nm reaches about 1.0. IPTG is added to a final concentration of 0.5 m*M* and the cells are cultured for another 15 to 18 hr. The cells are pelleted by centrifugation and resuspended in 10 volumes of lysis buffer [20 m*M* Tris-HCl (pH 8.0), 0.5 *M* NaCl, 10 m*M* 2-mercaptoethanol, 5 m*M* imidazole, 0.5% (v/v) Triton X-

100, and 100 μg/ml each of benzamidine and phenylmethylsulfonyl fluoride]. Cells are lysed by sonication and the resulting homogenates centrifuged at 20,000g for 30 min at 4°. The soluble fraction is applied to a 10-ml nickel-NTA (Qiagen) column previously equilibrated in the same buffer. The column is washed with 5 volumes of buffer containing 20 mM imidazole, 5 volumes of buffer containing 40 mM imidazole, and 5 volumes of buffer containing 60 mM imidazole. The histidine-tagged protein is eluted with buffer containing 200 mM imidazole. Elution fractions containing HIBADH activity are pooled, brought to 20% (v/v) glycerol, and stored at −70°. The advantage of the pET28 expression system is that the protein can be purified in one step, as opposed to the GST system which requires a second affinity chromatography step.

Site-Directed Mutagenesis of 3-Hydroxyisobutyrate Dehydrogenase

For site-directed mutagenesis, the *Bam*HI–*Xba*I fragment of pGEX-HIBADH is isolated and subcloned into the *Bam*HI and *Xba*I sites of the phagemid pBluescript II KS$^+$. Single-stranded DNA is isolated from this clone by using VCS-M13 helper phage according to the manufacturer (Stratagene) protocol. Synthetic oligonucleotides are designed to direct the synthesis of desired mutations, and site-directed mutagenesis is performed with the Sculptor system (Amersham, Arlington Heights, IL), according to the protocol suggested by the manufacturer. All mutant DNAs are analyzed by double-stranded sequencing, using dideoxynucleotide DNA sequencing methods. Mutant HIBADH cDNAs are excised from pBluescript with *Bam*HI and *Xba*I and religated to the *Bam*HI and *Xba*I sites of either pGEX-KG or pET28a.

Circular Dichroic Spectrapolarimetry of Recombinant 3-Hydroxyisobutyrate Dehydrogenase

Mutant and wild-type enzymes are analyzed for differences in secondary structure by circular dichroic (CD) spectrapolarimetry. For such analysis, wild-type and mutant forms of HIBADH are concentrated to 1 mg of protein per milliliter, using Centricon 30 (Amicon, Danvers, MA) concentrators. The concentrated samples are exhaustively dialyzed against a buffer containing 138 mM NaCl, 2.7 mM KCl, 1.2 mM KH$_2$PO$_4$, 8.1 mM Na$_2$HPO$_4$, and 1 mM 2-mercaptoethanol. CD spectra are recorded with a Jasco (Easton, MD) J-720 spectrapolarimeter. Concentrated, dialyzed enzyme is diluted to a protein concentration of 100 μg/ml. Spectra are recorded at 21° with a cell path length of 0.1 cm and a wavelength range of 300–190 nm. Secondary structure contents are estimated with the reference spectra of

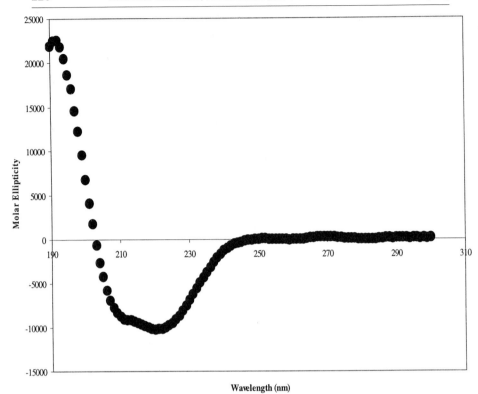

FIG. 2. CD spectra of purified wild-type HIBADH.

Yang et al.,[10] and the SSE 338 program (Japan Spectroscopic, Tokyo, Japan). Secondary structure contents of wild-type and mutant proteins are also compared by examination of the ratio of molar ellipticities at 208 and 222 nm. Figure 2 shows the CD spectra of wild-type HIBADH. Mutant and wild-type enzymes were analyzed for similarities in secondary structure by CD spectrapolarimetry. None of the mutant HIBADHs produced so far exhibit significant deviation from the wild-type spectrum. Secondary structure contents of wild-type HIBADH are estimated to be 33% α helix and 45% β sheet.

Characteristics of Native, Wild-type Recombinant, and Mutant 3-Hydroxyisobutyrate Dehydrogenase

Specific activity of purified native HIBADH at saturating substrate concentrations is about 10 μmol/min/mg of protein at 30°. The K_m for (S)-

[10] J. T. Yang, C. S. C. Wu, and H. M. Martinez, *Methods Enzymol.* **130,** 208 (1986).

3-hydroxyisobutyrate is approximately 61 μM and the K_m for NAD$^+$ is 23 μM. The native enzyme was also found to be active with 3-hydroxypropionate. HIBADH is particularly sensitive to product inhibition by NADH (K_i of 5.7 μM), and this may play a role in the physiological regulation of this enzyme. 3-Hydroxyisobutyrate can be released by certain tissues such as muscle and subsequently used for hepatic gluconeogenesis.[11] This process may be promoted under conditions that produce an elevated mitochondrial NADH/NAD$^+$ ratio such as starvation and diabetes.

Each form of recombinant HIBADH was assayed spectrophotometrically and characterized by bisubstrate kinetics as previously described.[2,9] Wild-type HIBADH produced as the GST-tagged protein displayed a V_{max} of 10 μmol/min/mg of protein, a K_m for (S)-HIBA of 59 μM, and a K_m for NAD$^+$ of 24 μM. Histidine-tagged HIBADH has enzymatic properties and stability similar to the protein expressed and purified from the GST fusion construct.[2,9] The recombinant enzyme has a clearly defined substrate specificity. The enzyme is active with (S)-HIBA (K_m = 59 μM), L-glycerate (K_m = 910 μM), and L-serine (K_m = 21 μM). Each of these substrates are 3-hydroxyacids with three carbon backbones. Although the K_m for each of these substrates differs, each reacts with a V_{max} of 8 to 10 units/mg of protein.

The purity and fidelity of the expressed recombinant enzyme are also analyzed by mass spectrometry. Figure 3 shows the deconvoluted electrospray ionization (ESI) mass spectrum of purified wild-type HIBADH with a measured mass of 33,790.0 compared with the calculated mass of 33,797.8 (a 0.023% difference).

Native and recombinant forms of HIBADH are sensitive to inactivation by cysteine modifiers.[4,9] Cysteine scanning mutagenesis has revealed that five of the six cysteine residues are dispensable whereas C215 appears to be critically important to catalysis by HIBADH. C215A and C215D mutant HIBADHs are enzymatically inactive under all conditions tested, whereas C215S HIBADH is active but displays a fivefold decrease in V_{max}.[9] Alanine substitution of C39, located near the NAD$^+$-binding domain, results in a small decrease in V_{max}. Alanine substitution of C163 results in a sevenfold increase in the K_m for S-HIBA.

HIBADH is also sensitive to treatment with tetranitromethane and N-acetylimidazole, chemical modifiers of tyrosine residues, suggesting the importance of a tyrosine residue (or residues) in catalysis.[9] Alanine substitution of Tyr-162 results in inactive enzyme, whereas a phenylalanine substitution of this residue has no effect on the activity or kinetic parameters of the enzyme. Phenylalanine substitution of Tyr-32, a residue in the NAD$^+$-binding domain, does not affect enzymatic activity or kinetic parameters. Two of these residues (Y162 and C163) directly precede a stretch of highly

[11] S.-H. C. Lee and E. J. Davis, *Biochem. J.* **233,** 621 (1986).

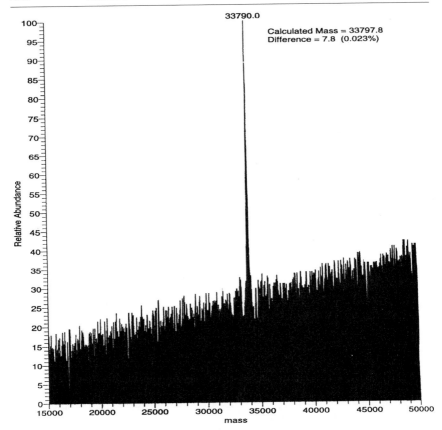

FIG. 3. Electrospray ionization mass spectrum of purified wild-type HIBADH. A sample of 250 μg of protein was desalted by reversed-phase HPLC, using a divinylbenzene–polystyrene copolymer column. The spectrum was recorded with a Finnigan MAT LCQ spectrometer with nitrogen as sheath gas, a spray voltage of 4 V, a capillary voltage of 12 V, and capillary temperature of 200°. The spectrum was deconvoluted with Bioworks software (Finnigan MAT).

conserved amino acid sequence that was proposed as the catalytic domain of HIBADH and 6-phosphogluconate dehydrogenase.[2]

Acknowledgments

This work was supported by grants from the U.S. Public Health Services (NIH DK40441 to R.A.H.), the Diabetes Research and Training Center of Indiana University School of Medicine (AM 20542), and the Grace M. Showalter Trust.

[23] 3-Hydroxyisobutyryl-CoA Hydrolase

By Yoshiharu Shimomura, Taro Murakami, Naoya Nakai,
Boli Huang, John W. Hawes, and Robert A. Harris

Introduction

The mitochondrial enzyme 3-hydroxyisobutyryl-CoA (HIB-CoA) hydrolase (3-hydroxy-2-methylpropanoyl-CoA hydrolase, EC 3.1.2.4) catalyzes the hydrolysis of (*S*)-HIB-CoA, an intermediate in the pathway of valine catabolism. The reaction catalyzed by this particular enzyme makes the valine catabolic pathway unique in that it produces a readily diffusible monocarboxylic acid, (*S*)-3-hydroxyisobutyric acid, from a CoA ester. This seems paradoxical because subsequent steps of the pathway again involve CoA ester intermediates. However, hydrolysis of HIB-CoA may be an important strategy to protect cells against toxic effects of methacrylyl-CoA,[1,2] an intermediate in the valine pathway occurring immediately upstream of HIB-CoA. Methacrylyl-CoA is a thiol-reactive molecule that would probably inactivate numerous enzymes in the absence of a mechanism designed to minimize its intramitochondrial concentration.[1-3] This likely accounts for why the activities of methacrylyl-CoA hydratase (crotonase) and HIB-CoA hydrolase are much higher in rat,[2,4] dog,[5] and human liver[4] than the branched-chain α-keto acid dehydrogenase complex, the rate-limiting enzyme of valine carabolism.

Cloning studies have revealed that HIB-CoA hydrolase is not a member of the esterase family of enzymes.[6] The deduced amino acid sequence of HIB-CoA hydrolase shows no similarity to well-known thioesterases, such as the fatty acyl-CoA thioesterases, and the catalytic triad common to the esterase family of enzymes[7] is not present in HIB-CoA hydrolase. Instead,

---

[1] G. K. Brown, M. S. Hunt, R. Scholem, K. Fowler, A. Grimes, J. F. B. Mercer, R. M. Truscott, R. G. H. Cotton, J. G. Rogers, and D. M. Danks, *Pediatrics* **70**, 532 (1982).

[2] Y. Shimomura, T. Murakami, N. Fujitsuka, N. Nakai, Y. Sato, S. Sugiyama, N. Shimomura, J. Irwin, J. W. Hawes, and R. A. Harris, *J. Biol. Chem.* **269**, 14248 (1994).

[3] T. W. Speir and E. A. Barnsley, *Biochem. J.* **125**, 267 (1971).

[4] K. Taniguchi, T. Nonami, A. Nakao, A. Harada, T. Kurokawa, S. Sugiyama, N. Fujitsuka, Y. Shimomura, S. M. Hutson, R. A. Harris, and H. Takagi, *Hepatology* **24**, 1395 (1996).

[5] T. Ooiwa, H. Goto, Y. Tsukamoto, T. Hayakawa, S. Sugiyama, N. Fujitsuka, and Y. Shimomura, *Biochim. Biophys. Acta* **1243**, 216 (1995).

[6] J. W. Hawes, J. Jaskiewicz Y. Shimomura, B. Huang, J. Banting, E. T. Harper, and R. A. Harris, *J. Biol. Chem.* **271**, 26430 (1996).

[7] M. Pazirandeh, S. S. Chirala, and S. J. Wakil, *J. Biol. Chem.* **266**, 20946 (1991).

the protein shows considerable sequence similarity to members of the enoyl-CoA/isomerase family of enzymes.[6] Indeed, the highly active substrates for HIB-CoA hydrolase, HIB-CoA and 3-hydroxypropionyl-CoA, are structurally related to specific substrates of enoyl-CoA hydratase. However, neither native nor recombinant HIB-CoA hydrolase exhibits hydratase or isomerase activity.[2,6]

This chapter reports methods for the assay of HIB-CoA hydrolase, its purification from rat liver, and its generation as a recombinant protein, along with a descriptive summary of the properties of the enzyme.

Assay Methods for 3-Hydroxyisobutyryl-CoA Hydrolase

Principle

The activity of HIB-CoA hydrolase is assayed spectrophotometrically by monitoring the reaction of 5,5'-dithiobis(2-nitrobenzoic acid) (DTNB) with free CoA formed by the hydrolysis of HIB-CoA. DTNB reacts with aliphatic thiol compounds in slightly alkaline medium to produce 1 mol of p-nitrothiophenol anion (extinction coefficient, 13.6 liter mmol^{-1} cm^{-1} at 412 nm) per mole of thiol.[8] Crotonase is used to generate HIB-CoA from methacrylyl-CoA in the assay cuvette. Crotonase can also be used to generate 3-hydroxypropionyl-CoA, another good substrate for HIB-CoA hydrolase, from acrylyl-CoA in the assay cuvette.

Reagents

Assay buffer (1.25×): 125 mM Tris-HCl, 1.25 mM EDTA, and 0.125% (w/v) Triton X-100 (pH 8.0 at 30°)

DTNB: 1 mM in 0.1 M Tris-HCl, pH 8.0 (final concentration in assay, 0.1 mM) Crotonase from bovine liver (Sigma, St. Louis, MO), 1000 units/ml (final concentration in assay, 10 units/ml): One unit of crotonase hydrates 1.0 μmol of crotonyl-CoA to 3-hydroxybutyryl-CoA/min at pH 7.5 and 25°

Methacrylyl-CoA, 10 mM (final concentration in assay, 0.2 mM): Methacrylyl-CoA is prepared according to the method of Stern.[9] Approximately 50 mg of CoA is dissolved in 4 ml of ice-cold water plus 0.4 ml of 1 M KHCO$_3$ in a glass test tube with a ground glass stopper. The CoA concentration of this solution is determined by mixing 5 μl with 0.995 ml of 0.1 M Tris-HCl (pH 8.0) containing 0.2 mM DTNB. Methacrylic anhydride (or methacryloyl chloride) (Aldrich,

[8] G. L. Ellman, *Arch. Biochem. Biophys.* **82,** 70 (1959).
[9] J. R. Stern, *Methods Enzymol.* **1,** 559 (1955).

Milwaukee, WI), 20 μl, is added to the solution with vigorous mixing. After the solution has been left on ice for 10 min, free CoA is again determined. Less than 2% of the original CoA should remain. If more than that remains, further additions of methacrylic anhydride should be made until the residual CoA is nearly undetectable. The mixture is then acidified to pH 2–3 with concentrated HCl and extracted twice with 4 ml of water-saturated diethyl ether to remove free methacrylic acid. Ether is aspirated from the tube after centrifugation at 2500 rpm for 2 min. The tube is placed in a desiccator connected to an aspirator for approximately 20 min to remove much of the residual diethyl ether. The methacrylyl-CoA concentration is estimated from its absorbance at 260 nm, using an extinction coefficient of 16.4 liter mmol^{-1} cm^{-1}. The concentration is adjusted to 10 mM before dividing the solution into several tubes and storage at $-80°$. Acrylyl-CoA is prepared by the same method described above except that 50 μl of acryloyl chloride (Aldrich) and 1.2 ml of 1 M KHCO$_3$ are used.

Procedure

The final volume of the assay mixture is 1 ml. Eight volumes of 1.25$\times$ assay buffer, 1 volume of 1 mM DTNB, and 0.5 volume of water are mixed and prewarmed to 30°. The mixture (0.95 ml), 20 μl of 10 mM methacrylyl-CoA, and 10 μl of crotonase (1000 units/ml) are mixed in a cuvette. The baseline is monitored at 412 nm before starting the reaction with 20 μl of enzyme solution. The rate of the reaction is established by following the absorbance change at 412 nm for 2 min. One unit of hydrolase activity catalyzes the formation of 1 μmol of CoA per minute.

Purification of 3-Hydroxyisobutyryl-CoA Hydrolase from Rat Liver

Buffers and Reagents

Buffer A: 50 mM potassium phosphate containing 0.1 mM EDTA, 0.1 μM leupeptin, and trypsin inhibitor (10 μg/ml) (pH 7.5 at 4°).

Buffer B: 10 mM Tris-HCl containing 0.1 mM EDTA (pH 8.5 at 4°)

Buffer C: 10 mM potassium citrate containing 0.1 mM EDTA (pH 5.6 at 4°)

Buffer D: 25 mM potassium phosphate containing 0.1 mM EDTA (pH 7.5 at 4°)

Buffers A–D are prepared as 2$\times$ stock solutions.

Phenyl-Sepharose CL-4B (Pharmacia LKB Biotechnology, Piscataway, NJ)

Octyl-Sepharose CL-4B (Pharmacia LKB Biotechnology)
DEAE-Sephacel (Pharmacia LKB Biotechnology)
CM-Sepharose CL-6B (Pharmacia LKB Biotechnology)
Hydroxylapatite, Bio-Gel HTP (Bio-Rad, Hercules, CA), suspended
    in water for at least 1 day before use. Small particles are removed
    by decantation
Sephacryl S-200 HR (Pharmacia LKB Biotechnology)
CoA-Sepharose, prepared from CoA-SH and thiopropyl-Sepharose
    6B (Pharmacia LKB Biotechnology) according to the instructions
    provided by Pharmacia
Ultrafilter with a YM10 membrane (Amicon, Danvers, MA)

Procedure

Homogenization and Ammonium Sulfate Fractionation. All procedures
are performed at 0–4°. Frozen rat liver (180 g) is homogenized in an Oster
blender (household type) at full speed for 4 min in 720 ml of buffer A
containing 1 mM EDTA, 1% (v/v) bovine serum, 10 μM N-tosyl-L-phenylalanine chloromethyl ketone (TPCK), 0.1 mM phenylmethylsulfonyl fluoride
(PMSF), and 0.5%, (w/v) Triton X-100.[10] The homogenate is centrifuged
at 9500g for 15 min. The supernatant is passed through four layers of
cheesecloth and the pH is adjusted to 7.5 with 2 M Tris. The supernatant
is made to 30% saturation in ammonium sulfate (176 g/liter) by slow addition of the solid salt with constant stirring and allowed to stir for 20 min
before centrifugation at 9500g for 20 min. Fat is effectively removed by
this step. The supernatant is passed through four layers of cheesecloth,
made to 45% saturation in ammonium sulfate (addition of 94 g/liter) as
described above, and allowed to stir for 20 min before centrifugation at
9500g for 30 min. The supernatant obtained is made to 75% saturation in
ammonium sulfate (addition of 210 g/liter) as described above and allowed
to stir for 20 min before centrifugation at 9500g for 20 min. The pellet is
dissolved in approximately 100 ml of buffer A and stored at −80°. The
procedure is repeated with a second batch of 180 g of liver to give two
crude preparations of the enzyme before proceeding to the next step.

Phenyl-Sepharose and Octyl-Sepharose Column Chromatography. The
two preparations are thawed, combined, and applied to a phenyl-Sepharose
gel (350 ml) equilibrated with buffer A containing 1 M ammonium sulfate
on a sintered-glass filter funnel (diameter, 9 cm.) The gel is washed with
the same buffer and a fraction containing the HIB-CoA hydrolase activity
not bound to the gel is collected. Solid ammonium sulfate, 382 g/liter (final,

---

[10] Triton X-100 and bovine serum are added to this buffer first. TPCK and PMSF are subsequently added from 10 and 100 mM stock solutions in ethanol, respectively.

approximately 80% saturation), is added slowly to the fraction with constant stirring, followed by stirring for another 20 min before centrifugation at 9500g for 20 min. The pellet is dissolved in approximately 60 ml of buffer A and applied to an octyl-Sepharose gel (approximately 100 ml) equilibrated with buffer A containing 1 M ammonium sulfate on a sintered-glass filter funnel (diameter, 6 cm). The gel is washed with the same buffer, and a fraction containing the hydrolase activity not bound to the gel is collected and concentrated by ammonium sulfate precipitation as described above.

DEAE-Sephacel Column Chromatography. The pellet obtained is dissolved in approximately 35 ml of buffer B and dialyzed against 1 liter of the same buffer for 18 hr with two buffer changes. Aggregated protein formed during dialysis is removed by centrifugation at 9500g for 10 min. The clear dialysate is applied to a DEAE-Sephacel column (2.5 × 15 cm) equilibrated with buffer B. The column is washed with the same buffer at a flow rate of approximately 65 ml/hr until the absorbance of the eluate at 280 nm decreases to almost zero (approximately 300 ml). The enzyme is then eluted from the column with buffer B containing 0.1 M NaCl.

CM-Sepharose Column Chromatography. The fractions with the hydrolase activity are combined and dialyzed against 2 liters of buffer C overnight. After removal of aggregated proteins as described above, the dialysate is applied to a CM-Sepharose column (2.5 × 8 cm) equilibrated with the same buffer. The column is washed with the buffer at a flow rate of approximately 75 ml/hr until the absorbance at 280 nm of the eluate is close to zero (approximately 220 ml) and then eluted with buffer C containing 0.1 M NaCl.

Hydroxylapatite Column Chromatography. The fractions with hydrolase activity are combined and concentrated to approximately 5 ml by ultrafiltration (YM10 membrane). The concentrate is diluted to 25 ml with buffer D and again concentrated. This dilution and concentration cycle is repeated again to change the buffer. The concentrate obtained is applied to a hydroxylapatite column (1.5 × 4 cm) equilibrated with buffer D. The column is washed with the buffer at a flow rate of 20 ml/hr until the absorbance of the eluate at 280 nm decreases to almost zero (approximately 100 ml). The enzyme is then eluted from the column with buffer D containing 50 mM potassium phosphate (total, 75 mM potassium phosphate). The hydrolase is eluted after a major protein peak in this step (Fig. 1). The main activity fractions are combined, and Tween 20 is added to 0.01% (w/v). A nonionic detergent must be added to all buffers after this step to minimize loss of the enzyme by adsorption to plastic surfaces.

First Sephacryl S-200 Column Chromatography. After concentration to approximately 1.5 ml by ultrafiltration, the preparation is made 10% (v/v) in glycerol and applied to a Sephacryl S-200 column (2.5 × 46.5 cm)

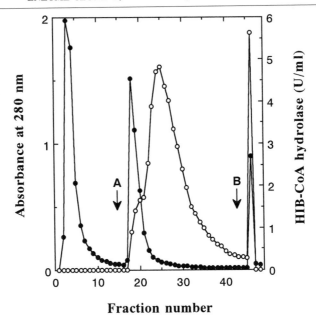

Fraction number

Fɪɢ. 1. Elution profile of HIB-CoA hydrolase purified by hydroxylapatite column chromatography. (○) HIB-CoA hydrolase activity, (●) absorbance at 280 nm. The column was eluted
with buffer B containing a total potassium phosphate concentration of 75 mM (arrow A)
followed by buffer B containing 500 mM potassium phosphate (arrow B). Fractions 22–44
were recovered for further purification. [Adapted with permission from Y. Shimomura, T.
Murakami, N. Fujitsuka, N. Nakai, Y. Sato, S. Sugiyama, N. Shimomura, J. Irwin, J. W. Hawes,
and R. A. Harris, *J. Biol. Chem.* **269,** 14248 (1994).]

equilibrated with 50 mM potassium phosphate containing 0.1 mM EDTA,
0.1 M KCl, and 0.05% (w/v) Tween 20 (pH 7.5 at 4°). The column is eluted
with the buffer at a flow rate of 17.5 ml/hr. The fractions with relatively
high specific activity of the hydrolase (fractions 38 and 39 in Fig. 2) are
combined and concentrated to approximately 2 ml by ultrafiltration. It is
important to combine and recover only fractions with ratios of hydrolase
activity to absorbance at 280 nm of greater than 200.

CoA-Sepharose Column Chromatography. The buffer of the preparation
is changed to buffer C containing 0.05% (w/v) Tween 20 by the ultrafiltration
method described above. The preparation is applied to a CoA-Sepharose
column (1 ml) equilibrated with the same buffer. The column is washed
with 14 ml of the buffer at a flow rate of 30 ml/hr and eluted with buffer
C containing 0.05% (w/v) Tween 20 and 0.1 M NaCl. The column is further
eluted with the same buffer containing 0.5 M NaCl to remove proteins
bound to the column and equilibrated again with buffer C containing 0.05%

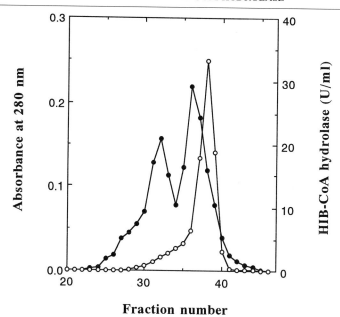

FIG. 2. Elution profile of HIB-CoA hydrolase purified by Sephacryl S-200 column chromatography. (○) HIB-CoA hydrolase activity; (●) absorbance at 280 nm. Fractions 38 and 39 were recovered for further purification. HIB-CoA hydrolase was eluted at a molecular weight of 36,000. [Adapted with permission from Y. Shimomura, T. Murakami, N. Fujitsuka, N. Nakai, Y. Sato, S. Sugiyama, N. Shimomura, J. Irwin, J. W. Hawes, and R. A. Harris, *J. Biol. Chem.* **269**, 14248 (1994).]

(w/v) Tween 20. Fractions with hydrolase activity that were eluted with buffer containing 0.1 M NaCl are combined. The NaCl concentration is decreased to less than 10 mM by the ultrafiltration method, and the sample is applied again to the CoA-Sepharose column. The column is washed with 10 ml of the equilibrating buffer and eluted with buffer D (pH 7.5) containing 0.05% (w/v) Tween 20.

Second Sephacryl S-200 Column Chromatography. The eluate is concentrated to approximately 1.5 ml by ultrafiltration, made 10% (v/v) in glycerol, and applied again to the Sephacryl S-200 column as described above. Protein peak fractions consisting of a single polypeptide (molecular weight of 36,000) as determined by sodium dodecyl sulfate–polyacrylamide gel electrophoresis (SDS–PAGE) are combined and stored at −80°.

HIB-CoA hydrolase is purified 7200-fold by the method described here (Table I).

TABLE I
PURIFICATION OF 3-HYDROXYISOBUTYRYL-CoA HYDROLASE FROM RAT LIVER[a]

Fraction	Protein (mg)	Total activity (units)	Yield (%)	Specific activity (units/mg)
Liver extract[b]	62,700	3,480	100	0.056
$(NH_4)_2SO_4$ (45–75%) precipitate	12,900	2,451	70	0.19
Phenyl-Sepharose-$(NH_4)_2SO_4$ precipitate	4,073	1,896	54	0.47
Octyl-Sepharose-$(NH_4)_2SO_4$ precipitate	2,934	1,828	53	0.62
DEAE-Sephacel	500	864	25	1.73
CM-Sepharose	82.1	522	15	6.36
Hydroxylapatite	6.51	360	10	55.4
Sephacryl S-200	0.85	188	5.4	222
CoA-Sepharose	0.40	161	4.6	403
Sephacryl S-200	0.29	126	3.6	427

[a] Adapted with permission from Y. Shimomura, T. Murakami, N. Fujitsuka, N. Nakai, Y. Sato, S. Sugiyama, N. Shimomura, J. Irwin, J. W. Hawes, and R. A. Harris, *J. Biol. Chem.* **269,** 14248 (1994).
[b] Refers to protein and activity of supernatant after centrifugation.

Generation of Recombinant 3-Hydroxyisobutyryl-CoA Hydrolase

Cloning of Human 3-Hydroxyisobutyryl-CoA Hydrolase

Twenty-four residues of the amino-terminal sequence of HIB-CoA hydrolase[6,11] were determined by Edman degradation, using a Porton PI 2090 microsequencing system. This sequence was matched with 75% identity to a translated human expressed sequence tag (EST) (GenBank accession number R20241) present in the database of the IMAGE Consortium. This clone contained an open reading frame of 1146 base pairs (bp) encoding a mature protein sequence of 352 amino acid residues with a calculated molecular weight of 39,398. The cDNA encodes an additional 28-amino acid sequence at the amino-terminal end of the protein that contains numerous basic residues characteristic of a mitochondrial leader sequence.

Construction of Vector for Prokaryotic Expression of Histidine-Tagged 3-Hydroxyisobutyryl-CoA Hydrolase

Polymerase chain reaction (PCR) primers (5′-AAAAGCGGCCGCG-CAGCAGAAGAGGTGCTATTGGAA-3′ and 5′-TAAAACGACGGC-CAGT-3′) were used to amplify the HIB-CoA hydrolase sequence with the purified EST DNA as template. The product was subcloned into pGEM-T vector (Promega, Madison, WI) to give pGEM-T-HHYD.[6]

---

[11] B. Huang, M.S. thesis. Indiana University, Indianapolis, Indiana, 1998.

Two primers with *Nde*I and *Xho*I restriction sites, respectively (5'-GGGAATTCC<u>CATATG</u>GCAGCAGAAGAGGTGCTATTG-3' and 5'-CTCCG<u>CTCGAG</u>TCAAAATTTCAAATCACTGCTTCC-3') were subsequently used to amplify the HIB-CoA cDNA of pGEM-T-HHYD. The amplified 1085-bp fragment was digested with *Nde*I and *Xho*I and ligated into pET28a to create an expression vector (pET28a-HHYD) encoding the entire mature protein with a six-histidine repeat and a thrombin cleavage site four amino acids upstream of the amino terminus.[11]

Expression and Purification of Histidine-Tagged 3-Hydroxyisobutyryl-CoA Hydrolase. One liter of TY medium supplemented with kanamycin (50 μg/ml) is inoculated with BL21(DE3) cells that has been previously transformed with the pET28a-HHYD plasmid. The cells are incubated at 37° with shaking until the A_{600} reaches 0.6 to 0.8. Isopropyl-1-thio-β-D-galactopyranoside (IPTG) is added to a final concentration of 1 mM and the incubation continued for 16–18 hr. The bacteria are harvested by centrifugation and resuspended in 30 ml of extraction buffer [20 mM Tris (pH 8.0), 5 mM imidazole, 0.5 M NaCl, 0.5% (w/v) Triton X-100]. The suspension is sonicated four times for 30 sec at 30-sec intervals. The homogenate is centrifuged before being applied to a 10-ml column of Ni^{2+}-NTA-agarose previously equilibrated in column buffer (20 mM Tris, 5 mM imidazole, 0.5 M NaCl). After sample loading, the columns are washed successively with 5 column volumes of column buffer plus a washing buffer (20 mM Tris, 60 mM imidazole, 0.5 M NaCl) before elution of the enzyme with elution buffer (20 mM Tris, 200 mM imidazole, and 0.5 M NaCl). Quantitative removal of the His$_6$ tag and all but four of the extra amino-terminal end amino acids is achieved by incubation of the enzyme (1 mg) with thrombin (3 μg) for 40 min at room temperature in the presence of 2.5 mM CaCl$_2$. Overall yield of the recombinant protein is approximately 6 mg of protein per liter of culture.[11]

Properties of Native and Recombinant 3-Hydroxyisobutyryl-CoA Hydrolase

Molecular Weight

The final preparation of native rat liver HIB-CoA hydrolase consists of a single polypeptide with a molecular weight of 36,000 as determined by SDS–PAGE.[2] The molecular weight of the native enzyme determined by gel filtration is also 36,000,[2] indicating that the enzyme exists as a monomer. The molecular weight of the recombinant human HIB-CoA hydrolase is estimated to be 38,000 by SDS–PAGE.[11]

TABLE II
SUBSTRATE SPECIFICITY OF 3-HYDROXYISOBUTYRYL-CoA HYDROLASE AND INHIBITION CONSTANTS
FOR VARIOUS CoA ESTERS[a]

CoA ester	Activity[b]		K_i^c (mM)
	Units/mg	%	
(S)-HIB-CoA (from methacrylyl-CoA + crotonase)	430	100	
3-Hydroxypropionyl-CoA (from acrylyl-CoA + crotonase)	244	57.1	
3-Hydroxy-2-methylbutyryl-CoA (from tiglyl-CoA + crotonase)	3.14	0.73	
L-3-Hydroxybutyryl-CoA (from crotonyl-CoA + crotonase)	1.75	0.41	
DL-3-Hydroxybutyryl-CoA	1.15	0.27	0.14
Acetoacetyl-CoA[d]	0.94	0.22	0.12
DL-Methylmalonyl-CoA	0.64	0.15	2.26
Isobutyryl-CoA	0.34	0.08	1.19
Malonyl-CoA	0.26	0.06	1.60
Acetyl-CoA	0.17	0.04	2.61
Propionyl-CoA	0.17	0.04	0.83
n-Valeryl-CoA	0.09	0.02	0.72

[a] Adapted with permission from Y. Shimomura, T. Murakami, N. Fujitsuka, N. Nakai, Y. Sato, S. Sugiyama, N. Shimomura, J. Irwin, J. W. Hawes, and R. A. Harris, J. Biol. Chem. 269, 14248 (1994).
[b] The enzyme activity was assayed under standard conditions except that the concentrations of CoA esters were at 0.5 mM and crotonase was omitted for CoA esters other than those indicated. For (S)-HIB-CoA and 3-hydroxypropionyl-CoA, 0.011 μg of enzyme was used, and for the other CoA esters 0.34 μg of enzyme was used.
[c] The type of inhibition was competitive for all CoA esters examined.
[d] Acetoacetyl-CoA was used at a concentration of 0.2 mM.

pH Optimum

It was reported in 1957 that the enzyme preparation purified eightfold from an alcohol-KCl extract of pig heart had a pH optimum of about 5.6.[12] However, the highly purified rat liver enzyme is active over a wide range (at least pH 5.1–10.3), with the optimum being about pH 8.[2]

Substrate Specificity, Kinetics, and Effects of Cations and Nucleotides

(S)-HIB-CoA, produced from methacrylyl-CoA by crotonase, is the best substrate for HIB-CoA hydrolase under standard conditions (Table II). 3-Hydroxypropionyl-CoA, produced from acrylyl-CoA by crotonase, is also a good substrate. Under the conditions of this assay crotonase converts about one-third of the methacrylyl-CoA and nearly all of the acrylyl-CoA to their respective hydroxy-CoA esters during the incubation that precedes initiation of the hydrolytic reaction by the addition of HIB-CoA hydrolase. Crotonase

[12] G. Rendina and M. J. Coon, J. Biol. Chem. 225, 523 (1957).

is added in great excess to ensure that this step is not rate limiting. V_{max} and K_m values for (S)-HIB-CoA are 443 units per mg of protein and 6 μM, respectively, while those for 3-hydroxypropionyl-CoA are 250 units/mg of protein and 25 μM, respectively. The enzyme has been found to hydrolyze 10 other CoA esters (Table II), but the rates of hydrolysis of these esters are low relative to that of (S)-HIB-CoA. Comparable results with respect to substrate specificity are obtained with human recombinant enzyme.[6] K_i values for eight CoA esters that have been examined are relatively high compared with the K_m value for (S)-HIB-CoA (Table II).

Hydrolysis of the following compounds by the native rat enzyme has not been detected: acrylyl-, benzoyl-, n-butyryl-, crotonyl-, glutaryl-, (R)-HIB-, 3-hydroxyisovaleryl- (produced from 3-methylcrotonyl-CoA by crotonase), 3-hydroxy-3-methylglutaryl-, (R)- and (S)-ibuprofenyl-, isovaleryl-, 3-methylcrotonyl-, methacrylyl-, palmitoyl-, phenylacetyl-, succinyl-, and tiglyl-CoA. Comparable results with respect to substrate specificity have been obtained with the recombinant human enzyme, with the exception that 3-hydroxyisovaleryl-CoA was found hydrolyzed by this enzyme, albeit slowly.[6]

The following are without effect on the enzyme activity: 60 mM KCl, 60 mM NaCl, 5 mM CaCl$_2$, 5 mM MgCl$_2$, and 3 mM ATP, ADP, NAD$^+$, or NADH.

Distribution of 3-Hydroxyisobutyryl-CoA Hydrolase in Tissues of Rat, Dog, and Human

The HIB-CoA hydrolase activities in several tissues of rat,[2] dog,[5] and human liver[4] are listed in Table III. All the tissues examined have high activity of HIB-CoA hydrolase relative to the branched-chain α-keto acid dehydrogenase complex, the rate-limiting enzyme in the pathway of valine catabolism. Western and Northern blot analyses indicate that heart, liver, and kidney express the greatest amounts of HIB-CoA hydrolase in humans and rats.[6]

Relationship of 3-Hydroxyisobutyryl-CoA Hydrolase to (S)-Methylmalonyl-CoA Hydrolase

DL-Methylmalonyl-CoA is hydrolyzed by HIB-CoA hydrolase, although slowly relative to the rate of (S)-HIB-CoA hydrolysis (Table II). The highest enzyme activity for methylmalonyl-CoA hydrolysis (approximately 3 units/mg of protein) is obtained at pH 6, indicating a much different pH optimum relative to that for (S)-HIB-CoA.[2] A D-(S)-methylmalonyl-CoA hydrolase has also been purified from rat liver and partially characterized.[13,14] The

[13] R. Kovachy, S. D. Copley, and R. H. Allen, *J. Biol. Chem.* **258**, 11415 (1983).
[14] R. Kovachy, S. P. Stabler, and R. H. Allen, *Methods Enzymol.* **166**, 393 (1988).

TABLE III
DISTRIBUTION OF 3-HYDROXYISOBUTYRYL-CoA HYDROLASE IN
TISSUES OF RAT, DOG, AND HUMAN LIVER[a]

| | Activity[a] (units/g wet weight) | | |
Tissue	Rat[b]	Dog[c]	Human[d]
Liver	9.9 ± 0.8	3.9 ± 0.2	7.3 ± 1.1
Kidney	6.2 ± 0.3	2.8 ± 0.3	—
Heart	8.3 ± 0.3	2.2 ± 0.1	—
Muscle	2.0 ± 0.4	1.3 ± 0.3	—
Pancreas	—	1.7 ± 0.3	—
Brain	2.7 ± 0.1	—	—

[a] HIB-CoA activity was measured under standard conditions. Samples from 3 rats, 5 dogs, and 10 humans were analyzed. Each value is the mean ± SEM.
[b] From Y. Shimomura, T. Murakami, N. Fujitsuka, N. Nakai, Y. Sato, S. Sugiyama, N. Shimomura, J. Irwin, J. W. Hawes, and R. A. Harris, *J. Biol. Chem.* **269**, 14248 (1994).
[c] From T. Ooiwa, H, Goto, Y. Tsukamoto, T. Hayakawa, S. Sugiyama, N. Fujitsuka, and Y. Shimomura, *Biochim. Biophys. Acta* **1243**, 216 (1995).
[d] From K. Taniguchi, T. Nonami, A. Nakao, A. Harada, S. Sugiyama, N. Fujitsuka, Y. Shimomura, S. M. Hutson, R. A. Harris, and H. Takagi, *Hepatology* **24**, 1395 (1996).

characteristics of this enzyme including pH profile, specific activity for the hydrolysis of methylmalonyl-CoA, molecular weight of the enzyme, and enzyme binding properties on ion-exchange columns are all quite similar to those of HIB-CoA hydrolase. Although (S)-HIB-CoA and 3-hydroxypropionyl-CoA were not tested as possible substrates in the work on (S)-methylmalonyl-CoA hydrolase,[13,14] it seems likely that it is the same enzyme purified as HIB-CoA hydrolase in this laboratory.[2] Thus, HIB-CoA hydrolase may have two functions, one involving HIB-CoA hydrolysis in the valine catabolic pathway and a second involving methylmalonyl-CoA hydrolysis in the event that this compound accumulates as a consequence of an inhibition or genetic defect in the catabolic pathway for propionyl-CoA.

Acknowledgments

This work was supported by grants from the U.S. Public Health Services (NIH DK40441 to R.A.H.), the Diabetes Research and Training Center of Indiana University School of Medicine (AM 20542), the Grace M. Showalter Trust (to R.A.H.), the Uehara Memorial Foundation (to Y.S.), and the University of Tsukuba Project Research Fund (to Y.S.).

[24] Mammalian Branched-Chain Acyl-CoA Dehydrogenases: Molecular Cloning and Characterization of Recombinant Enzymes

By Jerry Vockley, Al-Walid A. Mohsen, Barbara Binzak, Jan Willard, and Abdul Fauq

The acyl-CoA dehydrogenases (ACDs) are a family of related enzymes that catalyze the α,β-dehydrogenation of acyl-CoA esters, transferring electrons to electron-transferring flavoprotein (ETF).[1–4] Deficiencies of these enzymes are important causes of human disease. Biochemical and immunological studies have identified at least seven distinct members of this enzyme family, each with narrow substrate specificity.[1–5] Very long-, long-, medium-, and short-chain acyl-CoA dehydrogenases (VLCAD, LCAD, MCAD, and SCAD) catalyze the first step in the β-oxidation of straight-chain fatty acids with substrate optima of 16, 16, 8, and 4 carbon chains, respectively. These enzymes are encoded in the nuclear genome in precursor form. Precursor peptides are synthesized in the cytoplasm, transported into mitochondria, and processed to homotetramers (except VLCAD, which is processed to a homodimer), with each monomer containing a noncovalently, but tightly bound, FAD molecule.[2,3,6] Two ACDs are active in the metabolism of branched-chain amino acids. Isovaleryl-CoA dehydrogenase (IVD[1]; EC 1.3.99.10) and short/branched-chain acyl-CoA dehydrogenase (SBCAD; also known as 2-methyl branched-chain acyl-CoA dehydrogenase) catalyze the third step in leucine and isoleucine/valine metabolism, respectively.

cDNAs for each of the rat and human ACDs have been cloned by a variety of techniques.[5,7–14] DNA sequencing studies show that the amino

[1] F. Crane and H. Beinert, *J. Biol. Chem.* **218,** 717 (1955).
[2] Y. Ikeda, C. Dabrowski, and K. Tanaka, *J. Biol. Chem.* **258,** 1066 (1983).
[3] Y. Ikeda, K. O. Ikeda, and K. Tanaka, *J. Biol. Chem.* **260,** 1311 (1985).
[4] Y. Ikeda and K. Tanaka, *J. Biol. Chem.* **258,** 9477 (1983).
[5] K. Izai, Y. Uchida, T. Orii, S. Yamamoto, and T. Hashimoto, *J. Biol. Chem.* **267,** 1027 (1992).
[6] Y. Ikeda, K. Okamura-Ikeda, and K. Tanaka, *Biochemistry* **24,** 7192 (1985).
[7] T. Aoyama, M. Souri, I. Ueno, T. Kamijo, S. Yamaguchi, W. J. Rhead, K. Tanaka, and T. Hashimoto, *Am. J. Hum. Genet.* **57,** 273 (1995).
[8] Y. Matsubara, Y. Indo, E. Naito, H. Ozasa, R. Glassberg, J. Vockley, Y. Ikeda, J. Kraus, and K. Tanaka, *J. Biol. Chem.* **264,** 16321 (1989).
[9] Y. Matsubara, M. Ito, R. Glassberg, S. Satyabhama, Y. Ikeda, and K. Tanaka, *J. Clin. Invest.* **85,** 1058 (1990).

acid sequences of the various enzymes share 30–35% identical residues, indicating that the enzymes are encoded by separate but related genes. Interspecies identities between the same ACDs approach 90%. Thus, it is likely that these genes have evolved from a common ancestral gene and have attained their distinct substrate specificities through evolutionary divergence. The three-dimensional structures of porcine MCAD, *Megasphaera esdenii* butyryl-CoA dehydrogenase, rat SCAD, and human IVD have also been reported to be similar.[15–18]

Cloning of Mammalian Isovaleryl-CoA Dehydrogenases and Short/Branched-Chain Acyl-CoA Dehydrogenases

Mammalian IVDs and SBCADs have been reported in detail previously.[8,9,13,14] Rat IVD and human SBCAD cDNAs were identified by immunoselective techniques. The corresponding human IVD and rat SBCAD cDNAs were isolated by cross-species hybridization to screen cDNA libraries. Human genomic IVD sequences and gene structure were characterized in clones from a human genomic library. Sequence analysis of purified native rat liver 2-methyl branched-chain acyl-CoA dehydrogenase confirms that it is the same enzyme as SBCAD.

Assay Methods

Three assays are routinely employed to evaluate ACD activity. The phenazine methosulfate/2,6-dichlorphenol–indophenol (PMS/DCIP) reduction assay has been described in detail in this series.[19,20] A second assay used interchangeably with the PMS/DCIP assay employs ferricenium

[10] E. Naito, H. Ozasa, Y. Ikeda, and K. Tanaka, *J. Clin. Invest.* **83,** 1605 (1989).

[11] Y. Indo, T. Yang-Feng, R. Glassberg, and K. Tanaka, *Genomics* **11,** 609 (1991).

[12] D. Kelly, J. Kim, J. Billadello, B. Hainline, T. Chu, and A. Strauss, *Proc. Natl. Acad. Sci. U.S.A.* **85,** 4068 (1988).

[13] R. Rozen, J. Vockley, L. Zhou, R. Milos, J. Willard, K. Fu, C. Vicanek, L. Low-Nang, E. Torban, and B. Fournier, *Genomics* **24,** 280 (1994).

[14] J. Willard, C. Vicanek, K. P. Battaile, P. P. Van Veldhoven, A. H. Fauq, R. Rozen, and J. Vockley, *Arch. Biochem. Biophys.* **331,** 127 (1996).

[15] J.-J. P. Kim, M. Wang, and R. Paschke, *Proc. Natl. Acad. Sci. U.S.A.* **90,** 7523 (1993).

[16] J.-J. P. Kim, M. Wang, R. Paschke, S. Djordjevic, D. W. Bennett, and J. Vockley, *in* "Flavins and Flavoproteins 1993" (K. Yagi, ed.), p. 273. Walter de Gruyter, New York, 1994.

[17] J. M. Kirk, I. A. Laing, N. Smith, and W. S. Uttley, *J. Inher. Metab. Dis.* **19,** 370 (1996).

[18] S. Djordjevic, C. P. Pace, M. T. Stankovich, and J.-J. P. Kim, *Biochemistry* **34,** 2163 (1995).

[19] Y. Ikeda and K. Tanaka, *Methods Enzymol.* **166,** 360 (1988).

[20] Y. Ikeda and K. Tanaka, *Methods Enzymol.* **166,** 374 (1988).

hexafluorophosphate ($FcPF_6$) as an artificial electron acceptor.[21] The final assay utilizes ETF, the physiologic electron acceptor for these enzymes, directly as an electron acceptor. The PMS/DCIP and $FcPF_6$ assays are suitable for use with purified enzyme and are reported to be of similar sensitivity. They are not specific enough, however, for use with more crude samples such as tissue extracts. The ETF fluorescence reduction assay is the standard for determining ACD activity in crude extracts from tissue and cell cultures,[22] and is at least several-fold more sensitive than either of the other assays. Finally, a spectral scanning assay shows interaction of substrate with a purified ACD without the need for an electron acceptor to be present.[23,24]

$FcPF_6$ Assay

Reagents

$FcPF_6$, 2 mM in 10 mM HCl (prepared as described below)

Potassium phosphate buffer (pH 7.6), 100 mM, containing 0.1 mM EDTA

Substrate (50 μM): Most of the acyl-CoA substrates are commercially available from Sigma (St. Louis, MO). (R)- and (S)-2-Methylbutyryl-CoA are synthesized as described below. 2-Methylhexanoyl and 2-methylpalmitoyl-CoA are the kind gift of P. P. Van Veldhoven (Katholieke Universiteit, Leuven, Belgium). Substrates are dissolved in 2 mM sodium acetate, pH 5.0

Preparation of $FcPF_6$. Ferrocene (0.5 g, 2.7 mmol) is dissolved in 10 ml of concentrated sulfuric acid and incubated for 1 hr at room temperature. This is then slowly diluted into 150 ml of distilled water and filtered, using a coarse glass sintered funnel. Five milliliters of a saturated solution of $NaFcPF_6$ is added to the filtrate and incubated on ice for 30 sec. The precipitate is collected on a medium glass sintered funnel, washed with cold distilled water, then dried under vacuum. The product is recrystallized by resuspension in a minimal volume of cold distilled water, filtration on a medium glass sintered funnel, and incubation for 1 hr on ice, followed by lyophilization of the filtrate. A small amount of approximately 20 mM $FcPF_6$ is made by dissolving a small amount of crystals in 0.5 ml of 10 mM HCl. This can be stored at $-20°$ for several weeks between use. Immediately before use, the dye is thawed and vortexed, and a 2 mM working solution of $FcPF_6$ is made by adjusting the absorbance at 617 nm ($\Delta\varepsilon = 410\ M^{-1}$

[21] T. C. Lehman, D. E. Hale, A. Bhala, and C. Thorpe, *Anal. Biochem.* **186,** 280 (1990).

[22] F. E. Frerman and S. I. Goodman, *Biochem. Med.* **33,** 38 (1985).

[23] A.-W. A. Mohsen and J. Vockley, *Biochemistry* **34,** 10146 (1995).

[24] K. P. Battaile, J.-J. P. Kim, and J. Vockley, *Am. J. Hum. Genet.* **57,** A175 (1995).

cm^{-1}) of a 1:10 dilution of stock solution in 10 mM HCl to 0.82. Unused diluted FcPF$_6$ is discarded.

Assay Procedure. All solutions are prewarmed to 32°. To a 1-ml semi-UV plastic disposable cuvette, sample and 50 μl of 2 μM FcPF$_6$ are added along with phosphate buffer to give a final volume of 475 μl. Baseline reduction of the dye is measured in a spectrophotometer via continuous monitoring of absorbance at 300-nm wavelength in a heated block at 32° for 1 min. Twenty-five microliters of 50 μM substrate is then added and reduction in absorbance at 300-nm wavelength at 32° is monitored for an additional 1 min. Activity is calculated from the change in absorbance ($\Delta\varepsilon = 4300\ M^{-1}\ cm^{-1}$) and expressed as micromoles of FcPF$_6$ reduced per minute per milligram of enzyme protein. Sample concentration should be adjusted to give a pseudolinear reaction.

Anaerobic Electron-Transferring Flavoprotein Fluorescence Reduction Assay

Reagents

Tris (pH 8.0), 200 mM
D-Glucose, 20% (w/v)
Glucose oxidase, type V from *Aspergillus niger* (Sigma)
Catalase, bovine liver (Sigma)
ETF, porcine liver (must be purified as described below)
Acyl-CoA substrates, 50 μM (as noted previously)

Purification of Electron-Transferring Flavoprotein

Several methods have previously been published for purification of ETF from various sources, primarily from pig liver. The following procedure incorporates modifications to these methods that provide greater convenience, consistency, and ease.

Mitochondria Extract Preparation. Freshly harvested pig liver (500–600 g) is cut into small pieces, washed with ice-cold phosphate-buffered saline (pH 7.4), and homogenized in a blender with 50 mM KPO$_4$ (pH 8.0), 1 mM EDTA, 2.5% (v/v) glycerol, and 250 mM sucrose. The homogenate is filtered through cheesecloth to remove connective and tissue and larger tissue remnants, then transferred to 1-liter bottles and centrifuged at 1500g for 20 min. The supernatant is transferred to 250-ml bottles and centrifuged at 10,000g for 20 min. The soft pellets from three bottles are combined and resuspended in 250 mM sucrose in 50 mM KPO$_4$ (pH 8.0), 1 mM EDTA, and 2.5% (v/v) glycerol. The pellets are washed twice with the same buffer and can be stored at −80°. Between 200 and 250 g of pellet is processed at one time, and if frozen, are thawed overnight at 4°. A solution consisting of 100 mM KPO$_4$ (pH 8.0), 10% (v/v) glycerol, and FAD (0.1 mg/ml) is added to the pellets at 0.3 : 1 (v/w), and the sample is sonicated for five 45-sec bursts at maximum power, using a macrotip. The sample is centrifuged at 43,000g for 1 hr, the pH of the supernatant is adjusted to pH 8.0 with 5 N KOH, and FAD (0.1 mg/ml of sample) is added. A 40–60% ammonium sulfate fraction is prepared, and the final pellet is resuspended in 100 mM KPO$_4$ (pH 8.0)–10% (v/v) glycerol and dialyzed extensively (two or three buffer changes of 5 liters each) in 10 mM dibasic potassium phosphate (unbuffered) containing FAD at 10 mg/liter. The dialyzed sample is concentrated with polyethylene glycol powder at 4° with constant gentle shaking to avoid formation of local areas of aggregated or precipitated protein. The final sample size should be 30 ml or less.

Chromatography of Electron-Transferring Flavoprotein. Sixty grams of Whatman (Clifton, NJ) DE52 (DEAE-cellulose) resin is freshly prepared for each experiment by rinsing the resin three times with 4 liters of H$_2$O, and then equilibrating it overnight in 100 ml of 0.5 M potassium phosphate dibasic (unbuffered). The resin is packed into a Pharmacia (Piscataway, NJ) XK 50/20 column and washed with 1.2 to 1.5 liters of 10 mM potassium phosphate dibasic (unbuffered) at 10 ml/min, using a low-pressure chromatography system. The column bed height is approximately 8 cm after packing. The concentrated ammonium sulfate fraction is cleared by centrifugation at 43,000g for 15 min, and loaded onto the column at 2 ml/min. ETF is eluted with 10 mM potassium phosphate dibasic (unbuffered), and should begin to elute between approximately 60 and 120 ml as a yellow solution. If some red elutes from the column at the tail of the yellow fractions, it is a result of insufficient sample dialysis. All yellow fractions are combined immediately while they are eluting and FAD is added at 20 mg/50 ml. After all yellow fractions are combined, the sample should quickly be adjusted to pH 8.5 with 1 M potassium phosphate monobasic. If fractions are cloudy, they are centrifuged at 10,000g for 30 min. The final pooled sample is loaded immediately onto a CM-Sepharose FF column (XK 16/40; Pharmacia) that has been preequilibrated with 10 mM Tris, pH 8.5, and eluted with a 400-ml gradient of 10–100 mM NaCl in 10 mM Tris, pH 8.5 ETF elutes as a fluorescent green fraction toward the end of the gradient. The A_{270}/A_{436} ratio is then determined. Fractions with ratios below 6.5 can be used directly after concentration and dialysis as described below. Fractions with ratios between 6.5 and 10 should be pooled and further purified on a 20 μM ceramic hydroxyapatite column (Bio-Rad, Hercules, CA). These fractions are loaded immediately onto the column, which has been equilibrated in 10 mM potassium phosphate, pH 7.6, and eluted with a 300-ml gradient of 10–85 mM potassium phosphate, pH 7.6. Fractions with an A_{270}/A_{436} ratio below 6.5 are pooled and then concentrated and dialyzed in 10 mM Tris, pH 8.0, at 4° for 5 hr. The final preparation is brought to 20% (v/v) glycerol concentration and can be stored at −20° for extended periods of time.

Electron-Transferring Flavoprotein Assay Procedure

All solutions are prewarmed to 32°. The reaction mixture contains 50 mM Tris (pH 8.0), 0.5% (w/v) glucose, and 50 μM acyl-CoA substrate calculated to give a final volume of 0.8 ml after subsequent additions. The mixture is deaerated by repeated vacuum and layering with oxygen-free argon (10 cycles of 60 sec each) in a tightly sealed quartz semimicrocuvette. Five microliters containing 40 units of glucose oxidase and 1 unit of catalase is added to remove any remaining dissolved oxygen, followed by addition of purified pig ETF to give a final concentration of 1 μM. The reaction is initiated by adding the desired amount of enzyme sample, and activity is determined by exciting the ETF flavin at 342 nm and monitoring quenching of fluorescence at 496 nm, using an LS 50 B luminescence spectrometer with a heated sample block (Perkin-Elmer, Oak Brook, IL). Sample concentration should be adjusted to give a pseudolinear reaction over a time course of 1 min. The slope of the line is calculated by linear regression analysis, and 1 unit is defined as the amount of enzyme needed to completely reduce 1 μmol of ETF per minute at 32°.

Synthesis of Stereospecific (R)- or (S)-2-Methylbutyryl-CoA

General. Proton nuclear magnetic resonance (^{1}H NMR) spectra are measured with a Bruker (Billerica, MA) WH-300 instrument (^{1}H frequency at 300 MHz) in the solvent noted. ^{1}H chemical shifts are expressed in parts per million downfield from Me$_4$Si used as an internal standard. Elemental analyses are performed by Oneida Research Services (Whiteboro, NY). Silica gel 60 (230-400 mesh; Merck, Rahway, NJ) is used for column chromatography. Thin-layer chromatography is performed on Merck silica gel F-254 plates.

Synthesis of N^1-[(1S)-1-Phenylethyl]-(2R)-2-methylbutanamide and N^1-[(1S)-1-Phenylethyl]-(2S)-2-methylbutanamide. To a solution of L-(−)-phenylethylamine (Fig. 1, **1**) (10.12 ml, 79.3 mmol) and (±)-2-methylbutyric acid (Fig. 1, **2**) (8.56 ml, 79.3 mmol) in anhydrous tetrahydrofuran (THF, 150 ml) at 0° under nitrogen are added hydroxybenzotriazole (HOBT, 19.95 g, 83.2 mmol), dicyclohexyl carbodiimide (DDC), and diisopropylethylamine (14.4 ml, 83.2 mmol) in that order. The resulting mixture is stirred for 24 hr at room temperature. After addition of ether (300 ml), the mixture is filtered and sequentially washed with 1 N HCl, water, and

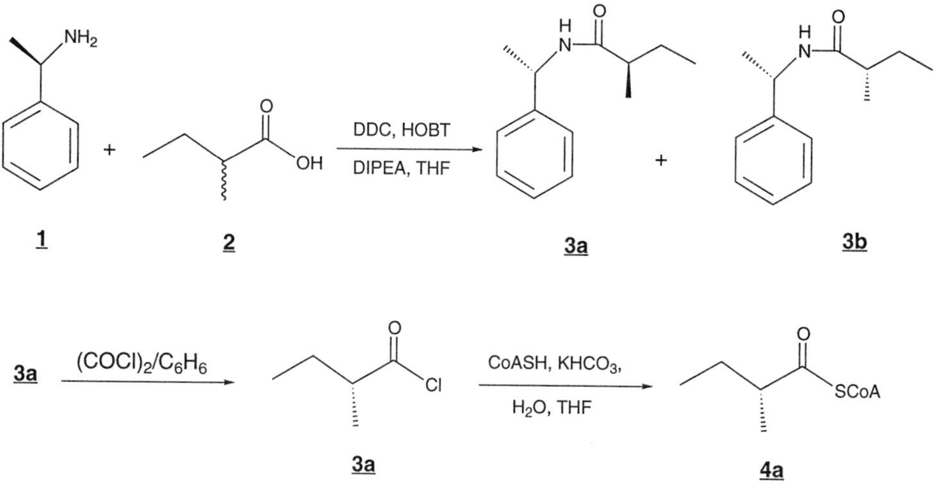

Fig. 1. Scheme for synthesis of (R)- and (S)-2-methylbutyryl-CoA. Steps indicated by boldface numbers are described in text.

saturated sodium chloride, and dried (MgSO$_4$). Thin-layer chromatography (TLC) analysis shows that two new closely spaced but distinct spots are obtained that can be successfully separated by silica gel chromatography using 30% (v/v) ether in petroleum ether to furnish N^1-[1S)-1-phenylethyl]-(2R)-2-methylbutanamide (Fig. 1, 3a) (higher R_f) and N^1-[(1S)-1-phenyl-ethyl]-(2S)-2-methylbutanamide (Fig. 1, 3b) (lower R_f) in 80% combined yield: ^{1}H NMR (CDCl$_3$) δ 7.37–7.28 (5H, m), 5.81 (br m, 1H), 5.16 (1H, m), 2.06 (1H, m), 1.68 (1H, m), 1.48 (3H, d, J = 6.3 Hz), 1.41 (1H, m), 1.12 (3H, d, J = 6.2 Hz), 0.89 (3H, t, J = 7 Hz). MS (ESI) m/z 206 (M$^+$ + 1).

Synthesis of (R)- and (S)-2-Methylbutyryl-CoA. The higher R_f amide (Fig. 1, 3a) (1.62 g, 7.9 mmol) is suspended in 35 ml of 6 N HCl and heated at 110° for 7 hr. A drop of phenolphthalein is added and the reaction mixture is neutralized with 35% (w/v) NaOH to a light pink color. The basic solution is then extracted three times with ether to remove the amine and the aqueous phase is reextracted three times with 50-ml aliquots of DCM. To a solution of the (R)-2-methylbutyric acid (80 mg) in benzene (4 ml) is added oxalyl chloride (75 μl) dropwise at 0°. After stirring at room temperature for 1 hr, the mixture is diluted with 4 ml of freshly distilled THF. This solution of (R)-2-methylbutyryl chloride (Fig. 1, 3a) is then added slowly to a mixture of CoASH (25 mg; Sigma), KHCO$_3$ (30 mg), THF (3 ml), and water (2.5 ml). During the addition, the pH of the clear reaction mixture should be closely monitored and maintained at approximately pH 8 by adding more $KHCO_3$ as needed. The stirring is continued at 0° for another 0.5 hr. After acidification with 2 N HCl to pH 3, the reaction mixture is concentrated under rotary evaporation. The aqueous phase is washed two times with DCM, filtered, and then lyophilized to a white powder. Purification is accomplished by high-performance liquid chromatography (HPLC) [Beckman (Fullerton, CA)] pump 126, Beckman Ultrasphere C_{18} column; linear gradient of 10% B–45% B in 25 min, where B is acetonitrile, A is 50 mM ammonium acetate, pH 5.5; λ_{max} = 254 nm). The product (1S)-1-phenylethyl-(2R)-2-methylbutanamide (Fig. 1, step **3a**) elutes at 15.7 min at room temperature. The fractions containing the product are lyophilized, dissolved in 5 ml of water, and relyophilized to give product **3a** (Fig. 1) as a white powder. ^{1}H NMR (D_2O) δ 8.55 (1H, s), 8.27 (1H, s), 6.16 (1H, d, J = 6.0 Hz), 4.70 (1H, hidden under D_2O peak), 4.56 (1H, m), 4.22 (2H, br m), 4.02 (1H, s), 3.78 (2H, m), 3.55 (1H, m), 3.43 (2H, t, J = 6.0 Hz), 3.30 (2H, t, J = 6.1 Hz), 2.96 (2H, t, J = 6.0 Hz), 2.61 (1H, m), 2.40 (2H, t, J = 6.2 Hz), 1.95 (2H), 1.57 (1H, m), 1.43 (1H, m), 1.07 (3H, d, J = 6.3 Hz), 0.88 (3H, s), 0.82 (3H, J = 7.5 Hz), 0.75 (3H, s).

Expression and Purification of Recombinant Human Isovaleryl-CoA Dehydrogenase

Construction of Escherichia coli Expression Vector for Human Isovaleryl-CoA Dehydrogenase

A cDNA for the mature coding region of human IVD is amplified by polymerase chain reaction (PCR) with a 5′-primer containing an *Eco*RI site and an ATG juxtaposed to 20 nucleotides beginning with the codon for the first amino acid of the mature protein. The 3′ primer contains a *Hin*dIII site, an antisense stop codon, and nucleotides complementary to the last 20 coding bases (antisense direction) of human IVD. The amplification product is cloned into an *Escherichia coli* expression vector (pKK223-3) that contains a strong *Tac* fusion promotor and a Shine–Dalgarno sequence upstream of a multicloning site, and transfected into *E. coli* strain JM105. The level of expression obtained with this vector, however, is extremely low. To enhance expression, the nucleotide sequence at the 5′ end of the coding region is altered to mimic codon usage of highly expressed proteins in *E. coli*. A total of 22 codons is changed within the first 111 nucleotides. IVD activity measured by the electron-transferring flavoprotein (ETF) fluorometric assay in the crude extract of cells expressing the altered IVD cDNA is increased 165-fold relative to the wild-type sequence. After a 24-hr induction, the amount of IVD is estimated to represent approximately

9% of the total protein in the crude cell-free extract and 29 nmol/g of wet cells.

Purification of Recombinant Isovaleryl-CoA Dehydrogenase

Eight-liter cultures of *E. coli* containing the human IVD high-expression vector pKKm*HIVD* are incubated for 20 hr at 37° in the presence of isopropyl-β-D-thiogalactopyranoside (IPTG, 1 mg/liter). Cells are harvested via centrifugation, resuspended in 100 ml of 20% (w/v) sucrose, 25 mM Tris, 0.5 mM EDTA, pH 8.0 containing lysozyme (0.5 mg/ml), and incubated for 30 min on ice. The cells are then pelleted and resuspended in 15 ml of 10 mM Tris, pH 7.5. EDTA is added to a final concentration of 0.15 M. The cell suspension is then sonicated four to six times for 60 sec each, and cell debris is removed by centrifugation first at 43,000g, and then at 250,000g (60 min each). The final supernatant is diluted 10 times with water and then loaded on a 16 × 40 mm DEAE Sepharose FF column (Pharmacia, Uppsala, Sweden) preequilibrated in 10 mM Tris, 10 mM EDTA, pH 7.5. IVD is eluted with a 400-ml gradient from 0 to 100 mM NaCl in the same buffer. The light green fractions containing IVD with the lowest A_{270}/A_{445} ratio (<12) are pooled and dialyzed against 10 mM potassium phosphate, pH 7.6, using potassium phosphate that has been treated with Chelex 100 resin (Bio-Rad). This sample is then loaded onto a 10-μm ceramic hydroxyapatite column (10 × 30 mm) and eluted with a gradient as described above. The protein-bound resin is washed with 150 ml of 50 mM potassium phosphate, pH 7.4, at a flow rate of 2 ml/min, and the enzyme is eluted with a 625-ml linear gradient to a final concentration of 175 mM potassium phosphate, pH 7.4. Enzyme elutes from the ceramic hydroxyapatite column as a broad peak. The A_{270}/A_{434} ratio of the pooled sample should be approximately 6.7. Sample is then concentrated to greater than 8 mg/ml. For storage, EDTA, dithiothreitol (DTT), and glycerol are added to the sample at a final concentration of 0.1 M, 1 mM, and 20% (v/v), respectively, and stored at −80°.

Preparation of Isovaleryl-CoA Dehydrogenase for Enzyme Analyses ("degreening"). Buffer (final concentration, 200 mM Tris, pH 8.0) is added to a small thawed sample, diluting it four times in a 2-ml glass tube with a rubber stopper. The sample is then deaerated with alternating cycles of oxygen and argon as for the ETF assay. Sodium hydrosulfite is added to the sample to a final concentration of about 0.6 M (100 mg/ml) and incubated for 3 hr at 37° under argon. During this time, the sample loses most of its green color. The enzyme is then dialyzed anaerobically under argon against 20 mM Tris, 10 mM sodium hydrosulfite, 50 mM EDTA, pH 7.5, for 5 hr. Final traces of green color disappear from the enzyme preparation after 3 hr of dialysis. The colorless enzyme solution is dialyzed aerobically against 20 mM Tris, pH 8.0, at 4° for another 5 hr, after which time it takes on a bright yellow color with an A_{270}/A_{445} ratio of 5.4. The "degreened" IVD is prone to precipitation; therefore reduced enzyme samples are always freshly prepared and used immediately for spectral and enzyme activity assays.

Characteristics of Recombinant Human Isovaleryl-CoA Dehydrogenase

Spectral analysis and X-ray crystallography of the recombinant IVD indicate that the green color is related to the presence of CoA persulfide bound to the purified enzyme.[25] This presumably acts as an inhibitor leading to the observed reduction in kinetic parameters seen in the "green" enzyme. With high EDTA concentrations, the specific activity of the enzyme prior to degreening is usually less than 3 μmol of ETF reduced min^{-1} mg^{-1}, while without the inclusion of the high EDTA concentrations, it is in the range of 8 to 12 μmol of ETF reduced min^{-1} mg^{-1}. After degreening of an enzyme sample prepared in the presence of high EDTA concentrations, specific activity has been measured to be as high as 90 μmol of ETF reduced min^{-1} mg^{-1}. This suggests tighter binding of the CoA persulfide resulting from the EDTA treatment.

Kinetic Constants and Spectral Properties

Kinetic constants of purified recombinant human IVD are shown in Table I. The catalytic efficiency is clearly in favor of isovaleryl-CoA as a substrate. The apparent absorbancy maxima (A_{max}) of recombinant IVD are at 269, 370, and 445 nm at ambient temperature, showing a slight shift compared with those of FAD in water (Fig. 2A; A_{max} 265, 377, and 449). The ratios of absorption maxima of the enzyme are 5.4 : 0.7 : 1.0. This A_{267}/A_{445} ratio is less than one-fourth that of the IVD isolated from human liver.[6] The experimental estimate of the FAD content of the purified IVD is 0.9 mol per mole of monomer, indicating little or no loss of FAD in the recombinant enzyme.

Catalytic Base in Isovaleryl-CoA Dehydrogenase

Site-directed mutagenesis has been used to replace Glu-254 with glycine, aspartate, and glutamine and construct a Glu254Gly/Ala375Glu IVD dou-

---

[25] K. A. Tiffany, D. L. Roberts, M. Wang, R. Paschke, A.-W. A. Mohsen, J. Vockley, and J.-J. P. Kim, *Biochemistry* **36**, 8455 (1997).

TABLE I

KINETIC CONSTANTS OF RECOMBINANT HUMAN ISOVALERYL-CoA
DEHYDROGENASE PURIFIED IN 0.1 M EDTA AND MEASURED WITH
ELECTRON-TRANSFERRING FLAVOPROTEIN FLUORESCENCE QUENCHING ASSAY[a]

Substrate	K_m (μM)	V_{max} (nmol min^{-1})	k_{cat}/K_m per mole of FAD[b] (μM^{-1} min^{-1})
Isovaleryl-CoA[c]	2.5	1.03	952 (170)[d]
Butyryl-CoA	49.5	4.45	10
Valeryl-CoA	19.3	4.70	28
Hexanoyl-CoA	30.3	2.96	11

[a] Adapted from A.-W. A. Mohsen and J. Vockley, *Biochemistry* **34**, 10146 (1995).

[b] Expressed as active monomer.

[c] The amount of enzyme used to determine the constants for isovaleryl-CoA and for the remaining substrates was 0.015 and 0.2 μg of IVD, respectively. The IVD specific activity measured with isovaleryl-CoA was 79.2 μmol of ETF reduced min^{-1} mg^{-1}.

[d] Value in parentheses was previously published [A.-W. A. Mohsen, B. Anderson, S. Volchenboum, K. P. Battaile, K. A. Tiffany, D. Roberts, J.-J. P. Kim, and J. Vockley, *Biochemistry* **37**, 10325 (1998)].

ble mutant.[23] The cDNAs for the mature mutant IVD proteins have been cloned into pKK223-3, expressed in *E. coli,* and purified. No enzymatic activity is observed for the glycine, aspartate, and glutamine mutants. The Glu254Gly/Ala375Glu mutant protein had 3.2% of the activity of the wild-type IVD for isovaleryl-CoA and 11.8% of the activity of the wild-type IVD for hexanoyl-CoA. These results suggest that Glu-254 of IVD interacts with the isoalloxazine moiety of FAD and support the hypothesis that the Glu-254 carboxylate plays an indispensable role in catalysis. In addition, it is interesting to note that the substrate specificity of the Glu254Gly/Ala375Glu mutant is altered compared with the wild-type enzyme, with the double mutant being more active toward hexanoyl-CoA relative to isovaleryl-CoA. Thus, it appears that the position of the catalytic residue in the active site plays a role in determining substrate specificity. Absorption spectra of the single mutant IVDs are altered compared with that of the wild-type (especially for the Asp mutant), suggesting perturbation in the binding of the isoalloxazine ring of the FAD, while the wild-type pattern is restored in the double mutant (Fig. 2A–D). Quenching of the FAD-related absorbance peaks with substrate binding and the establishment of a new maximum that is characteristic of the formation of an enzyme/substrate charge transfer complex are both disrupted in the single mutant

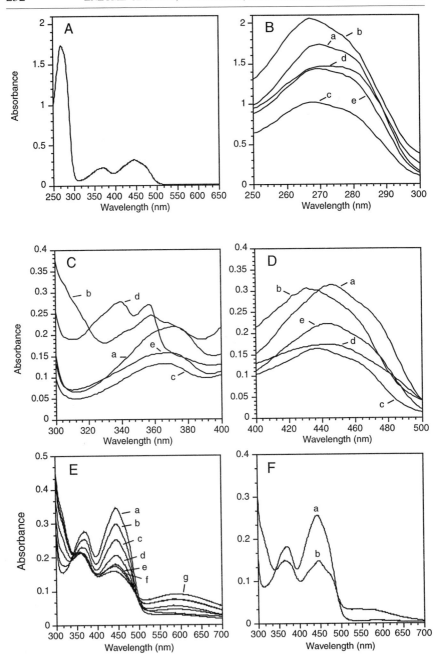

enzymes, while normal patterns are restored in the double mutant (Fig. 2E and F).

Isovaleric Acidemia

Deficiency of IVD results in the inborn error of metabolism isovaleric acidemia. A number of point mutations in the IVD gene have been identified in patients with isovaleric acidemia.[26,27] The positions of nine patient mutations are shown in the IVD crystal structure in Fig. 3. Kinetic properties of mutant enzymes that can be stably expressed in *E. coli* are shown in Table II.

Expression and Characterization of Rat and Human Short/Branched-Chain Acyl-CoA Dehydrogenase

cDNAs for the predicted mature coding regions of rat and human SBCADs have been constructed by PCR mutagenesis as for IVD and cloned into the pKK223-3 prokaryotic expression vector. An 8-liter culture of JM105 containing the expression vector is induced in mid-log phase with 0.5 mM IPTG and incubated overnight at 37°. Crude cellular extract prepared as described above is allowed to bind in batch to 25 g of DEAE-cellulose (Whatman DE52; Whatman Laboratories, Maidstone, England) in 25 mM Tris, pH 7.6. The beads are washed extensively with the same buffer and packed into a Pharmacia XK 16/40 column (Pharmacia, Piscataway, NJ), and the column is eluted with a 250-ml gradient of 0–300 mM

---

[26] J. Vockley, B. Parimoo, and K. Tanaka, *Am. J. Hum. Genet.* **40,** 147 (1991).

[27] A.-W. A. Mohsen, B. Anderson, S. Volchenboum, K. P. Battaile, K. A. Tiffany, D. Roberts, J.-J. P. Kim, and J. Vockley, *Biochemistry* **37,** 10325 (1998).

FIG. 2. (A–D) Ultraviolet/visible light absorption scans of IVD proteins at ambient temperature in 0.1 M potassium phosphate, pH 7.8. Scan of wild-type recombinant human IVD (7.2 μM) (A) and different regions of IVD protein spectra (B–D); wild-type IVD (7.2 μM) (a) and the E254G (7.8 μM) (b), E254Q (4.4 μM) (c), E254D (3.6 μM) (d), and E254G/A375E (6.6 μM) (e) mutant IVD proteins. (E and F) Ultraviolet/visible light absorption scans of IVD proteins in the presence of isovaleryl-CoA. Five nanomoles of wild-type recombinant human IVD sample was scanned in the presence of various amounts of substrate (A): 0 (a), 2.4 (b), 5.9 (c), 10.6 (d), 16.4 (e), 20.0 (f), and 32.9 (g) nmol of isovaleryl-CoA. (D) Scan of E254G/A375E mutant IVD proteins in the absence (a) and presence (b) of a 16-fold molar excess of isovaleryl-CoA. [Adapted from A.-W. A. Mohsen and J. Vockley, *Biochemistry* **34,** 10146 (1995).]

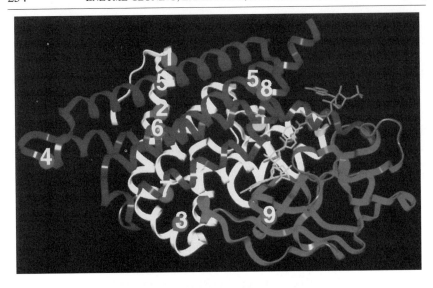

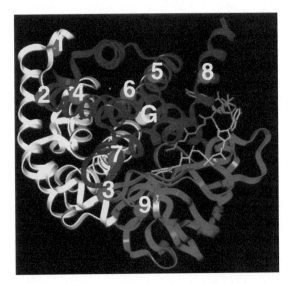

Fig. 3. IVD point mutations identified in patients with isovaleric acidemia. Numbering is as follows: 1, Leu13Pro; 2, Arg21Pro; 3, Asp40Asn; 4, Ala282Val; 5, Cys328Arg; 6, Val342Ala; 7, Arg363Cys; 8, Arg382Leu; 9, Gly170Val. The catalytic base Glu-375 is labeled with a "G."

TABLE II

KINETIC PARAMETERS OF RECOMBINANT HUMAN ISOVALERYL-CoA DEHYDROGENASE MUTANTS MEASURED BY ELECTRON-TRANSFERRING FLAVOPROTEIN FLUORESCENCE REDUCTION OR PHENAZINE METHOSULFATE/2,6-DICHLORPHENOL–INDOPHENOL ASSAYS[a]

Parameter	Wild-type		Ala282Val		Val342Ala		Arg382Leu	
	ETF	DCIP	ETF	DCIP	ETF	DCIP	ETF	DCIP
Enzyme used (μg)	0.07	17.0	1.0	10.4	1.1	22.0	0.55	11.0
K_m (μM)	3.1	29.6	27.0	410	2.8	24.0	6.9	4.5
V_{max} (nmol ETF min^{-1})	0.82	9.3	2.3	30.6	0.61	4.3	0.77	5.6
K_{cat} per mole FAD (sec^{-1})	8.8	10.0	2.2	2.81	0.81	0.29	2.0	0.73
Tetramer catalytic efficiency (μM^{-1} min^{-1})	520	61.8	11.4	1.0	27	1.1	28	15.5
Catalytic efficiency per mole FAD (μM^{-1} min^{-1})	170	20.4	4.8	0.4	7	0.73	17	9.6

[a] Adapted from A.-W. A. Mohsen, B. Anderson, S. Volchenboum, K. P. Battaile, K. A. Tiffany, D. Roberts, J.-J. P. Kim, and J. Vockley, *Biochemistry* 37, 10325 (1998).

NaCl in 25 mM Tris, pH 7.6. Fractions containing expressed SBCAD protein are identified by activity assay. Fractions with significant activity are pooled and used for subsequent substrate specificity assays. Protein concentration in the samples is determined with the Bio-Rad protein assay reagent. Because of a low level of expression, purification of SBCAD has not yet been achieved.

Substrate Specificity

The relative activities of rat and human SBCAD as measured with the ETF assay are shown in Table III.[13,14] No activity is obtained with extracts from cultures of *E. coli* containing the starting vector with no insert. The activity of the enzyme preparations is not inhibited by antibodies to purified human SCAD, and this antiserum identifies no specific bands on Western blotting in the SBCAD extracts. In addition to the substrates in Table III, no detectable activity is observed with the following compounds: isovaleryl-CoA, decanoyl-CoA, dedecanoyl-CoA, tetradecanoyl-CoA, glutaryl-CoA, 3-hydroxymethylglutaryl-CoA, and 3-methylcrotonyl-CoA. The pattern of substrate utilization is relatively conserved between rat and human for the other ACDs. In contrast, there are several significant differences in substrate utilization between rat and human SBCAD. Rat SBCAD is less active toward straight-chain substrates relative to 2-methylbutyryl-CoA

TABLE III
RELATIVE SPECIFIC ACTIVITY OF MATURE RAT SHORT/BRANCHED-CHAIN ACYL-CoA
DEHYDROGENASE EXPRESSED IN *Escherichia coli* FOR VARIOUS SUBSTRATES[a]

Substrate	Relative activity of rat SBCAD[b,c] (%)	Relative activity of human SBCAD[b,c] (%)
(S)-2-Methylbutyryl-CoA	100	100
(R)-2-Methylbutyryl-CoA	61	74
Isobutyryl-CoA	37	10
Butyryl-CoA	25	61
2-Ethylhexanoyl-CoA	30	0
2-Methylhexanoyl-CoA	8.6	86
Hexanoyl-CoA	8.4	109
Valprolyl-CoA	37	1.3

[a] Adapted from J. Willard, C. Vicanek, K. P. Battaile, P. P. Van Veldhoven, A. H. Fauq, R. Rozen, and J. Vockley, *Arch. Biochem. Biophys.* **331**, 127 (1996).

[b] Wild-type rat SBCAD activity, 16.6 ± 1.1 (7) as measured with (S)-2-methylbutyryl-CoA; wild-type human SBCAD activity, 120 ± 9.0 (6) as measured with (S)-2-methylbutyryl-CoA (mean ± standard deviation). The total number of determinations is in parentheses.

[c] Percent activity of SBCAD compared with (S)-2-methylbutyryl-CoA.

than its human counterpart, and while both are active against 2-methylhexanoyl-CoA, only the rat enzyme is active against 2-ethylhexanoyl-CoA. In addition, rat SBCAD is relatively more active toward valprolyl-CoA than the human enzyme. Thus it appears that the rat SBCAD substrate-binding site is tolerant of more bulky side groups, while that of human SBCAD can accommodate longer primary carbon backbones. The structural basis for these differences can be investigated by molecular modeling of SBCAD, using the known structures of other ACDs.

Molecular Modeling of Short/Branched-Chain Acyl-CoA Dehydrogenase

Using the Insight II family of molecular modeling software (Biosym Technologies, Mountain View, CA), amino acid sequence alignments of human SBCAD, human IVD, and porcine MCAD are optimized for structural modeling over a continuous set of overlapping stretches. Two techniques are employed to identify regions of structural homology. In the first, the root mean square (RMS) difference between the backbone atom coordinates of the segment is calculated iteratively while the range of residues being compared is adjusted in both extent and position. In the second, an interatomic distance matrix is constructed from the known ACD α-carbon coordinates. A continuous range of peptide segments from SBCAD is then overlaid on the matrix of each of the ACDs, and the minimum RMS value and corresponding residue match are stored. Comparison of the segments below a defined RMS value is performed in an iterative fashion, and a final map of conserved segments is formed. The coordinates of the areas of conservation are then mapped onto the amino acid sequence of SBCAD, using the Homology module of the Insight II software. Finally, the newly defined structure is imported into the Discover module of Insight II, where it is minimized for 300 iterations. To support the validity of this technique in predicting ACD structures, a similar calculation for human IVD based on the known structure of MCAD and a bacterial butyryl-CoA dehydrogenase has been performed and then compared with the actual structure for IVD. The carbon backbone of the predicted IVD structure

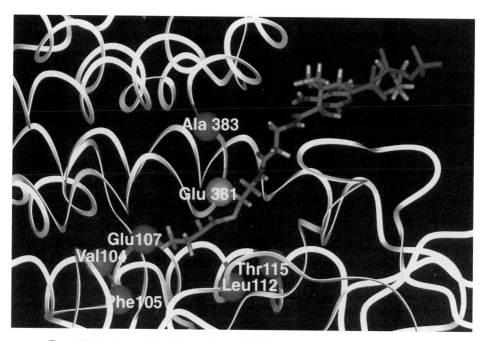

FIG. 4. Molecular model of human SBCAD. Carbon backbone is shown in white. Amino acid residues predicted to differ between the human and rat enzyme are solid rendered and numbered. A molecule of (S)-2-methylbutyryl-CoA is modeled into the substrate-binding pocket.

TABLE IV
RELATIVE SPECIFIC ACTIVITY FOR VARIOUS SUBSTRATES OF PARTIALLY PURIFIED WILD-TYPE AND
MUTANT HUMAN SHORT/BRANCHED-CHAIN ACYL-CoA DEHYDROGENASE PROTEINS
EXPRESSED IN *Escherichia coli*[a]

Substrate	Wild-type[c] (mU/mg)	E381G/G260E[c] (mU/mg)	E381D[c] (mU/mg)
(S)-2-Methylbutyryl-CoA	18.0 ± 1.5	4.6 ± 0.4 (26%)	12.5 ± 0.4 (70%)
(R)-2-Methylbutyryl-CoA	13.3 ± 2.1 (74%)	1.3 ± 0.07 (10%)	1.7 ± 0.2 (13%)
Butyryl-CoA	10.9 ± 3.4 (61%)	0 (0%)	0.47 ± 0.01 (4%)
Isobutyryl-CoA	0 (0%)	1.3 ± 0.4	0.49 ± 0.03
(R/S)-2-Methylhexanoyl-CoA	15.5 ± 1.7 (86%)	7.1 ± 0.2 (46%)	0.47 ± 0.01 (3%)
Hexanoyl-CoA	19.5 ± 4.4 (109%)	10.5 ± 1.0 (54%)	0 (0%)

[a] Adapted from B. Binzak, J. Willard, and J. Vockley, *Biochim. Biophys. Acta* **1382,** 137 (1998).

[b] Determined with the ETF fluorescence reduction assay.

[c] Mean activity ± the standard deviation of three determinations. Values in parentheses represent percentage of wild-type activity.

correlates quite well with that of the actual structure except in the very amino and carboxy termini, where some deviation is seen (not shown). In particular, the substrate-binding pocket and the relative position of the side chain of the catalytic base are remarkably closely approximated. The molecular model for the substrate-binding pocket of human SBCAD is shown in Fig. 4. Amino acid residues that differ between human and rat SBCAD are solid rendered. These residues are likely to be important in determining differences in substrate specificity between the two enzymes.

Catalytic Base in Short/Branched-Chain Acyl-CoA Dehydrogenase. On the basis of sequence homology and molecular modeling, it is likely that the catalytic residue in SBCAD is Glu-381. To examine this, human SBCADs mutated at the codon corresponding to this residue have been constructed and expressed in *E. coli,* and crude cell-free extracts have been partially purified by chromatography over a DEAE-Sepharose Fast Flow column. Loss of activity in the glycine, arginine, and glutamine mutant proteins, and restoration of activity in the Glu381Gly/Gly260Glu (which places a glutamate in a position analogous to the catalytic base in IVD) protein, confirm the identify of Glu-381 as the catalytic base in SBCAD (Table IV).[28]

[28] B. Binzak, J. Willard, and J. Vockley, *Biochim. Biophys. Acta* **1382,** 137 (1998).

[25] 3-Hydroxy-3-methylglutaryl-CoA Reductase

By Victor W. Rodwell, Michael J. Beach, Kenneth M. Bischoff,
Daniel A. Bochar, Bryant G. Darnay, Jon A. Friesen,
John F. Gill, Matija Hedl, Tuajuanda Jordan-Starck,
Peter J. Kennelly, DongYul Kim, and Yuli Wang

1. Reactions Catalyzed by 3-Hydroxy-3-methylglutaryl-CoA Reductase*

Biosynthesis of the isoprenoid precursor mevalonate employs NADPH and the enzyme HMG-CoA reductase (EC 1.1.1.34, 3-hydroxy-3-methylglutaryl-coenzyme A reductase) to reductively deacylate HMG-CoA:

$$(S)\text{-HMG-CoA} + 2\,\text{NADPH} + 2\,\text{H}^+ \longrightarrow (R)\text{-Mevalonate} + 2\,\text{NADP}^+ + \text{CoASH} \tag{1}$$

This four-electron oxidoreduction proceeds via two successive reductive stages in which mevaldehyde is believed to participate as an enzyme-bound intermediate:

$$(S)\text{-HMG-CoA} + \text{NADPH} + \text{H}^+ \longrightarrow [\text{Bound } (R)\text{-mevaldehyde}] + \text{NADP}^+ + \text{CoAS}^- \tag{1a}$$

$$[\text{Bound } (R)\text{-mevaldehyde}] + \text{NADPH} + \text{H}^+ \longrightarrow (R)\text{-Mevalonate} + \text{NADP}^+ \tag{1b}$$

Attempts to trap the putative intermediate mevaldehyde during the course of the overall reaction have been unsuccessful. HMG-CoA reductase nevertheless readily catalyzes two reactions of free mevaldehyde, the reduction of mevaldehyde to mevalonate [reaction (2)], and the oxidative acylation of mevaldehyde to HMG-CoA [reaction (3)]. These reactions appear to parallel stage (1b) and reverse of stage (1a) of the overall reaction, respectively.

---

* Section 1 by Daniel A. Bochar, Jon A. Friesen, and Victor W. Rodwell.

METHODS IN ENZYMOLOGY, VOL. 324

TABLE I
HMG-CoA Reductase Genes That Have Been Expressed in *Escherichia coli*

HMG-CoA reductase gene	GenBank accession number	Ref.
Homo sapiens (human)	M11058	1
Mesocricetus auratus (Syrian hamster)	M12705	2
Cricetulus griseus (Chinese hamster)	X00494, L00165	3
Blatella germanica (cockroach)	X70034	4
Arabidopsis thaliana isoform 1	X15032	5
Raphanus sativus 1 (radish)	M21329	6
Ustilago maydis (maize fungal pathogen)	L19262	7
Schistosoma mansoni (schistosome)	M22255	8
Haloferax volcanii (archaeon)	L19349	9
Sulfolobus solfataricus (archaeon)	M22002	10
Pseudomonas mevalonii (eubacterium)	M83531	11

(*R*)-Mevaldehyde   (*R*)-Mevalonate   (2)

(*R*)-Mevaldehyde   (*S*)-HMG-CoA   (3)

Several HMG-CoA reductases have been purified from their natural sources. In addition, an ever-increasing number of the *hmgr* genes that encode HMG-CoA reductases have been cloned and sequenced. Table I[1-11]

[1] R. J. Mayer, C. Debouck, and B. W. Metcalf, *Arch. Biochem. Biophys.* **267,** 110 (1988).
[2] K. Frimpong, B. G. Darnay, and V. W. Rodwell, *Protein Expression Purif.* **4,** 337 (1993).
[3] Y. P. Ching, S. P. Davies, and D. G. Hardie, *Eur. J. Biochem.* **237,** 800 (1996).
[4] J. Martínez-González, C. Buesa, M. D. Piulachs, X. Bellés, and F. G. Hegardt, *Eur. J. Biochem.* **213,** 233 (1993).
[5] S. Dale, M. Arro, B. Becerra, N. G. Morrice, A. Boronat, D. G. Hardie, and A. Ferrer, *Eur. J. Biochem.* **233,** 506 (1995).
[6] A. Ferrer, C. Aparicio, N. Nogues, A. Wettstein, T. J. Bach, and A. Borant, *FEBS Lett.* **266,** 67 (1990).
[7] R. Croxen, M. W. Goosey, J. P. Keon, and J. A. Hargreaves, *Microbiology* **140,** 2363 (1994).
[8] A. Rajkovic, J. N. Simonsen, R. E. Davis, and F. M. Rottman, *Proc. Natl. Acad. Sci. U.S.A.* **86,** 8217 (1989).
[9] K. M. Bischoff and V. W. Rodwell, *J. Bacteriol.* **178,** 19 (1996).
[10] D. A. Bochar, J. R. Brown, W. F. Doolittle, H.-P. Klenk, W. Lam, M. E. Schenk, C. V. Stauffacher, and V. W. Rodwell, *J. Bacteriol.* **179,** 3632 (1997).
[11] M. J. Beach and V. W. Rodwell, *J. Bacteriol.* **171,** 2994 (1989).

lists *hmgr* genes that have been expressed in *Escherichia coli*. Phylogenetic analysis has revealed the existence of two distinct classes of HMG-CoA reductase, termed class I and class II.[12] With one exception, the listed genes encode class I forms of the enzyme. The exception is the class II HMG-CoA reductase from *Pseudomonas mevalonii,* whose crystal structure has been solved.[13]

2. Preparation of Substrates*

The substrate for both HMG-CoA reductase and HMG-CoA lyase is (*S*)-HMG-CoA. (*R*)-HMG-CoA does not inhibit HMG-CoA lyase,[14] but inhibits HMG-CoA reductase competitively with respect to (*S*)-HMG-CoA.[14,15] (*R,S*)-HMG-CoA, but neither (*R*)- nor (*S*)-HMG-CoA, is commercially available. (*S*)-HMG-CoA has previously been prepared by the HMG-CoA synthase-catalyzed condensation of acetyl-CoA with acetoacetyl-CoA.[16]

It has often been assumed that the (*R*)-diastereomer is biologically inert, and consequently that the K_m for (*S*)-HMG-CoA is half that determined using (*R,S*)-HMG-CoA. However, as first noted for rat liver HMG-CoA reductase,[15] (*R*)-HMG-CoA can inhibit competitively with respect to the (*S*)-diastereomer. Many reported K_m values for HMG-CoA thus may be in error by a factor of two. (*R,S*)-HMG-CoA may, however, be used to estimate the K_m for (*S*)-HMG-CoA providing one knows whether the (*R*)-diasteromer inhibits, and if so, the relationship of K_i to K_m. In the special case where $K_i = K_m$ a plot of $1/v$ versus $1/[(R,S)\text{-HMG-CoA}]$ should give an apparent K_m equal to the true K_m without further correction.[14]

(S)-3-Hydroxy-3-methylglutaryl-CoA

Principle

(*R*)-Mevalonate + CoASH + 2NAD$^+$ → (*S*)-HMG-CoA + 2NADH + 2H$^+$

Synthesis of (*S*)-HMG-CoA employs the class II HMG-CoA reductase from *P. mevalonii* [EC 1.1.1.88][12] and an NAD$^+$-regenerating system to convert the (*R*)-diastereomer of (*R,S*)-mevalonate to (*S*)-HMG-CoA. Py-

[12] D. A. Bochar, C. V. Stauffacher, and V. W. Rodwell, *Mol. Genet. Metab.* **66,** 122 (1999).
[13] C. M. Lawrence, V. W. Rodwell, and C. V. Stauffacher, *Science* **268,** 1758 (1995).
* Section 2 by Kenneth M. Bischoff, Peter J. Kennelly, and Victor W. Rodwell.
[14] K. M. Bischoff and V. W. Rodwell, *Biochem. Med. Metab. Biol.* **48,** 149 (1992).
[15] A. Pastuszyn, C. M. Havel, T. J. Scallen, and J. A. Watson, *J. Lipid Res.* **24,** 1411 (1983).
[16] J. A. Watson and C. M. Havel, *Chem. Abstr.* **107,** 38114f (1987).

ruvate plus L-lactate dehydrogenase (EC 1.1.1.27) is included to reoxidize NADH and hence to favor formation of (S)-HMG-CoA. The conditions listed yield about 50 mg of 95% (or better) pure (S)-HMG-CoA. The course of the reaction and product purity are monitored by high-performance liquid chromatography (HPLC) on a C_8 reversed-phase column [(SynChrom, Lafayette, IN) RP-P-100, 250 × 4.6 mm] using 25 mM sodium phosphate, pH 4.6, in 15% (v/v) methanol as the mobile phase.

Reagents

Reversed-phase HPLC column (C_8)
DEAE-cellulose
(R,S)-Mevalonic acid
Coenzyme A
NAD$^+$
Ammonium formate
Formic acid
Sodium pyruvate
L-Lactate dehydrogenase
Pseudomonas mevalonii HMG-CoA reductase

Biosynthesis of (S)-3-Hydroxy-3-methylglutaryl-CoA. Combine and incubate at room temperature 22 ml of 100 mM KCl, 100 mM Tris (pH 8.1), 160 mg of NAD$^+$, 95 mg of coenzyme A, 80 mg of sodium pyruvate, 3.0 ml of 500 mM potassium mevalonate (pH 8.0), 10 enzyme units (eu) of *P. mevalonii* HMG-CoA reductase, and 50 eu of L-lactate dehydrogenase. For both HMG-CoA reductase and lactate dehydrogenase 1 eu corresponds to the reduction of 1 μmol of NAD$^+$ per minute. Monitor the course of the reaction by HPLC for approximately 3 hr. When equilibrium has been reached, i.e., when there is no further increase in the area of the HMG-CoA peak, add concentrated (88%, v/v) formic acid to pH 4.4. Maintain at room temperature for 30 min to allow complete conversion of unreacted mevalonate to its uncharged lactone. The sample may then be processed immediately or stored at 4°.

Ion-Exchange Chromatography. Apply the acidic reaction mixture to a 4.2 × 15 cm column of DEAE-cellulose equilibrated in 0.1 M ammonium formate, pH 4.4. Wash the sample with the same buffer, and then elute the column with 4 liters of 0.22 M ammonium formate, pH 4.4. Measure the absorbance of fractions at 260 nm. Compounds elute in the order NAD$^+$, CoASH, (S)-HMG-CoA. Assess the homogeneity of fractions that contain HMG-CoA by HPLC, inspecting the peak symmetry and area for HMG-CoA relative to other UV-absorbing components.

Product Workup. On the basis of HPLC data, combine fractions that contain HMG-CoA of the desired purity and freeze-dry to concentrate the sample and remove ammonium formate. A second lyophilization may be necessary to remove residual ammonium formate. Alternatively, or should the sample not stay frozen, considerable ammonium formate may be removed prior to freeze-drying by Sephadex G-25 gel filtration or by concentration over a YC05 Amicon (Danvers, MA) filter in a stirred cell concentrator and subsequent dilution with water.

Quantitation of Thiols and Thioesters. Concentrations of CoASH and of HMG-CoA may be determined by the procedure of Grunert and Phillips.[17] Various thiols may be used as standards. Thiol esters are determined after base-catalyzed hydrolysis.

FREE THIOLS. Combine 400 μl of sample, 400 μl of 5 M NaCl, 50 μl of 67 mM KCN, and 50 μl of 1.5 M Na$_2$CO$_3$. Mix. Add 100 μl of 67 mM sodium nitroprusside. Determine the A_{524} immediately after mixing.

THIOESTERS. Decrease sample volume to 300 μl. Add NaCl, KCN, and Na$_2$CO$_3$ as described above, followed by 50 μl of 5 M NaOH. After allowing 15 min to complete thioester hydrolysis, add 50 μl of 5 M HCl. Next, and nitroprusside as described above and measure the A_{524}.

(R)-3-Hydroxy-3-methylglutaryl-CoA

Principle

$$(S)\text{-HMG-CoA} + H_2O \rightarrow \text{acetoacetate} + \text{acetyl-CoA}$$

Synthesis of (R)-HMG-CoA uses *P. mevalonii* HMG-CoA lyase [EC 4.1.3.4] to selectively cleave the (S)-diastereomer of (R,S)-HMG-CoA. (R)-HMG-CoA is then isolated by ion-exchange chromatography and product purity is assessed by HPLC as described above for (S)-HMG-CoA.

Reagents

Dithiothreitol
(R,S)-HMG-CoA
Pseudomonas mevalonii HMG-CoA lyase

Procedure. Conduct the reaction in a volume of 10 ml that contains 10 mM MgCl$_2$, 2 mM dithiothreitol, 130 mM Tris (pH 8.0), 13 mg of (R,S)-HMG-CoA, and 0.1 eu of reactive blue fraction HMG-CoA lyase[18] or

[17] R. R. Grunert and P. H. Phillips, *Arch. Biochem.* **30,** 217 (1951).
[18] D. S. Scher and V. W. Rodwell, *Biochim. Biophy. Acta* **1003,** 321 (1989).

recombinant HMG-CoA lyase.[19] When HPLC analysis indicates that the selective cleavage of (S)-HMG-CoA has reached equilibrium, acidify the reaction mixture to pH 4.4 with ammonium formate and purify the product by chromatography on DEAE-cellulose as described above.

(R,S)-Mevaldehyde

Principle. The ability of HMG-CoA reductase to catalyze the reduction or the oxidative acylation of the catalytic intermediate mevaldehyde [reactions (2) and (3)] has proved useful in identifying the precise roles of active site residues in the catalysis of the overall reaction, reaction (1).[20] Mevaldehyde (mevaldic acid) exhibits limited stability in aqueous solution and should be prepared fresh just prior to use from "mevaldic acid precursor," *N,N'*-dibenzylethylenediammonium mevaldate methyl acetal. The precursor is first converted to the ammonium salt by adding NH_4OH and extracting the dibenzylethylenediamine into toluene. While the original procedure employs diethyl ether as the extraction solvent, even fresh reagent-grade ether contains contaminants that support the nonenzymatic oxidation of NAD(P)H and interfere with the accurate measurement of HMG-CoA reductase activity. The use of toluene stored and shipped in glass containers eliminates this problem. Mevaldic acid methyl acetal is then cleaved by acidifying the solution with HCl. The liberated methanol and any residual toluene are then removed by passing a stream of dry nitrogen over the solution for several minutes.

Reagents

Mevaldic acid precursor, the *N,N'*-dibenzylethylenediammonium salt of mevaldic acid methyl acetal (3-hydroxy-3-methyl-5,5'-dimethoxypentanoic acid) (Sigma, St. Louis, MO)

2,4-Dinitrophenylhydrazine

Synthesis. Dissolve 60 mg of mevaldic acid precursor, FW 312.4 g/mol, in 1.5 ml of ice-cold water. Chill on ice for a further 5 min, then add 1.5 ml of ice-cold 1 *N* NH_4OH. Mix. Extract the cloudy solution five times with 3-ml portions of toluene. Following each extraction, retain the aqueous (lower) phase. Subsequent operations are conducted at room temperature. Add concentrated HCl to a final concentration of 0.6 *M*. Mix. Pass a gentle

[19] J. R. Roberts, C. Narasimhan, P. W. Hruz, G. A. Mitchell, and H. M. Miziorko, *J. Biol. Chem.* **269,** 17841 (1994).

[20] D. A. Bochar, J. A. Friesen, C. V. Stauffacher, and V. W. Rodwell, Biosynthesis of mevalonic acid from acetyl-CoA. *In* "Comprehensive Natural Products Chemistry," Vol. 2: "Isoprenoids Including Carotenoids and Steroids" (D. Cane, ed.), Chap. 2. Pergamon Press, New York, 1999.

stream of dry nitrogen over the solution for approximately 15 min to remove methanol and residual toluene. Add 0.5 M K_xPO_4, pH 6.6, containing 10 mM EDTA to a final concentration of 25 mM K_xPO_4. Adjust the pH of the solution to pH 6.6 by adding 5 N KOH as needed. The final concentration of mevaldehyde should be approximately 90 to 100 mM.

Quantitation. Quantitation exploits the ability of aldehyde 2,4-dinitro-phenylhydrazones to form colored complexes in alkali.[21] Because the acidic conditions under which the 2,4-dinitrophenylhydrazones are formed cleave acetals, a fresh 1 mM aqueous solution of mevaldic acid precursor can be used to generate a standard curve that covers the range of 10 to 100 nmol of mevaldehyde. Place the sample in a glass test tube and dilute to 500 μl with water. Add 200 μl of 0.1% (w/v) 2,4-dinitrophenylhydrazine in 2 M HCl. Mix. Let stand for 10 min, then add 300 μl of 2.5 M NaOH. Mix. The solution will turn dark red, and then clarify. Measure the absorbance at 430 nm versus a reagent blank.

3. Assay of 3-Hydroxy-3-methylglutaryl-CoA Reductase Activity*

Two procedures, one radioisotopic, the other spectrophotometric, are used to measure the ability of HMG-CoA reductase to catalyze reaction (1). The radioisotopic method measures the formation of [^{14}C]mevalonate from [^{14}C]HMG-CoA. The spectrophotometric method measures the disappearance of NADPH at 340 nm. Each method has advantages and limitations. Advantages of the radioisotopic procedure, an "end-point" assay, include high sensitivity and applicability to simultaneous analysis of large numbers of samples. Ideal for analysis of microsomal preparations or limited quantities of enzyme, its drawbacks include the use of radioisotopes and their relatively high cost. Because radioactive mevaldehyde is not commercially available, the radioisotopic assay also is unsuitable for assay of reaction (2) or (3).

The spectrophotometric assay, a kinetic assay that follows the disappearance of NADPH, is most suitable for analysis of enzyme that has been expressed at a high level and that has undergone at least partial purification. Spectrophotometric assays typically require only 15–30 sec, avoid the use of radioisotopes, and are applicable to reactions (1), (2), and (3). Disadvantages include unsuitability for the analysis of microsomal preparations and the necessity of sequential rather than simultaneous assay of large numbers of samples.

---

[21] T. Friedman and G. Haugen, *J. Biol. Chem.* **147**, 415 (1943).

* Section 3 by Kenneth M. Bischoff, Matija Hedl, and Victor W. Rodwell.

Radioactive Assay of Mammalian 3-Hydroxy-3-methylglutaryl-CoA Reductase

A known initial concentration of NADPH may be used to assay purified preparations of HMG-CoA reductase. However, for crude preparations such as microsomes it is preferable to use $NADP^+$, glucose, and glucose-6-phosphate dehydrogenase to maintain a constant level of NADPH. Tritiated mevalonate is included as an internal standard to permit correction for loss due to manipulations, conversion of mevalonate to other products, incomplete sample recovery, etc.

Reagents

Mylar-backed 6 × 6 inch silica gel thin-layer chromatography (TLC) sheets without fluorescent indicator
Glucose
Glucose-6-phosphate dehydrogenase
$[3-^{14}C]-(R,S)$-HMG-CoA
$[5-^3H]-(R,S)$-Mevalonate
$NADP^+$
Buffer A: 30 mM EDTA, 250 mM NaCl, 1.0 mM dithiothreitol, 50 mM K_xPO_4, pH 7.5
Cofactor–substrate solution: A 50-μl portion of cofactor–substrate solution suitable for a single analysis contains 3.6 μmol of EDTA, 4.5 μmol of glucose 6-phosphate, 0.45 μmol of $NADP^+$, 0.3 international unit (IU) of glucose-6-phosphate dehydrogenase, 450 nmol of $NADP^+$, 50 nmol of $[3-^{14}C]-(R,S)$-HMG-CoA (specific activity ~1.5 Ci/mol), and 3×10^4 dpm of $[5-^3H]$-mevalonate (specific activity ~7 Ci/mol) in buffer A

Analysis. The conditions given are suitable for assay of microsomal or soluble HMG-CoA reductase.[22,23] While they may be conducted in as little as 35 μl, incubations typically employ a final volume of 150 μl. Mix enzyme (100–500 μg of microsomal protein, 1–30 μg of solubilized enzyme, typically in buffer A) with 50 μl of cofactor–substrate solution and incubate at 37°. After 15 min, add 15 μl of 6 N HCl to arrest the reaction. Continue incubation at 37° for an additional 15 min to allow complete lactonization of mevalonate. Centrifuge to pellet denatured protein, then apply 100-μl or smaller portions of supernatant liquid to the origin of an activated silica gel sheet that has been previously scored vertically into eight channels to prevent cross-over of radioactivity between channels. Develop the chromatogram in toluene–acetone (1 : 1, v/v). HMG-CoA, HMG, and mevalonate

[22] D. J. Shapiro, R. L. Imblum, and V. W. Rodwell, *Anal. Biochem.* **31**, 383 (1969).
[23] D. J. Shapiro, J. L. Nordstrom, J. J. Mitschelen, V. W. Rodwell, and R. T. Schimke, *Biochim. Biophys. Acta* **370**, 369 (1974).

phosphates remain at the origin whereas mevalonolactone migrates to an R_f of 0.5–0.9. Mobility can, however, vary significantly between different lots of silica gel or with changes in binder composition. Therefore, for each new batch of silica gel chromatograph a mevalonolactone standard to determine the region to count. Air dry the chromatogram, cut out the region containing mevalonolactone, transfer to scintillation fluor, and count for ^{14}C and [3]H. Using the ^{3}H disintegrations per minute (dpm) to correct for product loss, calculate the quantity of [^{14}C]-mevalonate formed.

Radioactive Assay of Haloferax volcanii
3-Hydroxy-3-methylglutaryl-CoA Reductase

Reaction (1) catalyzed by *Haloferax volcanii* HMG-CoA reductase may be assayed by a modification[9] of the radioisotopic assay described above for the mammalian enzyme. Assays, 30-μl final volume, contain 3 M KCl, 5.0 mM dithiothreitol, 0.5 mM (*R,S*)-[3-^{14}C]HMG-CoA (1.3 mCi/mmol), 1.0 mM NADPH, 1.5 nCi of (*R,S*)-[5-^{3}H]-mevalonate, and 25 mM K$_x$PO$_4$, pH 7.3. Incubations at 37° are initiated by adding HMG-CoA. Reactions are terminated after 30 min by adding 8 μl 6 N HCl. Mevalonolactone is then separated from HMG-CoA and HMG by silica gel thin-layer chromatography and quantitated as described above for mammalian HMG-CoA reductase.

Spectrophotometric Assay of Mammalian
3-Hydroxy-3-methylglutaryl-CoA Reductase

The spectrophotometric assay is applicable to all reactions catalyzed by HMG-CoA reductase.[20] Assays employ a spectrophotometer equipped with a cell holder maintained at constant temperature, generally 37°, to monitor the oxidoreduction of NADP(H) at 340 nm. The use of half-height 1-cm path spectrophotometer cells and a final reaction volume of 200 μl are recommended to minimize assay cost. For all assays, 1 enzyme unit represents the turnover, in 1 min, of 1 μmol of NADP(H). For reaction (1), this corresponds to the turnover of 0.5 μmol of HMG-CoA, but for reactions (2) and (3) this corresponds to the turnover of 1 μmol of mevaldehyde. All assays are initiated by the addition of substrate, HMG-CoA or mevaldehyde.

Reagents

Coenzyme A
(*R,S*)- or (*S*)-HMG-CoA
(*R,S*)-Mevaldehyde
NADPH
NADP[+]

Buffers I, II, and III: 100 mM NaCl, 10 mM dithiothreitol, 20 mM Na$_x$PO$_4$, adjusted to pH 6.5 (buffer I), pH 6.0 (buffer II) or pH 9.0 (buffer III). For all buffers, add dithiothreitol just prior to use.

The conditions given below are optimal for assay of the expressed catalytic domain of Syrian hamster HMG-CoA reductase.[2] Cold enzyme should be maintained at 37° for 12 min prior to assay. Monitor the background rate of NADP(H) oxidoreduction in the absence of substrate. Initiate the reaction by adding HMG-CoA or mevaldehyde, and then monitor the rate of change in absorbance at 340 nm, typically for 15–30 sec.

Reaction (1): Reductive deacylation of HMG-CoA to mevalonate: Assays contain, in 200 μl, 0.2 mM (R,S)- or (S)-HMG-CoA and 0.2 mM NADPH in buffer I, pH 6.5.

Reaction (2): Reduction of mevaldehyde to mevalonate: Assays contain, in 200 μl, 5 mM (R,S)-mevaldehyde and 0.2 mM NADPH in buffer II, pH 6.0.

Reaction (3): Oxidative acylation of mevaldehyde to HMG-CoA: Assays contain, in 200 μl, 2 mM coenzyme A, 4 mM NADP$^+$, and 5 mM (R,S)-mevaldehyde in buffer III, pH 9.0.

Spectrophotometric Assay of Haloferax volcanii 3-Hydroxy-3-methylglutaryl-CoA Reductase

The assay parallels that described above for hamster HMG-CoA reductase, with the indicated modifications. The conditions given below are optimal for assay of expressed *H. volcanii* HMG-CoA reductase. Monitor the background rate of NADP(H) oxidoreduction in the absence of substrate. Initiate the reaction by adding HMG-CoA or mevaldehyde, and then monitor the rate of change in absorbance at 340 nm, typically for 15–30 sec.

Additional Reagents

Buffer IV: 3.0 M KCl, 5.0 mM dithiothreitol, 50 mM K$_x$PO$_4$, pH 7.3
Buffers V and VI: 3.0 M KCl, 5.0 mM dithiothreitol, 50 mM Tris, 50 mM K$_x$PO$_4$, 50 mM glycine, adjusted to pH 6.5 (buffer V) or pH 8.5 (buffer VI).

Note: For all buffers, add dithiothreitol just prior to use.

Reaction (1): Reductive deacylation of HMG-CoA to mevalonate: Assays contain, in 200 μl, 0.2 mM NADPH and 0.5 mM (R,S)-HMG-CoA in buffer IV, pH 7.3.

Reaction (2): Reduction of mevaldehyde to mevalonate: Assays contain, in 200 μl, 0.2 mM NADPH, 3.0 mM (R,S)-mevaldehyde, and 1.0 mM coenzyme A in buffer V, pH 6.5. While not a reactant, coenzyme A stimulates mevaldehyde reduction.

Reaction (3): Oxidative acylation of mevaldehyde to HMG-CoA: Assays contain, in 200 μl, 3.5 mM NADP$^+$, 5.0 mM coenzyme A, and 3.0 mM (R,S)-mevaldehyde in buffer VI, pH 8.5.

Spectrophotometric Assay of Sulfolobus solfataricus 3-Hydroxy-3-methylglutaryl-CoA Reductase

The acidic pH and high temperature optimal for the activity of *Sulfolobus solfataricus* enzyme dictate conditions that differ from those for assay of a mesophilic HMG-CoA reductase. Assays are conducted at 55° and acidic pH. For assay of reactions (1) and (2), the instability of NADPH under these conditions requires an initial concentration of NADPH too great to measure accurately at 340 nm ($\varepsilon_{340} = 6220\ M^{-1}\mathrm{cm}^{-1}$). Disappearance of NADPH is therefore monitored at 366 nm ($\varepsilon_{366} = 3300\ M^{-1}\ \mathrm{cm}^{-1}$).[10] Initiate all assays by adding substrate, HMG-CoA, or mevaldehyde, and then monitor the absorbance at 366 nm, typically for 15–30 sec.

Additional Reagent

2-(N-Morpholino)ethanesulfonic acid (MES)

Reaction (1): Reductive deacylation of HMG-CoA to mevalonate: Assays contain, in 200 μl, 0.2 mM NADPH and 0.5 mM (R,S)-HMG-CoA in 100 mM MES, 100 mM potassium acetate, pH 5.5.

Reaction (2): Reduction of mevaldehyde to mevalonate: Assays contain, in 200 μl, 0.25 mM NADPH, and 4.0 mM (R,S)-mevaldehyde in 100 mM MES, 100 mM potassium acetate, pH 4.5.

Reaction (3): Oxidative acylation of mevaldehyde to HMG-CoA: Assays contain, in 200 μl, 4.5 mM NADP$^+$, 6.0 mM coenzyme A, and 3.0 mM (R,S)-mevaldehyde in 100 mM MES, 100 mM potassium acetate, pH 5.5.

Spectrophotometric Assay of Pseudomonas mevalonii 3-Hydroxy-3-methylglutaryl-CoA Reductase

Unlike other HMG-CoA reductases, the class II enzyme from *P. mevalonii* utilizes NAD(H) rather than NADP(H) as oxidoreductant for catalysis of reactions (1), (2), and (3). In addition, *P. mevalonii* HMG-CoA reductase readily catalyzes the reverse of reaction (1), the oxidative acylation of mevalonate to HMG-CoA.

Additional Reagents

2-(*N*-Cyclohexylamino)ethanesulfonic acid (CHES)
3-(*N*-Morpholino)propanesulfonic acid (MOPS)
NAD$^+$
NADH

Initiate reactions by adding substrate, HMG-CoA or mevaldehyde, and then monitor the rate of change in absorbance at 340 nm for 15–30 sec. For all three reactions, 1 enzyme unit represents the turnover in 1 min of 1 μmol of NAD(H).

Reaction (1): Reductive deacylation of HMG-CoA to mevalonate: Assays contain, in 200 μl, 0.4 m*M* NADH and 0.2 m*M* (*R,S*)-HMG-CoA in 100 m*M* MOPS, pH 7.7.

Reaction (2): Reduction of mevaldehyde to mevalonate: Assays contain, in 200 μl, 0.4 m*M* NADH and 5.0 m*M* (*R,S*)-mevaldehyde in 100 m*M* MES, pH 5.5.

Reaction (3): Oxidative acylation of mevaldehyde to HMG-CoA: Assays contain, in 200 μl, 3.0 m*M* NAD$^+$, 1.0 m*M* coenzyme A, and 5.0 m*M* (*R,S*)-mevaldehyde in 100 m*M* CHES, pH 8.7.

Reaction (−1): Oxidative acylation of mevalonate to HMG-CoA: Assays contain, in 200 μl, 2.6 m*M* (*R,S*)-mevalonate, 3.0 m*M* NAD$^+$, and 1.0 m*M* coenzyme A in 100 m*M* CHES, pH 8.9.

4. Syrian Hamster 3-Hydroxy-3-methylglutaryl-CoA Reductase*

The procedure is for the expression in *Escherichia coli* and the subsequent purification of HMGR$_{cat}$EEF, the modified catalytic domain of Syrian hamster HMG-CoA reductase. Expression vector pKFT721 contains the catalytic domain, residues 433 to 887, plus four residues (M-A-R-I-) that were added to the N-terminus during subcloning. Three additional residues (-E-E-F) subsequently were added to the C-terminus to permit purification on a EEF antibody affinity support.[2] However, affinity chromatography proved unnecessary, because in contrast to the heterogeneous product resulting from expression of the gene that did not encode the EEF epitope, expression of HMGR$_{cat}$EEF yielded essentially homogeneous protein. The purification protocol is modeled on that for purification of the catalytic domain of the human enzyme.[1]

---

* Section 4 by Bryant G. Darnay, Kenneth Li, Yuli Wang, and Victor W. Rodwell.

Materials

Blue Sepharose Cl-6B (Pharmacia, Piscataway, NJ)

LB_{amp} medium: 10 g of NaCl, 5 g of yeast extract, 10 g of tryptone per liter containing ampicillin (50 μg/ml) added just prior to use

Phenylmethylsulfonyl fluoride (PMSF), 100 mM in 2-propanol

Buffer A: 20 mM Na$_x$HPO$_4$, 50 mM NaCl, 10 mM dithiothreitol, 10% (v/v) glycerol, 100 mM sucrose, pH 7.3. Add dithiothreitol just prior to use

Buffer B: Buffer A containing 1.0 mM PMSF and 5 mM EDTA

Ammonium sulfate solution: Saturated at room temperature, pH 7.4 when measured at a 1:20 dilution

Expression plasmid pKFT721

Escherichia coli BL21(DE3): The genomic DNA of this *E. coli* strain contains a single copy of the gene that encodes T7 RNA polymerase.[24] The ϕ10 promoter of pKFT721[2] is the target for T7 RNA polymerase

Growth and Harvesting of Cells

Transform competent BL21(DE3) cells with pKFT721 and add directly to 100 ml of LB_{amp} medium. Shake at 37°, 300 rpm overnight, and use to inoculate 1 liter of LB_{amp} medium. Grow to a cell density of 1.0 to 1.5 g dry cell equivalent per liter. Harvest the cells by centrifugation and discard the spent medium. Suspend the cells in cold 0.9% (w/v) saline and centrifuge as described previously. Decant and discard the wash fluid. Retain the washed cells.

Preparation of Cell-Free Extract

Unless otherwise stated, conduct all subsequent operations at about 4°. Suspend the washed cells in 15 ml of buffer B per gram of dry cells. Rupture the cells by passage three times through a French pressure cell. Centrifuge the cell lysate at 30,000 rpm, 4°, for 60 min, then decant and retain the supernatant liquid as the cytosol.

Fractionation with Ammonium Sulfate

Add to the cytosol, with gentle stirring, 0.54 volume of room temperature saturated ammonium sulfate solution. This adjusts the concentration of ammonium sulfate to 35% saturation, Centrifuge. Discard the 0–35% precipitate. To the 35% supernatant liquid add 0.44 volume of saturated

---

[24] F. W. Studier, A. H. Rosenberg, J. J. Dunn, and J. W. Dubendorf, *Methods Enzymol.* **185**, 60 (1990).

ammonium sulfate solution. After centrifugation, dissolve the 35–55% precipitate in 5–10 ml of buffer A to give the ammonium sulfate fraction. The procedure may be halted at this point. When frozen in liquid nitrogen, the ammonium sulfate fraction retains full activity for several months.

Affinity Chromatography on Blue Sepharose

Conduct subsequent operations at about 4°. Because dilution of the ammonium sulfate fraction causes rapid loss of activity, small portions of this fraction are diluted as needed just prior to application to Blue Sepharose. To determine the dilution that will ensure binding of the enzyme to Blue Sepharose, add 0.5 ml of the ammonium sulfate fraction to 10 ml of cold buffer A. Determine the conductance, then dilute with cold buffer A to a conductance of 3–4 mmho. Assume that an equivalent dilution of the ammonium sulfate fraction will achieve the same conductivity. Apply sucessive 1-ml portions of diluted ammonium sulfate fraction, conductance 3–4 mmho, to a 2.6 × 3.5 cm (18-ml bed volume) column of Blue Sepharose equilibrated in buffer A until the entire fraction has been applied. Wash the column with 150 ml buffer A. Elute with a 150 × 150 ml gradient of 0 to 1.2 M NaCl in buffer A. Collect 5-ml fractions. Combine fractions that contain the majority of the activity and precipitate the protein with 1.5 volumes of saturated ammonium sulfate solution. Centrifuge, dissolve the precipitate in 2–4 ml of buffer A, divide in portions, and store the Blue Sepharose fraction in liquid nitrogen. $R_{cat}EEF$ protein is recognized both by rabbit anti-rat HMG-CoA reductase antibody and by anti-yeast γ-tubulin antibody. Tables II and III summarize the purification and kinetic properties of the enzyme, respectively.

TABLE II
PURIFICATION OF SYRIAN HAMSTER HMG-CoA REDUCTASE, $R_{cat}EEF^a$

Fraction	Total activity (eu)	Total protein (mg)	Specific activity (eu/mg)	Enrichment (fold)	Recovery of activity (%)
Crude extract	1250	570	2.2	(1.0)	(100)
(NH$_4$)$_2$SO$_4$	910	180	5.1	2.3	73
Blue Sepharose	830	22	38	17	66

^{a} The data are for purification of $R_{cat}EEF$, the modified catalytic domain of Syrian hamster HMG-CoA reductase, from 1 liter of culture. Assays were conducted at pH 6.75. One enzyme unit (eu) corresponds to the oxidation of 1 μmol of NADPH per minute.

TABLE III
KINETIC PARAMETERS FOR SYRIAN HAMSTER HMG-CoA REDUCTASE

Reaction	Optimal pH	V_{max}	K_m (mM)					
			HMG-CoA	Mevaldehyde	Mevalonate	CoASH	NADP$^+$	NADPH
HMG-CoA → mevalonate	6.2	64	0.02					0.08
Mevaldehyde → mevalonate	6.0	14		1.6				0.16
Mevalonate → HMG-CoA	8.5	0.6			0.02	0.01	0.51	
Mevaldehyde → HMG-CoA	9.0	2.0		0.09		0.05	0.60	

5. *Pseudomonas mevalonii* 3-Hydroxy-3-methylglutaryl-CoA Reductase*

Recombinant *Pseudomonas mevalonii* HMG-CoA reductase is produced in *Escherichia coli* transformed with plasmid pHMGR, which contains the *P. mevalonii hmgr* gene under the control of the *tac* promoter.[11] The protocol for subsequent isolation of the expressed enzyme[11,25] draws on that for isolation of the enzyme from its native host.[26]

Solutions

LB$_{amp}$ medium: 10 g of NaCl, 5 g of yeast extract, 10 g of tryptone per liter, to which ampicillin (50 μg/ml) is added just prior to use
Buffer KPEG: 1.17 g of K$_2$HPO$_4$, 0.46 g of KH$_2$PO$_4$, 0.37 g of Na$_2$EDTA, and 100 ml of glycerol per liter, pH 7.3
Phenylmethylsulfonyl fluoride (PMSF), 100 mM in 2-propanol
Buffer KPEG/PMSF: Buffer KPEG containing 1 mM phenylmethylsulfonyl fluoride added just prior to use
Ammonium sulfate solution saturated at 4°, pH 7.4, when measured at room temperature as a 1:20 dilution

Growth and Harvesting of Cells

Inoculate 10 ml of LB$_{amp}$ with BL21(DE3) cells that have been freshly transformed with pHMGR. Shake at 37°, 300 rpm overnight, and then transfer to 100 ml of LB$_{amp}$. Grow for about 5 hr, and then transfer to 1

* Section 5 by Michael J. Beach, Bryant G. Darnay, Jon A. Friesen, John F. Gill, Matija Hedl, Tuajuanda Jordan-Starck, and Victor W. Rodwell.
[25] T. C. Jordan-Starck and V. W. Rodwell, *J. Biol. Chem.* **264**, 17913 (1989).
[26] J. F. Gill, Jr., M. J. Beach, and V. W. Rodwell, *J. Biol. Chem.* **260**, 9393 (1985).

liter of LB$_{amp}$. Shake at 37°, 300 rpm. After approximately 10 hr, record the culture density and estimate the cell yield in grams dry cell equivalent per liter. Cell yields average 1.0 to 1.5 g/liter. Harvest the cells by centrifugation, discard the spent medium, and suspend the cells in cold 0.9% (w/v) NaCl. Centrifuge, discard the wash fluid, and retain the washed cells.

Cell-Free Extract

Unless otherwise stated, all subsequent operations are conducted at about 4°. Suspend the washed cells in about 15 ml of KPEG/PMSF per gram dry cell equivalent. Pass the cell suspension three times through a French pressure cell to give the crude extract. Centrifuge (30,000 rpm, 60 min, 4°), and then decant and retain the supernatant liquid as the cytosol.

Fractionation with Ammonium Sulfate

Adjust the cytosol to 40% saturation with ammonium sulfate by adding, with stirring, a volume of saturated ammonium sulfate solution equal to two-thirds the volume of the cytosol. Stir for 10 min. Centrifuge, discard the supernatant liquid, and dissolve the precipitate in 10–20 ml of KPEG/PMSF to give the ammonium sulfate fraction. This may be stored overnight in liquid nitrogen.

Chromatography on DEAE-Sepharose

Because the enzyme loses activity at low ionic strength, successive small portions of the ammonium sulfate fraction are diluted appropriately with cold KPEG prior to application to the DEAE column. Measure the conductance of 1 ml of chilled ammonium sulfate fraction plus 7 ml of cold buffer KPEG. Continue to add buffer KPEG to a conductivity of 2.0 to 2.5 mmho. Assume that an equivalent dilution of the ammonium sulfate fraction will have a conductivity of 2.0–2.5 mmho. Apply successively diluted portions of the ammonium sulfate fraction to a 2.5 × 8.5 cm (45 ml) column of DEAE-Sepharose (Pharmacia DEAE-Sepharose Fast Flow) in KPEG until all of the enzyme has been loaded. Wash the column with 150 ml of KPEG. A small quantity may fail to bind to the support. Elute the column with 150 ml of 150 mM KCl in KPEG. Assay fractions for activity and combine active fractions, except for those that emerge well after the major peak of activity, to give the DEAE fraction. Add KCl to a final concentration of 400 mM KCl and store in liquid nitrogen. Tables IV and V summarize the purification and kinetic properties of the enzyme, respectively.

TABLE IV

PURIFICATION OF *Pseudomonas mevalonii* HMG-CoA REDUCTASE[a]

Fraction	Total activity (eu)	Total protein (mg)	Specific activity (eu/mg)	Enrichment (fold)	Recovery (%)
Cell lysate	5190	680	7.6	(1.0)	(100)
Cytosol	5060	556	9.1	1.2	97
$(NH_4)_2SO_4$	4690	227	20.6	2.7	90
DEAE-Sepharose	4560	111	40.9	5.4	87

[a] The data are for the purification of enzyme from 1 liter of culture.

6. *Haloferax volcanii* 3-Hydroxy-3-methylglutaryl-CoA Reductase*

Expression in *E. coli* of the gene that encodes the HMG-CoA reductase of the halophilic archaeon *Haloferax volcanii* results in an initially inactive enzyme. Activation is achieved by exposure to NaCl or other salts, an environment more favorable for correct folding than that of *E. coli* cytosol.[9]

Reagents

DEAE-cellulose
Dithiothreitol
Phenylmethylsulfonyl fluoride
Isopropylthiogalactoside
Buffer A: 25 mM K_xPO_4, pH 6.6, containing 5.0 mM dithiothreitol added just prior to use

* Section 6 by Kenneth Bischoff and Victor W. Rodwell.

TABLE V

KINETIC PARAMETERS FOR *Pseudomonas mevalonii* HMG-CoA REDUCTASE

Reaction	Optimal pH	V_{max}	K_m (mM)					
			HMG-CoA	Mevaldehyde	Mevalonate	CoASH	NAD$^+$	NADH
HMG-CoA → mevalonate	7.7	66	0.02					0.08
Mevaldehyde → mevalonate	5.5	140		8.1				1.8
Mevalonate → HMG-CoA	8.7	42			0.26	0.06	0.30	
Mevaldehyde → HMG-CoA	8.9	2		0.08		0.11	0.12	

LB$_{amp}$ medium: 10 g of tryptone, 5 g of yeast extract, 10 g of NaCl per liter, containing ampicillin (75 μg/ml)

Ammonium sulfate solution: Saturated at room temperature, pH 7.4, when measured as a 1:20 dilution

Expression plasmid pT7-Vred

Growth of Cells

Grow cells harboring expression plasmid pT7-Vred at 37° in 1 liter of LB$_{amp}$ medium. At a culture density of about 80 Klett units (red filter), add isopropylthiogalactoside to a final concentration of 0.5 mM. Allow growth to continue to a culture density of about 200 Klett units. Harvest the cells by centrifugation, suspend them in 0.9% (w/v) NaCl, centrifuge, and suspend the washed cells in 150 mM KCl, 5 mM dithiothreitol, 1.0 mM phenylmethylsulfonyl fluoride, 10% (v/v) glycerol, 20 mM Tris, pH 8.3 (15 ml/g dry cell equivalent). Rupture the cells by passage twice through a French pressure cell. Following centrifugation of the cell lysate (100,000g, 4°, 30 min), retain the supernatant liquid as the cytosol. Note that the cytosol is catalytically inactive and must be activated before proceeding with purification.

Activation of the Cytosol

Activation of the cytosol is achieved by adding solid KCl to a concentration of 3.0 M followed by incubation at 4° for 6 hr. NaCl or (NH$_4$)$_2$SO$_4$ also activate the enzyme. After centrifugation to remove precipitated protein, decant and retain the supernatant liquid as the activated cytosol.

Precipitation with Ammonium Sulfate

Dialyze the activated cytosol against 20 volumes of 1.9 M (NH$_4$)$_2$SO$_4$ in buffer A. Centrifuge to remove precipitated protein. Decant the supernatant liquid and retain it as the ammonium sulfate fraction.

Heat Fractionation

Heat the ammonium sulfate fraction to 70°. After 10 min, allow it to cool to room temperature and centrifuge to sediment denatured protein. Decant and retain the supernatant liquid as the heat fraction.

Ion-Exchange Chromatography

Apply the heat fraction to a 1.5 × 6.5 cm column of DEAE-cellulose equilibrated with 1.9 M (NH$_4$)$_2$SO$_4$ in buffer A. Wash the column with 100 ml of the same solution. Most of the HMG-CoA reductase, but few $E.$ $coli$ proteins, bind to the support under these conditions. Elute the column with 100 ml of 1.5 M (NH$_4$)$_2$SO$_4$, in buffer A. Combine the active fractions and

TABLE VI
PURIFICATION OF *Haloferax volcanii* HMG-CoA REDUCTASE[a]

Fraction	Total activity (eu)	Total protein (mg)	Specific activity (eu/mg)	Enrichment (fold)	Recovery of activity (%)
Cytosol	0	209			
Activated cytosol	23.1	160	0.14	(1.0)	(100)
(NH$_4$)$_2$SO$_4$	21.6	40	0.54	4	93
Heat	31.4	14	2.2	16	136
DEAE	14.8	0.6	24	170	64

[a] The data are for the purification of enzyme from 1 liter of culture. Because no activity is detectable in unactivated cytosol, recovery of enzyme units is based on the activated cytosol.

concentrate by ultrafiltration through a PM30 membrane to give the DEAE fraction. Tables VI and VII summarize the purification and kinetic properties of the enzyme, respectively.

7. *Sulfolobus solfataricus* 3-Hydroxy-3-methylglutaryl-CoA Reductase*

Recombinant *Sulfolobus solfataricus* HMG-CoA reductase is expressed in *E. coli* transformed with plasmid pET-21b(Sol)HMGR, a derivative of pET-21b (Novagen, Madison, WI) that contains the *S. solfataricus* HMG-CoA reductase gene[10] under the control of the T7 promoter. Purification to homogeneity of *S. solfataricus* HMG-CoA reductase is achieved by

* Section 7 by Daniel A. Bochar, Dongyul Kim, and Victor W. Rodwell.

TABLE VII
KINETIC PARAMETERS FOR *Haloferax volcanii* AND *Sulfolobus solfataricus* HMG-CoA REDUCTASE

Reaction	Optimal pH	V_{max}	K_m (mM) HMG-CoA	Mevaldehyde	Mevalonate	CoASH	NADP$^+$	NADPH
1. HMG-CoA → mevalonate								
Sulfolobus solfataricus	5.0	7.6	0.045					0.055
Haloferax volcanii		12	0.06					0.066
2. Mevaldehyde → mevalonate	6.0							
Sulfolobus solfataricus		5.8		7.0				0.045
Haloferax volcanii		11		0.55				0.032
3. Mevalonate → HMG-CoA	9.5							
Sulfolobus solfataricus		0.60			1.5	0.11	0.090	
4. Mevaldehyde → HMG-CoA	9.0							
Sulfolobus solfataricus		1.4		4.3		2.6	0.90	
Haloferax volcanii		2.1						

successive heat treatment, ion-exchange chromatography, hydrophobic interaction chromatography, and affinity chromatography.

Materials

LB$_{amp}$medium: 10 g of NaCl, 5 g of yeast extract, 10 g of tryptone per liter containing ampicillin (75 μg/ml) added just prior to use

Phenylmethylsulfonyl fluoride, 100 mM in 2-propanol

Buffer A: 1.0 mM EDTA in 20 mM K$_x$PO$_4$, pH 7.0, with 1 mM phenylmethylsulfonyl fluoride added just prior to use

SP Sepharose FF, Q Sepharose FF, butyl Sepharose 4 FF, and Sephadex G-25 (Pharmacia)

Reactive Red-120 agarose (Sigma)

Escherichia coli strain BL21 (DE3)

Expression vector pET-21b(Sol)HMGR

Growth and Harvesting of Cells

Grow 10 liters of *E. coli* BL21 (DE3) cells transformed with pET-21b(Sol)HMGR at 37°, with shaking, on LB$_{amp}$ medium to a culture density of 400–425 Klett units (red filter). Harvest the cells by centrifugation.

Cell-Free Extract

Suspend the cells in 100 ml of buffer A that contains 10% (v/v) glycerol and rupture them by passage twice through a French pressure cell. After centrifugation of the cell lysate at 105,000g for 60 min at 4°, decant and retain the supernatant liquid as the cytosol.

Heat Treatment

Maintain approximately 40-ml portions of the cytosol at 80° for 25 min. Cool on ice and centrifuge to sediment denatured protein. Decant and retain the supernatant liquid. Suspend the precipitate in approximately 4 volumes of buffer A, centrifuge, and combine the supernatant liquid with the initial supernatant liquid to give the heat fraction.

Ion-Exchange Chromatography

The following procedure exploits the inability of either anion or cation exchangers to retain the enzyme. Apply the heat fraction to a 1.5 × 6.0 cm column of SP Sepharose FF in buffer A linked in series to a 1.5 × 6.0 cm column of Q Sepharose FF in buffer A. Elute the linked columns with 100 ml of buffer A. Combine the liquid emerging during sample loading with the eluate. Adjust to 0.82 M in (NH$_4$)$_2$SO$_4$ by adding 0.25 ml of

TABLE VIII
PURIFICATION OF *Sulfolobus solfataricus* HMG-CoA REDUCTASE[a]

Fraction	Total activity (eu)	Total protein (mg)	Specific activity (eu/mg)	Enrichment (fold)	Recovery of activity (%)
Cytosol	1330	3700	0.36	(1.0)	(100)
Heat	870	440	2.0	6	65
Ion	850	290	3.0	8	64
Butyl	690	110	6.2	17	52
Red	500	29	17.5	49	38

[a] The data are for the purification of the enzyme from 10 liters of culture.

saturated $(NH_4)_2SO_4$ solution (saturated at room temperature, pH 7.0 when diluted 1 : 20) per milliliter of combined eluate. Centrifuge to remove precipitated protein and retain the supernatant liquid as the ion fraction.

Hydrophobic Interaction Chromatography

Apply the ion fraction to a 2.5 × 12 cm column of butyl Sepharose 4 FF in 0.82 M $(NH_4)_2SO_4$ in buffer A and wash it in with 200 ml of 0.82 M $(NH_4)_2SO_4$ in buffer A. Elute the column, first with a decreasing gradient of 0.82 to 0 M $(NH_4)_2SO_4$ in buffer A (400 ml total), and subsequently with 300 ml of buffer A. Continue to elute with additional buffer A until the eluate contains no significant activity. Combine the active fractions and adjust to 65% saturation in $(NH_4)_2SO_4$ by adding 1.86 ml of saturated $(NH_4)_2SO_4$ per milliliter. Recover the precipitated protein by centrifugation and suspend it in buffer A to give the butyl fraction.

Affinity Chromatography

To remove $(NH_4)_2SO_4$, pass the butyl fraction through a 2.5 × 19 cm column of Sephadex G-25 equilibrated in buffer A. Combine the active fractions and apply to a 1.5 × 12 cm column of Reactive Red-120 agarose equilibrated in buffer A. Wash the column with 150 ml of buffer A, and then elute with a gradient of 0 to 1.5 M KCl in buffer A (400 ml total). Continue to elute with 100-ml portions of 1.5 M KCl in buffer A until no further activity emerges. Combine the active fractions and adjust to 65% saturation in $(NH_4)_2SO_4$. After centrifugation, suspend the pellet in 10% (v/v) glycerol in buffer A to give the red fraction. Store the red fraction, which contains homogeneous *S. solfataricus* HMG-CoA reductase, at −70°. Tables VIII and VII summarize the purification and kinetic properties of the enzyme, respectively.

Kinetic Parameters

Sulfolobus solfataricus HMG-CoA reductase exhibits a substrate specificity typical of biosynthetic HMG-CoA reductases. NADH cannot replace NADPH, nor can (*R*)-HMG-CoA replace (*S*)-HMG-CoA. Optimal activity for catalysis of the reductive deacylation of HMG-CoA occurs at about pH 5.5. V_{max} is 17 eu/mg at 50°. Optimal activity for catalysis of reaction (1) occurs at pH 5.5, well below the optimum pH for the only other characterized archaeal HMG-CoA reductase, that of the halophile *H. volcanii* (pH 7.3) and of the optimal pH for the hamster enzyme (pH 6.1).

Temperature Profile for Activity and for Stability

Significant features of *S. solfataricus* HMG-CoA reductase include high thermal stability, high optimal temperature, and low optimal pH. As might be anticipated for an enzyme from a thermophile, concentrated solutions of homogeneous *S. solfataricus* HMG-CoA reductase are stable at elevated temperatures ($t_{1/2} = 3.2$ hr at 90°). The activity of dilute solutions is optimal at 85° and ΔH for catalysis of reaction (1) is about 47 kJ (11 kcal)/mol.

[26] Characterization of 3-Methylcrotonyl-CoA Carboxylase from Plants

By Eve Syrkin Wurtele and Basil J. Nikolau

Introduction

3-Methylcrotonyl-CoA carboxylase (MCCase, EC 6.4.1.4) is a biotin-containing enzyme that catalyzes the ATP-dependent carboxylation of 3-methylcrotonyl-CoA to form 3-methylglutaconyl-CoA.[1] As with all biotin-containing enzymes, the reaction catalyzed by MCCase takes place in two steps (biotin carboxylation and carboxyl transfer), with the enzyme-bound carboxybiotin being an intermediate of the reaction. MCCase was initially purified and characterized from bacterial[2,3] and animal[4] sources, and was

[1] J. Moss and M. D. Lane, *Adv. Enzymol.* **35,** 321 (1971).
[2] U. Schiele and F. Lynen, *Methods Enzymol.* **71,** 781 (1981).
[3] R. Fall, *Methods Enzymol.* **71,** 791 (1981).
[4] E. P. Lau and R. R. Fall, *Methods Enzymol.* **71,** 800 (1981).

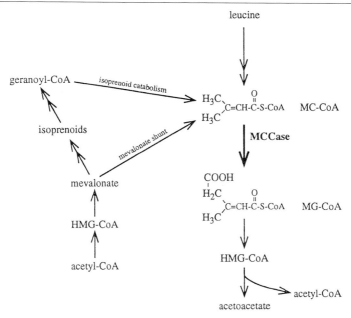

FIG. 1. Metabolic roles of MCCase. The reaction catalyzed by MCCase, the carboxylation of 3-methylcrotonyl-CoA (MC-CoA) to form 3-methylglutaconyl-CoA (MG-CoA), is required in the mitochondrial catabolism of leucine to acetoacetate and acetyl-CoA. This reaction is required in two additional metabolic processes: (1) the catabolism of acyclic isoprenoids, such as geranoyl-CoA, to acetyl-CoA, and (2) the mevalonate shunt, by which mevalonate is diverted from isoprenoid biosynthesis to acetyl-CoA. HMG-CoA, 3-Hydroxy-3-methylglutaryl-CoA.

in fact the enzyme via which the biochemical function of biotin as a enzyme-bound cofactor was discovered. Relative to this earlier work, the identification and characterization of MCCase in plants is recent.[5] Hence, the metabolic role of MCCase in plants is only just starting to be elucidated.

The reaction catalyzed by MCCase is required in a number of metabolic pathways, which may be interconnected (Fig. 1). These include the catabolism of leucine and isoprenoids, and the mevalonate shunt. In animals the catabolism of leucine to acetoacetate and acetyl-CoA is a mitochondrial process,[6] as is the subcellular location of MCCase.[7] However, in plants,

[5] E. S. Wurtele and B. J. Nikolau, *Arch. Biochem. Biophys.* **278**, 179 (1990).

[6] D. J. Danner and L. J. Elsas, in "The Metabolic Basis of Inherited Disease," 6th Ed. (C. R. Scriver, A. L. Beaudet, W. S. Sly, and D. Valle, eds), p. 671. McGraw-Hill, New York, 1989.

[7] M. L. Hector, B. C. Cochran, E. A. Logue, and R. R. Fall, *Arch. Biochem. Biophys.* **199**, 28 (1980).

leucine catabolism occurs via two independent pathways: one pathway catabolizes leucine to acetyl-CoA and acetoacetate, requires MCCase, and is located in mitochondria[8]; the second, MCCase-independent pathway occurs in peroxisomes and catabolizes leucine to acetyl-CoA.[9]

In addition to leucine catabolism, MCCase is also required for the operation of the mevalonate shunt,[10] which diverts mevalonate from isoprenoid biosynthesis to primary metabolism (acetyl-CoA). This shunt appears also to operate in plants.[11] Finally, in some bacteria, MCCase has a role in the catabolism of acyclic isoprenoids, such as geranoyl-CoA, to acetyl-CoA.[12] This pathway also requires the biotin-containing enzyme geranoyl-CoA carboxylase.[3] The identification of geranoyl-CoA carboxylase in plants[13] indicates that this process may also be operating in these organisms.

This chapter discusses the characterization of MCCase from plant sources, which since the discovery of this enzyme in these organisms has led to a clearer understanding of the structure and genetic regulation of this mitochondrial biotin-containing enzyme.

Purification and Assay Methods

MCCase has been purified from a number of plant sources by two basic protocols. One of these protocols relies on the specific reaction of avidin with biotin. The other protocol uses more conventional protein purification procedures. Both of these protocols are described below.

Plant Materials

MCCase has been purified from carrot,[14] maize,[15] soybean,[16] barley,[17] pea, and potato.[18] Studies of the distribution of the enzyme indicate that it

---

[8] D. Anderson, P. Che, J. Song, B. J. Nikolau, and E. S. Wurtele, *Plant Physiol.* **118,** 1127 (1998).

[9] H. Gerbling and B. Gerhardt, *Plant Physiol.* **91,** 1387 (1989).

[10] J. Edmond and G. Popjak, *J. Biol. Chem.* **249,** 66 (1974).

[11] W. D. Nes and T. J. Bach, *Proc. R. Soc. London B* **225,** 425 (1985).

[12] S. G. Cantwell, E. P. Lau, D. S. Watt, and R. R. Fall, *J. Bacteriol.* **135,** 324 (1978).

[13] X. Guan, T. Diez, K. T. Prasad, B. J. Nikolau, and E. S. Wurtele, *Arch. Biochem. Biophys.,* in press (1999).

[14] Y. Chen, E. S. Wurtele, X. Wang, and B. J. Nikolau, *Arch. Biochem. Biophys.* **305,** 103 (1993).

[15] T. A. Diez, E. S. Wurtele, and B. J. Nikolau, *Arch. Biochem. Biophys.* **310,** 64 (1994).

[16] J. Song, "Molecular Cloning and Characterization of 3-Methylcrotonyl-CoA Carboxylase from Soybean." Ph.D. Thesis, Iowa State University, Ames, Iowa, 1993.

[17] A. Maier and H. K. Lichtenthaler, *J. Plant Physiol.* **152,** 213 (1998).

[18] C. Alban, P Baldet, S. Axiotis, and R. Douce, *Plant Physiol.* **102,** 957 (1993).

is prevalent in most plant tissues, but is most abundant in nonphotosynthetic tissues.[8] However, because of the lack of interfering substances and the ease of obtaining large quantities of tissue, young expanding seedlings (8- to 14-day-old maize seedlings or 5- to 12-day-old soybean seedlings) are a convenient source of plant material for the purification of MCCase.

Soybean seeds (*Glycine max* cv. Corsoy 79) can be germinated in a sterile mixture of 30% (v/v) black soil, 30% (v/v) peat moss, and 40% (v/v) perlite in $50 \times 30 \times 6$ cm flats, in a greenhouse maintained at 22 to 25°, under a cycle of 15 hr of illumination and 9 hr of darkness, with a maximum daily irradiance of about 1200 μmol photons m^{-2} sec^{-1}. Maize seeds can be germinated in sand in a greenhouse maintained at 25 to 30°, under a cycle of 14 hr of illumination and 10 hr of darkness, with a maximum daily irradiance of about 1200 μmol photons m^{-2} sec^{-1}. Seedlings are watered daily.

3-Methylcrotonyl-CoA Carboxylase Activity Assay

As with most biotin-containing enzymes MCCase activity is routinely assayed by a radiochemical assay. MCCase activity is determined as the 3-methylcrotonyl-CoA-dependent incorporation of radioactivity from NaH^{14}CO$_3$ into the acid-stable product.[5] The standard assay is carried out in a total volume of 0.2 ml, which is composed of 0.1 M Tricine–KOH (pH 8.0), 5 mM MgCl$_2$, 1 mM ATP, 2 mM dithiothreitol (DTT), 5 mM NaH^{14}CO$_3$ (5 mCi/mmol), 0.2 mM 3-methylcrotonyl-CoA, and up to 0.01 unit of enzyme activity.

The assay is mixed, on ice, as follows:

Tricine-KOH (pH 8.0, 0.4 M), 20 mM MgCl$_2$, 8 mM DTT	50 μl
ATP (10 mM), adjusted to pH 7.0 with KOH	20 μl
NaH^{14}CO$_3$ (5 mCi/mmol), 50 mM: This solution is prepared just before use, by dissolving 3.8 mg of NaHCO$_3$ in 0.875 ml of water and adding 0.125 ml of NaH^{14}CO$_3$ (2 mCi/ml; specific radioactivity, 55 mCi/mmol)	20 μl

Enzyme and water are added to adjust the volume to 180 μl, and the mixture is incubated at 37° for 30 sec. The assay is then initiated by adding of 20 μl of 2 mM 3-methylcrotonyl-CoA, and further incubated at 37° for 10 min. The reaction is terminated by the addition of 50 μl of 6 N HCl. An aliquot (100 μl) is spotted on a strip of Whatman (Clifton, NJ) 3 MM paper, which is dried in a fume hood on a hot plate maintained at about 50°. The acid-stable radioactivity is determined by liquid scintillation counting.

This assay is susceptible to interference when used to assay MCCase activity in crude plant extracts. This interference occurs because such extracts contain abundant carboxylases (i.e., ribulose-1,5-bisphosphate carboxylase and/or phospho*enol*pyruvate carboxylase) and their substrates. This interference can be eliminated by passing the extract through a small gel-filtration column (Sephadex G-25 or Bio-Gel P6), which will remove low molecular weight molecules. These "desalting columns" can be obtained from commercial suppliers or can be made by pouring preswollen matrices. The column bed volume needs to be at least fivefold larger than the volume of extract that is being desalted.

Preparation of Monomeric Avidin Affinity Column

Although monomeric avidin affinity columns are available from a number of commercial sources, we have not obtained reliable results with these matrices. Synthesis of a monomeric avidin affinity matrix is carried out on the basis of previously published procedures.[19] Twenty grams of cyanogen bromide-activated Sepharose 4B is suspended in 100 ml of 1 mM HCl. This suspension is sequentially washed with 1 liter of 1 mM HCl and 500 ml of 10 mM potassium phosphate buffer (pH 7.0). Avidin (200 mg), dissolved in 200 ml of 10 mM potassium phosphate buffer (pH 7.0), is added to the activated Sepharose suspension (200 ml), and the mixture is allowed to react overnight at 4°. The reacted Sepharose is then sequentially washed with 1 liter of 1 M ethanolamine hydrochloride (pH 7.0) and 2 liters of 10 mM potassium phosphate buffer (pH 7.0). The resulting Sepharose–avidin conjugate is poured into a 2.5 × 50 cm column (72-ml bed volume). To dissociate the avidin subunits, the column is washed with 5 bed volumes of 6 M guanidine hydrochloride and maintained in the presence of this solution at room temperature for 20 hr. The column is then washed with 2–3 additional bed volumes of 6 M guanidine hydrochloride, followed by 10 bed volumes of 10 mM potassium phosphate buffer (pH 7.0). The elution of the dissociated avidin subunits is monitored by the A_{280} of this last wash. Once the A_{280} of the eluate is less than 0.01, the column is washed with 100 ml of 10 mM potassium phosphate buffer (pH 7.0) containing 2 mM biotin. The column is then extensively washed with 0.1 M glycine hydrochloride (pH 2.0) in order to remove biotin bound to exchangeable binding sites. (*Note:* It is important to wash the column extensively at this stage because the solubility of biotin at pH 2.0 is low and any precipitated biotin that remains on the column will subsequently interfere with enzyme purification.) The column is finally equilibrated with buffer A. Buffer A is composed of 10 mM HEPES–KOH (pH 7.0), 20 mM 2-mercaptoethanol, and 20% (v/v) glycerol.

MCCase is purified by affinity chromatography with such monomeric

---

[19] K. P. Anderson, S. H. G. Allen, and W. L. Maloy, *Anal. Biochem.* **94**, 366 (1979).

avidin columns. Enzyme preparations are bound to the matrix in the presence of high concentration of salts, and MCCase is specifically eluted with a solution containing biotin. Between uses the affinity matrix can be regenerated by washing the column with 3–5 bed volumes of 0.1 M glycine hydrochloride (pH 2.0), and then equilibrating with buffer A. Once every three or four uses, the column should be regenerated by washing with 6 M guanidine hydrochloride, followed by sequential washes with 10 mM potassium phosphate buffer (pH 7.0) containing 2 mM biotin, and with 0.1 M glycine hydrochloride (pH 2.0). Although such regeneration of a monomeric avidin matrix by denaturation–renaturation cycles extends the life of the matrix, its binding capacity is reduced. Hence, in our hands such columns have a limited lifetime and need to be remade after about 10 uses.

Purification of 3-Methylcrotonyl-CoA Carboxylase

Two methods for the purification of MCCase are presented below. The first method (method A) is illustrated by the purification of the enzyme from soybean seedlings (Table I), and the second method (method B) is illustrated by the purification of the enzyme from maize seedlings (Table II). Both methods give similarly pure preparations of MCCase, with approximately similar yields. In the case of the maize MCCase, recovery of the enzyme is somewhat difficult to judge, as activity in the extract is underestimated, probably because of the presence of some interfering substance, which is removed in the first purification step (see Table II).

The long-term stability of MCCase in extracts is enhanced if the enzyme is extracted at neutral pH. MCCase extracted with 0.1 M Tris-HCl (pH 8.0) is relatively unstable; the enzyme is completely inactive 2–3 days after extraction. Either addition of 1% (w/v) bovine serum albumin (BSA) to the extract or changing the pH of the extract fails to stabilize the enzyme.

TABLE I
PURIFICATION OF 3-METHYLCROTONYL-CoA CARBOXYLASE FROM SOYBEAN SEEDLINGS[a]

Fraction	Protein (mg)	Total activity (units[b])	Specific activity (units/mg protein)	Purification (fold)
Extract	714	663	0.9	1
PEG	595	595	1.0	1.1
Cibacron Blue	106	724	6.8	7.6
Q-Sepharose	80	549	6.9	7.7
Monomeric avidin	0.9	303	336.7	374.0

[a] Purification from 350 g of soybean seedlings.
[b] One unit equals 1 nmol of HCO_3^- incorporated per minute.

TABLE II
PURIFICATION OF 3-METHYLCROTONYL-CoA CARBOXYLASE FROM MAIZE SEEDLINGS[a]

Fraction	Protein (mg)	Total activity (units[b])	Specific activity (units/mg protein)	Purification (fold)
Extract	5040	192	0.04	1
4–16% PEG	1295	723	0.56	14
Propyl-agarose	438	362	0.83	20
Cibacron Blue	42	520	12.5	312
Q-Sepharose	2.8	566	206.2	5155

[a] Purification from 1 kg of maize seedlings.
[b] One unit equals 1 nmol of HCO_3^- incorporation per minute.

This problem is overcome by changing the extraction buffer to 0.1 M HEPES–KOH (pH 7.0) and by including 20% (v/v) glycerol. To prevent loss of enzyme activity due to proteolysis, protease inhibitors are added to the extraction buffer [Phenylmethylsulfonyl fluoride (PMSF) or *trans*-epoxysuccinyl-L-leucylamido(4-guanidino)butane]. In addition, to enhance the extractability of the enzyme, Triton X-100 is included in the extraction buffer.

Purification Method A. The purification method described below relies on the fact that MCCase contains biotin, which can specifically bind to immobilized avidin. Hence purification of MCCase can be readily achieved via affinity chromatography with immobilized monomeric avidin. Variations of this method have been used to purify MCCase from carrot,[14] barley,[17] pea, and potato.[18]

The method detailed below is that used for the purification of the soybean MCCase. In this method MCCase is purified by sequential use of three chromatographic procedures: Cibacron Blue (Amersham Pharmacia Biotech, Piscataway, NJ) affinity chromatography, Q-Sepharose anion-exchange chromatography, and monomeric avidin affinity chromatography. Although the latter procedure achieves the purification of the enzyme, the first two procedures are required to separate MCCase from the other biotin-containing proteins present in the extract (e.g., acetyl-CoA carboxylase isozymes). In addition, the Q-Sepharose column has the effect of concentrating the enzyme preparation. A summary of a typical purification achieved by this method is presented in Table I.

Five-day-old soybean seedlings (300 g) are pulverized with a mortar and pestle in the presence of liquid N_2. The frozen, ground tissue is homogenized with 1 liter of buffer A containing 1 mM EDTA, 0.1% (v/v) Triton X-100, PMSF (100 μg/ml), 10 μM *trans*-epoxysuccinyl-L-leucylamido(4-guanidino)butane. Subsequent purification steps are performed at 4°. The extract is filtered through four layers of cheesecloth, and the filtrate is centrifuged at 22,100g for 15 min. Polyethylene glycol 8000 (PEG 8000) is added to the resulting supernatant to a final concentration of 18% (w/v). Precipitated proteins are collected by centrifugation at 22,100g for 15 min, and the pellet is dissolved in the minimal volume of buffer A.

The enzyme preparation obtained from PEG precipitation is applied to an agarose-Cibacron Blue 3GA column (2 × 10 cm). The column is washed with 1 liter of buffer A, and MCCase is eluted with a linear gradient of 0–1 M KCl in buffer A. Peak MCCase activity is eluted at about 0.5 M KCl. All the fractions containing MCCase activity are pooled and dialyzed against 50 volumes of buffer A.

The resulting preparation is applied to a Q-Sepharose column (3 × 20 cm), preequilibrated with buffer A. The column is then washed with 500 ml of buffer A to remove proteins that do not bind to the column. MCCase is then eluted with a linear gradient of 0–0.8 M KCl in buffer A. Peak MCCase activity is eluted at about 0.6 M KCl. All the fractions containing MCCase activity are pooled and dialyzed against 50 volumes of buffer A.

The resulting preparation is applied to an agarose-monomeric avidin affinity column (3 × 20 cm), preequilibrated with buffer A. The column is washed with 300 ml of buffer A containing 0.25 M KCl and MCCase is then eluted with buffer A containing 0.25 M KCl and 0.4 mM biotin. MCCase activity elutes as a sharp peak immediately on the addition of biotin to the column. This preparation is homogeneous and can be stored at −20° (although we routinely store these preparations at −70°). Stored in buffer A, under these conditions, purified MCCase is stable for at least 1 year with minimal loss of activity.

Purification Method B. Purification of MCCase from maize seedlings has been achieved by using the method outlined below.[15] This method achieves purification of MCCase by the sequential application of three chromatographic procedures: hydrophobic chromatography on propyl-agarose, Cibacron Blue affinity chromatography, and Q-Sepharose anion-exchange chromatography. A summary of the purification achieved by this method is presented in Table II.

Maize leaves (1 kg) are frozen in liquid N_2 and pulverized with a mortar and pestle. The resulting powder is homogenized at 4° in a stainless steel Waring blender for 1–3 min with 3 liters of buffer A containing PMSF (0.1 mg/ml), 0.1% (v/v) Triton X-100, and 1 mM EDTA. The mixture is filtered through several layers of cheesecloth, and the filtrate is immediately centrifuged at 12,200g for 30–40 min at 4°. The supernatant (extract) is retained, and the pellet discarded.

Finely powdered PEG 8000 is slowly added to the crude extract to a final concentration of 4 g of PEG/100 ml. The solution is stirred until the PEG is completely dissolved and the mixture is then centrifuged at 12,200g for 30 min at 4°. The supernatant is retained, and more PEG 8000 is dissolved, to a final concentration of 16 g of PEG/100 ml. The precipitate (4–16% PEG fraction) is collected by centrifugation and resuspended in buffer A.

The 4–16% PEG fraction is applied to a propyl-agarose column (2.5 × 45 cm) previously equilibrated with buffer A. The column is washed with buffer A until the A_{280} of the eluate is less than 0.05. Elution of MCCase is achieved with a linear gradient of 0–0.5 M KCl in buffer A, at a flow rate of 0.5–1 ml/min. The fractions containing MCCase activity are pooled.

The pooled MCCase-containing fractions are applied to a column of Cibacron Blue-agarose (2.5 × 10 cm) previously equilibrated with buffer A. The column is washed with 5 column volumes of buffer A to remove unbound proteins. Elution of MCCase is achieved with an 800-ml linear gradient of 0–1.5 M KCl in buffer A at a flow rate of 1–1.5 ml/min. Fractions containing MCCase activity are pooled.

The pooled MCCase preparation obtained from the Cibacron Blue-agarose column is dialyzed against 4 liters of buffer A for 3–6 hr. The dialyzed sample is then applied to a Q-Sepharose column (2.6 × 18 cm) that has been equilibrated with buffer A. The column is washed with 10 volumes of buffer A. The enzyme is then eluted by using an 800-ml linear gradient of 0–0.75 M KCl in buffer A. Fractions containing MCCase are pooled and stored at −20°. Under these conditions, the purified maize MCCase can be stored for up to 6 months with minimal loss of activity.

Characterization of 3-Methylcrotonyl-CoA Carboxylase

Electrophoresis and Western Blot Analyses

Polyacrylamide gel electrophoresis (PAGE), performed under denaturing[20] and nondenaturing[21,22] conditions, is used to characterize the purified MCCase preparations. After electrophoresis, proteins can be stained with Coomassie blue or can be silver stained. Alternatively, proteins can be electrophoretically transferred to a nitrocellulose membrane, using a semidry transfer apparatus.[23] Biotin-containing proteins are specifically detected with ^{125}I-labeled streptavidin.[24] Immunological detection can be performed by sequential incubation of the nitrocellulose filter with polyclonal antisera

[20] U. K. Laemmli, *Nature (London)* **227,** 680 (1970).
[21] J. L. Hedrick and A. J. Smith, *Arch. Biochem. Biophys.* **126,** 155 (1968).
[22] P. Lambin and J. M. Fine, *Anal. Biochem.* **98,** 160 (1979).
[23] J. Kyhse-Andersen, *J. Biochem. Biophys. Methods* **10,** 203 (1984).
[24] B. J. Nikolau, E. S. Wurtele, and P. K. Stumpf, *Anal. Biochem.* **149,** 448 (1985).

(diluted about 1 : 1000), and with ^{125}I-labeled protein A [100 ng/ml, at a specific radioactivity of 1×10^{10} disintegrations per minute (dpm)/mg]. Antiserum directed against the biotin-containing[25,26] (MCC-A) and nonbiotinylated[27,28] (MCC-B) subunits of MCCase have been prepared against recombinantly expressed proteins.

Structural Characterization of 3-Methylcrotonyl-CoA Carboxylase

Sodium dodecyl sulfate (SDS)–PAGE analyses of purified MCCase indicate that the enzyme is composed of two subunits (Fig. 2). The larger, MCC-A subunit contains covalently bound biotin and has a molecular mass of about 80 kDa. The smaller, nonbiotinylated MCC-B subunit has a molecular mass of about 60 kDa. The holoenzyme appears to be composed of an equal molar ratio of the two subunits.

The quaternary organization of MCCase has been inferred from the molecular weight of the holoenzyme. Gel-filtration chromatography on Superdex 200 yields a molecular weight of about 500,000 for the pea and potato enzyme,[18] which has been interpreted as indicating an octomeric enzyme with an A_4B_4 subunit stoichiometry, analogous to the bacterial MCCase.[2,3]

However, because of the dissociation of MCCase on chromatography through Sephacryl S400, the molecular weight of the carrot,[14] maize,[15] soybean,[16] and tomato[29] MCCase could not be ascertained by gel-filtration chromatography. Rather, the molecular masses of these latter enzymes were determined to be between 800 and 900 kDa by nondenaturing PAGE.[15,16,29] These molecular weight determinations indicate that the MCCase holoenzyme is dodecameric, with an A_6B_6 subunit stoichiometry; such a quaternary organization is analogous to the animal MCCase.[4]

The amino acid sequences of the MCC-A[25,26,30] and MCC-B[27,28] subunits have been determined by translation of the nucleotide sequence of the cloned cDNAs. Sequence similarities between MCC-A and other biotin-containing enzymes indicate that the MCC-A subunit is composed of two functional domains, which are sequentially arranged in the primary se-

[25] X. Wang, E. S. Wurtele, G. Keller, A. L. McKean, and B. J. Nikolau, *J. Biol. Chem.* **269,** 11760 (1994).

[26] J. Song, E. S. Wurtele, and B. J. Nikolau, *Proc. Natl. Acad. Sci. U.S.A.* **91,** 5779 (1994).

[27] A. L. McKean, "Molecular Biology of the 3-Methylcrotonyl-CoA Carboxylase Subunits." M.Sc. Thesis, Iowa State University, Ames, Iowa, 1996.

[28] A. L. McKean, J. Ke, J. Song, P. Che, S. Achenbach, B. J. Nikolau, and E. S. Wuretele, *J. Biol. Chem.* **275,** 5582 (2000).

[29] X. Wang, "Characterization of β-Methylcrotonyl-CoA Carboxylase of Tomato, a Newly Identified Biotin Enzyme in Plants." Ph.D. Thesis, Iowa State University, Ames, Iowa, 1993.

[30] L. M. Weaver, L. Lebrun, E. S. Wurtele, and B. J. Nikolau, *Plant Physiol.* **107,** 1013 (1995).

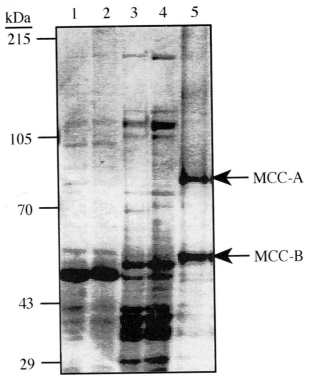

FIG. 2. SDS–PAGE analysis of the purification of MCCase from soybean seedlings. Samples from the fractions obtained during the purification of the soybean MCCase (see Table I) were subjected to SDS–PAGE and the resulting gel was stained with Coomassie blue. Each lane was loaded as follows: 1, extract; 2, PEG fraction; 3, Cibacron Blue fraction; 4, Q-Sepharose fraction; 5, monomeric avidin fraction. Positions of molecular weight standards are indicated. The two MCCase subunits (MCC-A and MCC-B) are arrowed.

quence: the biotin carboxylase domain and the biotin carrier domain. Hence, the MCC-A subunit contains the active site for the catalysis of the first half-reaction catalyzed by MCCase. The sequence of the MCC-B subunit is most similar to carboxyltransferase subunits of the biotin-containing enzymes methylmalonyl-CoA decarboxylase (35% identity), propionyl-CoA carboxylase (30% identity), and transcarboxylase (33% identity). These findings imply that the MCC-B subunit contains the active site for catalysis of the second half of the reaction catalyzed by MCCase.

Both the MCC-A and MCC-B subunits are initially synthesized as pre-

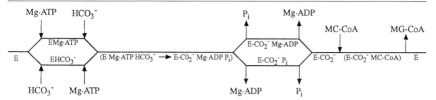

Fig. 3. Schematic representation of the random bi–bi uni–uni ping–pong kinetic mechanism of MCCase.

cursors, each containing an N-terminal extension that targets the subunits to mitochondria.

Enzymological Characterization of 3-Methylcrotonyl-CoA Carboxylase

The enzymological properties of MCCase have been investigated with the purified enzyme from carrot,[14] maize,[15] pea,[18] potato,[18] barley,[17] and soybean.[16] These investigations indicate that the kinetic behavior of MCCase can be described by the Michaelis–Menten equation. Detailed studies of the initial velocities and inhibition patterns of competitive inhibitors of the soybean MCCase indicates that the kinetic mechanism of the reaction is as shown in Fig. 3. In this mechanism ATP and bicarbonate react as the first two substrates to generate the first two products (ADP and P_i). Subsequently, 3-methylcrotonyl-CoA reacts to form 3-methylglutaconyl-CoA.

The Michaelis constants (K_m) for the substrates methylcrotonyl-CoA and bicarbonate are in the range of 0.01–0.05 and 0.8–2 mM, respectively. As with most other biotin-containing enzymes that require ATP, ATP reacts with MCCase as MgATP, and the K_m for this substrate is about 0.02 mM. The optimum pH for MCCase activity is between 8.0 and 8.5. Tricine is used to buffer MCCase assays. It is important to use a buffer that does not complex Mg^{2+} ions (i.e., do not use Tris, phosphate, etc.), because MCCase requires free Mg^{2+} ions for activation (in excess of that required to complex ATP). The requirement for free Mg^{2+} ions can be partially replaced by Mn^{2+} or Co^{2+} ions, but not by Zn^{2+}. The monovalent cations K^+, Cs^+, Rb^+, and NH_4^+ are activators of MCCase, but Li^+ and Na^+ are inhibitors.

In addition to carboxylating methylcrotonyl-CoA, MCCase will also carboxylate crotonyl-CoA, but the latter is a much poorer substrate. In contrast to bacterial and animal MCCases, which can also carboxylate acetoacetyl-CoA, MCCase from plant sources cannot do so. In fact, acetoacetyl-CoA is a potent inhibitor of plant MCCases.

Physiological Role and Regulation of 3-Methylcrotonyl-CoA Carboxylase in Plant Metabolism

In animals and bacteria, the primary role of this enzyme is considered to be in leucine catabolism. In addition, MCCase has been implicated in the metabolism of mevalonate via the mevalonate shunt, and in the catabolism of noncyclic isoprenoids, such as geranoyl-CoA (Fig. 1). Whether these metabolic processes are also operating in plants is still unclear. Hence, the discovery of MCCase in plants has opened a new avenue of research into plant metabolism. It is now becoming clear that plants catabolize leucine via at least two physically separated pathways: an MCCase-requiring pathway in mitochondria,[8] and an MCCase-independent pathway in peroxisomes.[9] How the operation of these two catabolic pathways is coordinated and regulated must still be explored.

Interestingly, MCCase activity in plants can be regulated by a mechanism that is specific to biotin-containing enzymes, namely biotinylation.[31] Although biotinylation is obviously essential for the function of every biotin-containing enzyme, biotinylation has not been extensively studied as a potential regulatory mechanism by which the activity of such enzymes can be controlled. In the case of MCCase in tomato plants the developmental pattern of MCCase distribution can be regulated by the biotinylation status of the enzyme. Specifically, whereas the MCC-A subunit accumulates to equal levels in leaves and roots of tomato plants, and this subunit is completely biotinylated in roots, only 10% of the MCC-A subunits that accumulate in leaves are biotinylated. Hence, the 10-fold difference in MCCase activity between leaves and roots of tomato plants is due to the differential biotinylation of the enzyme between these two organs. The biochemical mechanism that leads to the differential biotinylation of MCCase in roots and leaves requires additional research.

However, the presence of a pool of apo-MCC-A in tomato leaves leads to a number of questions as to the nature of the apoenzyme. MCCase is present in plant mitochondria[32]; therefore, it will be interesting to ascertain where in the cell the apo-MCC-A subunit accumulates. If it accumulates within the mitochondria, does it occur within assembled MCCase molecules? In this latter case, a mixture of apo and holo subunits may be present in tomato leaf mitochondria. Studies to ascertain the effect of this subunit heterogeneity on MCCase will undoubtedly lead to new and interesting information about the regulation and structure of biotin-containing enzymes.

---

[31] X. Wang, E. S. Wurtele, and B. J. Nikolau, *Plant Physiol.* **108**, 1133 (1995).
[32] P. Baldet, C. Alban, S. Axiotis, and R. Douce, *Plant Physiol.* **99**, 450 (1992).

[27] Purification of D-Hydroxyisovalerate Dehydrogenase from *Fusarium sambucinum*

By RAINER ZOCHER

D-Hydroxyisovalerate dehydrogenase is an NADPH-dependent enzyme that catalyzes the reversible reduction of 2-ketovaleric acid (KIV) to D-2-hydroxyisovaleric acid (D-HIV) and plays a key role in depsipeptide synthesis in fungi.[1] D-HIV is the hydroxy acid constituent of many cyclopetides and peptolides of prokaryotic and eukaryotic origin, such as enniatins, beauvericin, destruxin, and valinomycin.[2-5] Enniatins are produced by several strains of the genus *Fusarium*[6] and exhibit antibiotic properties due to their ionophoric activity. They are composed of three residues each of D-HIV and an N-methylated branched-chain amino acid, which are arranged in an alternating fashion. Enniatins are synthesized by a 350-kDa multifunctional enzyme (enniatin synthetase) from their primary precursors D-HIV and a branched-chain amino acid under the consumption of adenosyl-L-methionine (AdoMet) and ATP.[7] Thus the D-hydroxy acid is an intermediate in the biosynthetic pathway of enniatins in *Fusarium*. In mammals, the branched-chain amino acids leucine, isoleucine, and valine are usually transaminated to form 2-keto acids. These 2-keto acids are oxidatively decarboxylated by mitochondrial branched-chain 2-keto acid dehydrogenase[8-10] and converted to acyl-CoA. However, in some fungal species, namely the enniatin producers of the genus *Fusarium*, there is another pathway leading from L-Val to D-HIV via the keto acid catalyzed by D-HIV dehydrogenase.

Stereospecific D- and L-lactate dehydrogenases are well known from

[1] C. Lee, H. Görisch, H. Kleinkauf, and R. Zocher, *J. Biol. Chem.* **267**, 11741 (1992).
[2] T. K. Audhya and D. W. Russel, *Can. J. Microbiol.* **19**, 1051 (1973).
[3] H. Peeters, R. Zocher, N. Madry, P. B. Oelrichs, H. Kleinkauf, and G. Kraepelin, *J. Antibiot.* **36**, 1762 (1983).
[4] M. Pais, B. C. Das, and P. Ferron, *Phytochemistry* **20**, 715 (1981).
[5] O. D. Smith, W. L. Duax, D. A. Langs, O. T. Detitta, J. W. Edmonds, D. C. Rohrer, and C. M. Weeks, *J. Am. Chem. Soc.* **97**, 7242 (1975).
[6] P. A. Plattner, U. Nager, and A. Bauer, *Helv. Chim. Acta* **41**, 594 (1948).
[7] R. Zocher and U. Keller, *Adv. Microb. Physiol.* **38**, 85 (1997).
[8] O. Livesey and P. Lund, *Methods Enzymol.* **166**, 3 (1988).
[9] A. L. Gasking, W. T. E. Edwards, A. Hobson-Frohock, M. Elia, and G. Livesey, *Methods Enzymol.* **166**, 20 (1988).
[10] O. P. L. Crowell, R. H. Miller, and A. E. Harper, *Methods Enzymol.* **166**, 39 (1988).

microorganisms, as is L-lactate dehydrogenase from mammalian tissues.[11,12] L-Lactate dehydrogenase catalyzes predominantly the reduction of pyruvate to lactate among the various 2-keto acids. Pyruvate showed the highest V_{max}/K_m and other 2-keto acids, e.g., 2-KIV and 2-ketoisocapronate, were poor substrates. New NADH-dependent stereospecific D- and L-hydroxyisocapronate dehydrogenases were reported from *Lactobacillus* species, which preferentially reduced 2-ketoisocapronate to D- and L-hydroxyisocapronate.[13,14] However, these enzymes exhibited low substrate affinity with respect to 2-KIV. D-HIV dehydrogenase differs from other NADPH-dependent oxidoreductases with broad substrate specificity by its high affinity for 2-KIV and to a lesser extent for 2-keto-3-methyl-*n*-valerate. These findings indicate clearly that the branching CH_3 group in the β position is essential for the substrate to enter the binding site of the enzyme. The high specificity of D-HIV dehydrogenase may also explain the fact that D-HIV is the exclusive hydroxy acid component in enniatins isolated from *Fusarium*.[6]

Growth of Organisms

Strains

Fusarium sambucinum BBA 63933 is obtained from the collection of the Biologische Bundesanstalt Berlin (Berlin, Germany). Strain 16-4R is a D-HIV dehydrogenase-overproducing variant obtained as described previously.[15] Generally all enniatin-producing *Fusarium* strains can be used as a source for D-HIV dehydrogenase isolation.

Media

FCM liquid medium: 3% (v/v) molasses, 1% (v/v) cornsteep liquor
FDM medium by Madry *et al.*[15]: Contains 12.5 g of glucose, 4.25 g of NaNO$_3$, 5 g of NaCl, 2.5 g of MgSO$_4 \cdot 7H_2O$, 1.36 g of KH$_2$PO$_4$, 0.01 g of FeSO$_4 \cdot 7H_2O$, 0.002 g of ZnSO$_4 \cdot 7H_2O$, and water to 1 liter

Maintenance of Strains and Fermentation

Fusarium sambucinum is maintained on FCM agar slants [1.5% (w/v) agar]. For precultures we use chemically defined medium (FDM). Spore suspensions of precultures are obtained by filtration of 4-day-old submerged cultures through a double layer of Cleenex cloth. Spores are inoculated in

[11] A. Meister, *J. Biol. Chem.* **184,** 117 (1950).
[12] W. Hummel, H. Schütte, and M.-R. Kula, *Eur. J. Appl. Microbiol. Biotechnol.* **18,** 75 (1983).
[13] H. Schütte, W. Hummel, and M.-R. Kula, *Appl. Microbiol. Biotechnol.* **19,** 167 (1984).
[14] W. Hummel, H. Schütte, and M.-R. Kula, *Appl. Microbiol. Biotechnol.* **21,** 7 (1985).
[15] N. Madry, R. Zocher, and H. Kleinkauf, *Eur. J. Microbiol. Biotechnol.* **17,** 75 (1983).

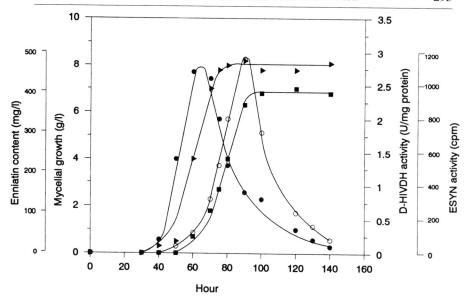

Fig. 1. Growth curve of *F. sambucinum* BBA 63933 (16-4R) and activities of D-HIV dehydrogenase and enniatin synthetase in FCM liquid medium. (○) Activity of D-HIV dehydrogenase per milligram of protein crude extract; (●) activity of enniatin synthetase per milligram of protein crude extract; (▲) mycelial growth; (■) enniatin content per liter of culture.

500-ml Erlenmeyer flasks containing 100 ml of FCM liquid medium and cultivated on a rotary shaker (105 rpm, 27°). As shown in Fig. 1, during the growth of *F. sambucinum* 16-4R in submerged FCM culture, production of enniatins occurs in the log phase of mycelial growth and remains at a constant level when the stationary phase is reached (80–100 hr). To obtain optimal amounts of active enzyme, cultures are harvested at various times and the activities of D-HIV dehydrogenase are measured in crude extracts. D-HIV dehydrogenase activity reaches its maximum after 90 hr and decreases rapidly after 100 hr. On the basis of these results, 90-hr-old cultures are used for the procedure of enzyme isolation. Cultures for enzyme preparation are harvested by suction filtration about 90 hr after inoculation. The mycelial cake is washed with 0.3 M KCl in 50 mM potassium phosphate buffer (buffer A) and shock-frozen at −80°.

Comments on Expression of D-2-Hydroxyisovalerate Dehydrogenase

As reported previously by Billich and Zocher,[16] enniatin synthetase in *Fusarium scirpi* behaves like a constitutive protein of primary metabolism

[16] A. Billich and R. Zocher, *Appl. Environ. Microbiol.* **54**, 2504 (1988).

and it is also present in the stationary phase of growth. This seems to be the case with D-HIV dehydrogenase as well. The growth curve of *F. sambucinum* 16-4R shows that maximal activity of soluble D-HIV dehydrogenase in cell-free extracts does not coincide with that of enniatin synthetase during the fermentation. The reason for this phenomenon is unknown. The increase in soluble D-HIV dehydrogenase activity in the end phase of enniatin production may be explained by the assumption that D-HIV dehydrogenase is membrane-bound during the production phase and is released in the stationary phase as a soluble enzyme. If one considers the concentration of D-HIV dehydrogenase and enniatin synthetase in crude extracts of the fungus it is noteworthy that the latter enzyme is present in much higher amounts[17] compared with D-HIV dehydrogenase described in this chapter. This discrepancy can be explained by the different turnover numbers of both enzymes, which have been calculated to be 7890/sec in the case of D-HIV dehydrogenase and 0.1–0.5/sec in the case of enniatin synthetase.

Assay Methods

Spectrophotometric Assay

Principle. D-HIV dehydrogenase catalyzes the reversible reduction of 2-ketoisovaleric acid (KIV):

$$2\text{-KIV} + \text{NADPH} + \text{H}^+ \rightleftharpoons \text{D-HIV} + \text{NADP}^+$$

The reaction is monitored by recording the change in absorbance at 340 nm due to oxidation of NADPH. Alternatively, the reverse reaction (oxidation of HIV) can be measured.

Reagents

Buffer A: 0.05 M potassium phosphate, pH 7.0
2-KIV (pH 7.0), 0.1 M
NADPH, 0.01 M

Assay Procedure. The standard D-HIV dehydrogenase assay mixture contains buffer A, 0.7 mM 2-KIV, 0.29 mM NADPH, and enzyme in a final volume of 0.35 ml. The reaction is initiated by the addition of substrate, and the decrease in absorbance at 340 nm is measured at 35° by using a Uvicon 930 spectrophotometer. A molar extinction coefficient of 6.22 cm^2/g per mole of NADPH is used for the calculation of enzyme activity. One unit is defined as the amount of enzyme catalyzing the oxidation of 1 μmol of NADPH per minute under the standard assay conditions. Specific activity

---

[17] R. Zocher, U. Keller, and H. Kleinkauf, *Biochemistry* **21**, 43 (1982).

is expressed as units of enzyme activity per milligram of protein. The reaction mixture for the assay of the reverse reaction contains 50 mM Tris-HCl buffer (pH 8.9), 2.85 mM NADP$^+$, 5.7 mM D-HIV, and enzyme in a final volume of 0.35 ml. The increase in the rate of reduction of NADP$^+$ due to oxidation of D-HIV is measured at 340 nm (45°).

Alternative Radioactive Assay

As an alternative assay for D-HIV dehydrogenase a coupled reaction can be used.[1] It is based on the enzymatic synthesis of enniatin in the presence of enniatin synthetase and the necessary substrates (ATP, adenosyl-L-methionine and L-[U-^{14}C]valine) with the exception of D-HIV. The latter compound is formed from 2-KIV in the presence of D-HIV dehydrogenase and NADPH, yielding a complete reaction mixture that allows synthesis of radioactive enniatin B, which is extracted with ethyl acetate and counted in a liquid scintillation counter.

Purification of D-2-Hydroxyisovalerate Dehydrogenase

All operations are carried out at 4°. Buffer A containing 1 mM EDTA and 4 mM dithioerythritol is used for all operations.

Step 1: Preparation of Crude Extract

The mycelial cake is lyophilized and subsequently homogenized with sand in a mortar. After the extraction of soluble proteins with 0.3 M KCl in buffer A by stirring for 40 min, the homogenate is centrifuged for 20 min at 20,000g.

Step 2: Polyethyleneimine Precipitation

A solution of polyethyleneimine [8.5% (w/v), pH 7] is then added to the supernatant to give a final concentration of 0.2% (w/v). After being allowed to stand on ice for 30 min with occasional stirring, the precipitate is removed by centrifugation as described above.

Step 3: Ammonium Sulfate Precipitation

Saturated ammonium sulfate solution in buffer A is gradually added to the supernatant to give 35% saturation. After being allowed to stand on ice for 30 min the pellet obtained after centrifugation (20 min, 20,000g) is discarded. Enough saturated ammonium sulfate solution is added to the supernatant to give 45% saturation. After being allowed to stand on ice for 30 min the precipitate is collected by centrifugation and dissolved in 3–4 ml of buffer A.

Step 4: Ultrogel AcA-44 Chromatography

The enzyme solution is applied to an Ultrogel AcA-44 column (1.5 × 120 cm) previously equilibrated with buffer A. The fraction size is 3 ml. The enzyme elutes with a V_e/V_0 ratio of 2.1. Active fractions are collected and give a total volume of about 10 ml.

Step 5: Mono Q HR Ion-Exchange Chromatography

Ultrogel AcA-44 fractions are applied to a Mono Q HR 5/5 ion-exchange FPLC (fast protein liquid chromatography) column (Pharmacia LKB Biotechnology, Piscataway, NJ) equilibrated with the same buffer. The flowthrough fraction contains most of the D-HIV dehydrogenase activity whereas the bulk of protein is adsorbed to the column.

Step 6: S-Sepharose Cation-Exchange Chromatography

The flowthrough fraction of the last step is applied to an S-Sepharose Fast Flow HR 5/5 cation-exchange FPLC column (Pharmacia). After washing the column with the same buffer, the enzyme is eluted by a linear NaCl gradient from 0 to 0.25 M in buffer A. Active fractions (0.14–0.16 M NaCl) are collected and concentrated by lyophilization.

The entire purification procedure of D-HIV dehydrogenase involving ammonium sulfate and polyethyleneimine precipitation, Ultrogel AcA-44 column chromatography, and Mono Q and S-Sepharose ion-exchange chromatography is summarized in Table I. As can be seen, the enzyme is

TABLE I

PURIFICATION OF D-HYDROXYISOVALERATE DEHYDROGENASE FROM *Fusarium sambucinum* BBA 63933 (16-4R)[a]

Fraction	Total activity (units)	Total protein (mg)	Specific activity (units/mg)	Purification (fold)	Recovery (%)
Crude extract	2001	712.00	3	1.0	100
Polyethyleneimine precipitate	1977	472.00	4	1.3	99
Ammonium sulfate precipitate	1262	102.00	12	4.4	63
Ultrogel AcA-44	252	0.83	302	101	13
Mono Q	190	0.21	896	299	10
S-Sepharose	180	0.02	9000	3000	9

[a] Eighteen grams of lyophilized mycelium was used.

purified about 3000-fold from crude extracts to apparent homogeneity in a 9% overall yield. Mono Q and S-Sepharose chromatography proved to be the most effective purification steps, leading to a strong increase in specific activity (about 10-fold). S-Sepharose cation-exchange chromatography, which yields a single protein band in sodium dodecyl sulfate–polyacrylamide gel electrophoresis (SDS–PAGE), is the last step of the purification procedure.

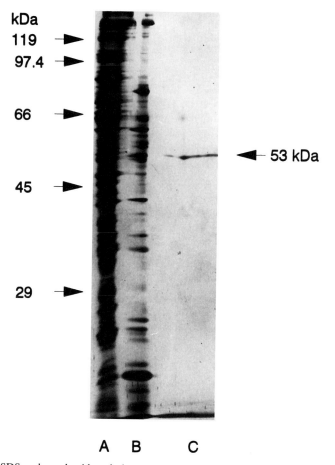

FIG. 2. SDS–polyacrylamide gel electrophoresis of protein samples from different steps in the purification of D-HIV dehydrogenase. Lane A, ammonium sulfate step (48 μg); lane B, pooled active fractions of AcA-44 gel chromatography (20 μg); lane C, purified D-HIV dehydrogenase S-Sepharose step (2 μg).

TABLE II
KINETIC CONSTANTS OF D-HYDROXYISOVALERATE DEHYDROGENASE WITH
VARIOUS SUBSTRATES

Substrate	K_m (mM)	V_{max} (mmol/min · l)	V_{max}/K_m (1/min)
2-Ketovalerate	4.2	1.25	0.3
2-Ketoisovalerate	0.2	4.76	23.8
2-Keto-3-methylvalerate	0.4	1.00	2.5
2-Ketocapronate	4.40	0.5	0.11
2-Ketoisocapronate	5.0	0.33	0.07

Properties of D-2-Hydroxyisovalerate Dehydrogenase

Molecular Weight and Subunit Structure

SDS–PAGE shows that D-HIV dehydrogenase migrates as a single protein band with a molecular mass of about 53 kDa compared with standard proteins [β-galactosidase (116 kDa), phosphorylase G (97.4 kDa), bovine serum albumin (66 kDa), egg albumin (45 kDa), carbonic anhydrase (29 kDa)] (see Fig. 2). Molecular mass determinations of the native enzyme are carried out by means of size-exclusion chromatography, using a Sephadex G-150 column that has been previously calibrated with standard proteins [aldolase (150 kDa), albumin (66 kDa), carbonic anhydrase (29 kDa), cytochrome c (12.4 kDa)]. The elution volume of D-HIV dehydrogenase indicates a molecular mass of 53.7 kDa for this enzyme, assuming a globular structure. The fact that both methods give similar molecular masses for D-HIV dehydrogenase suggests that this enzyme is composed of a single polypeptide chain.

Substrate Specificity and Kinetic Measurements[18]

Various substrate analogs of 2-ketoisovaleric acid including pyruvate, 2-ketobutyrate, 2-ketovalerate, 2-ketocapronate, 2-ketoisocapronate, 2-keto-3-methyl-*n*-valerate, and 2-ketooctanate are used to investigate the substrate specificity of D-HIV dehydrogenase. Michaelis constants are estimated from double-reciprocal plots by plotting the reciprocal concentrations of the second substrate versus the corresponding intercepts of the *1/V* axis (not shown). The results are summarized in Table II. The enzyme exhibits high substrate specificity with respect to 2-ketoisovalerate ($V_{max}/K_m = 24$) and 2-keto-3-methyl-*n*-valerate ($V_{max}/K_m = 2.5$).

[18] C. Lee, Ph.D. dissertation, Technical University of Berlin (1992).

The homologous compounds 2-ketovalerate, 2-ketoisocapronate, and 2-ketocapronate are poor substrates. The enzyme is specific for NADP as a coenzyme (K_m = 333 mM), which cannot be replaced by NADH. The K_m values for the reverse reaction have been found to be 9 mM for D-HIV and 350 mM for NADP$^+$. The reductive reaction is inhibited competitively by the end products D-HIV and NADP$^+$.

Optimal Temperature and pH of D-2-Hydroxyisovalerate Dehydrogenase

For determination of the optimal temperature for the D-HIV dehydrogenase reaction, mixtures containing enzyme and NADPH are preincubated at various temperatures for 1 min. The reaction is started by addition of 2-KIV. An optimal temperature of 35° is found for the reductive reaction; for the oxidative reaction it is 45°.

The effect of pH on the activity of the enzyme has been determined over a wide range between pH 3.7 and 10.5, using acetate, phosphate, and Tris-HCl buffer. A sharp optimum is observed for the reductive reaction at pH 7.0, and a plateau-like optimum between pH 8 and 9 is observed for the oxidative reaction.

Isoelectric Focusing

The isoelectric point of D-HIV dehydrogenase is determined by using Sevalyt precotes (Serva, Heidelberg, Germany) according to the instructions of the manufacturer.

A relative isoelectric point of the enzyme, pI 7.0, is determined from its mobility relative to those of the Serva protein test mixture.

[28] Purification and Characterization of Recombinant 3-Isopropylmalate Dehydrogenases from *Thermus thermophilus* and Other Microorganisms

By YOKO HAYASHI-IWASAKI and TAIRO OSHIMA

Introduction

3-Isopropylmalate dehydrogenase (EC 1.1.1.85; IPMDH) is the third enzyme in the leucine biosynthetic pathway (Fig. 1), and catalyzes dehydrogenation and decarboxylation of 3-isopropylmalate (3-IPM) in the presence of NAD$^+$ and a divalent cation, such as Mn^{2+} or Mg^{2+}. This pathway is analogous to the first three reactions of the tricarboxylic acid (TCA) cycle.

0076-6879/00 $30.00

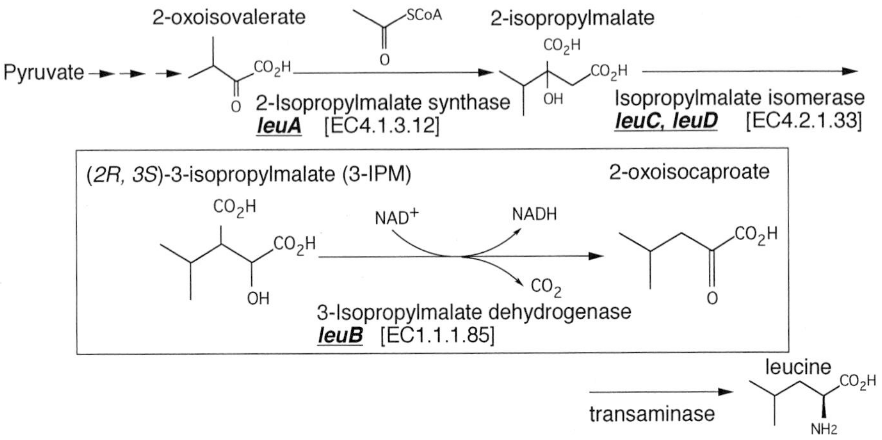

Fig. 1. Leucine biosynthetic pathway in microorganisms, plants, and yeast.

Isocitrate dehydrogenase (EC 1.1.1.42; ICDH) is the third enzyme in the TCA cycle, and IPMDH shows a marked similarity to ICDH both in three dimensional (3-D) structure and catalytic mechanism. Many dehydrogenases (such as lactate dehydrogenase, malate dehydrogenase, and alcohol dehydrogenase) belong to a well-known enzyme family called the NAD-dependent dehydrogenase group, which has a typical NAD-binding motif called the Rossman folds. However, IPMDH and ICDH do not have the motif (Fig. 2), and are classified into a unique enzyme family called "decarboxylating dehydrogenases." These enzymes catalyze chemically equivalent reactions, i.e., dehydrogenation at C-2 and decarboxylation at C-3 of (2R,3S)-2-hydroxy acids, and produce 2-keto acids.

Genes encoding IPMDH, *leuB* and *leu2*, have been cloned and sequenced from various sources such as plants, yeast, and microorganisms, including weakly, moderately, extremely, and hyperthermophilic bacteria and some thermophilic archaea (Fig. 3). The amino acid sequences of the enzymes from various microorganisms show a high degree of homology. *Thermus thermophilus leuB* gene is the first gene cloned and sequenced

Fig. 2. (A) Three-dimensional structure of a subunit of *T. thermophilus* IPMDH. α Helices and β strands are indicated as cylinders and arrows, respectively. These secondary structural elements are labeled. (A) was generated by Insight II. (B) Folding topologies of the polypeptide chains of (1) a subunit of *T. thermophilus* IPMDH, (2) a subunit of *E. coli* ICDH, and (3) NAD-binding motif of alcohol dehydrogenase. Open rectangles and arrows represent α helices and β strands, respectively. [K. Imada, M. Sato, N. Tanaka, Y. Katsube, Y. Matsuura, and T. Oshima, *J. Mol. Biol.* **222**, 725 (1991).]

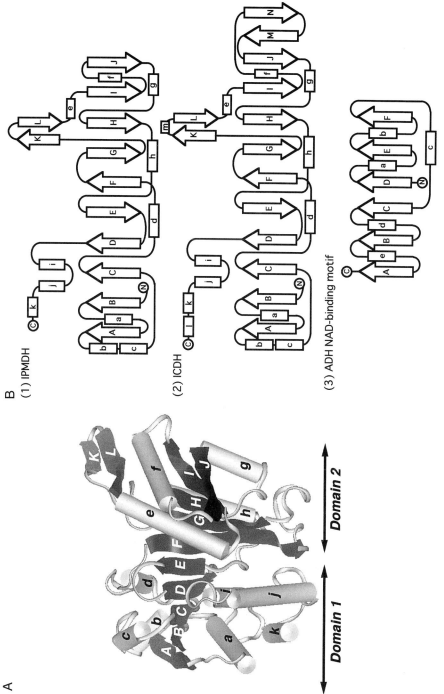

A

Domain 1   Domain 2

B

(1) IPMDH

(2) ICDH

(3) ADH NAD-binding motif

```
                1          20                                    60        80              94
Tth     ---MKVAVL  PGDGIGPEVT  ERAIKVLRAL  DEAEG-IGLA  YEVFPPGGAA  IDAFGEPFPE  PTRKGVEEAE  AVLLGSVGGP  KWDGLPRKIS  PETG-LLSLR
Tfer    --MKKIAIF   AGDGIGPEIV  AARRQVLDAV  DQRAH-LGLR  FENALIGGAA  LDASDDPLPA  ASLQLAMAAD  AVIIGAVGGP  RWDAYPPAKR  PEQG-LRLR
Bcoag   -MKUKLAVL   PGDGIGPEVL  DAAIRVLKTV  LDNRG-HEAV  FENALIGGAA  ETLDICRRSD  AILLGAVGGP  KWDHNPASLR  PEKG-LLGLR
Bsub    --MKKRIALL  PGDGIGPEVL  ESAKDVLKSV  AERFN-HEFE  FEVGLIGGAA  IDEHNPLPE   ETVAACKNAD  AVLFGSVGGP  KWDQNLSELR  PEKG-LLSIR
Ecoli   MSKNTHIAVL  PGDGIGPEVM  RNRFA-MRIT  TSHYDVGGAA  IDNHQPLPP   ATVECCEQAD  AVLFGSVGGP  KWENLPPDQ   PERGALLPLR
Sty     MSKNTHIAVL  PGDGIGPEVM  AQALKVLDAV  RSRFD-MRIT  TSHYDVGGIA  IDNHGHPLPK  ATVEGCEQAD  AILLGSVGGP  KWENLPESQ   PERGALLPLR
Vsp     -MSNQILLL   PGDGIGPEIM  AEAVKVLNLA  NEKYS-LGFE  LSFDDLGGAA  IDRYGVPLAD  ETLARAKAAD  AILLGAVGGP  KWDAIDPAIR  PERG-LLKIR
Scer    MSAPKKIVVL  PGDHVGQEIT  AERIKVLKAI  SDVRSNVKFD  FENHLIGGAA  IDATGVPLPD  EALEASKKVD  AVLLGAVGGP  KWGTG--SVR  PEQG-LLKIR
Aaeo+   -MKKYKIAVL  KGDGIGPEIV  EQALRVLDKI  GEIIG-VEFE  VREGLIGGEA  IDKTGDPLPE  ETLKICKESD  AILLGAVGGP  KWDNLFTDKR  PEKG-LLKIR
Sul#    --MGFTVALI  QGDGIGPEIV  SKSKRILAKI  NELYS-LPIE  YIEVEAGDRA  LARYGEALPK  DSLKIIDKAD  IILKGPVG--  -------ESA  AD-V-VVKLR
Aful#   --MKKIVVI   PGDGIGKEVM  EAAMLIEEKL  D-------LPFE  YSYYDAGDEA  LEKYGKALPD  ETLEACRKSD  AVLFGRAG--  -------ETA  AD-V-IVRLR

d

                                                   140              160            180
Tth     KSODLFANLR  PAKVFPG-LE  RLSPLKEIA   R-GVDVLIVR  ELIGGIYFGK  PRGMSEA---  -EAWNTERI   SKPEVRVAR   VAFEAARKR   KH--VVSVDK
Tfer    KGLDLFANLR  PAQIFPQ-LL  DASPLRPELV  R-DVDILVVR  ELIGDIYFGO  PRGLEVI-DG  KRRGFNTMVI  DEDEIRRIAH  VAFRAAQGR   KQ--LCSVDK
Bcoag   KENGLFANLR  PVKAIAT-LL  NASPLRKERV  R-NVDLVIVR  ELIGGLFGR   PSERRGP--G  ENEVVDTLAY  TREEIERIIE  KAFQLAQIRR  KK--LASVDK
Bsub    KQLDLFANLR  PVKVPES-LS  DRSPLKKEYI  D-NVDFVIVR  ELIGGIYFGO  PSKRIVNTEG  EQEAVDTLF   KRTEIERVIK  EGFKMAATRK  GK--VTSVDK
Ecoli   KHFKLFSNLR  PAKLYQG-LE  AFCPLRADIA  ANGFDILCVR  ELIGGIYFGO  PKGREGS-GQ  YEKAFDTEVY  HRFEIERIAR  IAFESARKR   HK--VTSIDK
Sty     KHFKLFSNLR  PAKLYQG-LE  AFCPLRADIA  ANGFDILCVR  ELIGGIYFGO  PKGREGS-GQ  YEKAFDTEVY  HRFEIERIAR  IAFESARKR   RK--VTSIDK
Vsp     SQLGLFGNLR  PAILYQG-LA  DASSLKREIV  A-GLDILIVR  ELIGGIYFGO  PRESKVLESG  ERMADTLPY   SESEIRRIAR  VGFDMARVRG  KK--LCSVDK
Scer    KELQLYANLR  PCNFASDSLL  DLSPIKPQFA  K-GTDFVVVR  ELVGGIYFGK  RKEDD-----  DGVAWDSEQY  TVFEVQRITR  MGAFWALQHE  PPLPIWSLDK
Aaeo+   KELDLIANLR  PAKVWDA-LI  SSSPLKEEVV  K-GTDHIVIR  ELTSGIYYGE  PRGIFPE-NG  KRYAINTMKY  TEDEIRRIVR  KGFEAINRKK  KK--LVSVDK
Sul#    QIIDMIANIR  PAKSIPG---  -------IDTK  YGNVDILVR   ENTEDLYKG-  FEHI-----VS  DGVAVGMKII  TRFASERIAK  VGLNFALRRR  KK--VTCVHK
Aful#   RELGTFANVR  PAKAIEG---  -------IECL  VPGLDIVVR   ENECIIMG-   FEF-------  GDVTEAIRVI  TREASERIAR  YAFELAKREG  RKK-VTALHK

e

                                         240              260           277
Tth     ANVLEVGEFW  RKTVEEVGR-  GYPDVALEHQ  VVDAMAMQLI  RSPARFD-VV  VTGNIFGDII  SDLASVLPGS  LGLLPSASLG  R-------GTP  VFEPVHGSAP
Tfer    ANVLETRLM   REVVEVAR--  DYPDVRLSHM  VVDNAAMQLI  EPWAQRV     LTGNMFGDII  SDEASQLPGS  IGMLPSASLG  E-------GRA  MKEPIHGSAP
Bcoag   ANVLESSRRW  REIAEETAK-  KYPDVELSHM  LVDSTSMQLI  EKERAAI     VENNAFGDII  SDEASVLPGS  SDEASVTGS   SD-------RFG  LFEPVHGSAP
Bsub    ANVLESSRRW  REVAEVAQ-   EFPDVELERM  LVDNARMOLI  YARNOWF-VV  VENNAFGDII  SDEASHLTGS  LGMLPSASLS  SS-------GLH  LFEPVHGSAP
Ecoli   ANVLQSSILW  REIVNEIAT-  EYPDVELAHM  VIDNATMQLI  KDPSQFD-VL  LCSNLFGDII  SDECAMITGS  MGMLPSASLN  EQ-------GFG  LYEPAGGSAP
Sty     ANVLQSSILW  REIVNDVAK-  TYPDVELERM  VIDNATMQLI  KDPSQFD-VL  LCSNLFGDII  SDECAMITGS  MGMLPSASLN  EQ-------GFG  LYEPAGGSAP
Vsp     RTVVIEVAK-  DYPDVELSHM  YVDNRAMQLV  RAPKQFD-VM  VTDNLFGDII  SDEASALTGS  IGMLPSASLD  AN-------NKG  MYEPCHGSAP
Scer    DLP-KNKVDP  RKTVEETIKN  EFPTLKVOHQ  LIDSAAMILV  RRPTHLNGII  ITSNMFGDII  SDEASVIPGS  LGLLPSASLA  SLPDKNTAFG  LYEPCHGSAP
Aaeo+   ROLVEEEKE-  NVPDVELEHL  VIDNCAMQLV  RREFSRFD-VI  VTGNIVGDII  SDEAGVVVGS  LGMLPSASIG  D-------RYA  LYEPVHGSAP
Sul#    FAEACRSVLK  G--KYEISEM  VVDAAAANLV  RNPOMFD-VI  VTENVIGDII  SDEASQIAGS  LGIAPSANIG  D-------KKA  LFEPVHGAAF
Aful#   FRDVCREVAK  DYPEIQYNDI  YIDAACNYLV  MDFFRFD-VI  VTNNMFGDIV  SDLAAGLVGG  LGLAPSANVG  E-------RTA  IFEVHGAAF

h

              280          300            320              340    345
Tth     DIAGKGIANP  TAAILSAAWM  LEHAFGL---  ---VELARKV  EDAVAKALLE  TP--PPDLGG  SA----GTEA  FTATVLRHLA  R---------
Tfer    DIAGQDKANP  LATILSVARM  LRHSLNA---  ---EPWAQRV  EANVQRVLDD  GLR-TADIAA  POTPVIGTRA  MGAAVVNALN  LKD-------
Bcoag   DIAGQGKANP  LGTVLSAALM  LRYSFGL---  ---EKERAAI  EKAVDDVLQD  GYC-TGDLQV  ANGKVVSTIE  LTDRLIEKLN  NSAAGPRIFQ
Bsub    DIAGKGMANP  FAAILSAAML  LRTSFGL---  ---EEEAKAV  EDAVNKVLAS  GKR-TRDLAR  SEEFS-STQA  ITEEVKAAIM  SENTISNV--
Ecoli   DIAGKNIANP  IAQILSLALL  LRYSLDA---  ---DDAACAI  ERAINRALEE  GIR-TGDLAR  GAAAV-STDE  MGDIIARVA   EGV-------
Sty     DIAGKNIANP  IAQILSLALL  LRYSLDA---  ---NDRATAI  EQAINRALEE  GVR-TGDLAR  GAAAV-STDE  MGDIIARVA   EGV-------
Vsp     DIAGQGIANP  LATILSVSKM  LRYSFNQ---  ---TAAADAI  ELAVSNVIDQ  GIR-TGDLGG  AGTVKVGTTA  AGDAVVEALR  SL--------
Scer    DLP-KNKVDP  IATILSAAWM  LKLSLNL---  ---PEEGKAI  EDAVKKVLDA  GIR-TPDIYS  SNSTTEVGDA  VAEEVKKILA  ---------
Aaeo+   DIAGKGIANP  IATILSAAWM  LKYSFNM---  ---DKAHDLI  ERAILYLYK   GYR-TPDIVS  EGCIKVGDKE  ITDKIELNLE  RLKDAYT---
Sul#    DIAGKNIGNP  TAFLLSVSKN  YERMELSND   DRYIKASRAL  ENAILVLYEK  KRALTFPDVG  NL----KIME  FANEVASLID  ---------
Aful#   DIAGKGIANP  TAMILTACNM  LRHFGTV---  ---EEAVEKTIKE  ---------
```

from this extreme thermophile.[1,2] The gene is located in the leucine operon of the thermophile, linked with other leucine genes, *leuC* and *leuD*.[3] It has been reported that these genes are expressed under the control of the thermophile leucine promoter in *Escherichia coli*.[4] The wealth of sequence information allows IPMDH to be an ideal model protein for the purpose of protein engineering.

In this chapter, we describe the purification of recombinant *T. thermophilus* IPMDH produced in *E. coli,* and its modified procedures for several homologous enzymes from different organisms. The structural properties, the catalytic properties, and the thermostability of these enzymes are also described in the following sections.

Assay Methods

IPMDH catalyzes dehydrogenation and decarboxylation of 3-IPM with concomitant conversion of NAD^+ to NADH in the presence of a divalent cation such as Mn^{2+} and Mg^{2+}. The reaction is monitored spectrophotometrically by following the increase in NADH at 340 nm. For a routine assay of IPMDH from *T. thermophilus,* the activity is measured at 60° in a buffer solution containing 100 mM potassium phosphate (pH 7.6), 0.4 mM $MnCl_2$ (or $MgCl_2$), 200–1000 mM KCl, 0.4 mM DL-*threo*-3-isopropylmalic acid (DL-3-IPM), and 0.8 mM β-NAD^+. For detailed kinetic measurements, *N*-2-hydroxyethylpiperazine-*N'*-2-ethane sulfonate (HEPES) buffer [50 mM HEPES (pH 8.0), 100–1000 mM KCl, 5 mM $MgCl_2$ or $MnCl_2$, 1 mM

---

[1] T. Tanaka, N. Kawano, and T. Oshima, *J. Biochem.* **89,** 677 (1981).

[2] Y. Kagawa, H. Nojima, N. Nukima, M. Ishizuka, T. Nakajima, T. Yasuhara, T. Tanaka, and T. Oshima, *J. Biol. Chem.* **259,** 2956 (1984).

[3] M. Tamakoshi, A. Yamagishi, and T. Oshima, *Gene* **222,** 125 (1998).

[4] J. E. Croft, D. R. Love, and P. L. Bergquist, *Mol. Gen. Genet.* **210,** 490 (1987).

FIG. 3. Multiple amino acid sequence alignments of IPMDHs from various sources. Tth, the extremely thermophilic bacterium *T. thermophilus;* Tfer, the mesophilic bacterium *Thiobacillus ferrooxidans;* Bcoag, the moderately thermophilic bacterium *Bacillus coagulans;* Bsub, the mesophilic bacterium *Bacillus subtilis;* Ecoli, the mesophilic bacterium *E. coli;* Sty, the mesophilic bacterium *Salmonella typhimurium;* Vsp, the psychrotrophic bacterium *Vibrio* sp. 15; Scer, the yeast *Saccharomyces cerevisiae;* Aaeo+, the hyperthermophilic bacterium *Aquifex aeolicus;* Sul#, the extremely thermoacidophilic archaeon *Sulfolobus* sp. strain 7; Aful#, the extremely thermophilic archaeon *Archaeoglobus fulgidus.* The sequence data were retrieved from databases, and the multiple sequence alignments were performed using a CLUSTAL X graphical interface [J. D. Thompson, T. J. Gibson, F. Plewniak, F. Jeanmougin, and D. G. Higgins, *Nucleic Acids Res.* **25,** 4876 (1997)] with minor manual adjustments. The sequence numbers and the secondary structure of *T. thermophilus* IPMDH are indicated.

3-IPM, and 5 mM β-NAD$^+$] is preferable because of the requirement for a sufficient amount of Mn^{2+} or Mg^{2+} ion in the assay mixture and the chelating by potassium phosphate. The reaction is initiated by addition of the enzyme (10 μl) to the assay mixture (490 μl) in a cell cuvette prewarmed for 10 min at 60°. The change in absorbance at 340 nm is recorded, and the initial rate is calculated.

The substrate, DL-3-IPM can be chemically synthesized according to the literature.[5] The classic method, using mutant fungus,[6] is somewhat laborious. The chemically synthesized substrate is also commercially available from Wako Pure Chemical Industries (Osaka, Japan).

Purification of Recombinant *Thermus thermophilus*
3-Isopropylmalate Dehydrogenase

Expression of Recombinant Thermus thermophilus 3-Isopropylmalate Dehydrogenase in Escherichia coli

In earlier studies, the expression of the GC-rich *leuB* gene encoding 3-isopropylmalate dehydrogenase of *T. thermophilus* in *E. coli* was difficult, and a leader open reading frame has been found to be important for sufficient expression.[7,8] Two kinds of plasmid vectors for expression of *T. thermophilus* IPMDH are routinely used in our laboratory for the purpose of spectroscopy, microcalorimetry, and crystallization, as briefly described below.

The *Bam*HI fragment (1.1 kbp) containing the *T. thermophilus leuB* gene was cloned into *Bam*HI sites of pUC 118 and 119, and the resulting plasmids, pUTL 118 and pUTL 119, respectively, are utilized for the purpose of protein preparation. For protein expression, *E. coli* JA221 (F$^-$, *hsdR trpE5 leuB6 lacY RecA1*) or OM 17 (*leuB supE endA sbcB15 hsdR4 rpsL thi* Δ(*lac-proAB*), F′[*traD36 proAB$^+$ lacI^q lacZΔM15*]; a derivative of JM105)[9] is used as the host strain. A typical yield, using these plasmids, is about 2–3 mg of IPMDH from 10–15 g (wet weight) of cells (2-liter cultivation).

[5] H. Terasawa, K. Miyazaki, T. Oshima, T. Eguchi, and K. Kakinuma, *Biosci. Biotech. Biochem.* **58**, 870 (1994).

[6] J. M. Calvo and S. R. Gross, *Methods Enzymol.* **17A**, 791 (1971).

[7] M. Ishida, M. Yoshida, and T. Oshima, *Extremophiles* **1**, 157 (1997).

[8] M. Ishida and T. Oshima, *J. Bacteriol.* **176**, 2767 (1994).

[9] O. V. Mortensen, A. Yamagishi, and T. Oshima, Tokyo Institute of Technology, unpublished results (1995).

A much higher level of expression of *T. thermophilus* IPMDH can be achieved with the pET vector system.[10] The pET-21c (Novagen, Madison, WI) expression construct (a kind gift of A. Motojima, Tokyo Institute of Technology, Yokohama, Japan) contains the *T. thermophilus leuB* gene cloned into the *Nde*I and *Bam*HI restriction sites of pET-21c, which gives rise to expression of the enzyme without any extra tags. The most notable feature of this high-level expression plasmid is substitution of the first six codons of the *leuB* gene to the equivalent codons preferably used in *E. coli* by site-directed mutagenesis. This plasmid is transformed into *E. coli* BL21(DE3), which contains the T7 RNA polymerase gene under the control of the *lac*UV5 promoter. The expression of the *leuB* gene is induced with 0.2–1 mM isopropyl-β-D-thiogalactopyranoside (IPTG) at the late log phase of growth. Cells are cultivated for 2–4 hr after induction. About 30–50 mg of protein can be obtained from 4–5 g of cells (1-liter cultivation).

Reagents

DE52 is purchased from Whatman (Clifton, NJ), butyl- and DEAE-Toyopearl 650S prepacked columns from Tosoh (Tokyo, Japan), and DEAE-Sephacel, DEAE-Sepharose, Resource Q (6 ml), and HiLoad 26/10 Q Sepharose HP from Amersham Pharmacia Biotech (Piscataway, NJ). DL-3-IPM is obtained from Wako Pure Chemicals. β-NAD$^+$ is obtained from Oriental Yeast (Tokyo, Japan). All other reagents are of reagent grade, and water purified by the Milli-Q system (Millipore, Bedford, MA) is used.

Purification Procedure

Step 1: Disruption of Cells. *Escherichia coli* cells, which produce the recombinant enzyme, are harvested by centrifugation and stored at −80° until use. Frozen cells are thawed and suspended in buffer A [20 mM potassium phosphate buffer, pH 7.6, containing 0.5 mM ethylenediaminetetraacetic acid (EDTA); approximately 2–3 ml of buffer A per gram of wet cells]. The sample should be kept on ice to avoid proteolysis, up to the step of butyl column chromatography (step 4). All other steps are conducted at room temperature for the thermophilic enzyme. The cells are disrupted by ultrasonication with a flat chip at 50% duty cycle with an output level of 5–7 for 7–10 min (ultrasonic disrupter; Tommy). The solution is kept on ice throughout the sonication step to prevent the sample from heating.

[10] F. W. Studier, A. H. Rosenberg, J. J. Dunn, and J. W. Dunbendorff, *Methods Enzymol.* **185**, 60 (1991).

When the solution exceeds 100 ml, 50–80 ml is sonicated at one time. The lysate is then centrifuged (65,000g, 20 min, 4°) to remove debris.

Step 2: Heat Treatment. The crude extract thus obtained is incubated for 10–15 min in a circulating water bath at 75° with periodic gentle stirring. This heat treatment step precipitates unstable proteins originated from the *E. coli* host cells, and is one of the most effective steps for purification of *T. thermophilus* IPMDH. The solution is centrifuged again (65,000g, 20 min, 4°) to remove precipitated proteins.

Step 3: DEAE Chromatography and Ammonium Sulfate Precipitation. We perform a simple anion-exchange chromatography with stepwise elution to remove remaining lipids and other contaminated compounds, and to keep the following prepacked columns clean. Because another anion-exchange chromatography is performed later, this procedure is not crucial for the final purity. The sample solution is diluted with 3 volumes of buffer A, and then applied to a small DEAE column (DE-52 or DEAE-Sephacel; ~1 ml of gel per gram of cells) preequilibrated with the same buffer. The column is washed with 3 column volumes of buffer A, and then eluted with 2 column volumes of 400 mM KCl in buffer A. The eluted sample is roughly fractionated and the active fractions are pooled. Solid ammonium sulfate is added to 60% saturation (390 g/liter). At this point, the sample solution can be kept at 4° for a short time. This ammonium sulfate fractionation step can also be omitted.

Step 4. Hydrophobic Butyl Chromatography. The solution, in buffer A to which ammonium sulfate has been added to 60% saturation, is centrifuged (65,000g, 20 min, 4°) to collect the precipitate. The pellet is dissolved in buffer A, and solid ammonium sulfate is added to 20% saturation (114 g/liter) (when step 3 is omitted, solid ammonium sulfate is directly added to the sample solution after step 2 to 20% saturation). The solution is then centrifuged again (65,000g, 20 min, 4°) to remove insoluble materials.

The sample is filtered with a disk membrane filter (0.22-μm pore size) and loaded onto a prepacked butyl-Toyopearl 650S column (70 ml; Tosoh, Tokyo, Japan) preequilibrated with 3 column volumes of buffer A containing ammonium sulfate to 20% saturation. The column is washed with 1 column volume of the equilibration buffer, and then eluted with a linear gradient in 3 column volumes of ammonium sulfate (20–0% saturated) in buffer A (240 ml) at a flow rate of 4 ml/min. The IPMDH is eluted at about 5% saturation of ammonium sulfate. The active fractions are pooled and dialyzed against buffer A for at least 4 hr.

Step 5. Anion-Exchange Chromatography. Several different anion-exchange columns can be used. We use Resource Q (6 ml), HiLoad 26/10 Q

TABLE I
PURIFICATION OF RECOMBINANT *Thermus thermophilus*
3-ISOPROPYLMALATE DEHYDROGENASE[a]

Fraction	Total protein (mg)	Specific activity (units/mg protein)	Purification (fold)	Recovery (%)
Crude extract	560	6	1	100
Heat treatment	114	25	4.2	85
DE-52	75	34	5.7	77
Butyl-Toyopearl 650S	36	61	10	66
HiLoad Q HP	20	71	12	42

[a] From 6 g of *E. coli* BL21 (DE3) harboring the plasmid, pET-21-c-*T. thermophilus leuB*. Protein was measured with the BCA (Pierce, Rockford, IL) protein assay reagent, using bovine serum albumin as standard.

Sepharose High Performance (about 60 ml), or DEAE Toyopearl 650S (70 ml) depending on the amount of enzyme to be purified. A resource Q column can be used at a high flow rate (4–6 ml/min), and allows purification of up to 10 mg of IPMDH in one run, in about 60 min. HiLoad 26/10 Q may be applicable for up to 100 mg of protein.

Resource Q is washed with 10 column volumes of 1 M KCl in buffer A followed by equilibration with 10 column volumes of buffer A before applying the sample. The dialysate is filtered with a disk membrane (0.22-μm pore size) and applied to the column. The column is washed with 5 column volumes of buffer A, and the porteins are eluted with 20–25 column volumes of a linear gradient from 0 to 400 mM KCl in buffer A. When HiLoad Q is used, the column is equilibrated with 5 column volumes of buffer A, and the protein protein is eluted with 5–10 column volumes of a linear gradient of 0–300 mM KCl in buffer A. The purity of the active fractions is analyzed by sodium dodecyl sulfate–polyacrylamide gel electrophoresis (SDS–PAGE) [13% (w/v) acrylamide].[11] Protein concentration is determined by the molar absorption coefficient at 280 nm for *T. thermophilus* IPMDH (30,400 M^{-1} cm^{-1} per subunit).[12] A summary of the purification is presented in Table I.

[11] U. K. Laemmli, *Nature (London)* **227**, 680 (1970).
[12] T. Yamada, N. Akutsu, K. Miyazaki, K. Kakinuma, M. Yoshida, and T. Oshima, *J. Biochem.* **108**, 449 (1990).

Purification Procedures for 3-Isopropylmalate Dehydrogenase from
Other Sources

Escherichia coli 3-Isopropylmalate Dehydrogenase

Expression. High-level expression plasmids using the pET system have
been described for *E. coli* IPMDH.[13] One of the expression plasmids,
pE1EP1, is a pET-21c-based plasmid containing the *E. coli leuB* gene and
was constructed in our laboratory[14]; a typical yield of purified enzyme is
about 5 mg/g of wet cells. The expression procedure is the same as that
for *T. thermophilus* IPMDH.

Purification. Essentially the same purification procedures described
above are applicable with slight modifications, but attention should be paid
to proteolysis and denaturation of the less stable IPMDH. To avoid the
proteolysis in the disruption step (step 1), 1 mM phenylmethylsulfonyl
fluoride (PMSF) and 5–10 µg/ml each of leupeptin and pepstatin A are
added to buffer A as protease inhibitors. The thermal stability of the
enzyme was investigated prior to the purification,[13] and the heat treatment
temperature for the crude extract was determined from the denaturation
profiles to be 55° (10 min) in step 2.

In step 3, the sample should be applied to a larger size of anion-exchange
column (DE-52, or DEAE-Sephacel; about 5–10 ml of resin per gram of
cells) at 4°. The column is washed with at least 3 column volumes of buffer
A, and the protein is eluted with a linear gradient of 0–400 mM KCl in 3
to 4 column volumes of buffer A. The active fractions are pooled, and then
solid ammonium sulfate is added to 60% saturation. Because *E. coli* IPMDH
has a tendency to aggregate on ammonium sulfate fractionation, gentle
mixing is required in this step. No modifications are applied to steps 4 and
5, and the purity is >95% after step 5. Additional gel-filtration chromatography may be used to obtain a higher purity of the enzyme.

Bacillus subtilis 3-Isopropylmalate Dehydrogenase

Because *Bacillus subtilis* IPMDH shows cold denaturation at temperatures below 4°, most steps should be done at room temperature. However,
the recombinant *E. coli* cells that produce the enzyme can be stored at
−80°. To avoid the proteolysis, the cell disruption step (step 1) should be
performed on ice or at 4°. Little loss of the activity is observed at these

---

[13] G. Wallon, K. Yamamoto, H. Kirino, A. Yamagishi, S. T. Lovett, G. A. Petsko, and T.
Oshima, *Biochim. Biophys. Acta* **1337**, 105 (1997).
[14] C. Motono and T. Suzuki, Tokyo University of Pharmacy and Life Science, unpublished
results (1998).

steps. Buffer A to suspend the cells should also contain PMSF and some other protease inhibitors as described in the preceding section. Step 2 should be omitted, because no heat treatment is applicable to *B. subtilis* IPMDH. Instead, the diluted sample solution is directly applied to a DEAE column, as described above for *E. coli* IPMDH in step 3. After the precipitation with ammonium sulfate at 60% saturation, the pellet should be dissolved in buffer A, and solid ammonium sulfate is then added to 25% saturation (144 g/liter) prior to hydrophobic column chromatography in step 4.

In step 4, butyl-Toyopearl column chromatography is not applicable to this enzyme because of a low recovery of *B. subtilis* enzyme. We use a phenyl column [Phenyl-5PW, Phenyl-Toyopearl 650S (Tosoh), or Phenyl Sepharose Fast Flow (Amersham Pharmacia Biotech)] in place of a butyl-Toyopearl column. The protein is eluted at a relatively higher ammonium sulfate concentration (~10% saturation). No modification is applied to step 5. If the purity is less than 95%, an additional anion-exchange chromatography step is performed at a different pH (6.5 of 8.5) after the dialysis of the enzyme.

Sulfolobus Species Strain 7 3-Isopropylmalate Dehydrogenase

Expression. Cloning, sequencing, purification, and characterization of IPMDH from a thermophilic archaeon, *Sulfolobus* sp. strain 7, have been reported elsewhere.[15] The expression plasmid, pE7-SB6, is constructed by inserting the fragment containing *Sulfolobus* sp. *leuB* gene into plasmid pET-17b (Novagen) at the *Nde*I and *Eco*RI sites. The host cells and the expression procedures are the same as those described for *T. thermophilus* IPMDH (the pET-21c construct).

Purification. The buffer used in the purification of the archaeal recombinant enzyme is 20 mM potassium phosphate, pH 7.8, containing 0.5 mM EDTA (buffer A'). The crude extract obtained by the same procedures as described for *T. thermophilus* enzyme (step 1) is heat treated at 83° for 20 min. After centrifugation to remove denatured proteins, an equal amount of 2 M ammonium sulfate in buffer A' is added to the solution. The sample is then applied to a butyl-Toyopearl 650S column preequilibrated with 1 M ammonium sulfate in buffer A'. The protein is eluted with a linear gradient of 0.7–0.3 M ammonium sulfate in buffer A'.

Because the isoelectric point of the archaeal IPMDH is about pH 6.9,[16]

[15] T. Suzuki, Y. Inoki, A. Yamagishi, T. Iwasaki, T. Wakagi, and T. Oshima, *J. Bacteriol.* **179,** 1174 (1997).

[16] E. Yoda, Y. Anraku, H. Kirino, T. Wakagi, and T. Oshima, *FEMS Microbiol. Lett.* **131,** 243 (1995).

the enzyme does not bind to an anion-exchange resin at neutral pH. Thus, the pooled fractions obtained from a butyl-Toyopearl column chromatography step are passed through a DEAE-Sepharose Fast Flow column, to obtain the purified enzyme.

Saccharomyces cerevisiae 3-Isopropylmalate Dehydrogenase

Purification and enzymatic properties of *Saccharomyces cerevisiae* IPMDH have been reported previously.[17,18] An expression plasmid in *E. coli* for this IPMDH, pETL2, was constructed in our laboratory on the basis of pET-21c.[19] The *leu2* gene encoding *S. cerevisiae* IPMDH was cloned into the *Nde*I and *Sal*I sites of pET-21c. The expression procedure is the same as that for *T. thermophilus* IPMDH.

Purification. To purify,[19] *E. coli* cells are suspended in buffer B [50 mM Tris-HCl (pH 8.3), 10% (v/v) glycerol, 1 mM EDTA] containing 1 mM PMSF and pepstatin A and leupeptin (10 μg/ml) (buffer B should be used throughout the purification in place of buffer A). The disruption step (step 1) is the same as that described for *T. thermophilus* IPMDH. The heat treatment step should be omitted. The crude extract is directly applied to a DE-52 column as described for *T. thermophilus* IPMDH, except for the elution buffer (0.1 M NaCl in buffer B). The active fractions are pooled, and solid ammonium sulfate is added to 50% saturation (313 g/liter). After centrifugation to remove precipitates, the supernatant is applied to a butyl-Toyopearl 650S column preequilibrated with buffer B containing ammonium sulfate (30% saturation; 176 g/liter). The column is washed with the same buffer, and the protein is eluted by a linear gradient of 30–0% saturation of ammonium sulfate in buffer B. After dialysis of active fractions against buffer B, the sample is applied to an anion-exchange column (e.g., HiLoad Q) and then eluted by a linear gradient of 0–0.1 M NaCl in buffer B. At this step, the enzyme is almost pure (when required, rechromatography with an anion-exchange column is effective to remove small amounts of contaminating proteins).

Chimeric 3-Isopropylmalate Dehydrogenases

Several chimeric enzymes between *T. thermophilus* and *B. subtilis* IPMDH have been constructed and purified in our laboratory.[20] The same

[17] Y.-P. Hsu and G. B. Kohlhaw, *J. Biol. Chem.*, **255**, 7255 (1980).
[18] G. B. Kohlhaw, *Methods Enzymol.* **166**, 429 (1988).
[19] S. Kakizawa and M. Tamakoshi, Tokyo University of Pharmacy and Life Science, unpublished results (1998).
[20] K. Numata, M. Muro, N. Akutsu, Y. Nosoh, A. Yamagishi, and T. Oshima, *Protein Eng.* **8**, 39 (1995).

purification procedures as described for *E. coli* IPMDH can be applied to these enzymes, except for the heat treatment step (step 2). An appropriate temperature for the treatment should be determined according to the thermostability analysis of each enzyme.

Storage

Purified IPMDH can be stored at 4° as a suspension in buffer A containing ammonium sulfate (60% saturation). IPMDH ($\sim$1 mg/ml or higher) can also be stored in buffer A without any ammonium sulfate, although some chimeric proteins and *B. subtilis* enzyme were unstable in the absence of ammonium sulfate. When used, the precipitate is collected by centrifugation (15,000–65,000g, 20–30 min, 4°) and dissolved in an appropriate buffer followed by overnight dialysis against the same buffer. *Saccharomyces cerevisiae* IPMDH has been reported to be cold labile. Ammonium sulfate (1.24 *M*) and glycerol (30%, v/v) should be added to the protein solution before storage at 4°.[18]

Properties of 3-Isopropylmalate Dehydrogenase

General Properties

Enzymatic properties of *T. thermophilus*, *B. subtilis*, *E. coli*, *Thiobacillus ferrooxidans* (an acidophilic chemolithouautotrophic bacterium), and *Sulfolobus* sp. IPMDHs are summarized in Table II. The amino acid sequences of the enzymes from various sources show a high degree of homology (Fig. 3). However, *Sulfolobus* IPMDH also shows a relatively high sequence homology to mitochondrial NAD-dependent isocitrate dehydrogenases (EC 1.1.1.41), and it was suggested that the archaeal enzyme is in one of the lowest branches of the decarboxylating dehydrogenase family.[15] Most IPMDHs reported so far are homodimers, with each subunit consisting of $\sim$350 amino acid residues, while the *Sulfolobus* enzyme has been reported to be a homotetramer as judged by the gel-filtration column chromatography[16] and sedimentation equilibrium analysis.[15]

The first 3-D structure of IPMDH was determined by high-resolution X-ray crystallographic analysis using *T. thermophilus* enzyme.[21] The 3-D structures of *E. coli*,[22] *Salmonella typhimurium*,[22] *Bacillus coagulans*,[23] and

[21] K. Imada, M. Sato, N. Tanaka, Y. Katsube, Y. Matsuura, and T. Oshima, *J. Mol. Biol.* **222**, 725 (1991).

[22] G. Wallon, G. Kryger, S. T. Lovett, T. Oshima, D. Ringe, and G. A. Petsko, *J. Mol. Biol.* **266**, 1016 (1997).

[23] D. Tsuchiya, T. Sekiguchi, and A. Tekenaka, *J. Biochem.* **122**, 1092 (1997).

TABLE II

KINETIC PARAMETERS AND ENZYMATIC PROPERTIES OF 3-ISOPROPYLMALATE DEHYDROGENASES

Organism	Amino acid residue no.	Molecular mass[a] (Da)	Identify vs. Tth[b] (%)	pI	K_m (μM) 3-IPM	K_m (μM) NAD$^+$	k_{cat}[c] (sec^{-1})	Optimum concentration of KCl (mM)	T_m[d] (°C)
Thermus thermophilus	345[e]	36,780 × 2[e]	—	4.67[f]	1.26[g]	40.9[g]	13.6[g]	100–1000[h,i]	85[i]
Bacillus subtilis	365[i]	39,757 × 2[j]	58[j]	4.39[f]	30.5[k]	333[k]	10.6[k]	100–1000[f]	45[f]
Escherichia coli	363[e]	39,517 × 2[e]	57[e]	—	105[h,l]	321[h,l]	35[h,l]	300[h]	63[h]
Thiobacillus ferrooxidans	358[m]	38,462 × 2[m]	50[m]	—	26[m,n]	800[m,n]	—	—	~60[m]
Sulfolobus sp.	337[o]	36,832 × 4[o]	37[o]	6.9[p]	1.2[o,q]	150[o,q]	3.6[o,q]	200[p]	97[p]

[a] Molecular mass calculated from each *leuB* gene sequence.
[b] Sequence identity against *T. thermophilus* IPMDH.
[c] Per α subunit (an active site).
[d] Half-denaturation temperatures obtained from the remaining activity profile.
[e] H. Kirino, M. Aoki, M. Aoshima, Y. Hayashi, M. Ohba, A. Yamagishi, T. Wakagi, and T. Oshima, *Eur. J. Biochem.* **220**, 275 (1994).
[f] K. Numata, M. Muro, N. Akutsu, Y. Nosoh, A. Yamagishi, and T. Oshima, *Protein Eng.* **8**, 39 (1995).
[g] Activities were assayed in 50 mM HEPES (pH 8.0), 100 mM KCl, 5 mM MgCl$_2$, 5 mM β-NAD$^+$, 1 mM 3-IPM, at 60°; K. Miyazaki and T. Oshima, *FEBS Lett.* **332**, 37 (1993).
[h] T. Yamada, N. Akutsu, K. Miyazaki, K. Kakinuma, M. Yoshida, and T. Oshima, *J. Biochem.* **108**, 449 (1990).
[i] G. Wallon, K. Yamamoto, H. Kirino, A. Yamagishi, S. T. Lovett, G. A. Petsko, and T. Oshima, *Biochim. Biophys. acta* **1337**, 105 (1997).
[j] R. Imai, T. Sekiguchi, Y. Nosoh, and K. Tsuda, *Nucleic Acids Res.* **15**, 4988 (1987).
[k] Activities were assayed in 100 mM potassium phosphate buffer (pH 7.6), 1 M KCl, 0.2 mM MnCl$_2$, 4 mM β-NAD$^+$, 0.4 mM 3-IPM, at 40°; S. Akanuma, A. Yamagishi, N. Tanaka, and T. Oshima, *Protein Sci.* **7**, 698 (1998).
[l] Activities were assayed in 20 mM potassium phosphate buffer (pH 7.6), 300 mM KCl, 0.2 mM MnCl$_2$, 0.8 mM β-NAD$^+$, 0.4 mM 3-IPM, at 40°.
[m] H. Kawaguchi, K. Inagagaki, Y. Kuwata, H. Tanaka, and T. Tano, *J. Biochem.* **114**, 370 (1993).
[n] Activities were assayed in 0.1 M Tris-HCl (pH 9.0), 50 mM KCl, 0.5 mM MgCl$_2$, 6.7 mM β-NAD$^+$, 0.67 mM 3-IPM.
[o] T. Suzuki, Y. Inoki, A. Yamagishi, T. Iwasaki, T. Wakagi, and T. Oshima, *J. Bacteriol.* **179**, 1174 (1997).
[p] E. Yoda, Y. Anraku, H. Kirino, T. Wakagi, and T. Oshima, *FEMS Microbiol. Lett.* **131**, 243 (1995).
[q] Activities were assayed in 50 mM HEPES (pH 8.0), 100 mM KCl, 5 mM MgCl$_2$, 5 mM β-NAD$^+$, 1 mM 3-IPM, at 70°.

T. ferrooxidans[24] (complex with 3-IPM; see below) are also available, and are similar to that of *T. thermophilus* enzyme. Here we summarize the structural features of IPMDH based on the structure of *T. thermophilus* enzyme.

The folding topology (Fig. 2) is categorized as a parallel α/β doubly

[24] K. Imada, K. Inagaki, H. Matsunami, H. Kawaguchi, H. Tanaka, N. Tanaka, and K. Namba, *Structure* **6**, 971 (1998).

A   Subunit   B

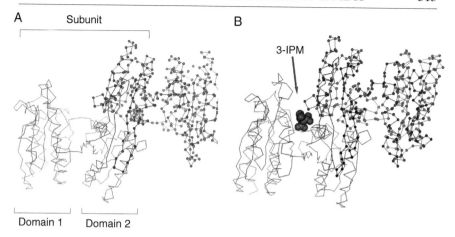

Domain 1   Domain 2

FIG. 4. Three-dimensional structures of IPMDHs with one subunit by line, the other by line and dots. (A) The backbone trace of *T. thermophilus* IPMDH. Domains and a subunit are indicated. [From K. Imada. M. Sato, N. Tanaka, Y. Katsube, Y. Matsuura, and T. Oshima, *J. Mol. Biol.* **222**, 725 (1991).] (B) The backbone trace of *T. ferrooxidans* IPMDH complexes with 3-IPM. The bound 3-IPM is shown as space-filling model for one subunit. (A) and (B) were generated by Insight II [K. Imada, K. Inagaki, H. Matsunami, H. Kawaguchi, H. Tanaka, N. Tanaka, and K. Namba, *Structure* **6**, 971 (1998).]

wound β-sheet motif,[25] as observed for the *E. coli* ICDH structure.[26] The enzyme is a homodimer, and each subunit consists of two domains (Fig. 4A). The first domain (or domain 1) of *T. thermophilus* IPMDH consists of residues 1–99 and 252–345, and includes the N and C termini. The second domain (or domain 2) consists of residues 100–251, and contains a subunit interface of the homodimeric structure. The active site is located at the cleft between these two domains, and residues of the other subunit are also involved in the formation of the region. Thus, each IPMDH dimer has two active sites. At the subunit interface, two α helices (α helices h and g) contributed from domain 2 of each subunit form a tight four-helix hundle structure (Fig. 5). Moreover, the amino acid residues 134–158 (β-strands K and L) form a long "armlike" structure that protrudes from the domain, forming an intersubunit β sheet with the same region of the other subunit (Fig. 5).

[25] J. S. Richardson, *Methods Enzymol.* **115**, 341 (1985).
[26] J. H. Hurley, P. E. Thorsness, V. Ramalingam, N. H. Helmers, D. E. J. Koshland, and R. M. Stroud, *Proc. Natl. Acad. Sci. U.S.A.* **86**, 8635 (1989).

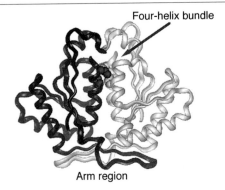
Four-helix bundle

Arm region

FIG. 5. Schematic representation of the subunit interface of *T. thermophilus* IPMDH with one subunit in light gray, the other in dark gray. The four-helix bundle structure and the arm region consisting of β strands are involved in the subunit–subunit interactions. The residues shown as space-filling models, L246 and V249, are involved in the intersubunit hydrophobic interaction, and contribute critically to the thermostability. Figure was generated by Insight II. [From H. Kirino, M. Aoki, M. Aoshima, Y. Hayashi, M. Ohba, A. Yamagishi, T. Wakagi, and T. Oshima, *Eur. J. Biochem.* **220,** 275 (1994).]

Catalytic Properties

Kinetic parameters of *T. thermophilus, B subtilis, E. coli, T. ferrooxidans,* and *Sulfolobus* sp. IPMDHs are summarized in Table II. The presence of a divalent cation is essential for the catalytic reaction of IPMDH. Most IPMDHs preferably require Mn^{2+} and Mg^{2+} (Mn^{2+} is more effective than Mg^{2+} for most of the enzymes).[12,13,17,27] In addition, the activity of IPMDH is remarkably enhanced by a monovalent cation such as K^+, NH_4^+, and Rb^+.[12,13,28] Among them, a potassium ion is the most effective for the activation of *T. thermophilus* IPMDH, which results in a 30-fold activation of the specific activity.[12] *Escherichia coli, B. subtilis,* and *T. ferrooxidans* enzymes show similar cation dependencies, although the effect of a monovalent cation is relatively small in *Sulfolobus* sp. IPMDH.[16] It has been reported on the basis of kinetic analyses[29,30] that the kinetic mechanism of *T. thermophilus* IPMDH is steady-state random, as is the case for the $NADP^+$-dependent ICDH.

---

[27] S. J. Parsons and R. O. Burns, *Methods Enzymol.* **17A,** 793 (1971).
[28] H. Kawaguchi, K. Inagagaki, Y. Kuwata, H. Tanaka, and T. Tano, *J. Biochem.* **114,** 370 (1993).
[29] A. M. Dean and D. E. J. Koshland, *Biochemistry* **32,** 9302 (1993).
[30] A. M. Dean and L. Dvorak, *Protein Sci.* **4,** 2156 (1995).

Substrate and Coenzyme Specificity

IPMDH and ICDH belong to the same decarboxylating dehydrogenase family as described in the Introduction. The substrates of these enzymes, 3-IPM and isocitrate, respectively, have a malate moiety, and the difference between these compounds is the γ moiety attached to (2R)-malate. Most residues involved in the catalytic reaction are conserved in both enzymes, but the two substrates are strictly recognized by their respective enzymes. A steady-state kinetic study showed that a variety of alkyl malates with various alkyl groups, such as a methyl, ethyl, isopropyl, isobutyl, *tert*-butyl, or isoamyl group, as the γ moiety can be recognized as the substrate for *T. thermophilus* IPMDH.[31] *Escherichia coli* and *Sulfolobus* IPMDHs have a similar broad specificity toward some alkyl malates.[13,15] Nevertheless, IPMDH exhibits no catalytic activity for isocitrate, which has a negatively charged carboxymethyl group as the γ moiety.[31]

Coenzyme specificities of *T. thermophilus* IPMDH and ICDH are also different. IPMDH utilizes NAD^+ as the preferred coenzyme, while ICDH utilizes $NADP^+$. Conversion of the coenzyme specificity between *T. thermophilus* IPMDH and ICDH has been achieved by a "module" replacement.[32,33]

The structure of a binary complex with 3-IPM for *T. ferrooxidans* IPMDH provides insight into the kinetic mechanism of IPMDH.[24] A divalent cation, Mg^{2+}, connects the malate backbone of 3-IPM and the side chains of two aspartate residues (Asp-222' and Asp-246 in the *T. ferrooxidans* enzyme), explaining the requirement for divalent cations in the catalytic reaction. A large domain movement of domain 1, i.e., a domain closure, is induced by substrate binding (Fig. 4B), which results in the formation of the hydrophobic pocket [Glu-88, Leu-91, Leu-92, and Val-193' [prime indicates the residues from the other subunit)] for the isopropyl group of 3-IPM (Fig. 6). Glu-88 (Glu-87 in *T. thermophilus* IPMDH) plays an important role in substrate recognition. The long side chain of Glu-88 (C_β and C_γ) provides a flexible hydrophobic surface for a variety of hydrophobic γ moieties of the alkyl malates. On the other hand, the electric repulsion between the negative charge of Glu-88 and γ-carboxylate of isocitrate suppresses the binding of isocitrate to IPMDH, as previously proposed by mutational analysis.[31,34] A structural model of the IPMDH–IPM–NAD^+ complex has suggested that a carboxyl group of Glu-88 is also likely to

[31] K. Miyazaki, K. Kakinuma, H. Terasawa, and T. Oshima, *FEBS Lett.* **332**, 35 (1993).
[32] M. Gō, *Adv. Biophys.* **19**, 91 (1985).
[33] T. Yaoi, K. Miyazaki, Y. Komukai, T. Oshima, and M. Gō, *J. Biochem.* **119**, 1014 (1996).
[34] J. H. Hurley, A. M. Dean, D. E. J. Koshland, and R. M. Stroud, *Science* **249**, 1012 (1990).

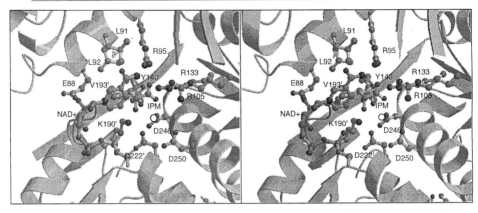

FIG. 6. Stereoview of the possible active site structure of IPMDH [K. Imada, K. Inagaki, H. Matsunami, H. Kawaguchi, H. Tanaka, N. Tanaka, and K. Namba, *Structure* **6,** 971 (1998)]. A ternary complex model of IPMDH–IPM–NAD⁺ was constructed on the basis of the structures of the *T. ferrooxidans* IPMDH binary complex with 3-IPM [K. Imada, K. Inagaki, H. Matsunami, H. Kawaguchi, H. Tanaka, N. Tanaka, and K. Namba, *Structure* **6,** 971 (1998)] and *E. coli* ICDH ternary complex with NADP⁺ and isocitrate [J. M. Bolduc, D. H. Dyer, W. G. Scott, P. Singer, R. M. Sweet, D. E. Koshland, Jr., and B. L. Stoddard, *Science* **268,** 1312 (1995)]. The nicotinamide mononucleotide moiety of NAD⁺, 3-IPM, and side chains involved in the active site are shown as ball-and-stick models. The residues E88, L91, L92, and V193′ in *T. ferrooxidans* IPMDH, labeled here, are involved in the formation of the hydrophobic pocket responsible for the recognition of the alkyl γ group of the substrate, and the equivalent residues in *T. thermophilus* IPMDH are E87, L90, L91, and V188′, respectively. The residues R95, R105, R133, Y140, K190′, D222′, and D246, D250 in the *T. ferrooxidans* enzyme, labeled here, are hydrogen bonded to the malate backbone of 3-IPM, and the equivalent residues in *T. thermophilus* IPMDH are R94, R104, R132, Y139, K185′, D217′, D241, and D245, respectively. The Mg²⁺ ion, indicated as an open circle, is present between 3-IPM and the side chains of D222′ and D246. This figure was kindly provided by K. Imada (Matsushita Electric Industrial Co., Ltd.), and was generated with MOLSCRIPT [P. J. Kraulis, *J. Appl. Crystallogr.* **24,** 946 (1991)] and RASTER3D [E. A. Merritt and M. Murphy, *Acta Crystallogr. D* **50,** 869 (1994)].

interact with nicotinamide mononucleotide ribose. The residues involved in the catalytic reaction are shown in Fig. 6, based on the model of the IPMDH–IPM–NAD⁺ complex. The equivalent amino acid residues are well conserved in IPMDHs known so far (Fig. 3).

Thermostability

IPMDH is an ideal model protein for studying the thermostability of thermophilic proteins because of the wealth of sequence and structural information. The dimeric structure and the domain arrangement of IPMDH show that it is a "regular" type of protein, having a certain degree of structural complexity. The 3-D structure and/or sequence comparisons of IPMDH provide possible structural bases of thermostability because enzymes from different sources have different denaturation temperatures (Table II). IPMDH was also utilized as a model protein for the stabilization of proteins by an evolutionary molecular engineering technique, using *T. thermophilus* as a host. The most remarkable feature of this technique is that one can obtain stabilizing mutations from a number of possibilities even in the absence of detailed structural information. The principle and the application of this technique have been reviewed elsewhere.[35] Designed and selected mutant enzymes are purified and evaluated for thermostability as described below, to reveal the mechanism of thermostability.

Analysis of Thermostability

Remaining Activity Analysis. Remaining activity analysis is a convenient way to evaluate the thermostability of purified (or crude) IPMDH. The enzyme solution (0.1–0.2 mg/ml) in an Eppendorf tube is incubated for 10 min at various temperatures. The heated enzyme is chilled quickly in an ice bath. The tube is centrifuged to remove insoluble aggregates, and the remaining activity is measured in the standard assay mixture (see Assay Methods). Each remaining activity is plotted against the heat-treated temperature, and the half-denaturation temperature, T_m (or T_h), can be obtained from the plot (Fig. 7A). The inactivation rates can also be obtained from a series of experiments in which the remaining activity is measured as a function of the incubation time, and the activation energy of the inactivation reaction can be calculated on the basis of the Arrhenius equation.[20] The remaining activity analysis cannot exclude the possible existence of refolding proteins, which may give rise to a higher shift in apparent denaturation temperatures, as compared with circular dichroic (CD) analysis (see the next section). It should be noted that the remaining activity method provides information on protein thermotolerance, rather than thermostability. However, denaturation temperatures obtained from these analyses do not differ greatly, probably because these IPMDHs show irreversible thermal denaturation. Therefore, this method can also be used for preliminary estimation of thermostability.

Circular Dichroic Melting Analysis. CD melting analysis is a more direct way to measure the unfolding process of a protein; it is done by monitoring the decrease in secondary structures on heating. The sample solution (0.1–0.5 mg/ml) in 20 mM potassium phosphate buffer, pH 7.6, or other appro-

---

[35] T. Oshima, *Curr. Opin. Struct. Biol.* **4**, 623 (1994).

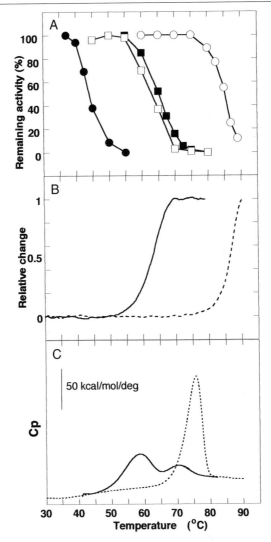

FIG. 7. The thermostability of IPMDHs. (A) Remaining activities are plotted against heat-treatment temperatures. The IPMDH solution in 20 mM potassium phosphate, pH 7.6, was heat treated at various temperatures for 10 min, and the remaining activity was measured in the standard assay mixture. (●) *Bacillus subtilis* IPMDH (T_m 45°); (□) *E. coli* IPMDH (T_m 63°); (■) a chimera between *T. thermophilus* and *B. subtilis* IPMDH (2T2M6T) (T_m 65°); (○) *T. thermophilus* IPMDH (T_m 85°). (B) Normalized CD denaturation curves of IPMDHs monitored at 222 nm. Solid trace, *E. coli* IPMDH (T_m 63°); dashed trace, *T. thermophilus* IPMDH (T_m 85°). The heating rate is ~1°/min. The buffer used is 20 mM potassium phosphate, pH 7.6, and the protein concentration is ~0.2 mg/ml. (C) DSC melting curves of IPMDHs. Dashed trace, *T. thermophilus* IPMDH [peak temperature (T_p) 76°]; solid trace, the chimera

priate buffer, is placed into a 1-mm cell. The temperature of the cell is increased, using a circulating bath with a programmable controller unit, and the CD value at 222 nm is monitored at the same time. A typical heating rate is 1°/min. When information at the near-UV region is required, the protein concentration and the cell length should be changed (e.g., 1.2 mg/ml in a 0.5-cm cell),[36] to improve the signal-to-noise (S/N) ratio. The temperature of the solution in the cell should be measured directly with a thermocouple. The melting profile thus obtained can be normalized, assuming the linear temperature dependence of the baseline of the native and unfolded states (Fig. 7B).

Differential Scanning Microcalorimetry. Differential scanning microcalorimetry (DSC) measurement is also applicable to IPMDH, and gives the accurate denaturation temperature of the enzyme. However, higher order aggregation in the unfolding of *T. thermophilus* IPMDH was observed when the pH of the solution was less than 9.5.[36] The protein solution in an alkaline buffer (e.g., 5–20 mM Na$_2$B$_4$O$_7$-NaOH buffer, pH 10–11) is filtered and degassed, and injected into the DSC cell. The endothermic peak corresponding to the protein unfolding is observed at about 75° depending on the pH (Fig. 7C). The peak temperature is relatively low as compared with the T_m obtained from the CD analysis, because the measurements are conducted under alkaline conditions. The irreversibility and the slight scan rate dependencies of the thermal unfolding process of *T. thermophilus* IPMDH make the quantitative analysis difficult. In spite of these problems, CD and DSC measurements provide valuable information. On the other hand, the equilibrium studies of the unfolding of *T. thermophilus* IPMDH using denaturants allow the quantitative analysis, which provides ΔG, ΔH, and ΔS values for the folding/unfolding reaction of the enzyme.[37]

Mechanism of Thermostability of 3-Isopropylmalate Dehydrogenase

The studies of the thermostability of IPMDH have revealed a number of residues that contribute to the thermostability of *T. thermophilus*

[36] Y. Hayashi-Iwasaki, K. Numata, A. Yamagishi, K. Yutani, M. Sakurai, N. Tanaka, and T. Oshima, *Protein Sci.* **5**, 511 (1996).
[37] C. Motono, A. Yamagishi, and T. Oshima, *Biochemistry* **38**, 1332 (1999).

2T2M6T (T_{p1} 59°, T_{p2} 71°) [Y. Hayashi-Iwasaki, K. Numata, A. Yamagishi, K. Yutani, M. Sakurai, N. Tanaka, and T. Oshima, *Protein Sci.* **5**, 511 (1996)]. The heating rate is 1°/min. The buffer used is 20 mM NaHCO$_3$-NaOH, pH 10.4. The chimeric IPMDH 2T2M6T has a mesophilic portion in the domain interface region of the *T. thermophilus* IPMDH sequence. It shows a biphasic transition, which has been suggested to be independent unfoldings of two domains [Y. Hayashi-Iwasaki, K. Numata, A. Yamagishi, K. Yutani, M. Sakurai, N. Tanaka, and T. Oshima, *Protein Sci.* **5**, 511 (1996)].

IPMDH.[20,38–41] One of the major contributors to the marked thermostability of *T. thermophilus* IPMDH consists of the intersubunit and interdomain interactions. The importance of hydrophobic interaction at the subunit interface (Fig. 5) has been confirmed by the results of a site-directed mutagenesis study.[38] In addition, replacements at the domain interface by the mesophilic sequence in *T. thermophilus* IPMDH caused the decooperativity of two domains (Fig. 7C), resulting in a decrease in the thermostability of the enzyme.[36] Moreover, several stabilized mutant IPMDHs selected by the evolutionary screening method, using *T. thermophilus,* have mutation sites at the interdomain hydrophobic contact regions, also supporting the importance of the interdomain interaction.[41,42] Thus, for the multidomain and multisubunit proteins, the interaction(s) between these structural units (i.e., domains or subunits) are indicated to be of particular importance for the overall thermostability.

Acknowledgments

The authors thank Dr. Katsumi Imada (International Institute for Advanced Research, Matsushita Electric Industrial Co., Ltd.) for valuable discussion. We also thank colleagues and collaborators. This investigation was supported in part by Grants-in-Aid for Scientific Research from the Ministry of Education, Science, Culture, and Sports of Japan (10044095, 09558081, and 09480154).

[38] H. Kirino, M. Aoki, M. Aoshima, Y. Hayashi, M. Ohba, A. Yamagishi, T. Wakagi, and T. Oshima, *Eur. J. Biochem.* **220,** 275 (1994).
[39] M. Tamakoshi, A. Yamagishi, and T. Oshima, *Mol. Microbiol.* **16,** 1031 (1995).
[40] T. Kotsuka, S. Akanuma, M. Tomuro, A. Yamagishi, and T. Oshima, *J. Bacteriol.* **178,** 723 (1996).
[41] S. Akanuma, A. Yamagishi, N. Tanaka, and T. Oshima, *Protein Sci.* **7,** 698 (1998).
[42] S. Akanuma, A. Yamagishi, N. Tanaka, and T. Oshima, *FEBS Lett.* **410,** 141 (1997).

[29] Wild-Type and Hexahistidine-Tagged Derivatives of Leucine-Responsive Regulatory Protein from Escherichia coli

By Rowena G. Matthews, Yuhai Cui, Devorah Friedberg, and Joseph M. Calvo

Leucine-responsive regulatory protein (Lrp) is a moderately abundant regulatory protein in *Escherichia coli* (about 3000 dimers per cell) that stimulates expression of more than a dozen operons and represses expression of another dozen (reviewed in Refs. 1 and 2). Proteins having nearly identical amino acid sequences exist in other enteric bacteria[3] and regulatory proteins that are clearly evolutionarily related to Lrp (but that likely have different functions) are found in a large number of microorganisms.[4] The acronym Lrp derives from the fact that leucine, when present in high concentrations in media, can modulate the effect of Lrp. In many cases, leucine has little or no effect on Lrp-mediated expression *in vivo*, but may show small effects *in vitro*.[5,6] In some well-documented instances, exogenous leucine reduces the (positive or negative) effect of Lrp, whereas in other instances it potentiates the effect of Lrp.[1,2] Alanine also modulates the effect of Lrp on some promoters.[2,7]

Lrp has a monomer molecular mass of 18.8 kDa (164 amino acids) and exists as a homodimer in solution (10 μM). It binds to DNA sequences that are a variation of the consensus YAGHAWATTWTDCTR, where Y = C or T, H = not G, W = A or T, D = not C, and R = A or G.[8] In members of the Lrp regulon that have been studied extensively, there are multiple Lrp-binding sites near the promoter and Lrp binds cooperatively to these sites *in vitro*.[2] Binding induces a bend in the DNA.[9] A mutational analysis suggests that Lrp contains three functional domains of about equal sizes: an N-terminal domain containing the site for DNA binding, a middle domain responsible for transcription activation, and a C-terminal domain that is required for the response to leucine.[10]

This chapter describes the purification of wild-type Lrp and of a hexahistidine (His$_6$)-tagged Lrp that is easy to purify and that is useful for a variety of *in vitro* experiments. The His$_6$ derivative behaves like the wild-type protein in a number of ways, both *in vitro* and *in vivo*.

Purification of Wild-Type Leucine-Responsive Regulatory Protein

Principle

Lrp is purified from the overexpressing *Escherichia coli* strain JWD3-1. Strain JWD3-1 is strain JM105 [*supE endA sbcB15 hsdR4 rpsL thi* Δ(*lac-*

[1] E. B. Newman, T. T. Lin, and R. D'Ari, *in* "*Escherichia coli* and *Salmonella*" (F. C. Neidhardt, ed.), p. 1513. ASM Press, Washington, D.C., 1996.

[2] J. M. Calvo and R. G. Matthews, *Microbiol. Rev.* **58**, 466 (1994).

[3] D. Friedberg, J. V. Platko, B. Tyler, and J. M. Calvo, *J. Bacteriol.* **177**, 1624 (1995).

[4] B. R. Belitsky, M. C. U. Gustafsson, A. L. Sonenshein, and C. V. Wachenfeldt, *J. Bacteriol.* **179**, 5448 (1997).

[5] B. R. Ernsting, M. R. Atkinson, A. J. Ninfa, and R. G. Matthews, *J. Bacteriol.* **174**, 1109 (1992).

[6] B. R. Ernsting, J. W. Denninger, R. M. Blumenthal, and R. G. Matthews, *J. Bacteriol.* **175**, 7160 (1993).

[7] E. Mathew, J. Zhi, and M. F. Freundlich, *J. Bacteriol.* **178**, 7234 (1996).

proAB)] carrying pJWD1.[6] Plasmid pJWD1 contains the *lrp* coding sequence inserted into the NcoI–XbaI sites of the polycloning region of the expression vector p*Trc*99A (Pharmacia, Piscataway, NJ) and confers ampicillin resistance on strain JWD3-1.

The preparation involves 3 days of cell growth and then purification, which takes up most of the following 3 days.

Reagents

STG medium (per liter)

Bacto-tryptone (Difco, Detroit, MI)	20 g
Yeast extract (Difco)	10 g
NaCl	5 g
Glycerol	2.52 g (2 ml; 0.2%, v/v)

Potassium phosphate to final concentration of 50 mM (pH 7.4)

K$_2$HPO$_4$ (dibasic), 0.5 M	40.4 ml
KH$_2$PO$_4$ (monobasic), 0.5 M	9.6 ml

Bring the volume to 1 liter with glass-distilled water. *Beware:* K$_2$HPO$_4$ and KH$_2$PO$_4$, when combined, have a smaller volume than the sum of their individual volumes. The STG broth should have a pH of about 7.4 after diluting to 1 liter

Ampicillin: 50 mg/ml in glass-distilled water. Store at 4°

Dithiothreitol: Electrophoretic grade (Bio-Rad, Hercules, CA)

TG$_{10}$ED buffer (per liter)

Tris-HCl buffer (pH 8.0), 0.1 M	100 ml
Glycerol, reagent grade	100 ml
EDTA (pH 7.4), 30 mM	3 ml
Dithiothreitol, 0.1 mM (add immediately	
before use)	15.4 mg

NaCl: 1 M in glass-distilled water

Phenylmethylsulfonyl fluoride (PMSF): 20 mg/ml (115 mM) in 2-propanol

Tosyl-L-lysyl chloromethyl ketone (TLCK): 1 mg/ml (2.7 mM) in glass-distilled water. Prepare no more than 1 hr before use and store at 4°

Bio-Rex70 weak cation-exchange resin: Analytical grade, 100–200 mesh, sodium form (Bio-Rad)

Isopropyl-β-D-thiogalactopyranoside (IPTG): 0.1 M; store at −20°

[8] Y. Cui, Q. Wang, G. D. Stormo, and J. M. Calvo, *J. Bacteriol.* **177,** 4872 (1995).
[9] Q. Wang and J. M. Calvo, *EMBO J.* **12,** 2495 (1993).
[10] J. V. Platko and J. M. Calvo, *J. Bacteriol.* **175,** 1110 (1993).

Procedure

Growth of Cell Cultures. Streak a Luria broth plate[11] containing ampicillin (100 μg/ml) with strain JWD3-1. After 24 hr of incubation at 37°, inoculate 4 ml of STG medium plus ampicillin (100 μg/ml) with a single colony and grow at 37° with shaking for 8 hr. Inoculate 46 ml of STG medium plus ampicillin (100 μg/ml) with the 4-ml culture and incubate with shaking at 37° for 8 hr. Inoculate 450 ml of STG medium (no ampicillin added) with the 50-ml culture and incubate at 37° with shaking. If desired the preparation can be scaled up by growing six 500-ml culture flasks. When the OD$_{550}$ reaches a value between 1.0 and 1.5, add isopropyl-β-D-thiogalactopyranoside (IPTG) to a final concentration of 0.5 mM. Incubate for 2 hr at 37°. Chill the cell cultures at 4° and pellet the cells by centrifuging for 15 min at 10,000g at 4°. A 500-ml culture will generally yield about 6 g (wet weight) of cells.

Preparation of Sonicate. Resuspend the cells in TG$_{10}$ED–0.2 M NaCl in a small metal beaker, adding 3.5 ml per g wet weight of cells. Add 100 μl of PMSF (~20 μM final concentration) and 20 μl of TLCK (~2 μM final concentration) per 500 ml of cultured cells. Immerse the beaker in ice, and sonicate the cells, alternating five 1-min sonication pulses [we use a setting of 8 on a Branson (Danbury, CT) sonicator] with 2-min cooling periods. Centrifuge the sonicate in an ultracentrifuge for 1 hr at 33,000g at 4°. Collect the supernatant.

Cation-Exchange Chromatography. Suspend Bio-Rex in TG$_{10}$ED–0.2 M NaCl to yield ~30 ml of settled resin and titrate to pH 8.0 with 1 M unneutralized Tris. Rinse the settled resin several times with TG$_{10}$ED until the pH of the suspension is 8.0. Pour the Bio-Rex70 resin in TG$_{10}$ED buffer into a 35 × 2 cm column; this amount of resin (~10 × 2 cm) is sufficient for preparation of Lrp from 3 liters of cultured cells. Load the sonicate supernatant on the column and wash with TG$_{10}$ED buffer until the OD$_{280}$ of the eluate drops to 0.05; typically this occurs after 80 ml of wash has eluted. Elute Lrp with a 200-ml linear gradient from 0.2 to 1.0 M NaCl in TG$_{10}$ED and then wash with 30 ml of TG$_{10}$ED–1 M NaCl. Collect 3- to 4-ml fractions throughout the elution. The Lrp should elute at about 0.4 M NaCl and can be located by measuring the A_{280} of each tube. Run a 15% (w/v) acrylamide gel in the presence of sodium dodecyl sulfate to determine which fractions contain Lrp and collect and pool the purest fractions. Use an Amicon (Danvers, MA) concentrator with a YM10 filter to concentrate

[11] T. Maniatis, E. F. Fritsch, and J. Sambrook, "Molecular Cloning: A Laboratory Manual." Cold Spring Harbor Laboratory Press, Cold Spring Harbor, New York, 1982.

the preparation to between 2.5 and 5 mg/ml; an Lrp concentration of 1 mg/ml (as determined by amino acid analysis) has an A_{280} of 0.6.

Roesch and Blomfield[12] have used a 30-ml bed volume column of Whatman (Clifton, NJ) cellulose phosphate for this step, eluting the Lrp with a 300-ml linear gradient of 0.2–1.5 M NaCl in $TG_{10}ED$ buffer.

Size-Exclusion Chromatography. Equilibrate a Superose-12 MR 10/30 FPLC (fast protein liquid chromatography) column with 50 ml of $TG_{10}ED$– 0.2 M NaCl. Add 250 μl of Lrp concentrate from the preceding step and develop with $TG_{10}ED$–0.2 M NaCl at a flow rate of 0.45 ml/min and a pressure below 3 mPa. Monitor the A_{280} of the eluate while collecting 1-ml fractions. The major peak is typically Lrp, and elutes in about four tubes. Continue to load 250-μl aliquots of the Lrp concentrate from the preceding step until all the concentrate has been chromatographed. The purity of each fraction is determined by sodium dodecyl sulfate–polyacrylamide gel electrophoresis (SDS–PAGE); during the analysis all fractions are stored at 4°. Homogeneous fractions are pooled and the protein concentration determined by the Bio-Rad protein assay. Concentrate the pooled fractions to ~6 mg/ml and dilute with an equal volume of 50% (v/v) glycerol. Aliquot the Lrp into small tubes (100 μl) and store at −80°. Typically, ~6 mg of homogeneous Lrp is obtained from 500 ml of cultured cells.

Roesch and Blomfield[12] have successfully used a Bio-Gel P30 polyacrylamide gel-filtration column for this step. They collected 3-ml fractions and analyzed peak fractions on denaturing polyacrylamide gels to determine purity.

Purification of Hexahistidine–Leucine-Responsive Regulatory Protein

Principle

His$_6$–Lrp (Lrp-40) has 12 additional amino acids at the N terminus (RGSHHHHHHGS inserted after the initiating M), including 6 adjacent histidine residues that facilitate purification of the protein by affinity chromatography on Ni-NTA columns. HIS$_6$–Lrp is purified from the overexpressing *Escherichia coli* strain CV1494. The preparation involves 3 days of cell growth and then purification, which takes 1 day.

Strains and Plasmids

Strain CV1494 [*supE thi* Δ(*lac-proAB*) *ilvIH*::MudI1734 *lrp-35*::Tn*10*/ F' *traD36 lacI*^{q} Δ(*lacZ*) *M15 proA*$^+$B$^+$/pCV294] was derived by successive

---

[12] P. L. Roesch and I. C. Blomfield, *Mol. Microbiol.* **27,** 751 (1998).

transductions into strain JM101 of *ilvIH*::MudI1734 from strain CV975 and *lrp-35*::Tn*10* from strain CV1008. Plasmid pCV294 carrying the *lrp-40* allele was created by amplifying the *lrp* gene from plasmid pCV168[10] by the polymerase chain reaction (PCR), using primers 5'-ACAATA<u>GGATCC</u>G-TAGATAGCAAGAAG and 5'-TTTGCA<u>AAGCTT</u>CCGTGTTAGCG-CGTC (*Bam*HI and *Hin*dIII sites underlined) and cloning the product into plasmid pQE30 (Qiagen, Valencia, CA) cut with *Bam*HI and *Hin*dIII.

Reagents

Prepare the following in Milli-Q water (Millipore, Bedford, MA), except where noted.

Luria broth[13]
Ampicillin: 50 mg/ml; store at $-20°$
Tetracycline: 5 mg/ml in absolute ethanol; store at $-20°$
IPTG: 0.1 *M*; stored at $-20°$
Buffer A (per liter)
  Na$_2$HPO$_4$ 7.1 g
  NaCl   17.5 g
  Adjust pH to 8.0
Buffer B (per liter)
  Na$_2$HPO$_4$ 7.1 g
  NaCl   17.5 g
  Glycerol   100 ml
  Adjust pH to 6.0
RNase A: 10 mg/ml in 10 m*M* Tris (pH 7.5), 15 m*M* NaCl; store at $-20°$
DNase I: 10 mg/ml in water; store at $4°$
Ni-NTA resin: 50% slurry in buffer A
  Pellet the product from the manufacturer [Qiagen: 50% (w/v) slurry in ethanol] by centrifugation at about 200*g* for 1–2 min, remove the supernatant, and resuspend it in 1 volume of buffer A. Repeat this five times
TG$_{50}$ED buffer: Same as described above, except 500 ml of glycerol per liter

Procedure

A single colony of strain CV1494 selected from a Luria broth plate containing ampicillin (100 μg/ml) and tetracycline (20 μg/ml) is used to prepare a stationary-phase culture grown in the same liquid medium and

[13] J. H. Miller, "Experiments in Molecular Genetics," pp. 1–352. Cold Spring Harbor Laboratory Press, Cold Spring Harbor, New York, 1972.

this is used to inoculate a larger volume of the same medium [2% (v/v) inoculum]. Cells are grown at 37°, with shaking, to an A_{600} of 0.7–0.8, and IPTG is added to a final concentration of 0.5 mM. After incubation for 6 hr, cells are harvested by centrifugation at 4000g for 10 min at 4°, resuspended in 2 volumes of buffer A with the aid of a Potter–Elvehjem homogenizer, and stored frozen at −70° until use. Samples are passed through a French pressure cell at 10,000 psi. For each milliliter of sample, 10 μg of RNase A and 5 μg of DNase I are added and after incubation for 15 min on ice, the sample is clarified by centrifugation at 15,000g for 20 min at 4° and the supernatent is mixed with a 50% (v/v) slurry of Ni-NTA resin in buffer A (8 ml of slurry per liter of cells). After gentle mixing in a rotating tube for 1 hr, the resin is added to a Qiagen column (5 ml, 1.5 × 8 cm for 8 ml of slurry), washed with buffer A until the A_{280} is less than 0.01 (0.5 ml min^{-1}), washed with buffer B until the A_{280} is less than 0.02, washed with 0.1 M imidazole in buffer B (20 ml/liter of cells), and eluted with 0.25 M imidazole in buffer B (10 ml/liter of cells). Fractions (1–1.5 ml), collected on starting the 0.25 M imidazole elution, are analyzed by SDS–PAGE through 15% (w/v) acrylamide. Those containing Lrp are pooled, dialyzed against TG$_{50}$ED buffer containing 0.1 M NaCl (Slide-A-Lyzer dialysis cassette, 10-kDa cutoff; Pierce, Rockford, IL), and stored at −70°.

Cells produce about 10% of their total protein as His$_6$–Lrp (Lrp-39; see Remarks, below) following induction with IPTG, but presumably most is formed in inclusion bodes because it is lost during the centrifugation step following breakage of the cells. A significant amount of Lrp derivative can be recovered by extracting the pellet with a homogenizer, but we have been unable to increase the fraction of the total that is soluble by altering the temperature (25 versus 37°) or composition (minimal medium, rich medium, presence of 0.45 M sucrose) of the medium. Without an additional extraction step, the yield of soluble Lrp-39 per gram wet weight cells after purification by Ni-NTA affinity chromatography is about 1.5 mg.

Remarks

The His$_6$–Lrp (Lrp-40) described above has, except for the 11 additional amino acids, a wild-type sequence. A closely-related derivative containing a K6R change was inadvertently prepared and studied before it was realized that the mutation had been introduced during the polymerase chain reaction. This derivative, called His$_6$–Lrp (Lrp-39), has properties that are similar to wild-type Lrp, and because the K6R change is conservative, it seems likely that His$_6$–Lrp (Lrp-40) behaves similarly. Lrp-39 and wild-type Lrp bound *in vitro* to a 285-bp fragment of DNA from the *ilvIH* promoter region[14] with binding constants in the nanomolar range that differed by less than 25%.

By following procedures and using strains provided by Blomfield *et al.*,[15] strain CV1468 [F⁻ *ara thi* Δ(*lac-pro*) *lrp-39*] was prepared with the wild-type *lrp* gene in the chromosome replaced by the *lrp-39* allele. *Escherichia coli* K-12 requires functional Lrp for growth in a medium containing arginine or glycine as sole nitrogen source.[5] His$_6$–Lrp (Lrp-39) must retain some function because strain CV1468 grows normally in these media. In addition, His$_6$–Lrp (Lrp-39) functioned *in vivo* as an activator of the *ilvIH* promoter as measured using a *lacZ* reporter gene, although levels of expression were reduced by a factor of 2.7. His$_6$–Lrp (Lrp-39) also functioned *in vivo* to repress its own synthesis, but again, its efficiency was somewhat less than that of the wild-type protein. These results suggest that His$_6$–Lrp is at least partially active for three of the known functions of the protein.

Acknowledgments

This work was supported by NIH Grant GM48861 (J.M.C.) and NSF grant MCB9807237 (R.G.M.). We thank Travis Tani, who tested the purification of the histidine-tagged Lrp-40 in the Matthews laboratory.

[14] Q. Wang and J. M. Calvo, *J. Mol. Biol.* **229**, 306 (1993).
[15] I. C. Blomfield, V. Vaughn, R. R. Rest, and B. I. Eisenstein, *Mol. Microbiol.* **5**, 1447 (1991).

[30] Purification of Branched-Chain Keto Acid Dehydrogenase Regulator from *Pseudomonas putida*

By KUNAPULI T. MADHUSUDHAN and JOHN R. SOKATCH

Introduction

Branched-chain keto acid dehydrogenase (Bkd) is an important member of the α-keto acid dehydrogenase family of enzymes. This multienzyme complex is induced in *Pseudomonas putida*[1] and *Pseudomonas aeruginosa*[2] by growth in media containing branched-chain amino acids or branched-chain keto acids. The enzyme has been characterized from several sources, including *P. putida*,[3] *P. aeruginosa*,[2] bovine kidney,[4]

[1] V. P. Marshall and J. R. Sokatch, *J. Bacteriol.* **110**, 1073 (1972).
[2] V. McCully, G. Burns, and J. R. Sokatch, *Biochem. J.* **233**, 737 (1986).
[3] J. R. Sokatch, V. McCully, and C. M. Roberts, *J. Bacteriol.* **148**, 647 (1981).
[4] F. H. Pettit, S. J. Yeaman, and L. J. Reed, *Proc. Natl. Acad. Sci. U.S.A.* **75**, 4881 (1978).

rabbit liver,[5] rat kidney,[6] and *Bacillus subtilis*.[7] The purified branched-chain keto acid dehydrogenase from *P. putida* and *P. aeruginosa* contains three components: E1 ($\alpha_2\beta_2$, the dehydrogenase–decarboxylase), E2 (the transacylase), and E3 (lipoamide dehydrogenase) or Lpd-Val.[2,8] The genes of the *bkd* operon are *bkdA1* and *bkdA2,* encoding E1α and E1β, respectively; *bkdB,* encoding the E2 component; and *lpdV,* encoding Lpd-Val.[9–11] The mRNA of the *bkd* operon of *P. putida* encoding subunits of the multienzyme complex is polycistronic and all four genes are tightly linked.[11]

BkdR is encoded in *P. putida* by *bkdR,* which is divergently transcribed from the *bkd* operon. Chromosomal inactivations of *bkdR* in *P. putida* result in loss of branched-chain keto acid dehydrogenase activity and are complemented by supplying BkdR in *trans.* Therefore, expression of the *bkd* operon is positively regulated by BkdR.[12,13] In addition, mutations in *bkdR* have no effect on either branched-chain amino acid transport or transamination in *P. putida.*[12] BkdR shares 37.5% amino acid identity with Lrp, the leucine-responsive protein of *Escherichia coli,*[12,14] and is the second member of this family to be characterized. Lrp is a global transcriptional regulator that regulates the expression of several operons either in a leucine-dependent or leucine-independent manner.[15] Lrp complements *bkdR* mutations in *P. putida.*[12] Anti-Lrp antibodies recognize BkdR but anti-BkdR antibody does not recognize Lrp on Western blots.

Materials and Reagents

Bacteria

Escherichia coli DH5α

Plasmids

pUC18[16]: Cloning vector
pCYTEXP1[17]: *E. coli* expression vector

[5] R. Paxton and R. A. Harris, *J. Biol. Chem.* **257,** 14433 (1982).
[6] R. Odessey, *Biochem. J.* **204,** 353 (1982).
[7] P. N. Lowe, J. A. Hodgson, and R. N. Perham, *Biochem. J.* **215,** 133 (1983).
[8] J. R. Sokatch, V. McCully, J. Gebrosky, and D. J. Sokatch, *J. Bacteriol.* **148,** 639 (1981).
[9] G. Burns, T. Brown, K. Hatter, J. M. Idriss, and J. R. Sokatch, *Eur. J. Biochem.* **176,** 311 (1988).
[10] G. Burns, T. Brown, K. Hatter, and J. R. Sokatch, *Eur. J. Biochem.* **176,** 165 (1988).
[11] G. Burns, T. Brown, K. Hatter, and J. R. Sokatch, *Eur. J. Biochem.* **179,** 61 (1989).
[12] K. T. Madhusudhan, D. Lorenz, and J. R. Sokatch, *J. Bacteriol.* **175,** 3934 (1993).
[13] K. T. Madhusudhan, N. Huang, and J. R. Sokatch, *J. Bacteriol.* **177,** 636 (1995).
[14] D. A. Willins, C. W. Ryan, J. V. Platko, and J. M. Calvo, *J. Biol. Chem.* **266,** 10768 (1991).
[15] J. M. Calvo and R. G. Matthews, *Microbiol. Rev.* **58,** 466 (1994).
[16] C. Yanisch-Perron, J. Vieira, and J. Messing, *Gene* **33,** 103 (1985).
[17] T. N. Belev, M. Singh, and J. E. G. McCarthy, *Plasmid* **26,** 147 (1992).

pJRS119[13]: *bkdR* expression plasmid derived from pCYTEXP1
pJRS146[13]: *Nae*I–*Nae*I fragment of 705 bp, containing a fragment of *bkdR*, the *bkdR–bkdA1* intergenic region, and a fragment of *bkdA1*

Growth Media

L-broth[18]: Ampicillin is used at a final concentration of 200 μg/ml for the growth of *E. coli* DH5α (pJRS119)

Chromatography Media

DEAE Sepharose CL-6B (Pharmacia, Piscataway, NJ)
Heparin Sepharose CL-6B (Pharmacia)

Solutions

TAE buffer: 40 mM Tris-acetate (pH 8.0), 1 mM EDTA
TMN buffer: 20 mM Tris-HCl (pH 7.5), 0.05% (v/v) 2-mercaptoethanol, 0.02% (w/v) sodium azide
Tris-HCl (pH 8.0), 10 mM, containing 1 μM EDTA, 0.2 M NaCl, 0.005% (v/v) 2-mercaptoethanol, and 50% (v/v) glycerol
Binding buffer (10$\times$): 200 mM Tris-HCl (pH 7.5), 10% (v/v) glycerol, 5 mM MgCl$_2$,
0.1 mM EDTA, 1 mM dithiothreitol (DTT), 50 mM NaCl, salmon sperm DNA (5 μg/ml)

Methods

Gel Mobility Shift Assay

BkdR binds specifically to the intergenic region between *bkdR* and *bkdA1* and can be identified by gel mobility shift assays using specific DNA. The DNA fragment used for gel mobility shift assays is prepared by digesting pJRS146 with *Sty*I and *Eco*RI, releasing a 378-bp DNA fragment containing the *bkdR–bkdA1* intergenic region. The *Sty*I site is located in *bkdR* and *Eco*RI is in the polylinker. A 435-bp DNA fragment is prepared for use as a negative control that contains *bkdR* sequence by digesting pJRS146 with *Sty*I and *Hind*III, which cuts in the polylinker. The 3' ends of the DNA fragments are labeled with [α-^{32}P]dCTP and Klenow fragment according to standard methods.[18] The labeled DNA is purified by electrophoresis through 5% (w/v) polyacrylamide gels before use.[18]
   Binding of BkdR and DNA is initiated by mixing 1 μl of binding buffer with 3 ng of labeled DNA and 109 nM BkdR tetramer in a total reaction

[18] J. Sambrook, E. F. Fritsch, and T. Maniatis, "Molecular Cloning." Cold Spring Harbor Laboratory Press, Plainview, New York, 1989.

volume of 10 μl. The reaction is incubated at room temperature for 5 min to facilitate binding of protein and DNA. The reaction is loaded on a 5% (w/v) polyacrylamide gel in TAE buffer. Electrophoresis is carried out at 100 V for 1.5 hr and the dried gels are autoradiographed.

Expression of bkdR from pJRS119 and Purification of BkdR

Because the copy number of BkdR in *P. putida* is low, it is necessary to hyperexpress *bkdR* from a plasmid in order to purify the protein. The plasmid is pJRS119, which contains *bkdR* cloned in pCYTEXP1, an *E. coli* expression vector.[17] The promoter for the expression of *bkdR* is λ P$_R$P$_L$, which is tightly regulated by a λ temperature-sensitive repressor, c*Its*857.[17] pCYTEXP1 also contains the *atpE* translation initiation region, which promotes expression. Initial purification of BkdR from cell-free extracts is achieved by anion-exchange chromatography after the protein is expressed in *E. coli*, with the final purification by heparin-Sepharose chromatography.

pJRS119 is relatively stable in *E. coli* when stock cultures are maintained on agar plates at 30° with monthly transfers. Two liters of L-broth containing ampicillin at 200 μg/ml is inoculated with 20 ml of an overnight culture containing *E. coli* DH5sα (pJRS119) grown in the same medium. The culture is grown at 30° with aeration until an A_{600} of 0.8 is reached. The temperature is quickly shifted to 42°, and the culture is grown at this temperature for 3 hr. Induction of *bkdR* expression beyond 3 hr does not significantly increase the amount of BkdR in the soluble fraction. The cells are harvested and suspended in 18 ml of TMN buffer. The cell suspension is placed in an ice bath, and the cells are broken by sonic oscillation[13] and then centrifuged for 1 hr at 90,000g at 4°. The expressed protein is found in both the soluble and pellet fractions of the cell-free extracts. The protein content of the crude extract is estimated by the biuret method.[19] All the purification procedures are done at 4°, starting with 250 to 300 mg of protein that is loaded on a DEAE-Sepharose CL-6B column (2 × 17 cm) equilibrated with TMN buffer containing 0.1 M NaCl. Unbound proteins are eluted with 100 ml of TMN buffer containing 0.1 M NaCl, and bound proteins are separated by a linear gradient of 0.1 to 0.3 M NaCl (400 ml) in buffer. Purification of BkdR is followed by sodium dodecyl sulfate-polyacrylamide gel electrophoresis (SDS–PAGE) to verify the characteristic protein band at about 20 kDa (Fig. 1). Fractions of 4 ml are collected, and the fractions containing BkdR are pooled. To facilitate the purification, the protein content of column fractions is determined by the dye-binding

[19] E. Layne, *Methods Enzymol.* **3**, 447 (1957).

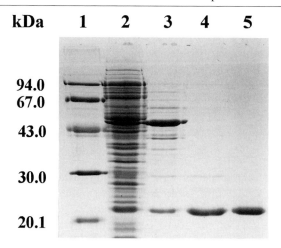

FIG. 1. SDS–PAGE analysis and purification of BkdR. The samples are electrophoresed in a 12% (w/v) separating gel and stained with Coomassie Brilliant Blue R250. Lane 1, molecular weight markers in descending molecular weight order: phosphorylase *b*, bovine serum albumin, ovalbumin, carbonic anhydrase, and soybean trypsin inhibitor; lane 2, cell extract of *E. coli* DH5α (pJRS119), 14 μg of protein; lane 3, DEAE-Sepharose CL-6B pooled fractions, 7 μg of protein; lane 4, pooled fractions from a heparin-Sepharose CL-6B column, 2.5 μg of protein; lane 5, precipitated and redissolved BkdR from the heparin-Sepharose CL-6B pool, 1.5 μg of protein.

assay using the Bio-Rad (Hercules, CA) microassay procedure[20] with bovine serum albumin as standard. Typically, fractions 49 through 64 from DEAE-Sepharose CL-6B chromatography contain BkdR as determined by SDS–PAGE.

The pool from DEAE-Sepharose CL-6B chromatography, containing 80 to 100 mg of protein, is loaded onto a heparin-Sepharose CL-6B column (1.6 × 11 cm). Unbound proteins are eluted by washing with 5 column volumes of TMN buffer containing 10 mM MgCl$_2$ and 0.1 M NaCl. Bound proteins are separated by a gradient of 0.1 to 0.4 M NaCl in 150 ml of TMN buffer containing 10 mM MgCl$_2$ and 0.1 M NaCl. Fractions of 2 ml are collected, and the fractions containing BkdR are pooled and dialyzed overnight against TMN buffer containing 2% (v/v) glycerol, causing precipitation of BkdR. Further purification of BkdR is achieved by dialyzing the heparin-Sepharose CL-6B pool against TMN buffer with 2% (v/v) glycerol.

[20] C. M. Stoscheck, *Methods Enzymol.* **182**, 50 (1990).

This procedure precipitates BkdR and effectively removes high molecular weight contaminants (Fig. 1). The retenate is centrifuged at 12,000g for 20 min. The pellet is redissolved in 10 mM Tris-HCl, pH 8.0, containing 1 μM EDTA, 0.2 M NaCl, 0.005% (v/v) 2-mercaptoethanol, and 50% (v/v) glycerol and stored at $-70°$ until used.

Properties of BkdR

The yield is 2 to 3 mg of BkdR that is about 98% pure, from 250 mg of crude extract. Purified BkdR is active for at least 2 years when stored at $-70°$. The ultraviolet absorption spectrum of BkdR from 240 to 320 nm exhibits a peak at 278 nm,[13] with a series of ripples in the region of 256 to 270 nm that are characteristic of phenylalanine.[21] The molar absorbance of BkdR at A_{280} is estimated to be 5240 M^{-1} cm^{-1} by the PeptideSort program of the Genetics Computer Group sequence analysis program.[22] The estimated molecular weight of 79,400 of BkdR by gel filtration is four times the computed molecular weight of the monomer, 18,500.[13] The molecular mass obtained by sedimentation equilibrium experiments is 84 $\pm$ 2 kDa, which means that BkdR is a tetramer.[23] Purified BkdR gives a single band on SDS–PAGE when fresh 2-mercaptoethanol is used in the sample buffer. However, when 2-mercaptoethanol is oxidized, two protein bands are found because of incomplete reduction of intermolecular disulfide bonds.

Discussion

A specific and simple biological assay of BkdR depends on its ability to bind to the *bkdR–bkdA1* intergenic region (Fig. 2), which takes place either in the presence or absence of branched-chain amino acids. However, BkdR does not bind to the *bkdR* structural gene (Fig. 2). No apparent effect on mobility and number of complexes formed between DNA and BkdR is observed by the addition of branched-chain amino acids.

Although Southern blotting of *P. putida* chromosome revealed a single copy of *bkdR,* analysis of the genomic sequence of *P. aeruginosa* revealed the presence of at least two genes that exhibited significant sequence similarity to *bkdR.* The released genomic sequence of *P. aeruginosa* is obtained from the Internet (http://www.pseudomonas.com/obtaining_data.html).

[21] D. B. Wetlaufer, *Adv. Protein Chem.* **17,** 303 (1962).
[22] J. Devereux, P. Haeberli, and O. Smithies, *Nucleic Acids Res.* **12,** 387 (1984).
[23] N. Huang, K. T. Madhusudhan, and J. R. Sokatch, *Biochem. Biophys. Res. Commun.* **223,** 315 (1996).

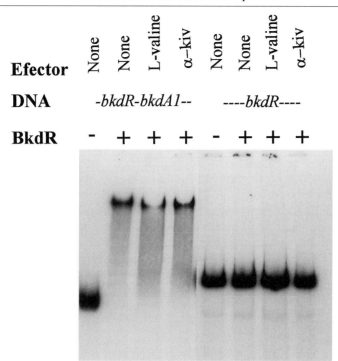

Efector	None	None	L-valine	α-kiv	None	None	L-valine	α-kiv
DNA		-bkdR-bkdA1--				----bkdR----		
BkdR	-	+	+	+	-	+	+	+

FIG. 2. Specific binding of BkdR to the *bkdR–bkdA1* intergenic region. Twenty-three nanograms of ^{32}P-labeled fragment that contains the *bkdR–bkdA1* intergenic region, or ^{32}P-labeled fragment that contains only the *bkdR* coding sequence, and 109 nM BkdR tetramer were incubated at room temperature and then electrophoresed in a 5% (w/v) polyacrylamide gel. The concentration of branched-chain amino acids or α-ketoisovaleric acid (α-kiv) was 50 nM.

Only one of the two *bkdR* homologs in *P. aeruginosa* is adjacent to genes encoding the branched-chain keto acid dehydrogenase operon.

Summary

BkdR can be isolated in nearly pure form as a tetramer by this procedure, which involves hyperexpressing *bkdR* from a plasmid, purification by chromatography on DEAE-Sepharose CL-6B, heparin-Sepharose CL-6B, and dialysis to precipitate BkdR. BkdR is relatively insoluble in aqueous buffers but can be kept in solution in buffer with 50% (v/v) glycerol and 0.2 M NaCl. Cultures of *E. coli* DH5α (pJRS119) should be maintained at 30° to promote plasmid stability. Because BkdR is prone to form intermolecular disulfide bonds, buffers for SDS–PAGE should contain fresh 0.5% (v/v) 2-mercaptoethanol.

[31] Mitochondrial Import of Mammalian Branched-Chain α-Keto Acid Dehydrogenase Complex Subunits

By Dean J. Danner

Many of the reactions catalyzed within mitochondria require a multiprotein complex to form in order for function to occur. Only the multienzyme complexes of the electron transport chain have subunits encoded by the mitochondrial genome and synthesized within the organelle. All other proteins in mitochondria are nuclear encoded and must be incorporated into the organelle after synthesis in the cytosol.[1,2] Those proteins destined for the inner membrane and/or matrix contain mitochondrial targeting sequences (MTS) at their amino-terminal ends. Import occurs at well-defined sites within the membranes, composed again of multiple proteins used as receptors and as components of the pore or channel through which the proteins transit.[3,4] The destination site of these proteins is determined by the peptide sequences within both the MTS and the mature portion of these proteins. Once inside the matrix the MTS is removed by a processing peptidase[5] and the remaining protein is further modified if necessary and folded into a functional conformation. Mitochondrial chaperones play important roles in the overall process from aiding in import to folding of the proteins.[4,6] Mammalian branched-chain α-keto acid dehydrogenase (BCKD) complex is an example of a nuclear encoded multienzyme complex found on the matrix side of the inner membrane.[7] Five enzymatic functions are associated with BCKD: three for the oxidative decarboxylation of the branched-chain α-keto acids and two involved in the regulation of this catalytic function, a complex-specific kinase and phosphatase.[8,9] Genes for all components save the phosphatase have been identified, cloned, and characterized.[10] Current information suggests that the catalytic complex is formed by 12

[1] A. Chomyn, M. W. Cleeter, C. I. Ragan, M. Riley, R. F. Doolittle, and G. Attaradi, *Science* **234**, 614 (1986).
[2] W. Neupert, *Annu. Rev. Biochem.* **66**, 863 (1997).
[3] M. Mori and K. Terada, *Biochim. Biophys. Acta* **1403**, 12 (1998).
[4] N. Pfanner, E. A. Craig, and A. Hönlinger, *Annu. Rev. Cell Dev. Biol.* **13**, 25 (1997).
[5] P. Luciano and V. Géli, *Experientia* **52**, 1077 (1996).
[6] B. Bukau and A. L. Horwich, *Cell* **92**, 351 (1998).
[7] S. J. Yeaman, *Biochem. J.* **257**, 625 (1989).
[8] R. Odessey, *Biochem. J.* **204**, 353 (1982).
[9] Z. Damuni and L. J. Reed, *J. Biol. Chem.* **262**, 5129 (1987).
[10] D. J. Danner and C. B. Doering, *Front. Biosci.* **3**, d517 (1998).

subunits of the E1 decarboxylase $\alpha_2\beta_2$ tetramer, 24 subunits of the E2 branched-chain acyltransferase, and 6 subunits of the E3 homodimer, lipoamide dehydrogenase.[11] The amount of kinase associated with the complex will vary with metabolic conditions and tissue.[12–14]

General Methods

Cell Culture

DG75, a human lymphoblastoid cell line, is maintained in RPMI 1640 medium supplemented with 15% (v/v) fetal bovine serum at 37° and 5% (v/v) CO_2 in a humidified environment. Mitochondria are isolated from ~3 × 10^8 cells that are growth-stimulated by the addition of a 20% volume of fresh medium 18 hr before harvest.

Plasmid Construction

cDNA encoding the full preproteins for the BCKD-specific subunits[15–18] are constructed in pGEM vectors (Promega, Madison, WI) so that the T7 promoter drives transcription. Experience has demonstrated that T7 is a stronger promoter than T3 for the BCKD subunits.

In Vitro Protein Synthesis

Proteins are prepared from the described plasmids with the Promega TNT-coupled transcription/translation system, according to the manufacturer directions, and T7 polymerase. Proteins are radiolabeled with [35S]methionine (E1α and E2) or [35S]cysteine (E1β) (10 mCi/ml; Amersham, Arlington Heights, IL). Aliquots from the reaction mixture are used as a source of the preproteins without further purification.

[11] L. J. Reed and M. L. Hackert, J. Biol. Chem. **265**, 8971 (1990).
[12] A. J. M. Wagenmakers, J. T. G. Schepens, J. A. M. Veldhuizen, and J. H. Veerkamp, Biochem. J. **220**, 273 (1984).
[13] J. Fujii, Y. Shimonura, T. Murakami, N. Nakai, T. Sato, M. Suzuki, and R. A. Harris, Biochem. Mol. Biol. Int. **44**, 1211 (1998).
[14] R. A. Harris, K. M. Popov, and Y. Zhao, J. Nutr. **125**, 1758S (1995).
[15] Y. Nobukuni, H. Mitsubuchi, F. Endo, I. Akaboshi, J. Asaka, and I. Matsuda, J. Clin. Invest. **86**, 2442 (1990).
[16] K. S. Lau, J. L. Chuang, W. J. Herring, D. J. Danner, R. P. Cox, and D. T. Chuang, Biochim. Biophys. Acta **1132**, 319 (1992).
[17] M. C. McKean, K. A. Winkeler, and D. J. Danner, Biochim. Biophys. Acta **1171**, 109 (1992).
[18] C. B. Doering, C. Coursey, W. E. Spangler, and D. J. Danner, Gene **212**, 213 (1998).

Preparation of Mitochondria

Mitochondria are prepared by the method of Greenawalt[19] from DG75 human lymphoblasts,[20] using a buffer that contains 220 mM mannitol, 70 mM sucrose, and 2 mM HEPES, pH 7.4. Mitochondrial protein concentration is estimated by Coomassie dye binding, using Bio-Rad (Hercules, CA) reagents, and adjusted to a final concentration of 10–15 mg of protein per milliliter with a buffer of 250 mM sucrose and 10 mM HEPES, pH 7.5, for use in import reactions.

Detection of Imported Proteins

Mitochondria are isolated from the import reaction by centrifugation at 10,000g for 3 min at 4° and washed twice with the sucrose–HEPES buffer by the same procedure. Mitochondrial proteins are resolved by 10% (w/v) sodium dodecyl sulfate–polyacrylamide gel electrophoresis (SDS–PAGE) under standard conditions except that 10 μl of 2-mercaptoethanol per 100-μl sample buffer is used to aid in resolution. The gel is then fixed for 5 min in 10% (v/v) acetic acid–40% (v/v) methanol, and rehydrated for 5 min in water prior to a 30-min soak in Fluoro-Hance (Research Products International, Mount Prospect, IL). The gel is dried and exposed to Hyperfilm (Amersham) or a phosphoimaging screen (Bio-Rad) for visualization of the radiolabeled proteins. Rainbow molecular weight markers (Amersham) are used for orientation.

Import and Processing of Preproteins by Mitochondria

Import reactions contain equal volumes of translation mix and mitochondria and are incubated for the indicated times at 30°. Because both E1α and E2 contain 11 methionine residues in their mature form, [^{35}S]methionine is used to label these proteins to essentially equal levels of radiolabel. E1β has only 3 methionine residues in he mature protein but 12 cysteine residues, and therefore [^{35}S]cysteine incorporation provides a radiolabel similar to that for methionine with the other 2 subunits. This equal labeling is evidenced by the similar intensity of the reaction products in the lysate lane for each preprotein (Fig. 1). Import occurs in a time-dependent manner but the extent of protein accumulated differs for each subunit.

The increase in mobility of the protein after import results from the removal of the MTS by a mitochondrial processing peptidase (MPP).[5] External protease treatment of mitochondria after the import reaction is

[19] J. W. Greenawalt, *Methods Enzymol.* **31**, 310 (1974).
[20] S. Litwer and D. J. Danner, *Am. J. Hum. Genet.* **43**, 764 (1988).

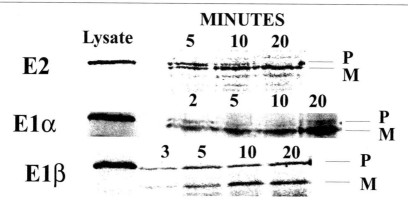

FIG. 1. *In vitro* time-dependent import of human BCKD subunits into mitochondria from cultured human lymphoblasts. Lysate lane contains 2 μl of the TNT reaction. P, Preprotein; M, mature protein. Shown is an autoradiographic analysis of ^{35}S-labeled proteins. All samples were treated with trypsin after import for the indicated times. Further details are found in text.

used to determine if the imported protein is sequestered within the matrix and protected from digestion by the inner membrane. Here trypsin is added (one-tenth reaction volume, 1.25-mg/ml solution of trypsin; Sigma, St. Louis, MO) to the import reaction mixture after the indicated times of incubation, followed by a 10-min incubation on ice. Digestion is stopped by the addition of a one-tenth reaction volume of soybean trypsin inhibitor (2.5 mg/ml; Sigma) solution.[20] The continued presence of the precursor protein after trypsin treatment likely reflects a slower rate of processing of that preprotein by the MPP. For the E2 protein, an intermediate mobility form of the protein is sometimes observed (Fig. 3). This intermediate represents the unlipoated apoprotein. The holoenzyme with lipoate covalently bound to Lys-43 is the fully matured form of the protein.[21–23]

Energy for Import

Movement of preproteins across the inner mitochondrial membrane may require energy in the form of the membrane potential [$\Delta\psi$]. To determine if $\Delta\psi$ is required, mitochondria are incubated with rhodamine 123 (60-ng/μl final concentration) for 5 min at room temperature prior to addition of preprotein to the reaction mixture.[20] This compound dissipates the membrane potential and therefore will eliminate import of any protein

[21] K. Fujiwara, K. Okamura-Ikeda, and Y. Motokawa, *J. Biol. Chem.* **271,** 12932 (1996).
[22] K. Fujiwara, K. Okamura-Ikeda, L. Packer, and Y. Motokawa, *J. Biol. Chem.* **27,** 19880 (1997).
[23] S. W. Jordan and J. Cronan, *J. Biol. Chem.* **272,** 17903 (1997).

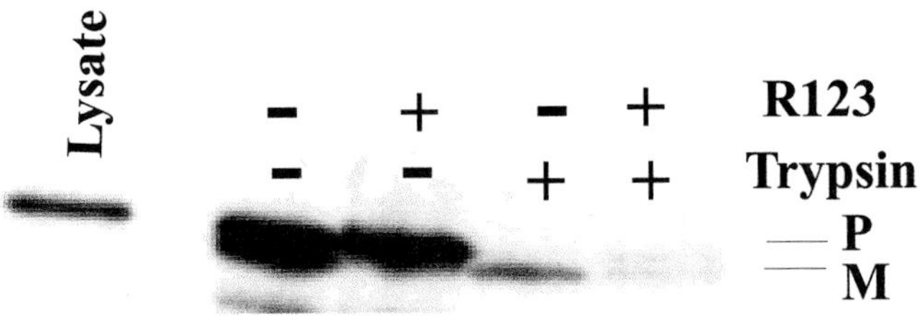

FIG. 2. Dependence of import of the E1α subunit on the membrane potential across the mitochondrial inner membrane. R123, Treatment of mitochondria with rhodamine 123 prior to import reaction.

that uses the $\Delta\psi$ for this process. Figure 2 illustrates the dependence of the E1α subunit on $\Delta\psi$. Identical results have been found for the other two subunits (data not shown). After rhodamine treatment and incubation with the preproteins all the radiolabeled preprotein is accessible to trypsin and digested (Fig. 2, right lane), confirming that the protein is not imported.

Requirement for ATP during Import

A more difficult task is to demonstrate whether ATP is needed for the import of a specific preprotein. The following conditions have been used by us, and others, with some success.[24,25] ATP is depleted from the lysate and mitochondria by treatment with 0.02 unit of apyrase (Sigma), 8 μM antimycin A, 20 μM oligomycin, 8 mM potassium ascorbate, and 0.2 mM N,N,N,N-tetramethylphenylenediamine (Sigma) as described.[26] ATP concentration is reestablished by addition of 5 μl of 440 mM ATP (Boehringer Mannheim, Indianapolis, IN), adjusted to pH 7.4 with KOH, at the beginning of import and again halfway through the incubation period. In some experiments mitochondria are treated with carboxyatractyloside (25 μg/ml; Sigma) to block the adenine nucleotide translocator. This inhibitor depletes internal ATP with time, by preventing ADP entry for renewing ATP production. However, this compound is no longer available from Sigma and a search of chemical catalogs on the Web returns no company as a supplier. Other compounds with these properties are not yet described.

ATP is restored to the matrix by washing away the carboxyatractyloside, followed by the addition of α-ketoglutarate at 5 mM for 3 min prior to addition of radiolabeled preprotein.[25] ATP is restored to the outside of

[24] T. L. Sitler, M. C. McKean, F. Peinemann, E. Jackson, and D. J. Danner, *Biochim. Biophys. Acta* **1404**, 385 (1998).

[25] C. Wachter, G. Schata, and B. S. Glick, *Mol Biol. Cell* **5**, 465 (1994).

[26] N. Pfanner, R. Pfaller, and W. Neupert, *FEBS Lett.* **209**, 152 (1986).

Fig. 3. Import of BCKD subunits when subunits are present in combination. Lane 1, lysate for E1α synthesis; lane 2, E1α import without trypsin treatment; lane 3, E1α import followed by trypsin treatment; lane 4, same as lane 2, with half of the lysate reaction; lane 5, lysate for E2 synthesis; lane 6, E2 import without trypsin treatment; lane 7, E2 import followed by trypsin treatment; lane 8, same as lane 5, with half the lysate reaction; lane 9, mixed import of E2 and E1α with lysate amounts as in lane 4 and 8. E2apo, Unlipolyated protein; E2m, lipoated mature protein; E1αm, mature protein.

the mitochondria by addition of an ATP-generating reaction that includes 7.5 mM creatine phosphate (Sigma) and 1.5 mg of creatine phosphokinase (Sigma).[25] As we reported previously,[24] only the ATP within the matrix can be demonstrated as necessary for import of all three BCKD subunits tested in this way. These finding are consistent with the need for ATP in the function of mitochondrial chaperones HSP70 and HSP60.[3,27] Also, ATP is presumed necessary for the covalent attachment of lipoate to the E2 protein during its maturation.[21–23] Without internal ATP processing would be slowed and could result in a blockage of import.

After import, the mature proteins must associate into an active complex with the hypothesized fixed stoichiometry described above. One way to achieve this desired ratio would utilize cooperative import of the subunits. To test for cooperative import the preproteins made *in vitro* are mixed prior to addition of the mitochondria and import of the subunits is compared with results of single protein import. As seen in Fig. 3, there does not appear to be an enhancement of import of any subunit when the others are present in the reaction. The presence of E2 does not change the import of E1α or E1β and E2 import is not altered. Even though E1α and E1β must form a tetrameric complex to function, there is no evidence that cooperativity occurs when both are present.

Conclusions

Although the *in vitro* import of proteins by mitochondria may not faithfully reflect the *in vivo* conditions, relative relationships between and among the proteins of a multiprotein complex can be assessed. Under

[27] G. Schatz, *J. Biol. Chem.* **271,** 31763 (1996).

identical *in vitro* conditions, the rates of entry and extent of accumulation for the components of the BCKD complex show a wide variation. Here we show that although the basic requirements are identical for all subunits, E1α imports rapidly and was readily processed to its mature form. E2 is slower in the overall process and may experience delays as the lipoate is added. Addition of lipoate is ATP dependent and thus adequate energy sources are required. For reasons yet to be explained, E1β does not import well and is only slowly converted to its mature form. The E1β subunit may thus regulate the formation of the full complex. Similar results were found with mouse liver mitochondrial import of the human subunits, indicating that there rates of import are not species specific. Factors controlling complex assembly still need to be defined. Likewise, factors that affect the import and replacement of BCKD components also need assessment. Mitochondria lacking a specific subunit because of inherited mutations in the gene for that subunit do not show altered *in vitro* import properties for the missing subunit.

Acknowledgments

Thank E. Jackson for technical assistance and C. Doering for editing suggestions. This work was supported in part by grants from the Emory University Research Committee and the National Institutes of Health (HD 38320).

[32] Cloning, Expression, and Purification of Mammalian 4-Hydroxyphenylpyruvate Dioxygenase/ α-Ketoisocaproate Dioxygenase

By Nicholas P. Crouch, Meng-Huee Lee, Teresa Iturriagagoitia-Bueno, and Colin H. MacKinnon

Introduction

The principal route for metabolism of leucine proceeds via transamination to its keto acid, α-ketoisocaproate, followed by degradation in the mitochondrial branched-chain keto acid dehydrogenase complex, a pathway that is common to both isoleucine and valine. However, unlike the other branched-chain amino acids (BCAAs), leucine also has a cytosolic catabolic pathway available for the degradation of its keto acid. For many years the

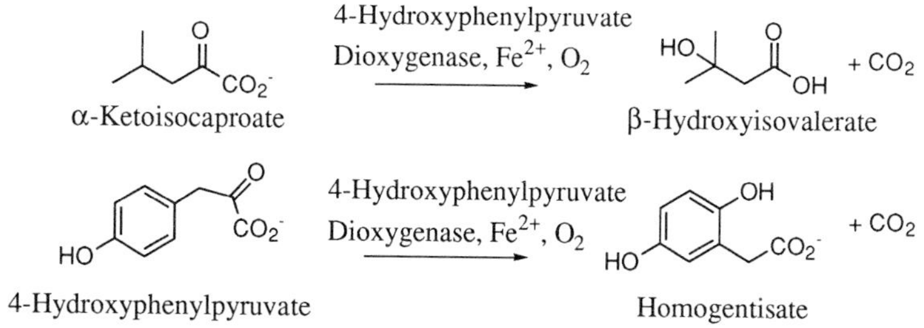

SCHEME 1. Conversion of α-ketoisocaproate to β-hydroxyisovalerate and of 4-hydroxyphenylpyruvate to homogentisate, catalyzed by 4-HPPD.

first enzyme in this cytosolic pathway was believed to be α-ketoisocaproate dioxygenase (α-KICD),[1] but work in our laboratory identified the enzyme responsible for the conversion of α-ketoisocaproate to β-hydroxyisovalerate as being 4-hydroxyphenylpyruvate dioxygenase (4-HPPD, EC 1.13.11.27), an enzyme involved in tyrosine catabolism.[2] The principal metabolic function of 4-HPPD is to convert 4-hydroxyphenylpyruvate (4-HPP) to homogentisate (Scheme 1). This duality between α-KICD and 4-HPPD was also independently reported.[3]

Of the "two" enzymes, 4-HPPD was the better characterized protein,[4–7] with sequence data available in the literature; α-KICD, on the other hand, had been purified to near homogeneity[8] but no sequence data had ever been obtained. We purified α-KICD to homogeneity and sequenced a tryptic digest of the N-terminally blocked protein and confirmed its identity as being 4-HPPD.[2] However, the confusion concerning 4-HPPD was not immediately resolved because other immunogenic proteins, the so-called F antigens, were known to have a high sequence homology with 4-HPPD, to the extent that they were considered to be species variants of the same protein.[9] The relationship between the F antigens and 4-HPPD has remained ambiguous.

[1] P. J. Sabourin and L. L. Bieber, *Methods Enzymol.* **166**, 288 (1988).
[2] J. E. Baldwin, N. P. Crouch, Y. Fujishima, M.-H. Lee, C. H. MacKinnon, J. P. N. Pitt, and A. C. Willis, *Bioorg. Med. Chem. Lett.* **5**, 1255 (1995).
[3] J. Jaskiewicz, K. M. Popov, and R. A. Harris, *FASEB J.* **9**(6), abstract 365 (1995).
[4] S. Lindstedt and B. Odelhog, *Methods Enzymol.* **142**, 139 (1987).
[5] D. J. Buckthal, P. A. Roche, T. J. Moorehead, B. J. R. Forbes, and G. A. Hamilton, *Methods Enzymol.* **142**, 132 (1987).
[6] J. H. Fellman, *Methods Enzymol.* **142**, 148 (1987).
[7] S. Lindstedt and B. Odelhog, *Methods Enzymol.* **142**, 143 (1987).
[8] P. J. Sabourin and L. L. Bieber, *J. Biol. Chem.* **257**, 7460 (1982).
[9] U. Ruetschi, A. Dellsen, P. Sahlin, G. Stenman, L. Rymo, and S. Lindstedt, *Eur. J. Biochem.* **213**, 1081 (1993).

Molecular Cloning of 4-Hydroxyphenylpyruvate Dioxygenase and Related Proteins

The rat F antigen (RFA) was the first protein related to 4-HPPD to be cloned.[10] At that time comparison of the nucleotide and amino acid sequences with data in the National Institutes of Health and European Gene Banks, using the FASTP program, revealed sequence homology only to a protein from *Escherichia coli* ribosomes, the so-called L28 component of the 50S subunit. The mouse F proteins[11,12] and the *Tetrahymena* F antigen[13] were subsequently cloned and the sequences shown to have high homology with the corresponding rat protein; however, the identity and function of the protein remained elusive. The issue was partially resolved when human 4-HPPD was cloned[14] and the sequence compared with that of the known F antigens.[9] The conclusion reached, as stated above, was that the F antigens were species variants of 4-HPPD. To date, 4-HPPD has been cloned from a wide range of sources, including mammalian,[9–12,15] protozoan,[13] bacterial,[16–18] and plant.[19,20] The methods used to clone 4-HPPD are fully described in the corresponding references, and consequently only the methodology adopted in our studies to resolve the identity of mammalian α-KICD/ 4-HPPD is described.[15,21]

[10] M. E. Gershwin, R. L. Coppel, E. Bearer, M. G. Peterson, A. Sturgess, and I. R. Mackay, *J. Immunol.* **139**, 3828 (1987).

[11] S. S. Teuber, R. L. Coppel, A. A. Ansari, P. S. C. Leung, P., Neve, I. R. Mackay, and M. E. Gershwin, *J. Autoimmun.* **4**, 857 (1991).

[12] J. P. Schofield, R. K. Vijayakumar, and D. B. G. Oliveira, *Eur. J. Immunol.* **21**, 1235 (1991).

[13] R. Hummel, P. Horgaard, P. H. Andreasen, S. Neve, K. Skjodt, D. Tornehave, and K. Kristiansen, *J. Mol. Biol.* **228**, 850 (1992).

[14] F. Endo, H. Awata, A. Tanoue, M. Ishiguro, Y. Eda, K. Titani, and I. Matsuda, *J. Biol. Chem.* **267**, 24235 (1992).

[15] N. P. Crouch, J. E. Baldwin, M.-H. Lee, C. H. MacKinnon, and Z. H. Zhang, *Bioorg. Med. Chem. Lett.* **6**, 1503 (1996).

[16] C. D. Denoya, D. D. Skinner, and M. R. Morgenstern, *J. Bacteriol.* **176**, 5312 (1994).

[17] C. Ruzafa, F. Solano, and A. Sanchez-Amat, *FEMS Microbiol. Lett.* **124**, 179 (1994).

[18] U. Ruetschi, B. Odelhog, S. Lindstedt, J. Barros-Soderling, B. Persson, and H. Jornvall, *Eur. J. Biochem.* **205**, 459 (1992).

[19] I. Garcia, M. Rodgers, C. Lenne, A. Rolland, A. Sailland, and M. Matringe, *Biochem. J.* **325**, 761 (1997).

[20] G. E. Bartley, C. A. Maxwell, W. S. Hanna, V. A. Wittenbach, and P. A. Scolnik, *Plant Physiol.* **114**(3), 1587 (1997).

[21] M.-H. Lee, Z. H. Zhang, C. H. MacKinnon, J. E. Baldwin, and N. P. Crouch, *FEBS Lett.* **393**, 269 (1996).

Materials and Methods

General

Assays for both α-KICD and 4-HPPD activity are based on trapping $^{14}CO_2$ released during decarboxylation of the corresponding 1-^{14}C-labeled substrates as previously described.[1,22] Alternatively, 4-HPPD activity can be monitored via a modification of a previously reported high-performance liquid chromatography (HPLC) assay.[23] Typical assay conditions consist of a final incubation volume of 5 ml containing 100 mM Tris-HCl (pH 7.0), 0.1 mM FeSO$_4$, 1 mM ascorbic acid, 5 mM dithiothreitol (DTT), catalase (0.8 mg/ml), 2 mM 4-HPP, and 3.5 mg of 4-HPPD per assay. The mixture is preincubated for 30 min at 27° in the absence of substrate, and then for 1 hr after the addition of 4-HPP. The reaction is quenched by the addition of methanol (5 ml) and the sample is centrifuged at 14,000 rpm for 15 min. An aliquot (100 μl) of the supernatant is analyzed by injection onto a Hypersil ODS column (250 × 7 mm, 5-μm particle size) with an aqueous mobile phase containing 10% (v/v) methanol and 0.05% (v/v) formic acid and a flow rate of 2.5 ml/min. Homogentisate is monitored at 220 and 290 nm and identified by comparison with an authentic standard, and its identity is confirmed by isolation and low-resolution mass spectrometry.

HPLC is performed with a Waters (Milford, MA) 600E system controller, a Waters 717 Autosampler, and a Waters 996 PDA detector.

Protein purification is carried out at 4–8° with the FPLC (fast protein liquid chromatography) or Biopilot FPLC systems supplied by Pharmacia Biotechnology (Piscataway, NJ).

Polymerase chain reactions (PCRs)[24] are performed with *Taq* polymerase (Pharmacia) over 30 cycles unless otherwise stated, with the first cycle at 94° for 5 min, 55° for 1 min, and then at 72° for 1 min. The subsequent 29 cycles are then performed at 94° for 1 min, 55° for 1 min, and 72° for 1 min.

Sequencing is performed with the Sequenase version 2.0 DNA sequencing kit and [α-^{35}S]dATP supplied by Amersham Life Sciences (Arlington Heights, IL).

cDNA Library Screening

Screening of a λ ZAP II rat liver cDNA library (Stratagene, La Jolla, CA; Sprague Dawley male rats, 6 months old) is carried out by PCR as

[22] B. Lindblad, *Clin. Chim. Acta* **34,** 113 (1971).
[23] C. Bory, R. Boulieu, C. Chantin, and M. Mathieu, *Clin. Chim. Acta* **189,** 7 (1990).
[24] T. Takumi and H. F. Lodish, *BioTechniques* **17**(3), 443 (1994).

described above. A total of 60,000 plaque-forming units (PFU) is initially aliquoted from the library and divided into 20 pools, each of approximately 3000 units. The pools are amplified and 2 μl of each pool used as a template for PCR screening. Phage are denatured by heating at 100° for 2 min in order to release the DNA.

As the complete RFA cDNA has never been isolated, primers used for screening are designed from the published[10] incomplete sequence:

5'-GTG GTC AGT CAT GTC ATC AAG CAA G-3' (forward) (169–193)

and

5'-ACA GAG AGT TGA AGT TAC CTG CTC C-3' (reverse) (1102–1078)

Amplified pools are screened and positive pools with a band of the correct molecular weight, as judged on an agarose gel, are further subdivided into subpools, each containing 200 PFU. Reamplification, screening, and subdivision are repeated to the level of a single plaque. A clone with an insert of 1.3 kb is selected for *in vivo* excision. Excision, plating, and single-stranded DNA rescue are performed according to the protocols provided by Stratagene.[25] The gene is then sequenced in both directions with a panel of synthetic primers, each of which is based on the published RFA sequence.[10] The full cDNA sequence as determined is given in Fig. 1.

Subcloning into Expression Vector

The insert is sequenced and the cDNA found to be incomplete at the 5' end. Comparison with the previously published murine alleles[11,12] reveals a high degree of homology. Because an intact rat F antigen gene with a complete 5' end has never been identified and attempted sequencing of wild-type RFA indicates that the N terminus is blocked, we have added the two N-terminal amino acid residues from mouse 4-HPPD,[12] namely, threonine and threonine. Subsequently the cDNA is amplified by PCR with Vent polymerase (New England BioLabs, Beverly, MA) with the addition of codons to introduce the additional two amino acids (boldface type) as well as *Nde*I [5'-end (underlined)] and *Bam*HI [3'-end (underlined)] restriction sites to assist subcloning. The forward primer

5'-GATCGA<u>CATATG</u>**ACCACCTACTGGGACAAAGGA**
CCAGAGCCTGAGAGAGGCCGGTTCCTC-3'

and reverse primer

[25] Stratagene, "Stratagene Lambda ZAP II Library Instruction Manual." Stratagene, La Jolla, California, 1994.

5'- **AA G**AGCCTGAGAGA GGCCGGTTCCTC   CATTTCCA**T**TCT GTGACCTTCTGG   50

GTTGGCAACGCC AAGCAGGCTGCC   TCCTTCTATTGC AACAAGATGGGC   98

TTCGAACCGCTG GCCTACAAGGGC   CTGGAGACGGGC TCCCGGGAGGTG   146

GTCAGTCATGTC ATCAAGCAAGGG   AAAATTGTGTTT GTTCTCTGCTCT   194

GCTCTCAATCCC TGGAACAAAGAG   ATGGGTGACCAC CTGGTGAAGCAT   242

GGTGATGGCGTA AAGGACATCGCA   TTCGAGGTGGAA GACTGTGAACAC   290

ATTGTGCAGAAA GCCCGAGAACGG   GGCGCTAAAATT GTACGGGAGCCA   338

TGGGTGGAGGAA GACAAATTCGGG   AAGGTGAAGTTC GCTGTGCTGCAG   386

ACGTATGGAGAT ACCACGCACACC   CTGGTGGAGAAG ATCAACTACACC   434

GGTCGTTTCTTA CCTGGATTCGAG   GCCCCAACATAC AAGGACACCCTA   482

CTTCCAAAACTA CCCAGCTGTAAC   CTGGAGATCATC GACCATATTGTA   530

GGCAACCAGCCC GACCAGGAAATG   GAGTCTGCCTCA GAATGGTACCTG   578

AAAAACCTGCAG TTCCACCGGTTC   TGGTCTGTAGAC GACACACAGGTG   626

CACACGGAGTAT AGCTCTCTGCGC   TCCATCGTGGTG GCCAACTATGAG   674

GAGTCCATCAAA ATGCCCATTAAT   GAACC**A**GCCCCG GGCAGGAAGAAG   722

TCTCAGATCCAG GAATATGTGGAC   TATAATGGGGGT GCTGGGGTCCAG   770

CACATCGCTCTC AGGACCGAAGAC   ATCATCACAACG ATCCGCCACTTG   818

AGGGAACGAGGC ATGGAGTTCTTG   GCTGTCCCGTCT TCTTACTACAGA   866

CTGCTTCGGGAG AATCTCAAGACC   TCCAAGATCCAA GTGAAGGAGAAC   914

ATGGATGTCTTG GAGGAGCTAAAA   ATCCTGGTAGAC TACGATGAGAAA   962

GGCTACCTCCTA CAGATCTTCACC   AAGCCCATGCAG GACCGGCCCACG   1010

CTCTTCTTGGAA GTCATCCAACGC   CACAACCACCAG GGCTTTGGAGCA   1058

GGTAACTTCAAC TCTCTGTTCAAA   GCTTTCGAAGAG GAGCAAGCCCTA   1106

CGGGGTAACCTC ACTGACCTGGAG   ACCAACGGTGTG AGGTCTGGAATG   1154

<u>taa</u>gctccacccacacccagaccgcgcaaggcgctggacaagtcaatcagctccaactggctgaaaggctggacctcagggctc

cacccacaccatggccacgcccctctacggcaaggcttctcctgatccgtttcgagtaaagatgccttcccag

AAAAAAAAAAAAAAAAA-3'

FIG. 1. The genetic sequence of 4-HPPD. [The stop codon (taa) is underlined and boldface letters represent differences between this clone and the RFA cDNA as reported by M. E. Gershwin, R. L. Coppel, E. Bearer, M. G. Peterson, A. Sturgess, and I. R. Mackay, *J. Immunol.* **139,** 3828 (1987).]

5'-GATCGA<u>GGATCC</u>TCATTACATTCCAGACCT
CACACCGTTGGTCTC-3'

are used. The PCR product after 30 cycles is analyzed by agarose gel electrophoresis, excised, and purified with a GeneClean II kit (Bio 101, La Jolla, CA). Purified DNA is then double digested with *Nde*I and *Bam*HI restriction enzymes in 1× KGB buffer (Stratagene), precipitated with 3 *M* sodium acetate, pH 6.0 (0.1 volume), and absolute ethanol (3 volumes), and then air dried for 30 min. p*trc* vector (pMAT-4; donated by D. Hart, New Chemistry Building, University of Oxford) is digested with *Nde*I and *Bam*HI restriction enzymes and a portion, 0.05 μg mixed with 0.1 μg of the insert in 1× ligation buffer (supplemented with ATP) and 1 μl of T4 DNA ligase (400 units/μl) in a total volume of 10 μl, is incubated for 16 hr to give the plasmid as shown in Fig. 2.

The 4-HPPD cDNA, once subcloned into the recombinant plasmid, is sequenced in both directions to confirm that no unwanted mutations have been introduced during PCR and the plasmid is then used to transform either BL21(DE3) or NM554 *E. coli* host cell lines.

Expression of 4-Hydroxyphenylpyruvate Dioxygenase

NM554 *E. coli* host cells, containing the pHPPD plasmid, are plated onto an LB agar plate containing chloramphenicol (34 μg/ml) and incubated overnight at 37°. In a sterile 50-ml conical tube, 5 ml of 2TY broth medium supplemented with chloramphenicol (34 μg/ml) is inoculated with one colony from the plate and then incubated at 37° for 6–8 hr. The starter material, four 100-ml portions of 2TY broth medium containing chloramphenicol (34 μg/ml), is prepared by inoculating each 100-ml portion with 1 ml of the cell culture. Cells are grown at 37° overnight. This starting culture is then used to inoculate 24 conical flasks (2-liter volume), each containing 750 ml of 2TY broth medium supplemented with chlorampheni-

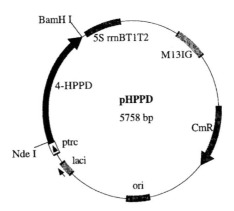

FIG. 2. 4-HPPD cDNA cloned into the p*trc* expression vector.

col (34 μg/ml). Cells are incubated at 37° until an OD_{600} of 2.00 is reached, at which point the incubation temperature is dropped to 27°. Protein expression is induced by the addition of isopropyl-β-D-thiogalactopyranoside (IPTG) to 0.5 mM, and incubation continued for a further 12–16 hr at 27°.

The resulting culture becomes a characteristic brownish color after overnight incubation, a typical observation reported for organisms producing high levels of 4-HPPD. The culture broth is concentrated to approximately 3 liters with a Sartorius tangential flow concentrator and the cells (80–90 g) harvested by centrifugation at 10,000g for 10 min at 4°. The cell pellet is then stored at −80° until required.

Purification of 4-Hydroxyphenylpyruvate Dioxygenase

To obtain recombinant 4-HPPD of ~98% purity, a four-step purification has been developed, consisting of the following: Q-Sepharose anion exchange → Superdex 75 gel filtration → Phenyl Resource hydrophobic interaction → Superdex 200 gel filtration. All purification steps are carried out at 4°. Purity is determined by sodium dodecyl sulfate–polyacrylamide gel electrophoresis (SDS–PAGE) analysis and protein concentration determined by the method of Bradford.[26] Generally, 350 mg of pure recombinant 4-HPPD is obtained from approximately 20–22 liters of cell culture.

Buffers

Lysis buffer
  Tris-HCl (pH 8.0), 50 mM
  EDTA, 5 mM
  Triton X-100, 0.1%(v/v)
  2-Mercaptoethanol, 1% (v/v): Add just before use
Q-Sepharose
  A: Tris-HCl (pH 8.0), 50 mM 2 liters
  B: Tris-HCl (pH 8.0), 50 mM
    EDTA, 2 mM 3 liters
  C: Tris-HCl (pH 8.0), 50 mM
    EDTA, 2 mM
    NaCl, 0.5 M 3 liters
Superdex 75
  Tris-HCl (pH 8.0), 50 mM
  EDTA, 2 mM 3 liters
Phenyl Resource
  A: Tris-HCl (pH 8.0), 50 mM

[26] M. M. Bradford, *Anal. Biochem.* **72,** 248 (1976).

EDTA, 2 mM
(NH$_4$)$_2$SO$_4$, 1 M 2 liters
B: Tris-HCl (pH 8.0), 50 mM
EDTA, 2 mM 2 liters
Superdex 200
Tris-HCl (pH 8.0), 50 mM 3 liters

Protein Extraction

Cells are lysed with a sonicator in the presence of lysis buffer (3 ml/g of cells). Polyethyleneimine [5% (w/v), pH 8.0] is added to a final concentration of 5% (v/v), and the cells centrifuged at 13,000 rpm for 15 min. The resulting supernatant is used in subsequent purification steps.

Q-Sepharose

A Q-Sepharose column (300-ml bed volume) is preequilibrated with 1 liter of Q-Sepharose buffer A and the protein extract is loaded onto the column at a flow rate of 20 ml/min. All unbound proteins are washed from the column with Q-Sepharose buffer B (800 ml) and the absorbed proteins are eluted by running a linear gradient from 1 to 60% Q-Sepharose buffer C (containing 1 M NaCl) in a 2-liter volume at a flow rate of 30 ml/min. Fractions of 25 ml are collected and analyzed by SDS–PAGE, which shows that 4-HPPD elutes as a single peak at approximately 10–15% Q-Sepharose buffer C. All fractions containing 4-HPPD are pooled and concentrated to 21 ml, using an Amicon (Danvers, MA) ultrafiltration concentrator.

Superdex 75

The sample containing 4-HPPD is loaded onto a Superdex 75 column (85.0 × 3.2 cm) preequilibrated with 800 ml of Superdex 75 buffer. The column is run three times overnight at a flow rate of 3 ml/min. For each run, 7 ml of sample is loaded and 10-ml fractions are collected. 4-HPPD elutes as a single peak at about 300 ml of buffer, as determined by SDS–PAGE.

Phenyl Resource

Fractions containing 4-HPPD are pooled and (NH$_4$)$_2$SO$_4$ added slowly to a final concentration of 1 M to avoid any protein precipitation. The protein is loaded onto a 50-ml Phenyl Resource column preequilibrated with 300 ml of buffer A containing 1 M (NH$_4$)$_2$SO$_4$. All unbound proteins are eluted with 200 ml of high-salt buffer and the remaining bound proteins are eluted with a linear gradient of salt, 1 to 0 M (NH$_4$)$_2$SO$_4$ in a 500-ml volume at a flow rate of 20 ml/min. Fractions of 20 ml are collected and 4-HPPD elutes as a single peak at 70–90% buffer B as determined by SDS–PAGE. Fractions containing 4-HPPD are pooled and concentrated to a final volume of 7 ml.

Superdex 200

A Superdex 200 column (60.0×3.5 cm) is equilibrated with 1 liter of Superdex 200 buffer containing no EDTA. The sample is loaded onto the column with a 10-ml Superloop and 4-HPPD elutes as a single peak at 300 ml of volume at a flow rate of 3 ml/min. Fractions of 5 ml are collected and the purity of 4-HPPD determined by SDS–PAGE analysis.

Protein Storage and Stability

Each fraction containing 4-HPPD is individually concentrated with a Millipore (Bedford, MA) Ultrafree-4 centrifugal filter unit to an approximate protein concentration of 40 mg/ml. The samples are then aliquoted into 75-μl portions in 0.5-ml Eppendorf tubes and stored at $-80°$. Purified 4-HPPD of incubation grade ($\sim$50% purity) appears to lose little activity even after several months of storage at $-80°$.

Comments on Cloning, Isozymes, Expression, and Purification of 4-Hydroxyphenylpyruvate Dioxygenase

Cloning

The relationship between the F antigens and 4-HPPD has still not been fully resolved. However, in the course of our studies we notice that the cDNA sequence of the RFA, as determined by Gershwin et al.,[10] contained four differences compared with the sequence obtained by us. Two of these differences were silent third-base substitutions: cytosine-60 and guanine-729 in RFA were replaced by thymine and adenine, respectively, in 4-HPPD. The third difference was noted in the first base of the codon encoding lysine (AAG) at position 28 in RFA, it being changed to guanine (GAG) and hence glutamic acid in 4-HPPD. Of these differences, it was the fourth that was the most significant. This entailed deletion of the thymine at position 1138 in RFA cDNA compared with 4-HPPD. As a result of this, a frame shift occurred and a further 14 amino acids were translated at the C terminus. Examination of the RFA cDNA[10] revealed that the rest of the sequence after position 1138 was identical to our sequence. Consequently, deletion of this thymine in the RFA cDNA would result in the transcription of the same 14 amino acids.

It was of considerable interest to us that in the case of the murine alleles cloned by Schofield et al.,[12] the authors reported that "the mouse F protein has an apparent deletion of the first base of the termination codon," which they believed to be a "PCR artifact." The absence of thymine-1138 was likewise reported at almost the same time by Teuber et al.,[11] who also sequenced the murine F alloantigens. It is likely therefore, that the RFA sequence obtained by Gershwin et al.[10] contains an error at position 1138 and hence the thymine at this position in RFA cDNA should be deleted. We have subsequently shown that the gene product obtained from the RFA sequence containing thymine at position 1138 generates a protein devoid of any activity toward 4-HPP or α-KIC.[21]

Isozymes

Observation of 4-HPPD isozymes has been a common feature in reports dealing with the isolation and purification of this protein from native sources. For example, it has been reported[27] that at least three isozymes of human 4-HPPD exist and that these isozymes could be resolved into three peaks by TEAE chromatography. Such behavior of highly purified native rat liver 4-HPPD was also observed during the course of our early work.[2] Consequently, the existence of 4-HPPD isozymes seems to be beyond doubt; however, the question that remains unresolved is whether these isozymes are encoded by separate genes or arise because of posttranscriptional modification.

In an attempt[28] to obtain a complete N-terminal sequence for rat liver 4-HPPD, we performed 5'-RACE (rapid amplification of cDNA ends) on the rat liver cDNA λ ZAP II library, using an internal primer that would anneal to bp 771–747 of 4-HPPD. T3 and T7 external primers were chosen because the insert cloned in the *Eco*RI sites could be of either orientation, but amplification would, of course, occur only when the internal and external primers annealed in opposite directions. To overcome the problem of nonspecificity of the T3 and T7 primers, which could prime every λ vector in the library, the 4-HPPD cDNA in the library was enriched prior to initiation of RACE. This was achieved by using only the internal primer in a pseudo-PCR.

The amplicons generated by this strategy were separated by agarose gel electrophoresis, excised, purified, and cloned into pCR-Script vectors (Stratagene). These were then sequenced to reveal the existence of at least

[27] M. Rundgren, *J. Biol. Chem.* **252**, 5085 (1977).
[28] M.-H. Lee, Ph.D. thesis. Oxford University, Oxford, UK, 1997.

1   5' GA CCA AAG CCT GAG AGA GGC CGG TTC CTC CAT TTC CAT TCT 3'

2   5'   AA GAG CCT GAG AGA GGC CGG TTC CTC CAT TTC CAT TCT 3'

3 & 4   5'   A GAG CCT GAG AGA GGC CGG TTC CTC CAT TTC CAC TCT 3'

4-HPPD   5'   AA GAG CCT GAG AGA GGC CGG TTC CTC CAT TTC CAT TCT 3'

RFA   5'   CA AAG CCT GAG AGA GGC CGG TTC CTC CAT TTC CAC TCT 3'

FIG. 3. 5' Sequences of four 4-HPPD clones isolated from a single rat liver cDNA library. [4-HPPD sequence from N. P. Crouch, J. E. Baldwin, M. H. Lee, C. H. MacKinnon, and Z. H. Zhang, *Bioorg. Med. Chem. Lett.* **6,** 1503 (1996). RFA sequence from M. E. Gershwin, R. L. Coppel, E. Bearer, M. G. Peterson, A. Sturgess, and I. R. Mackay, *J. Immunol.* **139,** 3828 (1987).]

three isoforms of 4-HPPD in the cDNA λ ZAP II library (see Fig. 3). These findings represent the first molecular evidence supporting the occurrence of 4-HPPD isoforms.

It is of interest to note that of the four murine F antigen/4-HPPD cDNAs isolated so far,[10–12,15] the regions containing the main differences occur at the extreme 5' and 3' ends and these are transcribed into the different peptides as shown in Fig. 4. The reason for such variation, e.g., location in high mutation "hot spots" (an assumption reasonable only for the C terminus) or differential splicing during mRNA maturation, is unclear.

Source	N-terminal	C-terminal
	11	374
DBA mouse	QHLSSVFKGPKPERGRF	..........FEEEPTYGATSLTWS PMV[*]
C57 mouse	DKKPERGRF	..........FEEEQALRGNLTDLE PNGVRSGM[*]
RFA	YWDKGPKPERGRF	.........FEEEQALRG[*]
4-HPPD	EPERGRF	..........FEEEQALRGNLTDLE TNGVRSGM[*]

FIG. 4. Regions of variation at N and C termini in different species of murine F antigens/4-HPPD. [DBA and C57 mouse sequences from S. S. Teuber, R. L. Coppel, A. A. Ansari, P. S. C. Leung, R. Neve, I. R. Mackay, and M. E. Gershwin, *J. Autoimmun.* **4,** 857 (1991). RFA sequence from M. E. Gershwin, R. L. Coppel, E. Bearer, M. G. Peterson, A. Sturgess, and I. R. Mackay, *J. Immunol.* **139,** 3828 (1987). 4-HPPD sequence from N. P. Crouch, J. E. Baldwin, M. H. Lee, C. H. MacKinnon, and Z. H. Zhang, *Bioorg. Med. Chem. Lett.* **6,** 1503 (1996).]

Further work is required to clarify the effects of these differences, not least because it is unknown whether the murine F antigens have 4-HPPD activity or not.

Expression

Initially, 4-HPPD subcloned into the *ptrc* vector was expressed in XL-1 Blue MRF' as host cell. Although this expression system yielded active enzyme, low expression levels were obtained (10–15% of the total soluble protein), with the majority of the expressed protein being deposited into inclusion bodies.[29] In addition, reproducibility of expression and the growth of the host cells were both poor.

To overcome these problems a new host cell (*E. coli* NM554) was used to express the enzyme and the incubation temperature raised from 27 to 37° before induction with IPTG. These alterations significantly improved the expression levels of 4-HPPD, as judged by SDS–PAGE analysis, which showed 4-HPPD production to be increased to 25–30% of the total soluble protein present in the cells. In addition, improved reliability in growth of the host was also achieved, as evidenced by the greater weight of harvested cells. The *ptrc* vector contains a pUC origin of replication, which carries a temperature-sensitive mutation resulting in the copy number of the plasmid increasing when the temperature is shifted from 27 to 37°. Such temperature mutation is inducible when using *E. coli* JM109 or NM554.[30]

Production of 4-HPPD in a large-scale culture was always performed in 2-liter flasks because attempts to grow the cells containing the 4-HPPD plasmid in fermenters was not successful.

Purification

We have described various purification protocols for production of recombinant 4-HPPD in the course of our work,[31,32] but the protocol described above has evolved in our laboratory as the most efficient procedure for the production of highly purified protein. It is clear that for enzymatic studies protein of a lower purity but higher overall total activity is sometimes preferable, and this quality of protein may be readily prepared with an

---

[29] M.-H. Lee and N. P. Crouch, unpublished observations (1996).

[30] J. E. Baldwin, J. M. Blackburn, J. D. Sutherland, and M. C. Wright, *Tetrahedron* **47**, 5991 (1991).

[31] N. P. Crouch, R. M. Adlington, J. E. Baldwin, M.-H. Lee, and C. H. MacKinnon, *Tetrahedron* **53**, 6993 (1997).

[32] N. P. Crouch, R. M. Adlington, J. E. Baldwin, M.-H. Lee, C. H. MacKinnon, and D. R. Paul, *Tetrahedron* **53**, 10827 (1997).

0–50% ammonium sulfate step followed by a DEAE-Sepharose purification step.[32]

Conclusions

The results of our studies have clearly shown that the enzyme responsible for the conversion of α-ketoisocaproate to β-hydroxyisovalerate is 4-HPPD and hence that the identity of α-KICD is really 4-HPPD.[1,15] In addition, comparison of the sequences obtained for mammalian 4-HPPD from a range of species (i.e., murine,[11,12] rat,[10] and human[14]) provides convincing evidence that the so-called F antigens are in fact 4-HPPDs; however, whether they are all functional proteins remains to be determined.

Many other key areas concerning 4-HPPD remain uninvestigated. For example, what is the significance of the various 4-HPPD isoforms? Perhaps an even more poignant question concerns the metabolic significance of the leucine cytosolic catabolic pathway, because for many years it has been recognized that leucine and/or its cytosolic metabolites (i.e., α-ketoisocaproate and β-hydroxyisovalerate) appear to have an anticatabolic effect.[33,34] Does this pathway play a regulatory role in protein synthesis and/or degradation? These are just some of the questions that are being addressed at present.

[33] R. C. Hider, E. B. Fern, and D. R. London, *Biochem. J.* **114**, 171 (1960).
[34] S. Nissen, R. Sharp, M. Ray, J. A. Rathmacher, D. Rice, J. C. Fuller, A. S. Connelly, and N. Abumrad, *J. Appl. Physiol.* **81**, 2095 (1996).

[33] Mammalian Branched-Chain Aminotransferases

By Myra E. Conway and Susan M. Hutson

The branched-chain aminotransferases (BCATs) catalyze reversible transamination of the short-chain aliphatic branched-chain amino acids (BCAAs) leucine, valine, and isoleucine to their respective α-keto acids (BCKAs) α-ketoisocaproate, α-ketoisovalerate, and α-keto-β-methylvalerate.

L-Leucine + α-ketoglutarate $\rightleftharpoons$ α-ketoisocaproate + L-glutamate

L-Valine + α-ketoglutarate $\rightleftharpoons$ α-ketoisovalerate + L-glutamate

L-Isoleucine + α-ketoglutarate $\rightleftharpoons$ α-keto-β-methylvalerate + L-glutamate

In mammals and other eukaryotes, there are two BCAT isoenzymes, a mitochondrial (BCATm) and a cytosolic (BCATc) form.[1] In humans and rodents BCATm is found in most tissues.[2,3] In contrast, BCATc is found almost exclusively in the brain,[2,4] and this isoenzyme may be a target for the anticonvulsant drug and leucine analog, gabapentin.[5] We have purified the rat isoenzymes,[4,6] and the cDNA sequences of the BCAT from rat,[7,8] human,[8,9] ovine,[10] and several other species have been cloned. On the basis of primary sequence comparisons, the mammalian BCAT have been placed in a separate folding class (fold type IV) of pyridoxal phosphate-dependent enzymes.[7,8] The only other enzymes in this folding class are the bacterial BCAT, and the bacterial enzymes D-amino-acid aminotransferase and 4-amino-4-deoxychorismate lyase.

In this chapter we report the assay and cloning of mammalian BCAT. Protocols for overexpression of the human recombinant BCATm and BCATc in *Escherichia coli* are described. The methods described here for purification of the human recombinant enzymes are designed to yield milligram quantities of highly purified proteins.

Enzyme Assay

BCAT activity is assayed with the BCKA/BCAA pair α-keto[1-^{14}C]isovalerate and isoleucine.[1] The radioactive α-keto acid is synthesized from [1-^{14}C]valine essentially as described by Rudiger *et al.*[11] Valine formation is measured as radioactivity remaining in the medium after chemical decarboxylation of the radioactive α-keto acid, using hydrogen peroxide.[1] Other BCAA/BCKA combinations can be used.

[1] S. M. Hutson, D. Fenstermacher, and C. Mahar, *J. Biol. Chem.* **263**, 3618 (1988).

[2] A. Ichihara, *in* "Transaminases" (P. Christen and D. E. Metzler, eds.), p. 430. John Wiley & Sons, New York, 1985.

[3] S. M. Hutson, R. Wallin, and T. R. Hall, *J. Biol. Chem.* **257**, 15681 (1992).

[4] T. R. Hall, R. Wallin, G. D. Reinhart, and S. M. Hutson, *J. Biol. Chem.* **268**, 3092 (1993).

[5] S. M. Hutson, D. Berkich, P. Drown, B. Xu, M. Aschner, and K. LaNoue, *J. Neurochem.* **71**, 863 (1998).

[6] R. Wallin, T. R. Hall, and S. M. Hutson, *J. Biol. Chem.* **265**, 6019 (1990).

[7] S. M. Hutson, R. K. Bledsoe, T. R. Hall, and P. A. Dawson, *J. Biol. Chem.* **270**, 30344 (1995).

[8] R. K. Bledsoe, P. A. Dawson, and S. M. Hutson, *Biochim. Biophys. Acta* **1339**, 9 (1997).

[9] O. Schuldiner, A. Eden, T. Ben-Yosef, O. Yanuka, G. Simchen, and N. Benvenisty, *Proc. Natl. Acad. Sci. U.S.A.* **93**, 7143 (1996).

[10] M. Faure, F. Glomot, R. Bledsoe, S. M. Hutson, and I. Papet, *Eur. J. Biochem.* **259**, 104 (1999).

[11] H. E. Rudiger, U. Langenbeck, and H. W. Goedde, *Biochem. J.* **126**, 446 (1972).

Reagents

Potassium hydroxide, 5 *M:* Store at room temperature

Potassium phosphate (pH 7.8), 100 m*M:* Store at 4°

Dithiothreitol (DTT), 100 m*M:* Make fresh

Pyridoxal phosphate (PLP), 5 m*M:* Prepare in 25 m*M* HEPES (pH 7.5) and store at −20° (in the dark)

Radioactive α-ketoisovalerate/isoleucine stock solution: Contains 10 m*M* α-keto[1-^{14}C]isovalerate (specific radioactivity, 150 dpm/nmol) and 120 m*M* isoleucine in deionized water; store at −20°

Acetic acid, 3 *N:* Store at 4°

Hydrogen peroxide (30%; Fisher, Pittsburgh, PA): Store at 4°

All solutions except DTT are prepared fresh on a monthly basis.

Twenty-milliliter glass scintillation vials are used for the assay. Serum stoppers (Kontes, Vineland, NJ) are fit with plastic center wells (Kontes). Fluted filter paper (3 MM; Whatman, Clifton, NJ) is placed in each well and 200 μl of 5 *M* potassium hydroxide is added. The potassium hydroxide acts as a trap for $^{14}CO_2$ that is released by chemical decarboxylation of the radioactive α-keto acid substrate at the hydrogen peroxide step (see below). To set up the assay vials, 360 μl of distilled water; 50 μl of 100 m*M* potassium phosphate, pH 7.8; 20 μl of 100 m*M* DTT; and 10 μl of 5 m*M* PLP are added to each vial.

A 10-μl aliquot of an appropriate dilution of the purified enzyme is added to each vial, and the vial is incubated at 37° in a shaking water bath for 3 min. At 3 min, the assay is started by adding 50 μl of the 10 m*M* α-keto[1-^{14}C]isovalerate/120 m*M* isoleucine solution, and the serum caps with center wells are placed securely over the assay vials. The reaction times are adjusted so that ≤20% of the radioactive substrate is used. Typical reaction times are 3–5 min. The reaction is stopped by injecting 500 μl of 3 *N* acetic acid into the sample through the serum stopper. After about 5 min, 500 μl of 30% hydrogen peroxide is injected to decarboxylate the remaining radioactive α-keto[1-^{14}C]isovalerate. The vials should be shaken for at least 30 min at 37°. [^{14}C]Valine formation is measured by counting 150 μl of the sample volume (one-tenth of the total volume), using a liquid scintillation counter. The sample disintegrations per minute (dpm) must be corrected for background radioactivity by running a separate sample that has been treated with acid before addition of the radioactive substrate. With the purified enzyme, it is sufficient to set up an assay vial containing all the assay components excluding enzyme protein.

One unit of enzyme activity is defined as 1 μmol of valine formed per minute. At each step in the purification protocol, protein is measured by the Amido Black assay with bovine serum albumin (BSA) as a standard.[12] Protein concentrations of the purified proteins can be calculated from the absorbance at 280 nm, using extinction coefficients of 67,600 and 86,300 M^{-1} cm^{-1} for human recombinant BCATm and BCATc, respectively.[13]

Cloning of Mammalian Mitochondrial and Cytosolic Branched-Chain Aminotransferases

A similar cloning strategy was used for the rat and ovine BCAT isoenzymes.[7,8,10] The rat BCATc isoenzyme was the first mammalian BCAT cDNA to be identified and cloned.[7] First, a partial cDNA sequence for the BCAT isoenzyme was obtained. The amino acid sequence of peptides obtained from a trypsin or endoproteinase Lys-C digest of the purified enzyme was used to design degenerate oligonucleotide primers that were used in a polymerase chain reaction (PCR) to obtain a partial cDNA probe. This partial cDNA was then used to screen a cDNA library.[7,8,10] The cloning of rat BCATc was facilitated by the discovery that the amino acid sequence of six of nine internal peptides from rat aligned with the deduced amino acid sequence of a mouse cDNA, ECA39.[14] Correction of two sequencing errors positively identified the ECA39 cDNA as the murine BCATc.[7] The rat, murine, and human BCATc cDNA sequences have been published, and the deduced amino acid sequences exhibit high sequence identity (>80% identical).[7,13] The murine and human BCATc cDNA sequences appear to contain a cMyc-binding sequence in the 5' untranslated region[15]; however, the physiological significance of this sequence remains to be established. The human,[8,14] rat,[8] ovine,[10] murine[8] (partial cDNA), and porcine[8] (partial cDNA) BCATm cDNA sequences are now available.

Insertion of Human Branched-Chain Aminotransferase Isoenzyme cDNAs into Expression Plasmid

Expression plasmids for the human BCAT isoenzymes have been constructed with the E. coli expression vector pET-28a (Novagen, Madison, WI), which carries a sequence encoding an N-terminal His · Tag/thrombin/ T7 · Tag. High levels of expression are achieved with this expression vector and the hexahistidine (His$_6$) residue tag on the N terminus facilitates purifi-

[12] W. Shaffner and C. C. Weismann, *Anal. Biochem.* **56,** 502 (1973).

[13] J. Davoodi, P. M. Drown, R. K. Bledsoe, R. Wallin, G. Reinhart, and S. M. Hutson, *J. Biol. Chem.* **273,** 4982 (1998).

[14] O. Niwa, T. Kumazaki, T. Tsukiyama, G. Soma, N. Miyajima, and Y. Yokoro, *Nucleic Acids Res.* **18,** 6709 (1990).

[15] A. Eden, G. Simchen, and N. Benvenisty, *J. Biol. Chem.* **271,** 20242 (1996).

cation of the recombinant protein. A PCR strategy is used to amplify the human BCAT cDNAs from cDNA prepared from the appropriate human tissue. The PCR product is designed to contain restriction sites that facilitate cloning of the cDNA into the expression plasmid. We first cloned the PCR product directly into the general purpose plasmid pT7Blue(R) for sequencing, and then removed the cDNA insert, purified it, and ligated it into the pET-28a vector. The procedure that we used is described below. Other restriction sites and strategies can be used.

For human BCATm, the sense primer 5'-AGC<u>CATATG</u>GCCTCCTC-CAGTTTCAAG-3' contains an *Nde*I restriction site (underlined). The last 18 nucleotides of the sense primer correspond to the first 18 bases encoding the mature BCATm protein.[8,13] The antisense primer 5'-GAGTGTCG-CAACCACAT-3' corresponds to nucleotides 1316–1332 of the full-length human cDNA.[13] For increased accuracy, a polymerase mixture of *Taq* and *Pwo* polymerases (Boehringer Mannheim, Indianapolis, IN) should be used in the PCR. The amplification conditions are 94° (1 min), 60° (2 min), and 72° (1 min), for 30 cycles. The amplified PCR product is cloned into a pT7Blue(R) vector (Novagen). The ligation mix is then used to transform DH5α cells (Life Technologies, Gaithersburg, MD), and individual colonies (8 to 10) are screened by restriction digest to determine the direction of the insert. Any restriction enzyme that will cut the vector and insert to produce fragments of different sizes for each orientation can be used. The insert is sequenced and the fidelity is verified by comparison with the original human BCATm cDNA nucleotide sequence. The purified cDNA is removed from the pT7 vector by digestion with the restriction enzymes *Nde*I and *Sal*I. The *Sal*I restriction site is found in the pT7Blue(R) vector. The cDNA is then purified on a 1.4% (w/v) agarose gel, and the BCATm insert is ligated into the pET-28a vector cut with *Nde*I and *Sal*I.

A similar strategy is used to construct the BCATc expression vector. Using the human BCATc cDNA, the sense primer contains an *Nhe*I site immediately preceding the codon for the initiator methionine followed by the nucleotides encoding the first five amino acids of the protein, 5'-TA<u>TGGCTAGC</u>ATGGATTGCAGTAACGGA-3'. The antisense primer 5'-TCAGGATAGCACAATTGTC-3' corresponds to nucleotides 1137–1155. Again the amplified product is cloned into the pT7Blue(R) vector and the fidelity of the sequence is verified. The purified cDNA is ligated into the pET-28a vector cut with *Sal*I and *Nhe*I.

Transformation Protocol

The expression vector containing either human BCATm or BCATc is transformed into competent BL21(DE3) cells. A number of different dilutions of the pET-28a vector are prepared with human BCATm or human BCATc insert containing 0, 1, 2, and 3 μg of DNA. Four Eppendorf tubes are assembled with the following:

pET-28a BCAT plasmid (ranging from 0 to 3 μg of DNA)	4 μl
Polyethylene glycol (PEG; 30%, w/v)	4 μl
KCM (10×), containing 1 M KCl, 0.5 M MgCl$_2$, 0.3 M CaCl$_2$	5 μl
Sterile water	to 15 μl

A 5-μl aliquot of competent BL21(DE3) cells purchased from Novagen is added to the preceding mixture and allowed to sit on ice for 20 min. The mixtures are removed from the ice, and the reaction is allowed to proceed at room temperature for 10 min. Five hundred microliters of LB (Luria-Bertani) medium (no antibiotic) is added and the tubes incubated at 37° for 1 hr with shaking. For each reaction tube, part of the reaction mixture is streaked on a plate containing kanamycin (30 mg/ml). The plates are allowed to dry for about 15 min, inverted, and incubated overnight at 37°.

White colonies should be present on the plates. To confirm that the colonies have the vector with the BCAT insert, a number of colonies are selected and one 3-ml volume of LB broth containing kanamycin (30 mg/ml) is inoculated. The culture is incubated at 37° overnight. The DNA from these cultures is purified as described in the Wizard Miniprep kit protocol (Promega, Madison, WI).

To verify that the plasmid DNA contains the cDNA insert, the purified DNA can be digested with either *Nde*I, *Nde*I and *Sal*I, or *Nhe*I (BCATc) and *Sal*I to release the insert from the vector. The digestion conditions for *Nde*I and *Sal*I are as follows:

Miniprep plasmid DNA (approximately 1–2 μg of DNA)	5 μl
Buffer D (Promega)	1 μl
*Nde*I	1 μl
*Sal*I	1 μl
Sterile water	2 μl

The mixture is incubated at 37° for 1 hr. A 2-μl aliquot of 6× loading dye (Promega) is added to each digest. The samples are loaded onto a 1% (w/v) agarose gel prepared in 1× TAE buffer (0.04 M Tris–acetate, 0.001 M EDTA) containing 5 μl of ethidium bromide (10-mg/ml solution). The DNA bands are visualized with ultraviolet light to determine which colonies contain the BCAT insert.

Glycerol stocks of the transformed cells should be prepared. Several colonies that express the highest levels of BCAT activity are inoculated in LB broth containing kanamycin (30 mg/ml), and the cultures are incubated overnight at 37°. Aliquots of 0.5 ml are transferred to microcentrifuge tubes

(1.5 ml) and glycerol is added to a final concentration of 50% (v/v). The glycerol stocks are stored at −70°.

Expression of Mitochondrial and Cytosolic Branched-Chain Aminotransferases

A glycerol stock is taken and LB medium (100 ml) containing kanamycin (30 mg/ml) is inoculated and incubated at 37° overnight. The cells are transferred to a 2-liter culture flask containing LB medium (1.0 liter) with kanamycin (30 mg/ml). The cells are grown at 37° with vigorous aeration until the culture reaches an absorbance OD_{600} between 0.6 and 0.9. Cells are then induced with 1 mM isopropyl-β-D-thiogalactopyranoside (IPTG). After 4 hr of induction, the cells are harvested by centrifugation at 7800g for 5 min at 4°. The pellets are stored on ice at all times. The pellets may be stored indefinitely at −70°. Approximately 4.5–5.5 g of wet cells is obtained per liter of medium.

Purification of Mitochondrial Branched-Chain Aminotransferase

All procedures are performed at 4° unless stated otherwise.

Buffers

Buffer A: 0.01 M Tris-HCl, 0.1 M sodium phosphate, 5 mM 2-mercaptoethanol (pH 8.0)

Buffer B: 0.01 M Tris-HCl, 0.1 M sodium phosphate, 5 mM 2-mercaptoethanol, 4 M urea (pH 8.0)

Buffer C: 0.01 M Tris-HCl, 0.1 M sodium phosphate, 0.5 M sodium chloride, 5 mM 2-mercaptoethanol, 20% (v/v) glycerol (pH 7.4)

Buffer D: 0.01 M Tris-HCl, 0.1 M sodium phosphate, 0.75 M sodium chloride, 5 mM 2-mercaptoethanol, 20% (v/v) glycerol (pH 6.0)

Buffer E: 0.01 M Tris-HCl, 0.1 M sodium phosphate, 0.75 M sodium chloride, 50 mM imidazole, 5 mM 2-mercaptoethanol, 10% (v/v) glycerol (pH 6.0)

Buffer F: 0.01 M Tris-HCl, 0.1 M sodium phosphate, 0.75 M sodium chloride, 0.5 M imidazole, 5 mM 2-mercaptoethanol, 10% (v/v) glycerol (pH 6.0)

Cell Extraction and Nickel Chromatography

The cells from 1 liter of culture are resuspended carefully in 15 ml of buffer A. The mixture is then placed on ice and sonicated at 1.6 output with a duty cycle of 70% at 1.5-min intervals for a total of 10 min, using a double-step microtip attached to a Branson sonifier model 250 (Branson Ultrasonics, Danbury, CT). The cells are centrifuged at 7800g for 10 min at 4°. The supernatant is decanted and measured into a clean conical flask. To permit efficient binding of the N-terminal histidine-tagged BCATm to the nickel-NTA resin, the extract is brought to 4 M urea and kept at 4°. The cell pellets are resuspended in 15 ml of buffer B, sonicated, and centrifuged as described. The two cell extracts are then combined.

Nickel-NTA resin (Qiagen, Chatsworth, CA) is used to purify the histidine-tagged recombinant protein. The nickel-NTA resin is equilibrated with 3 column volumes of buffer B. A 6-ml aliquot of resin is measured into a 15-ml conical tube, and the resin is equilibrated by inverting it in buffer B a total of three times.

Next, the total combined extract is incubated with 4 ml of nickel-NTA resin and stirred gently for 1 hr at 4°. The resin is then added to a glass column, using gravity flow. Buffers B through D are used to wash the column sequentially at a flow rate of 1.5 ml/min. In this and subsequent steps, protein in the column effluent is monitored by measuring absorbance at 280 nm. The BCATm-containing fraction is eluted with buffer E. Approximately 14–16 ml of a pale yellow-colored solution of partially purified BCATm is collected. Ethylenediaminetetraacetic acid (EDTA, 1 mM final concentration) is added to the BCATm to chelate any free nickel ions.

Thrombin Digestion and Gel-Permeation (Sephadex G-25) Chromatography I

Thrombin digestion is required to remove the histidine tag. A Sephadex G-25 column (13.5 × 3.0 cm) is equilibrated in buffer containing 50 mM Tris-HCl and 150 mM sodium chloride, pH 7.5. The BCATm-containing fraction from the nickel-NTA column is applied to the Sephadex G-25 column and eluted at a flow rate of 5 ml/min. The protein peak containing BCATm is collected. The histidine tag is removed by digestion with human α-Thrombin (7.5 NIH units/ml of eluate) for 1 hr at 28°.

Gel-Permeation (Sephadex G-25) II and Hydrophobic Interaction Chromatography

Reagents

Buffer X: 100 mM potassium phosphate, pH 7.5; store at 4°
Buffer Y: 35% saturated ammonium sulfate in 100 mM potassium phosphate, pH 7.5; store at 4°
Saturated ammonium sulfate: Prepare in 100 mM potassium phosphate, pH 7.5; store at −20°

α-Ketoisocaproate (KIC), 100 mM
Storage buffer: 50 mM Tris-HCl, 150 mM NaCl, 1 mM EDTA, 5 mM DTT, 5 mM glucose, 1 mM KIC (pH 7.5)

Procedure. After thrombin cleavage, the buffer containing the BCATm fraction is exchanged for buffer X. A Sephadex G-25 column (13.5 × 3.0 cm) is equilibrated with buffer X. The BCATm fraction is applied to the column at a flow rate of 5 ml/min. BCATm is converted to the pyridoxal phosphate form by adding 1 mM α-ketoisocaproate to the BCATm fraction.

A Hydropore-PK HIC 12U column (10 × 100 mm; Rainin, Woburn, MA) is equilibrated with 10 column volumes of buffer Y, at a flow rate of 5.0 ml/min. Saturated ammonium sulfate is added to the fraction eluted from the Sephadex G-25 gel permeation II column, to a final concentration of 35% (v/v). The sample is allowed to sit on ice for 15 min before it is applied to the column at a flow rate of 5.0 ml/min. Once the sample is loaded, the column is washed with 5 column volumes of buffer Y to remove any unbound proteins. The absorbance at 280 nm is monitored to ensure that BCATm binds to the column. BCATm is eluted with a descending gradient (35 to 0%) of ammonium sulfate in the potassium phosphate buffer, at a flow rate of 1.0 ml/min for a period of 10 min. BCATm will elute as a distinctive concentrated yellow peak at an ammonium sulfate saturation between 29 and 25%.

Because BCATm is unstable in the ammonium sulfate buffer, the purified BCATm fraction is dialyzed overnight against the storage buffer at 4°. The yield of purified BCATm from 1 liter of cells is usually about 4–6 mg. Glycerol is added to a final concentration of 30% (v/v) to purified BCATm, and the tube is flushed with argon. Stored at −20 or −70°, BCATm is stable for about 3 months. A summary of a typical BCATm purification is shown in Table I.

Purification of Cytosolic Branched-Chain Aminotransferase

Purification of human recombinant BCATc uses a similar protocol as for human recombinant BCATm, with the modifications detailed below.

Extraction. Because BCATc is more soluble than BCATm, urea extraction of the pellet is not necessary, and thus only buffer A is used in both steps. The protease inhibitor diisopropyl fluorophosphate (1 mM) must be added to buffer A.

Gel Permeation I: Hydroxylapatite Column Chromatography. The BCATc protocol includes a hydroxylapatite column chromatography step after the nickel-NTA resin step. One gram of hydroxylapatite is weighed into a beaker and the resin is washed three times with distilled water.

TABLE I
PURIFICATION OF RECOMBINANT HUMAN MITOCHONDRIAL
BRANCHED-CHAIN AMINOTRANSFERASE

Procedure	Volume (ml)	Total activity (units)	Protein (mg/ml)	Specific activity (units/mg)	Yield (%)	Purification (fold)
Cells	30	1110	10.0	3.7	100	1.0
Extraction	34	340	7.75	1.3	30	0.4
Nickel-NTA	17	434	0.51	51	38	14
Gel permeation I	22	594	0.38	71	53	19
Gel permeation II	27	788	0.40	73	70	20
Hydropore	5.0	410	0.97	85	37	23
Dialysis	4.8	510	0.85	125	46	34

To remove fines, excess water is decanted after each wash. The hydroxyl-apatite column is equilibrated in 10 mM potassium phosphate, pH 7.5. The BCATc-containing fraction from the nickel-NTA resin is loaded on the column after changing the buffer to 10 mM potassium phosphate (pH 7.5), using Sephadex G-25 gel-permeation chromatography as described above. BCATc is eluted with 200 mM potassium phosphate, pH 7.5. As described for BCATm, the histidine tag is removed by thrombin cleavage.

Mono Q Anion-Exchange Chromatography. In the final step of the BCATc purification protocol, anion-exchange chromatography is substituted for hydrophobic interaction chromatography.

TABLE II
PURIFICATION OF RECOMBINANT HUMAN CYTOSOLIC BRANCHED-CHAIN AMINOTRANSFERASE

Procedure	Volume (ml)	Total activity (units)	Protein (mg/ml)	Specific activity (units/mg)	Yield (%)	Purification (fold)
Cells	30	2178	33.0	2.2	100	1.0
Extraction	34	2493	14.1	5.2	114	2.4
Nickel-NTA	14	1115	2.43	33	51	15
Gel permeation	19	814	1.48	30	37	14
Hydroxylapatite	7	511	6.0	30	23	14
Gel permeation	12	864	1.92	38	40	17
Gel permeation	19	1378	0.91	82	63	37
Mono Q	4	730	2.15	85	34	39

Reagents

Buffer X: 10 mM potassium phosphate, pH 8.0
Buffer Y: 10 mM potassium phosphate with 0.5 M sodium chloride (pH 8.0)
Procedure. Using a Sephadex G-25 gel-permeation column, the buffer of the BCATc fraction is exchanged for buffer X. A mono Q HR 5/5 column (Pharmacia, Piscataway, NJ) is equilibrated with 5 column volumes of buffer X, at a flow rate of 1.0 ml/min. The sample is applied to the column, using the same flow rate, followed by 5 column volumes of buffer X to remove any unbound protein. Using a Mono Q column, BCATc is selectively eluted with a NaCl gradient from 0 to 0.5 M over 20 min at a flow rate of 1 ml/min. BCATc elutes as a distinctive concentrated yellow peak between 0.32 and 0.41 M NaCl. BCATc is then dialyzed into the storage buffer as described for BCATm. The average yield of purified BCATc from 1 liter of culture is 7.5 to 8.0 mg. A typical purification is summarized in Table II.

[34] Branched-Chain-Amino-Acid Transaminases of Yeast *Saccharomyces cerevisiae*

By CORINNA PROHL, GYULA KISPAL, and ROLAND LILL

Introduction

Genetic screens for components required for the biosynthesis of branched-chain amino acids (leucine, isoleucine, and valine) in the yeast *Saccharomyces cerevisiae* have led to the identification of almost the complete set of genes involved in this process.[1,2] Searches have failed, however, to identify branched-chain-amino-acid transaminases (Bat proteins) cata-

[1] E. W. Jones and G. R. Fink, *in* "The Molecular Biology of the Yeast *Saccharomyces: Metabolism and Gene Expression*" (J. N. Strathern, E. W. Jones, and J. R. Broach, eds.), p. 181. Cold Spring Harbor Laboratory Press, Cold Spring Harbor, New York, 1982.

[2] A. G. Hinnebusch, *in* "The Molecular and Cellular Biology of the Yeast *Saccharomyces*," Vol. II: "Gene Expression" (E. W. Jones, J. R. Pringle, and J. R. Broach, eds.), p. 319. Cold Spring Harbor Laboratory Press, Cold Spring Harbor, New York, 1992.

METHODS IN ENZYMOLOGY, VOL. 324

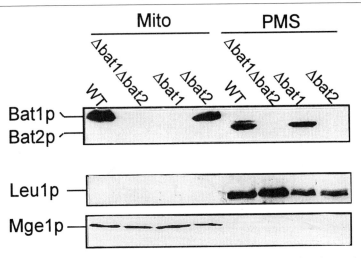

FIG. 1. Cellular localization of Bat1p and Bat2p. Mitochondria (Mito) and postmitochondrial supernatants (PMS) were isolated from the indicated yeast cells grown in YPD medium.[5,5a] Samples were analyzed by immunostaining, using antisera raised against Bat1p, the mitochondrial protein Mge1p, and the cytosolic protein Leu1p. Because of the high similarity to Bat1p, Bat2p is recognized in postmitochondrial supernatants by the anti-Bat1p antiserum. WT, Wild type.

lyzing the last step of biosynthesis or the first step of biodegradation. In part, this is due to the presence of two genes encoding Bat proteins in the yeast genome. The transaminase proteins termed Bat1p and Bat2p[3] have been identified by two independent approaches. In one study, *BAT1* encoding the mitochondrial isoform (Fig. 1) was identified as a high-copy suppressor of a temperature-sensitive mutant of the mitochondrial ABC transporter Atm1p.[4,5] The existence of Bat2p in the yeast cytosol was realized through its specific recognition by anti-Bat1p antibodies (Fig. 1[5,5a]). Cloning of the *BAT1* and *BAT2* genes revealed a high sequence similarity of the encoded proteins. In the second approach, Bat1p was discovered on the basis of its considerable homology to a mouse protein termed ECA39.[6,7] The mamma-

[3] The previous names for the genes *BAT1* and *BAT2* (*TWT1* or *ECA39* and *TWT2*, respectively) were introduced when the function of the encoded proteins as transaminases was still unknown.

[4] G. Kispal, H. Steiner, D. A. Court, B. Rolinski, and R. Lill, *J. Biol. Chem.* **271**, 24458 (1996).

[5] G. Kispal, P. Csere, B. Guiard, and R. Lill, *FEBS Lett.* **418**, 346 (1997).

[5a] F. Sherman, *Methods Enzymol.* **194**, 3 (1991).

[6] O. Schuldiner, A. Eden, T. Ben-Yosef, O. Yanuka, G. Simchen, and N. Benvenisty, *Proc. Natl. Acad. Sci. U.S.A.* **93**, 7143 (1996).

[7] A Eden, G. Simchen, and N. Benvenisty, *J. Biol. Chem.* **271**, 20242 (1996).

lian protein is highly expressed early in embryogenesis and in several tumors including mouse teratocarcinoma cells.[8,9] Expression of ECA39 has been proposed to be regulated by the c-Myc oncoprotein.[9] More recent investigations of a number of tumor cells including neuroblastoma cells have indicated that the control of ECA39 expression is much more complex and that the levels of the protein may not be under direct control of the c-Myc oncoprotein.[10]

Before the discovery of the transaminase activity of yeast Bat1p, a role for this protein in the control of the cell cycle has been suggested.[6] This contention was based on slight increases in growth rates of cells in which the *BAT1* gene was deleted (Δbat1 cells) as compared with wild-type cells and on somewhat higher numbers of Δbat1 cells in the G_1 phase. These findings were taken to propose a role for Bat1p in the regulation of the transition from G_1 to S phase.[6] However, subsequent findings rendered a direct involvement of the Bat proteins in cell cycle control rather unlikely. First, Bat1p was found to be located within mitochondria.[4] Second, both Bat1p and Bat2p were characterized as branched-chain-amino-acid transaminases.[4,7] Third, double-mutant cells lacking both *BAT1* and *BAT2* genes (Δbat1Δbat2 cells) displayed retarded growth on various growth media[4] (see below). Finally, *BAT1,* like other genes involved in branched-chain amino acid biosynthesis, was found to be under the control of the transcription factor Gcn4p, which regulates the general cellular utilization of nitrogen including the synthesis of amino acids.[11]

The amino acid sequences of Bat1p (43.6 kDa, including its mitochondrial presequence) and Bat2p (41.6 kDa) show 77% identity.[4] Both proteins share about 50% of their amino acid residues with the branched-chain-amino-acid transaminases from other species including human (BCAT), mouse (ECA39), rat, sheep and *Schizosaccharomyces pombe.*[9,12–15] Comparatively weak sequence identity (24%) exists with the branched-chain-amino-acid transaminase from *Escherichia coli,* IlvE.[16] For human Bat proteins it has been reported that these proteins function as dimers.[17]

[8] O. Niwa, T. Kumazaki, T. Tsukiyama, G. Soma, N. Miyajima, and K. Yokoro, *Nucleic Acids Res.* **18,** 6709 (1990).

[9] N. Benvenisty, A. Leder, A. Kuo, and P. Leder, *Genes Dev.* **6,** 2513 (1992).

[10] T. Ben-Yosef, O. Yanuka, D. Halle, and N. Benvenisty, *Oncogene* **17,** 165 (1998).

[11] T. Ben-Yosef, O. Yanuka, and N. Benvenisty, *Oncogene* **13,** 1859 (1996).

[12] S. M. Hutson, R. K. Bledsoe, T. R. Hall, and P. Dawson, *J. Biol. Chem.* **270,** 30344 (1995).

[13] R. K. Bledsoe, P. A. Dawson, and S. M. Hutson, *Biochim. Biophys. Acta* **1339,** 9 (1997).

[14] A. Eden and N. Benvenisty, *Yeast* **14,** 189 (1998).

[15] M. Faure, F. Glomot, R. Bledsoe, S. Hutson, and I. Papet, *Eur. J. Biochem.* **259,** 104 (1999).

[16] S. Kuramitsu, T. Ogawa, H. Ogawa, and H. Kagamiyama, *J. Biochem.* **97,** 993 (1985).

[17] J. Davoodi, P. M. Drown, R. K. Bledsoe, R. Wallin, G. D. Reinhart, and S. M. Hutson, *J. Biol. Chem.* **273,** 4982 (1998).

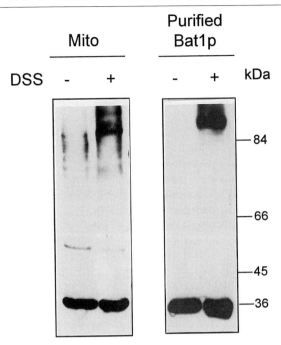

FIG. 2. Cross-linking of Bat1p. Isolated mitochondria that were lysed in 0.1% (v/v) Triton X-100 or purified Bat1p were incubated for 30 min at 0° in buffer G (250 mM sucrose, 1 mM EDTA, 15 mM MOPS–KOH, pH 7.2) without or with the homobifunctional cross-linker DSS (disuccinimidyl suberate; Pierce, Rockford, IL). The reaction was stopped by adding 5 mM Tris-HCl, pH 7.2. Samples were centrifuged and applied to SDS–PAGE. Immunostaining of Bat1p detected a band at 90 kDa in addition to the monomeric form of Bat1p. This band corresponds to the dimeric form of Bat1p. The positions of molecular mass markers are indicated on the right.

Cross-linking experiments performed with Bat1p derived from isolated mitochondria and with purified Bat1p demonstrated that the yeast protein forms a homodimer, suggesting that this is the functional form of Bat1p (Fig. 2).

Biochemical analysis of total cell extracts derived from wild-type or from single- or double-disrupted cells revealed that the transaminase enzyme activity of mitochondrial Bat1p is significantly higher than that of cytosolic Bat2p (Fig. 3[17a]). In mitochondria, Bat1p represents the major activity for transamination of branched-chain α-keto and amino acids, whereas in the cytosol other enzymes can efficiently replace the function of Bat2p as a transaminase.[4]

[17a] M. Woonter and J. A. Jaehning, *J. Biol. Chem.* **265,** 8979 (1990).

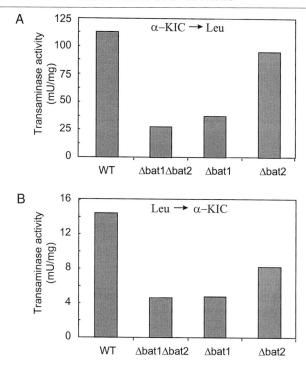

Fig. 3. Branched-chain-amino-acid transaminase activities of total yeast cell extracts. Wild-type (WT) yeast cells and the indicated *BAT* mutant cells were grown in YPD medium and collected by centrifugation. Cell extracts were prepared by vigorously vortexing cells with glass beads in 50 m*M* Tris-HCl buffer, pH 7.4, supplemented with 0.01 m*M* PMSF and 0.2% (v/v) Triton X-100.[17a] After pelleting the cell debris by centrifugation at 4000*g* for 5 min, the supernatants were used for measurement of transaminase activities. (A) Enzyme activities for the forward reaction, i.e., the conversion of α-ketoisocaproate (α-KIC) to leucine. (B) In the reverse reaction, the synthesis of α-ketoisocaproate from leucine was followed by the subsequent formation of its hydrazone.

Disruption of either *BAT1* or *BAT2* does not result in major consequences in cell growth, even on medium lacking leucine, isoleucine, and valine.[4,6] This suggests that transamination of branched-chain keto acids and amino acids can be performed either in the cytosol or the mitochondrial matrix. Apparently, these compounds can be exchanged between mitochondria and the cytosol fast enough to satisfy the biosynthetic needs of both cellular compartments. Disruption of both *BAT* genes causes the expected auxotrophy for leucine, isoleucine, and valine.[4] Surprisingly, the resulting ΔbatlΔbat2 cells exhibit retarded growth even after addition of these three amino acids. Impairment of growth is most significant for cultivation on rich medium containing glucose, where Δbat1Δbat2 cells grow at drastically reduced rates as compared with wild-type cells. In contrast, no detectable growth phenotype was noted for Δbat1Δbat2 cells in the presence of glycerol-containing rich medium. These observations have led to the proposal that the Bat proteins fulfill a second function in the cell, in addition to their involvement in transamination of branched-chain keto acids and amino acids. Apparently, the additional function of the Bat proteins is rate-limiting for growth under conditions of glucose repression, i.e., for cells containing functionally impaired mitochondria.

It has been demonstrated that the Bat proteins perform a function in the biosynthesis of cytosolic Fe/S proteins (Fig. 4).[18] Mitochondria play a crucial role in this process.[19] Synthesis of both mitochondrial and cytosolic Fe/S clusters is initiated in the mitochondrial matrix by the production of elemental sulfur, a reaction catalyzed by the cysteine desulfurase Nfs1p. A number of other mitochondrial proteins may cooperate with Nfs1p in the synthesis of Fe/S clusters, including two heat shock proteins and a mitochondrial ferredoxin.[19,20] Export of components required for Fe/S protein assembly in the cytosol is mediated by the mitochondrial ABC transporter Atm1p, a protein located in the mitochondrial inner membrane.[5,21] Findings indicate that the expression of at least one of the Bat proteins is necessary for efficient biosynthesis of extramitochondrial Fe/S clusters (Fig. 4). On the other hand, the presence of the Bat proteins is not required for the synthesis of intramitochondrial Fe/S cluster-containing proteins.[18] While the precise molecular role of the Bat proteins in cytosolic Fe/S cluster formation remains unknown, it may be noteworthy that a bacterial NifS-like protein was reported to exhibit transaminase enzyme activity.[22] It is thus conceivable that the two types of pyridoxal phosphate-dependent proteins perform a similar reaction in biosynthesis. The participation of Atm1p and Bat1p in the same biosynthetic process, namely the generation of cytosolic Fe/S proteins, provides a likely explanation for the functional complementation of a temperature-sensitive mutant of *ATM1* by high levels of Bat1p.[4] Further elucidation of the molecular basis of the functional interaction of these two proteins will aid studies addressing their precise molecular function in Fe/S cluster formation.

In the following, we first describe methods for the measurement of the branched-chain-amino-acid transaminase enzyme activity in either crude

[18] C. Prohl, G. Kispal, and R. Lill, unpublished observations.
[19] G. Kispal, P. Csere, C. Prohl, and R. Lill, *EMBO J.* **18**, 3981 (1999).
[20] J. Strain, C. R. Lorenz, J. Bode, S. Garland, G. A. Smolen, D. T. Ta, L. E. Vickery, and V. C. Culotta, *J. Biol. Chem.* **273**, 31138 (1998).
[21] J. Leighton and G. Schatz, *EMBO J.* **14**, 188 (1995).
[22] P. Leong-Morgenthaler, S. G. Oliver, H. Hottinger, and D. Soell, *Biochimie* **76**, 45 (1994).

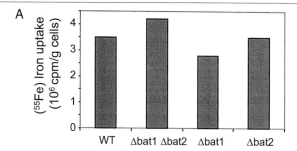

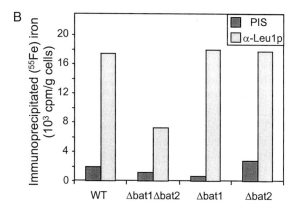

Fig. 4. Formation of the Fe/S cluster of cytosolic Leu1p *in vivo*. Wild-type (WT) and the indicated *BAT* mutant cells were grown overnight in medium A (minimal medium SD lacking added iron chloride) in the presence of the desired carbon source at 30°. At an optical density of 1–2 OD_{600}, cells (0.5 g wet weight) were collected and resuspended in 10 ml of medium A, and 0.1 mM ascorbate and 10 μCi of ^{55}Fe (Amersham) were added. Radiolabeling was for 1 hr at 30°. After addition of 100 μM iron sulfate and further incubation for 5 min at 30° cells were collected and washed with 50 mM citrate (pH 7.4)–1 mM EDTA. A cell lysate was prepared by breaking the cells with glass beads.[17a] Aliquots of the clarified lysate (10 min, 12,000g) were used (A) for determination of ^{55}Fe uptake by the various cells through liquid scintillation counting and (B) for immunoprecipitation, employing antibodies raised against Leu1p or antibodies derived from preimmune serum (PIS). Radioactivity associated with the immunoprecipitated material was quantitated by liquid scintillation counting. The standard error for detection of Leu1p-associated ^{55}Fe in a cell lysate was 10%, and the cellular uptake of ^{55}Fe varied by 20% in independent experiments.

cell extracts of *S. cerevisiae,* in isolated mitochondria, or in cytosolic fractions. Assays for both directions of the interconversion of branched-chain amino acids and α-keto acids are detailed. We then describe a simple method to purify the Bat proteins and determine their transaminase activity (Table I).

TABLE I
PURIFICATION OF Bat1p BY Ni-NTA AFFINITY CHROMATOGRAPHY[a]

Fraction	IPTG induction	Amount of protein (mg)	Total Bat activity (U)	Specific Bat activity (U/mg)
Crude cell extract	+	10	1	0.1
	−	10	0.15	0.015
Total Bat1p-containing fractions	+	0.12	0.5	4.2
Fraction of highest Bat1p activity	+	0.005	0.12	24
	−	ND	<0.005	ND

[a] Purification of Bat1p, taking advantage of a hexahistidinyl tag (for description of detailed procedure see Bacterial Expression of Bat1p^{His_6}). Bat1p^{His_6} was overexpressed in *E. coli* by the addition of the inducer isopropyl-β-D-thiogalactopyranoside (IPTG), and growth of cells was for 3 hr at 37°. The branched-chain-amino-acid transaminase (Bat) activity increased more than sixfold on induction by IPTG. The purification yielded a more than 40-fold increase in specific transaminase activity in the total Bat1p-containing fraction. The enzyme activity of the fraction containing the peak of Bat1p^{His_6} content was increased more than 200-fold. In the absence of induction hardly any activity was detectable in the fractions corresponding to the position of elution of Bat1p^{His_6}. ND, not detectable.

Assays

Assay of Enzyme Activity of Yeast Branched-Chain Amino Acid Transaminases

The transamination of branched-chain α-keto acids (α-ketoisocaproate, α-ketoisovalerate, and α-keto-β-methylvalerate) to the corresponding amino acids (leucine, valine, and isoleucine) represents an equilibrium reaction. Thus, the reactions can be measured in both directions. The forward reactions can be determined by coupling them to the glutamate dehydrogenase reaction [see reaction (1)]. Glutamate dehydrogenase converts α-ketoglutarate to glutamate and is dependent on NAD$^+$/NADH or NADP$^+$/NADPH. Accordingly, the formation of branched-chain amino acids can be recorded by the decrease in NAD(P)H concentration.[23]

Branched-chain α-keto acid + glutamic acid $\rightleftharpoons$
    branched-chain amino acid + α-ketoglutarate

α-Ketoglutarate + NH$_4^+$ + NAD(P)H $\underset{\text{dehydrogenase}}{\overset{\text{glutamate}}{\rightleftharpoons}}$

    glutamic acid + NAD(P)$^+$ + H$_2$O   (1)

[23] R. H. Collier and G. Kohlhaw, *J. Bacteriol.* **112**, 365 (1972).

In the reverse reaction [reaction(2)] the formation of a hydrazone adduct with the branched-chain keto acids is measured by its absorbance at 440 nm.[24] The irreversible formation of this adduct allows the quantitative determination of branched-chain α-keto acids.

Branched-chain amino acid + α-ketoglutarate $\rightleftharpoons$ branched-chain
α-keto acids + glutamic acid
Branched-chain α-keto acid + 2,4-dinitrophenylhydrazine $\rightarrow$ α-keto
acid 2,4-dinitrophenylhydrazone   (2)

Detailed procedures for both types of assays are provided below.

Forward Reaction. A 0.5-ml reaction mixture typically contains 0.25 ml of buffer A [200 mM Tris-HCl (pH 8.0), 100 mM NH$_4$Cl, 0.5 mM pyridoxal phosphate, and 2 mM NaN$_3$], 0.04 ml of 0.5 M glutamic acid, 0.01 ml of 10 mM NADH, 1 U of glutamate dehydrogenase, and 0.04 ml of sample in a quartz cuvette. The sample consists of either 30–50 μg of protein derived from yeast cell extracts or from cytosol, 10–30 μg of isolated mitochondria, or 1 μg of purified Bat1p. After reaching a stable signal of absorption at 340-nm wavelength, the transamination reaction is started by adding 0.02 ml of the desired α-keto acid (200 mM). The decrease in absorption is recorded for several minutes and expressed in units per milligram of protein. One unit is defined as the oxidation of 1 μmol of NADH per minute (extinction coefficient $\varepsilon_{340} = 6.22$ mM^{-1} cm^{-1}). A typical result using cell extracts derived from wild-type, Δbat1, Δbat2, or Δbat1Δbat2 cells is depicted in Fig. 3A for the transamination of α-ketoisocaproate to leucine. Deletion of *BAT1* results in a threefold reduction in total cellular branched-chain-amino-acid transaminase activity, whereas deletion of *BAT2* only weakly affects the total transaminase activity. These data indicate that Bat1p represents the major transaminase activity for branched-chain α-keto acids in the yeast cell. In the absence of both Bat proteins, 25% residual branched-chain-amino-acid transaminase activity is detected in yeast cell extracts. At least in these *in vitro* experiments, proteins other than Bat1p and Bat2p can perform the conversion of branched-chain α-keto acids to leucine (Fig. 3A), isoleucine, or valine.[4] Most likely, these activities are executed by other cellular transaminases with low specificity for α-keto acids with a branched aliphatic side chain.

Reverse Reaction. The reverse reaction is performed in a total volume of 1 ml containing 0.5 ml of buffer B (75 mM sodium pyrophosphate, pH 9.2), 10 μl of 6.7 mM isoleucine, leucine, or valine, 3.4 μl of 20 mM pyridoxal phosphate, and the sample to be analyzed. Typically, 300–500 μg of total cell extracts or of cytosolic fractions, 100–300 μg of isolated mitochondria,

[24] A. Ichihara and E. Koyama, *J. Biochem.* **59,** 160 (1966).

or 10 μg of purified Bat1p is used. The solution is preincubated for 5 min at 37°, and the reaction is started by adding 10 μl of 6.7 mM α-ketoglutarate. After 10 min at 37°, the transaminase reaction is terminated by the addition of 100 μl of 60% (w/v) trichloroacetic acid. The samples are centrifuged and the clarified supernatant is transferred to a 10-ml glass tube. After incubation for 5 min at 25° in a water bath, 2 ml of 2,4-dinitrophenylhydrazine [0.5% (w/v) in 2 N HCl] is added, and the incubation is continued for another 5 min. The mixture is supplemented with 5 ml of toluene and vortexed vigorously for 2 min. The bottom aqueous layer is aspirated off with a capillary pipette, and 5 ml of 0.5 N HCl is added to the organic phase. After shaking for 1 min, the insoluble hydrazone generated with α-ketoglutarate is removed by brief centrifugation. Two milliliters of the clarified toluene layer is combined with 2 ml of 10% (w/v) sodium carbonate in a fresh glass tube. After brief shaking and phase separation, 1 ml of the aqueous layer is mixed with 1 ml of 1.5 N NaOH, and the absorption is recorded at 440 nm. The levels of absorption at 440 nm of the hydrazone adducts of the three branched-chain α-keto acids are similar (extinction coefficients $\varepsilon_{440} = 14$ mM^{-1} cm^{-1}). For quantitation of the enzyme activity the background absorption (mainly due to the added 2,4-dinitrophenylhydrazine) must be subtracted. To this end, the reaction is performed in parallel with buffer instead of protein sample. A typical example of this assay is presented in Fig. 3B, using leucine as a substrate and extracts derived from wild-type, Δbat1, Δbat2, or Δbat1Δbat2 yeast cells. The results are comparable to those obtained for the forward reaction.

Transaminase Activity of Purified Bat1p from Saccharomyces cerevisiae

On deletion of the two *BAT* genes yeast cells contain residual activity of branched-chain-amino-acid transaminase (see Fig. 3). To establish unequivocally the function of the Bat proteins as transaminases, it is necessary to demonstrate that this enzyme activity is associated with the isolated Bat proteins. Rapid purification of the Bat proteins in active form can be achieved after expression of hexahistidinyl-tagged versions in *E. coli*. The procedure of isolation will be exemplified for the tagged version of mitochondrial Bat1p (termed Bat1p^{His_6}). The N-terminal mitochondrial targeting sequence (amino acids 1–16)[4] must be removed in the tagged fusion protein. The hexahistidinyl tag attached to the C terminus of Bat1p does not significantly affect the enzymatic function of the protein in transamination (not shown). Purification of Bat1p^{His_6} from bacterial cell extracts takes advantage of the specific binding of the histidine residues to a Ni-NTA (nickel nitrilotriacetic acid) resin. The one-step purification protocol yields

Bat1p^{His_6} of at least 60% purity (not shown). If higher purification is required, contaminating proteins can be removed by standard anion-exchange chromatography. The one-step purification procedure yields an enrichment of transaminase activity of more than 200-fold relative to the starting extract (Table I). Hardly any transaminase activity was detectable with extracts derived from bacteria that were not induced for expression of Bat1p^{His_6}. Thus, Bat1p functions as a branched-chain amino acid transaminase of yeast mitochondria. On purification, we noted that Bat1p may rapidly lose its transaminase activity. This may be efficiently prevented by inclusion of stabilizing agents such as glycerol, α-ketoglutarate, and pyridoxal phosphate in the buffer solutions.

Bacterial Expression of Bat1p^{His_6}. *Saccharomyces cerevisiae BAT1* lacking the coding information for amino acid residues 2–16 is cloned in plasmid pQE30 (Qiagen, Chatsworth, CA) and the resulting plasmid is used to transform *E. coli* strain DH5α. Bacteria are grown in LB medium containing ampicillin (100 μg/ml)[25] and the expression of the hexahistidinyl-tagged Bat1p fusion protein is induced by addition of isopropyl-β-D-thiogalactopyranoside (IPTG; 0.5 mg/ml) for 3 hr. Cells are harvested by centrifugation for 20 min at 7000g and resuspended in 0.4 ml of buffer C [10% (w/v) sucrose, 50 mM Tris-HCl (pH 8.0)]. Cell lysis is mediated by addition of lysozyme (10 mg/ml) and incubation for 30 min on ice. Five milliliters of buffer C containing 1% (v/v) Triton X-100 and 4 mM phenylmethylsulfonyl fluoride (PMSF) is added and the mixture is sonicated (8 times for 10 sec, duty cycle 50%, 30 W) to break the cells open. Finally, the cell debris is removed by centrifugation for 30 min at 15,000g at 4° and the supernatant is used for purification of Bat1p^{His_6}.

Purification of Bat1p^{His_6} Employing Ni-NTA Affinity Chromatography. The Ni-NTA affinity resin (Qiagen) is equilibrated with buffer D [50 mM sodium phosphate (pH 8.0), 300 mM NaCl, 10% (v/v) glycerol, 0.2 mM pyridoxal phosphate, 5 mM α-ketoglutarate, 0.2 mM EDTA, 5 mM freshly prepared 2-mercaptoethanol, and 0.1% (v/v) Triton X-100]. The bacterial extract (40 mg of protein) containing Bat1p^{His_6} is applied to a column containing 1 ml of resin. The column is washed once with 2 ml of buffer E (same as buffer D, but pH 6.0), and then three times with buffer F (same as buffer E, but without Triton X-100). The Bat1p^{His_6} is eluted with buffer F containing 150 mM EDTA. Fractions of 0.1 ml are collected and tested for protein content and transaminase activity (see above). The final yield of Bat1p^{His_6} protein is about 1–2% relative to the amount of total protein applied to the column.

[25] J. Sambrook, E. F. Fritsch, and T. Maniatis, "Molecular Cloning: A Laboratory Manual," 2nd Ed. Cold Spring Harbor Laboratory Press, Cold Spring Harbor, New York, 1989.

[35] Purification, Properties, and Sequencing of Aminoisobutyrate Aminotransferases from Rat Liver

By Nanaya Tamaki, Shigeko Fujimoto Sakata, and Koichi Matsuda

D-β-Aminoisobutyrate + pyruvate ⇌
D-methylmalonate semialdehyde + L-alanine

L-β-Aminoisobutyrate + 2-oxoglutarate ⇌
L-methylmalonate semialdehyde + L-glutamate

Introduction

Aminoisobutyrate is metabolized in the mitochondrial matrix by two characteristic aminoisobutyrate aminotransferases: D-β-aminoisobutyrate–pyruvate aminotransferase (D-AIBAT) (EC 2.6.1.40)[1,2] and L-β-aminoiso-butyrate–oxoglutarate aminotransferase (L-AIBAT) (EC 2.6.1.22).[3,4]

D-β-Aminoisobutyrate is formed by the degradation of thymine via dihydrothymine and N-carbamoyl-β-aminoisobutyrate in cytosol.[5,6] β-Aminoisobutyrate is transported into mitochondria,[7,8] where it is further metabolized to propionyl-CoA by D-AIBAT and methylmalonate semialde-hyde dehydrogenase in the mitochondrial matrix.[2,9] D-AIBAT is predomi-nantly distributed in the kidney and liver in the rat. The physicochemical, enzymological, and immunological properties of D-AIBAT are identical to those of alanine–glyoxylate aminotransferase 2 (EC 2.6.1.44),[10] aminolevu-

[1] Y. Kakimoto, K. Taniguchi, and I. Sano, J. Biol. Chem. 244, 335 (1969).
[2] N. Tamaki, M. Kaneko, C. Mizota, M. Kikugawa, and S. Fujimoto, Eur. J. Biochem. 189, 39 (1990).
[3] Y. Kakimoto, A. Kanazawa, K. Taniguchi, and I. Sano, Biochim. Biophys. Acta 156, 374 (1968).
[4] S. Fujimoto, N. Mizutani, C. Mizota, and N. Tamaki, Biochim. Biophys. Acta 882, 106 (1986).
[5] R. M. Fink, C. McCauchy, R. E. Cline, and K. Fink, J. Biol. Chem. 218, 1 (1956).
[6] K. Fink, R. E. Cline, R. B. Henderson, and R. M. Fink, J. Biol. Chem. 221, 425 (1956).
[7] O. W. Griffith, Annu. Rev. Biochem. 55, 855 (1986).
[8] N. Tamaki, S. Fujimoto, C. Mizota, and M. Kikugawa, J. Nutr. Sci. Vitaminol. 33, 439 (1987).
[9] G. W. Goodwin, P. M. Rougraff, E. J. Davis, and R. A. Harris, J. Biol. Chem. 264, 14965 (1989).
[10] Y. Kontani, M. Kaneko, M. Kikugawa, S. Fujimoto, and N. Tamaki, Biochim. Biophys. Acta 1156, 161 (1993).

linate aminotransferase (EC 2.6.1.43),[11] 2-aminobutyrate aminotransferase,[12] and dimethylarginine–pyruvate aminotransferase.[13]

L-β-Aminoisobutyrate is formed from L-valine via L-β-hydroxyisobutyrate.[14] L-AIBAT is mainly distributed in the brain, liver, and kidney in the rat. L-AIBAT is identical to aminobutyrate aminotransferase (β-alanine–oxoglutarate aminotransferase, EC 2.6.1.19).[15]

Assay Method

Principle

The activities of D-AIBAT and L-AIBAT are assayed at 37° by measuring the rate of formation of malonate semialdehyde from β-alanine in the presence of pyruvate and 2-oxoglutarate, respectively.

Reagents

Sodium borate buffer (pH 8.8), 0.5 M
Pyruvate, 0.1 M: Neutralize with 1 M NaOH
2-Oxoglutarate, 0.1 M: Neutralize with 1 M NaOH
Pyridoxal 5′-phosphate, 0.05 M
2-Mercaptoethanol, 0.5 M: Prepare fresh each day
β-[2-^{14}C]Alanine, 0.01 M: Specific activity, 37 GBq/mol

Procedure

D-AIBAT. To assay D-AIBAT,[2] the preceding reagents and water are added to a final volume of 1.0 ml: 0.1 ml of sodium borate buffer, 0.1 ml of pyruvate, 0.01 ml of pyridoxal 5′-phosphate, 0.01 ml of 2-mercaptoethanol, an appropriate amount of enzyme, and 0.1 ml of β-[2-^{14}C]alanine. The incubation is carried out in a shaking water bath for 30 min at 37°. The reaction is terminated by the addition of 0.5 ml of 2 M HCl, and the tube is immediately transferred to an ice bath. After the addition 0.02 ml of 1 M β-alanine and 2 ml of 0.2% (w/v) 2,4-dinitrophenylhydrazine (in 2 M HCl), the mixture is allowed to stand for 15 min at 37° and the dinitrophenylhydrazone formed is extracted by shaking with 5.0 ml of toluene. After

[11] T. Noguchi and R. Mori, *J. Biol. Chem.* **256**, 10335 (1981).
[12] E. Okuno, Y. Minatogawa, and R. Kido, *Biochim. Biophys. Acta* **715**, 97 (1982).
[13] T. Ogawa, M. Kimoto, and K. Sasaoka, *J. Biol. Chem.* **265**, 20938 (1990).
[14] D. A. Wolf and H. A. Akers, *Trends Biochem. Sci.* **11**, 390 (1986).
[15] N. Tamaki, S. Fujimoto, C. Mizota, and M. Kikugawa, *Biochim. Biophys. Acta* **925**, 238 (1987).

brief centrifugation, the radioactivity is measured with a liquid scintillation spectrometer.

L-AIBAT. L-AIBAT is assayed according to the preceding method for D-AIBAT activity, except that 0.1 ml of 2-oxoglutarate is used as the amino acceptor.[4]

Definition of Unit and Specific Activity

One unit of enzyme is defined as the amount that catalyzes the conversion of 1 μmol of β-alanine to malonate semialdehyde per minute at 37°. Specific activity is expressed in units per milligram of protein. The protein concentration is measured by the dye-binding method with albumin as protein standard.

Purification of D- and L-AIBATs

Buffers

Buffer A: 10 mM Potassium phosphate (pH 7.5), 1 mM EDTA, 2 mM 2-mercaptoethanol, 40 μM pyridoxal 5'-phosphate

Buffer B: 10 mM Potassium phosphate (pH 7.0), 1 mM EDTA, 2 mM 2-mercaptoethanol

Buffer C: 100 mM potassium phosphate (pH 6.8), 1 mM EDTA, 2 mM 2-mercaptoethanol, 40 μM pyridoxal 5'-phosphate

Buffer D: 100 mM potassium phosphate (pH 7.5), 1 mM EDTA, 2 mM 2-mercaptoethanol, 40 μM pyridoxal 5'-phosphate

Buffer E: 100 mM potassium phosphate (pH 7.0), 1 mM EDTA, 2 mM 2-mercaptoethanol, 40 μM pyridoxal 5'-phosphate

Buffer F: 10 mM potassium phosphate (pH 7.5), 1 mM EDTA, 2 mM 2-mercaptoethanol

Buffer G: 10 mM potassium phosphate (pH 6.0), 1 mM EDTA, 2 mM 2-mercaptoethanol

Purification Procedure

D-AIBAT

The activity of D-AIBAT[2] in the livers of rats is found to be 11.3 ± 1.2 mU/g of tissue, which is approximately one-tenth that of L-AIBAT (133.7 ± 15.0 mU/g tissue). The procedure outlined here requires a total of 70 g of rat liver. All subsequent steps are performed at about 4°.

Step 1. Preparation of Crude Extract. Rat livers are homogenized in a Waring blender in 350 ml of buffer A. The homogenate is centrifuged at

27,000g for 20 min. The precipitate is homogenized again in 140 ml of buffer A. The homogenate is centrifuged at 27,000g for 20 min. The supernatant is added to the preceding supernatant.

Step 2. Heat Treatment. The crude extract is heated to 55° for 3 min, cooled to 4°, and then centrifuged at 27,000g for 20 min.

Step 3. Ammonium Sulfate Precipitation. Solid ammonium sulfate is added (313 g/liter) slowly with gentle stirring to the preceding supernatant. After being left standing for 30 min, the precipitate is removed by centrifugation for 20 min. Additional solid ammonium sulfate is then added (137 g/liter of supernatant solution), and the precipitate is collected as described above. The precipitate is dissolved in a minimum volume of buffer A and desalted by dialysis against the same buffer overnight at 4°.

Step 4. CM-Sepharose Chromatography. After centrifugation at 27,000g for 20 min, the enzyme solution is applied to a CM-Sepharose CL-6B column (2.3 × 25 cm, 50 ml/hr) equilibrated with buffer B. D-AIBAT is passed through the column with buffer B.

Step 5. DEAE-Sepharose Chromatography. The fraction containing D-AIBAT from the CM-Sepharose CL-6B column is also applied directly to a DEAE-Sepharose CL-6B column (2.0 × 25 cm, 50 ml/hr) equilibrated with buffer B. The column is washed with approximately 100 ml of buffer B and enzyme is then eluted with 400 ml of the same buffer containing KCl in a continuous gradient of 0 to 150 m*M*. One peak of D-AIBAT appears between 60 and 100 m*M* KCl.

Step 6. Hydroxyapatite Chromatography. The active fractions are combined and applied directly to a hydroxyapatite column (1.5 × 20 cm, 20 ml/hr) equilibrated with buffer C. The column is washed with approximately 30 ml of buffer C, and the enzyme is eluted with 400 ml of the same buffer containing potassium phosphate in a continuous gradient of 100 to 500 m*M*. The fractions containing the enzyme activity are pooled and concentrated to about 1.5 ml with a membrane filter (Centriflo CF 25; Amicon, Beverly, MA).

Step 7. Gel Filtration. The enzyme preparation obtained by the hydroxyapatite chromatography is applied to a Sephacryl S-200 column (1.2 × 170 cm, 7 ml/hr) equilibrated with buffer D. The active fractions are combined and concentrated to about 1 ml.

Step 8. Chromatofocusing. The enzyme solution from gel filtration is then applied to a column (0.9 × 24 cm, 20 ml/hr) of polybuffer exchanger equilibrated with 25 m*M* imidazole hydrochloride, pH 7.4. D-AIBAT is separated with Polybuffer 74 (diluted 1:8 from the stock solution) at pH 6.7. The active fractions are combined, concentrated and desalted by dialysis against buffer E, overnight. The purified enzyme is stored at 4°.

A summary of the D-AIBAT isolation procedure is shown in Table I.

TABLE I
PURIFICATION OF D-β-AMINOISOBUTYRATE AMINOTRANSFERASE FROM RAT LIVER

Step	Total activity (milli units)	Total protein (mg)	Specific activity (milli units/mg protein)	Purification (fold)	Yield (%)
Extract	1340	8700	0.154	1.0	100
Heat treatment	958	7300	0.131	0.9	71
(NH$_4$)$_2$SO$_4$ fractionation	381	906	0.421	2.7	28
CM chromatography	387	557	0.694	4.5	29
DEAE chromatography	342	45.9	7.47	48.5	26
Hydroxyapatite chromatography	190	5.84	32.6	211.7	14
Sephacryl S-200 chromatography	70.4	0.44	161.0	1043	5.3
Chromatofocusing	18.3	0.056	326.8	2125	1.4

L-AIBAT

The activity of L-AIBAT[4,16] in the livers of rats is found to be 133.7 ± 15.0 mU/g of tissue, and those in the brain and kidney are found to be 95.8 ± 8.7 and 47.2 ± 8.7 mU/g of tissue, respectively. The procedure outlined here requires a total of 70 g of rat liver. All steps are performed at about 4°.

Step 1. Preparation of Crude Extract. Rat livers are homogenized in a Waring blender in 350 ml of buffer A. The homogenate is centrifuged at 27,000g for 20 min. The precipitate is rehomogenized in 140 ml of buffer A. The homogenate is centrifuged at 27,000g for 20 min. The supernatant is added to the preceding supernatant.

Step 2. Heat Treatment. The extract is heated to 55° for 3 min, cooled to 4°, and then centrifuged at 27,000g for 20 min. The precipitate is discarded.

Step 3. First Ammonium Sulfate Precipitation. Solid ammonium sulfate is added (277 g/liter) slowly with gentle stirring to the supernatant from step 2. After being left standing for 30 min, the precipitate is removed by centrifugation for 20 min. Additional solid ammonium sulfate is added (99 g/liter of supernatant solution), and the precipitate is collected as described above. The precipitate is dissolved in 350 ml of buffer A. The first ammonium sulfate fractionation procedure is followed by a second one.

Step 4. Second Ammonium Sulfate Precipitation. Solid ammonium sulfate is added (299 g/liter of the preceding solution). After the precipitate is removed, additional solid ammonium sulfate is added (101 g/liter of supernatant solution), and the precipitate is collected as described above.

[16] N. Tamaki, H. Aoyama, K. Kubo, T. Ikeda, and T. Hama, *J. Biochem. (Tokyo)* **92**, 1009 (1982).

The precipitate is dissolved in a minimum volume of buffer A and desalted by dialysis against the same buffer overnight at 4°.

Step 5. DEAE-Sepharose Chromatography. After centrifugation at 27,000g for 20 min, the enzyme solution is applied to a DEAE-Sepharose CL-6B column (2.3 × 25 cm, 50 ml/hr) equilibrated with buffer F. The column is washed with approximately 70 ml of buffer F and eluted with a linear gradient made with the equilibrating buffer (buffer F, 200 ml) and the same volume of buffer F containing 0.15 mM KCl. The active fractions are combined, and the enzyme protein is precipitated by the addition of solid ammonium sulfate. The dissolved pellet is dialyzed for 5 hr against buffer C.

Step 6. Hydroxyapatite Chromatography. The enzyme solution from step 4 is applied to a hydroxyapatite column (1.5 × 20 cm, 20 ml/hr) equilibrated with buffer C. The column is washed with approximately 30 ml of buffer C and eluted with 400 ml of the same buffer containing potassium phosphate in a continuous gradient of 100 to 500 mM. The fractions containing the enzyme activity are pooled and concentrated to about 1 ml.

Step 7. Gel Filtration. The enzyme preparation obtained by hydroxyapatite chromatography is applied to a Sephacryl S-200 column (1.2 × 170 cm, 7 ml/hr) equilibrated with buffer D. The active fractions are combined and concentrated to about 1 ml.

Step 8. CM-Sepharose Chromatography. The enzyme solution from gel filtration is then applied to a CM-Sepharose CL-6B column (1.0 × 10 cm) equilibrated with buffer G. The enzyme is eluted from the column with buffer G and concentrated and stored at 4° in buffer A.

A summary of the L-AIBAT isolation procedure is shown in Table II.

TABLE II

PURIFICATION OF L-β-AMINOISOBUTYRATE AMINOTRANSFERASE FROM RAT LIVER

Step	Total activity (milli units)	Total protein (mg)	Specific activity (milli units/mg protein)	Purification (fold)	Yield (%)
Extract	7198	9399	0.766	1.0	100
Heat treatment	4784	5324	0.899	1.2	66
First (NH$_4$)$_2$SO$_4$ fractionation	2994	684	4.38	5.7	42
Second (NH$_4$)$_2$SO$_4$ fractionation	1858	538	3.45	4.5	26
DEAE chromatography	2049	74.1	27.7	36.1	28
Hydroxyapatite chromatography	1544	6.23	246.1	321.3	21
Sephacryl S-200 chromatography	769	2.03	378.2	493.7	11
CM chromatography	560	1.37	408.4	533.1	7.8

Properties

D-AIBAT

See Ref. 2 for more details.

Stability. When the purified enzyme solution (50 μg/ml in buffer E) is stored at about 4°, a 50% decrease in the specific activity is observed within a few days and an 80% decrease is seen within 1 week.

Molecular Properties. On gel filtration with Sephacryl S-300, an enzyme activity peak is found at the same position as the protein peak, and no other peaks are found. From the reference proteins, the molecular mass of the purified enzyme is calculated to be 220 kDa. The subunit molecular mass of D-AIBAT is calculated to be 52 kDa by sodium dodecyl sulfate–polyacrylamide gel electrophoresis (SDS–PAGE). Cross-linking patterns obtained with D-AIBAT after treatment with dimethyl suberimidate or dimethyl adipimidate result only in the appearance of two main protein bands by SDS–PAGE: a monomer of 52 kDa and a dimer of 104 kDa. A concentration of dimethyl suberimidate higher than 70 mg/mg of protein and cross-linking reaction times longer than 1 hr reveal no association greater than dimer. The purified D-AIBAT from rat liver may be organized as a tetramer composed of two dimers.

Ultraviolet and Visible Light Absorption Spectra. At pH 7.0, the purified enzyme shows an absorption spectrum with peaks at 281 nm ($A_{0.1\%} = 1.08$) and 412 nm ($A_{0.1\%} = 0.19$) and a shoulder at 330 nm ($A_{0.1\%} = 0.17$).

pH Optimum and Isoelectric Point. Under standard assay conditions, the pH optimum for D-AIBAT is pH 9.5.

An analysis of chromatofocusing of the purified enzyme shows an active peak at the same position as the protein peak at pH 6.7.

Substrate Specificity. Table III shows the relative activity of the enzyme with different amino donors or amino acceptors, compared with those of L-AIBAT. When pyruvate is employed as an amino acceptor, D-β-aminoisobutyrate and β-alanine serve as good substrates, but the L-β-aminioisobutyrate is inert. γ-Aminobutyrate is slightly transaminated. It is found that among the β-alanine analogs and derivatives, δ-aminovalerate, ε-aminocaproate, taurine, anserine, carnosine, spermine, spermidine, putrescine, and propylamine cannot act as substrates for D-AIBAT. The enzyme is not active with the following L-amino acids and these amino acids do not inhibit the enzyme activity up to 2 mM: glycine, histidine, serine, and glutamate.

When β-alanine is employed as an amino donor, pyruvate, glyoxylate, and oxaloacetate are good amino acceptors, while 2-oxoglutarate is a poor substrate. Phenylpyruvate is inert as an amino acceptor.

K_m *Values.* The K_m values of amino donors at pH 7.3 are as follows:

TABLE III
Amino Donor and Amino Acceptor Specificity of D-β-Aminoisobutyrate
Aminotransferase and L-β-Aminoisobutyrate Aminotransferase from Rat Liver

| | Relative enzyme activity (%) | |
Group	D-β-Aminoisobutyrate aminotransferase	L-β-Aminoisobutyrate aminotransferase
Amino donor		
D-β-Aminoisobutyrate	100	1
L-β-Aminoisobutyrate	0	100
DL-β-Aminoisobutyrate	78	74
β-Alanine	60	154
γ-Aminobutyrate	1	154
δ-Aminovalerate	0	146
ε-Aminocaproate	0	25
Ornithine, spermine, spermidine, putrescine	0	0
Amino acceptor		
Glyoxylate	89	2
Pyruvate	100	1
Oxalacetate	63	1
2-Oxoglutarate	3	100
Phenylpyruvate	0	1

D-β-aminoisobutyrate, 0.12 mM; β-alanine, 1.4 mM; L-alanine, 2.2 mM; $N^G,N^{G'}$-dimethyl-L-arginine, 6.4 mM; L-α-aminobutyrate, 11.3 mM; δ-aminolevulinate, 2.1 mM. Among these amino acids, D-β-aminoisobutyrate has the highest affinity for D-AIBAT.

Inhibitors. D-AIBAT is competitively inhibited by pyrimidine derivatives against β-alanine.[17] The K_i values are as follows: 5-fluorouracil, 56 μM; 6-azauracil, 0.3 mM; α-fluoro-β-alanine, 8.0 mM. γ-Aminobutyrate is neither a substrate nor an inhibitor of D-AIBAT. However, gabaculine (5-amino-1,3-cyclohexadienylcarboxylate), an analog of γ-aminobutyrate, is a potent irreversible inhibitor of D-AIBAT, with a K_i of 8.3 μM.[18]

The D-β-aminoisobutyrate and β-alanine contents are increased 22- and 61-fold, respectively, in rat liver by a subcutaneous injection of 6-azauracil.[19] The incorporation of [*methyl*-14C]thymine into β-aminoisobutyrate is increased to 42-fold by 6-azauracil treatment.

[17] M. Kaneko, Y. Kontani, M. Kikugawa, and N. Tamaki, *Biochim. Biophys. Acta* **1112,** 45 (1992).
[18] M. Kaneko, S. Fujimoto, M. Kikugawa, and N. Tamaki, *FEBS Lett.* **276,** 115 (1990).
[19] N. Tamaki, S. Fujimoto, N. Mizutani, and C. Mizota, *FEBS Lett.* **191,** 113 (1985).

Immunological Properties. Antiserum to rat liver D-AIBAT produces a single connecting band of precipitin between purified D-AIBAT and crude extract from rat liver, but not with purified L-AIBAT by Ouchterlony double diffusion. The antiserum inhibits alanine–glyoxylate aminotransferase activity in rat liver in the same way as it inhibits that of D-AIBAT.

L-AIBAT

See Ref. 4 for more details.

Stability. It is found that the enzyme preparation can be stored in buffer E (1 mg/ml) at 4° for 1 week without loss of activity. When the enzyme is stored at −25°, a 50% decrease in the specific activity is observed within a few days.

Molecular Properties. On Sephacryl S-200 column chromatography, the molecular mass of purified L-AIBAT is found to be 105 kDa. The subunit molecular mass of purified enzyme is 56 kDa on SDS–PAGE. It is suggested from these results that the enzyme consists of two subunits of identical molecular weight.

Ultraviolet and Visible Light Absorption Spectra. At pH 7.0, the purified enzyme shows an absorption spectrum with peaks at 282 nm ($A_{0.1\%} = 1.22$), 330 nm ($A_{0.1\%} = 0.20$), and 414 ($A_{0.1\%} = 0.19$).

pH Optimum and Isoelectric Point. Under standard conditions, the pH optimum of L-AIBAT is pH 9.0–9.2.

An analysis of chromatofocusing of the enzyme shows an active peak at the same position as the protein peak at pH 6.0.

Substrate Specificity. The substrate specificity of L-AIBAT is different from that of D-AIBAT. The relative activity of the enzyme with different amino donors or amino acceptors is shown in Table III. L-β-Aminoisobutyrate, β-alanine, γ-aminobutyrate, and δ-aminovalerate serves as good amino donors, but D-β-aminoisobutyrate is inert and does not affect the enzyme activity, at least up to 10 mM. It is found that the β-alanine analogs and derivatives, taurine, anserine, carnosine, spermine, spermidine, and putrescine cannot act as a substrate for L-AIBAT. The enzyme is not active with the following L-amino acids, and these amino acids do not inhibit the enzyme activity up to 10 mM: glycine, alanine, ornithine, histidine, valine, leucine, isoleucine, aspartate, tyrosine, lysine, and tryptophan.

In contrast, 2-oxoglutarate is a specific amino acceptor.

K_m Values. The apparent K_m values for L-β-aminoisobutyrate and β-alanine are 2.7 and 1.1 mM, respectively, and the relative maximum velocity for L-β-aminoisobutyrate is about 80% of that for β-alanine.

Inhibitors. Among uracil derivatives, 6-azauracil, 6-azathymine, and 5-iodouracil are found to be potent inhibitors of purified L-AIBAT, while

6-azauridine and 6-azauridine 5'-phosphate are not. 6-Azauracil acts as a noncompetitive inhibitor with respect to β-alanine as well as 2-oxoglutarate, and has a K_i of 0.7 mM. 5-Fluorouracil acts as a competitive inhibitor of the enzyme with respect to β-alanine, with a K_i value of 1.9 mM, and is uncompetitive against 2-oxoglutarate, with a K_i of 1.8 mM.[17] Gabaculine acts as a potent irreversible inhibitor of L-AIBAT, with a K_i of 7.1 μM.[18]

Ethanol in the presence of disulfiram (N,N,N',N'-tetraethylthiuram disulfide, an inhibitor of aldehyde dehydrogenase) inhibits rat liver L-AIBAT activity in growing rats *in vivo*.[20] The L-AIBAT activity is reduced with a pseudo-first-order profile with time, and the half-life is calculated to be 12.3 ± 0.8 hr. The synthesis of L-AIBAT in rat liver is estimated to be 1.56×10^{-10} mol/g of wet tissue per hour at a steady state.

Immunological Properties. Antiserum to rat liver purified L-AIBAT gives precipitation lines against rat liver homogenate and rat brain homogenate. The antiserum inhibits L-AIBAT activity in rat liver in the same way it inhibits β-alanine–oxoglutarate aminotransferase activity.[15]

Interconversion between D- and L-Enantiomers
of β-Aminoisobutyrate

Both D- and L-enantiomers of β-aminoisobutyrate[21] occur in human urine and plasma.[22] Urinary β-aminoisobutyrate is present mostly in the D configuration, but the ratio of D-β-aminoisobutyrate to L-β-aminoisobutyrate in the urine of both tumor patients and patients without tumors is found to be nearly constant for the whole range of concentrations investigated.[23] These results suggest the interconversion of each enantiomer of β-aminoisobutyrate. The interconversion between D- and L-enantiomers of β-aminoisobutyrate is found in the presence of both aminoisobutyrate aminotransferases (Fig. 1).

Conversion from D- to L-β-Aminoisobutyrate

The assay mixture is as follows. The standard reaction mixture contains 50 mM triethanolamine (pH 7.3), 5 mM 2-mercaptoethanol, 1 mM D-β-aminoisobutyrate, 1 mM pyruvate, 10 mM L-glutamate, 0.5 mM pyridoxal

[20] Y. Kontani, S. Kawasaki, M. Kaneko, K. Matsuda, S. Fujimoto Sakata, and N. Tamaki, *J. Nutr. Sci. Vitaminol.* **44,** 165 (1998).

[21] N. Tamaki, M. Kaneko, M. Kikugawa, and S. Fujimoto, *Biochim. Biophys. Acta* **1035,** 117 (1990).

[22] E. Solem, E. Jellum, and L. Eldjarn, *Clin. Chim. Acta* **50,** 393 (1974).

[23] A. H. van Gennip, J. P. Kamerling, P. K. De Bree, and S. K. Wadman, *Clin. Chim. Acta* **116,** 261 (1981).

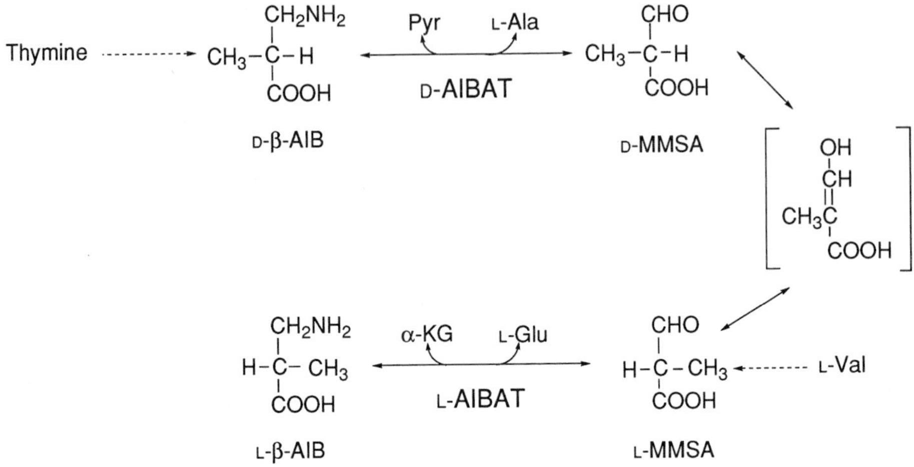

Thymine ---------▶ $\underset{\underset{\text{COOH}}{|}}{\overset{\overset{\text{CH}_2\text{NH}_2}{|}}{\text{CH}_3-\text{C}-\text{H}}}$ ◀—$\overset{\text{Pyr} \quad \text{L-Ala}}{\frown}$— $\underset{\underset{\text{COOH}}{|}}{\overset{\overset{\text{CHO}}{|}}{\text{CH}_3-\text{C}-\text{H}}}$

D-β-AIB   D-AIBAT   D-MMSA

$$\left[\ \underset{\underset{\text{COOH}}{|}}{\overset{\overset{\overset{\text{OH}}{|}}{\overset{\text{CH}}{\|}}}{\text{CH}_3\text{C}}}\ \right]$$

$\underset{\underset{\text{COOH}}{|}}{\overset{\overset{\text{CH}_2\text{NH}_2}{|}}{\text{H}-\text{C}-\text{CH}_3}}$ ◀—$\overset{\alpha\text{-KG} \quad \text{L-Glu}}{\frown}$— $\underset{\underset{\text{COOH}}{|}}{\overset{\overset{\text{CHO}}{|}}{\text{H}-\text{C}-\text{CH}_3}}$ ◀------- L-Val

L-β-AIB   L-AIBAT   L-MMSA

FIG. 1. Interconversion of D- and L-β-aminoisobutyrate. Pyr, Pyruvate; α-KG, 2-oxoglutarate; β-AIB, β-aminoisobutyrate; and MMSA, methylmalonate semialdehyde.

5′-phosphate, D-AIBAT (24 μg), and L-AIBAT (73 μg) in a final volume of 1.0 ml. The incubation is carried out in a shaking water bath for 3 hr at 37°. The reaction is terminated by the addition of 0.1 ml of 6 M perchloric acid, and the tube is immediately transferred to an ice bath. After the addition of 41.4 mg of K_2CO_3, the mixture is allowed to stand for 30 min. After centrifugation at 3000 rpm for 10 min, the supernatant is transferred onto a Sep-Pak C_{18} column. Amino acids are passed through the column with H_2O. The eluate is dried *in vacuo* and used as a sample for the amino acid analysis.

The analysis of D- and L-isomers of β-aminoisobutyrate is achieved by high-performance liquid chromatography (HPLC). The sample for amino acid analysis is dissolved in 2 ml of 1.7 M 2-propano l-HCl and heated for 60 min at 60°. The mixture is evaporated *in vacuo,* and dissolved in 2.5 ml of tetrahydrofuran. After the addition of 10 mg of 3,5-dinitrobenzoyl chloride and 25 μl of triethylamine, the solution is heated for 15 min at 60° in an oil bath. The solvent is evaporated *in vacuo* and the residue is dissolved in 1 ml of chloroform before being applied to a Sep-Pak silica column. N-3,5,-Dinitrobenzoylamino acid isopropyl esters are eluted with 2.5 ml of chloroform. The enantiomers of the esters are separated at room temperature by an HPLC system with a chiral resin (Sumipax OA-4100, Sumitomo Chemicals, Osaka).

By the application of HPLC, L-β-aminoisobutyrate, as well as L-alanine, are formed in the complete assay system. The ratio of the L to D form is

0.73. However, in the absence of L-AIBAT, L-alanine is found, while L-β-aminoisobutyrate is not. This suggests a lack of further reaction from D-methylmalonate semialdehyde to L-β-aminoisobutyrate. In the absence of L-glutamate in the assay mixture, L-β-aminoisobutyrate is not detected.

The hydrogen at the 2-position of methylmalonate semialdehyde is easily ionized by the facile enolization, and a rapid racemization is followed by the interconversion of the enantiomer of β-aminoisobutyrate (Fig. 1).

Conversion from L- to D-β-Aminoisobutyrate

The assay mixture for the conversion from L- to D-β-aminoisobutyrate contains 1 mM L-β-aminoisobutyrate, 1 mM 2-oxoglutarate, and 10 mM L-alanine instead of D-β-aminoisobutyrate, pyruvate, and L-glutamate, respectively, in the preceding reaction mixture.

Enzymatic reverse conversion from L- to D-β-aminoisobutyrate is also found in the presence of 2-oxoglutarate, L-alanine, L-AIBAT, and D-AIBAT. The ratio of the D to L form is 0.74. In the absence of D-AIBAT, only L-glutamate appears; D-β-aminoisobutyrate is not produced.

Cloning and Sequencing of Rat D-AIBAT and L-AIBAT

D-AIBAT[20] is purified from rat liver, and the N-terminal sequence is LHTKHNMPPCDFSPFKYQ-, showing identity with that of rat kidney AGT 2.[24] In the present study, based on the nucleotide sequence of rat AGT 2, two oligonucleotides (P1, 5'-CACCTTTCTCCTGTTGAATA-CAGCC-3'; and P2, 5'-GCATTCCTGACTGTTTGCACC-3') are synthesized. Complementary DNA (539 bp) is amplified with P1 and P2 primers from rat kidney cDNA. The cDNA encoding D-AIBAT is screened from a rat liver cDNA library with the preceding probes. The longest clone, 2172 bp, is sequenced. The nucleotide sequence data have been submitted to the DDJB/EMBL/GenBank DNA databases with the accession number AB002584 for D-AIBAT. The sequence includes a 1539-bp protein-coding region including the stop codon. The cDNA encoding mature D-AIBAT contains 1419 nucleotides for 473 amino acids with a molecular mass of 52,645 Da, which accords well with the subunit of the enzyme. A sequence of 117 nucleotides encoding 39 amino acids upstream from the amino-terminal leucine of the mature enzyme is the signal peptide of D-AIBAT (Fig. 2).

L-AIBAT[20] is purified from rat liver. The N-terminal sequence is

[24] I. S. Matsui Lee, Y. Muragai, T. Ideguchi, T. Hase, M. Tsuji, A. Ooshima, E. Okuno, and R. Kido, *J. Biochem. (Tokyo)* **117**, 856 (1995).

```
                                           ⇓
D-AIBAT   MSLAWRTLQKAFYLETSLRILQMRPSLSCASRIYVPKLTLHTKHNMPPCDFSPEKYQSLA    60

L-AIBAT        MAFLLTTRRLVCSSQKNLHLFTPGSRYISQAAAKVDFEFDYDGPLMKTEVP         51
                                            ⇑


D-AIBAT   YNHVLEIHKQHLSPVNTAYFQKPLLLHQGHMEWLFDSEGNRYLDFFSGIVTVGVGHCHPK   120
                          ....  ..* * .*** **..* * .* .*. **
L-AIBAT   GPRSQELMKQLNTIQNAEAVHFFCNYEESRGNYLVDVDGNRMLDLYSQISSVPIGYNHPA   111


D-AIBAT   VTA-VAKKQMDRLWHTSSVFFHSPMHEYAERL-SALL---PEPL-KVIFLV--NSGSEAN   172
          .. *.. *  .. .   .....    *  ......*  ..*.   *. ...* ..  ....*..
L-AIBAT   LAKLVQQPQNASTFINRPALGILPPENFVDKLRESLMSVAPKGMCQLITMACGSCSNENA   171


D-AIBAT   DLAMVM--------ARAYSNHTDIISFRGAYHGCSPYTLGLTNVGIYKMKVPSTIACQST   224
          .. *       .*..*.  .   .. ..**. *.. *. .*  ....*
L-AIBAT   FKTIFMWYRSKERGQRGFSKEELETCMVNQSPGCPDYSI-LSFMGAFHGRTMGCLATTHS   230


D-AIBAT   MCPDVFRGPWGGSHCRDSPVQTVRKCSCAPDGCQAKERYIEQFKD--TLNTSVATSIAGF   282
          . . . *  . . *   . . ...*. *...*..*. .* . . ..**.
L-AIBAT   KAIHKIDIPSFDWPIAPFPRLKYPLEEFVTDNQQEEARCLEEVEDLIVKYRKKKRTVAGI   290


D-AIBAT   FAEPIQGVNGVVQYPKEFLKEAFALVRERGGVCIADEVQTGFGRLGSHFWGFQTHDTMPD   342
          ..****. .* . ...*... ...*.* . ..****** * * .**. ..* .. *
L-AIBAT   IVEPIQSEGGDNHASDDFFRKLRDIARKHGCAFLVDEVQTGGGCTG-KFWA-HEHWGLDD   348


D-AIBAT   IVTMAKGIGNGFPMAAVVTTPEIASSLAKHLHHFSTFGGSPLACAIGSAVLEVIEEENLQ   402
          ... ..... .. ..    . *. .*  .. *.*. *.*   . ..*...*. *.*
L-AIBAT   PADV-MSFSKKMMTGGFFHKEEFRPSAPYRI--FNTWLGDPSKNLLLAEVINIIKREDLL   405


D-AIBAT   RNSQEVGTYMLLKFAKLRDEF-DIVGDVRGKGLMVGIEMVQDKI-SRQPLPKTEVNQIHE   460
          .* ...*. .* . .*..... ..*. ***.* . ...  . * .  * . . . .
L-AIBAT   NNVAHAGKTLLTGLLDLQAQYPQFVSRVRGRGTFCSFDTPDKAIRNKLILIARNKGVVLG   465


D-AIBAT   DCKDMGLLVGRGGNFSQTFRIAPPMRVTKLEVDFAFEVFRSALTQHMERRAK          512
          .* *.. .
L-AIBAT   GCGDKSIRFRPTLVFRDHHAHLFLNIFSGILADFK                         500
```

Fig. 2. Alignment of the amino acid sequences of D- and L-AIBAT. Asterisks and dots indicate a perfect match and a conservative replacement, respectively. Arrows indicate a processing site.

VDFEFDYDGPLMKTEVPGPR-, similar to those of pig[25] and human[26] γ-aminobutyrate (GABA) aminotransferase, except that Val-3 is substituted by Phe-3. In the present study, based on the nucleotide sequence of the pig GABA aminotransferase, two oligonucleotides are synthesized. The oligonucleotide sequences of the forward primer (P3) and reverse primer (P4) are 5'-ATGGTGGAGCCCATCCAGTCTGAGG-3' and 5'-CCTCCTCCTGTCTGGACCTCATC-3', respectively. The polymerase chain reaction (PCR) products (531 bp) amplified from the pig liver cDNA library with P3 and P4 primers are used as probes for L-AIBAT cDNA cloning. The cDNA encoding L-AIBAT is screened from a rat liver cDNA library. The longest clone, 1740 bp, is submitted to the DDBJ/EMBL/ GenBank DNA databases with the accession number D87839. The sequence includes a 1503-bp protein-coding region including the stop codon. The cDNA encoding mature L-AIBAT contains 1398 nucleotides for 466 amino acids, suggesting a molecular mass of 52,662 Da, which compares well with the subunits of the enzyme. A sequence of 102 nucleotides encoding 34 amino acids, upstream from the amino-terminal valine of the mature form of the enzyme, is the signal peptide of the precursor of L-AIBAT.

A sequence comparison between D-AIBAT and L-AIBAT is presented in Fig. 2. There is a fair degree of amino acid identity over the entire coding region, with 20.4% amino acid sequence identity. The leader peptide of D-AIBAT and L-AIBAT shows 26.7% amino acid sequence identity. A four-amino acid stretch of the sequence as -EPIQ- at positions 285–288 of D-AIBAT is a common sequence and is also found in ornithine δ-aminotransferase and lysine ε-aminotransferase.

[25] D. De Biase, B. Maras, F. Bossa, D. Barra, and R. A. John, Eur. J. Biochem. 208, 351 (1992).
[26] D. De Biase, D. Barra, M. Simmaco, R. A. John, and F. Bossa, Eur. J. Biochem. 227, 476 (1995).

[36] Branched-Chain Keto Acid Dehydrogenase of Yeast

By J. Richard Dickinson

Introduction

The precise physiological role of branched-chain α-keto acid dehydrogenase in the catabolism of branched-chain amino acids in the budding yeast *Saccharomyces cerevisiae* is unclear,[1] but its presence has never been in

[1] J. R. Dickinson, *Methods Enzymol* 324, Chap. 9, 2000 (this volume).

doubt since it was first demonstrated in homogenates.[2] The activity of the enzyme varies enormously from one strain to another,[2,3] hence when contemplating the study of this enzyme the investigator should ascertain that the strain has respectable levels of activity before commencing. The reason(s) for this variability is not known at this time, but, as with the ability to grow with leucine, valine, or isoleucine as sole nitrogen source, there are strains with either suitable or unsuitable genetic backgrounds. Since the entire genome of yeast has been sequenced, one would have expected database searches to reveal sequences that show homology to the known E1α, E1β, and E2 of branched-chain α-keto acid dehydrogenases from other organisms. This is not the case, and thus the genes that encode these functions in yeast remain unknown. Hence, it is not possible at this time to ascribe the presence or absence of branched-chain α-keto acid dehydrogenase activity to a specific genetic constitution. The screening of mutants that were defective in branched-chain amino acid catabolism did not reveal any with defects in branched-chain α-keto acid dehydrogenase.[4] Thus, apart from a lack of enzyme activity, a mutant specifically defective in branched-chain α-keto acid dehydrogenase presumably has a subtle phenotype. Knowledge of the molecular genetics of the E3 component, lipoamide dehydrogenase, is much more clear. There is one structural gene, *LPD1*, which encodes lipoamide dehydrogenase for pyruvate dehydrogenase, α-ketoglutarate dehydrogenase, and glycine decarboxylase as well as α-keto acid dehydrogenase. Mutation of the *LPD1* gene results in a loss of activity of all four complexes.[3,5–7] However, *lpd1* mutants can grow with a branched-chain amino acid as sole source of nitrogen,[8,9] confirming the idea that a mutant specifically lacking α-keto acid dehydrogenase will have a subtle phenotype.

Materials and Reagents

Acid-Washed Glass Beads

Glass beads (40-mesh), which are readily available from any chemical supplier as chromatography support material, should be prepared as fol-

[2] J. R. Dickinson and I. W. Dawes, *J. Gen. Microbiol.* **138,** 2029 (1992).

[3] M. M. Lanterman, J. R. Dickinson, and D. J. Danner, *Hum. Mol. Genet.* **5,** 1643 (1996).

[4] J. R. Dickinson and V. Norte, *FEBS Lett.* **326,** 29 (1993).

[5] J. R. Dickinson, D. J. Roy, and I. W. Dawes, *Mol. Gen. Genet.* **204,** 103 (1986).

[6] B. Repetto and A. Tzagoloff, *Mol. Cell. Biol.* **11,** 3931 (1991).

[7] D. A. Sinclair and I. W. Dawes, *Genetics* **140,** 1213 (1995).

[8] J. R. Dickinson, M. M. Lanterman, D. J. Danner, B. M. Pearson, P. Sanz, S. J. Harrison, and M. J. E. Hewlins, *J. Biol. Chem.* **272,** 26871 (1997).

[9] J. R. Dickinson, S. J. Harrison, and M. J. E. Hewlins, *J. Biol. Chem.* **273,** 25751 (1998).

lows. Place the beads into a large glass beaker and transfer to a fume cupboard. Carefully add sufficient concentrated nitric acid to cover all the glass beads, place a lid over the beaker, and leave it overnight. The next day, pour away as much as possible of the concentrated nitric acid and transfer the glass beads to a large conical flask, e.g., 2 or 5 liter. It will probably be necessary to use some tap water to "wash" the remaining beads that have stuck to the sides of the beaker. The beads must now be washed with copious amounts of tap water to remove all traces of acid. This is best done by approximately half-filling the large flask with water, and vigorously swirling and decanting the water. The beads, being considerably denser than water, quickly settle to the bottom of the flask. Repeat this procedure several times until pH test paper indicates that all the nitric acid has been removed. Wash the beads two or three times with distilled water to remove salts and ions present in the tap water. Decant the beads into a shallow container and place it in a hot oven overnight to allow it to dry. Usually drying leads to the formation of a "crust" on the uppermost beads; this can be easily broken up with a spatula. The dried beads can now be transferred to a screw-top jar and kept until required. The purpose of the washing procedure, which is recommended for all enzyme preparations from yeast, is to remove any contaminants that might act as inhibitors.

Buffers

> Buffer A: 50 mM potassium phosphate buffer, pH 7.4, containing 2 mM EDTA and 2 mM 2-mercaptoethanol
> Buffer B: 50 mM potassium phosphate buffer, pH 7.4, containing 0.1 mM EDTA, 0.2 mM thiamine pyrophosphate, and 0.2 mM phenyl-methylsulfonyl fluoride (PMSF)

Assay Method

Principle

The total activity of the branched-chain α-keto acid dehydrogenase complex can be assayed by monitoring the reduction of NAD$^+$ or the substrate analog acetylpyridine adenine dinucleotide at 340 or 366 nm (respectively), using a recording spectrophotometer.

Assay Reagents

> α-Ketoisovaleric acid, sodium salt (Sigma, St. Louis, MO), 200 mM
> Cysteine hydrochloride, 47 mg in 10 ml
> Coenzyme A, 2.6 mg in 1.5 ml

Thiamine pyrophosphate, 46.9 mg in 5 ml
3-Acetylpyridine adenine dinucleotide (APAD$^+$; Sigma), 22.8 mg in 2
  ml or NAD$^+$, 10 mg/ml
Sodium azide, 2% (w/v)

The substrates and cofactors are made up fresh each day in distilled water and stored in ice. The sodium azide is prepared in batches of 100 ml and stored at 4° between experiments.

Procedure

The assay mixture contains 600 μl of buffer A, 100 μl of cysteine hydrochloride, 100 μl of sodium azide, 40 μl of coenzyme A, 10 μl of thiamine pyrophosphate, 50 μl NAD$^+$ or APAD$^+$ and 100 μl of homogenate or purified enzyme preparation (diluted as appropriate with buffer A) in a 1.5-ml cuvette. The mixture is allowed to equilibrate to 30° for 3 min then the reaction is initiated by the addition of 50 μl of α-ketoisovaleric acid. The reaction blank has identical contents except that 50 μl of buffer A is added instead of α-ketoisovaleric acid. The sodium azide is included, and APAD$^+$ is used instead of NAD$^+$, to prevent NADH reoxidation in crude homogenates. With pure enzyme preparations sodium azide and the more expensive APAD$^+$ do not need to be employed.

Growth and Handling of Yeast

Branched-chain α-keto acid dehydrogenase can be assayed in crude extracts, isolated mitochondria, or in various stages of purification. The nature of the preparation that is to be assayed will determine the amount of yeast to be used. This in turn, will determine the scale and complexity of the yeast handling involved. To assay a homogenate 10^9 cells should be sufficient. This corresponds to 10 ml of a suitable laboratory strain grown to an optical density of 10. For a full-scale purification it is simpler to use fresh compressed baker's yeast, which can be purchased cheaply in bulk from a bakery. Before commencing a full-scale purification it is worthwhile to assay a crude homogenate of a small sample of the baker's yeast to verify that activity can be detected. If this proves negative then an alternative source must be sought. Often large bakeries purchase baker's yeast from more than one supplier, so it is worth enquiring whether "a different company's yeast" is available. A small bakery will likely use a single supplier, but they will probably tell you the source (if this is not apparent from the wrapper). Because of the enormous variability in enzyme levels in different yeast strains it is important to note the name of the producer.

(The author knew one unwary student who lost valuable time due to not checking this detail and subsequently learning that the bakery had changed their supplier of yeast between experiments.)

The growth medium in which the yeast has been cultivated has a profound effect on the activity of branched-chain α-keto acid dehydrogenase. Maximal activity is observed in cells that have been in YEP–glycerol medium[2] [yeast extract, 1% (w/v); bacteriological peptone, 2% (w/v); glycerol, 3% (v/v)]. Not all strains will grow with glycerol and baker's yeast will certainly not have been produced with glycerol as carbon source. Nevertheless, it has been observed that the specific activity of branched-chain α-keto acid dehydrogenase can be boosted significantly by incubating the yeast in this medium for 12 hr prior to harvesting and breaking of the cells. For compressed baker's yeast, 10 g of compressed yeast can be suspended in 100–150 ml of YEP–glycerol medium. Benzylpenicillin (100 μg/ml, final concentration) can be added to the medium to prevent the growth of unwanted bacteria if one harbors doubts about the microbiological integrity of the surface of the block of compressed baker's yeast.

Harvesting and Breaking Yeast

Irrespective of whether a laboratory strain of yeast has been grown in an experimental medium of choice or a mass of commercial baker's yeast is being used, the first step is to harvest the yeast cells. This can be done by centrifugation or filtration. The cells are then washed either in sterile distilled water or buffer A. If a homogenate is all that is required, the cells can be resuspended in buffer A. If a more elaborate preparation is required then buffer B should be used. Phenylmethylsulfonyl fluoride is included in buffer B to counter proteolysis and the thiamine pyrophosphate is included for enzyme protection. All subsequent steps take place at 0–4°. Next, the yeast cells are broken. This is most conveniently done with a Braun (Melsungen, Germany) MSK homogenizer and acid-washed glass beads. Equal volumes of glass beads and cell suspension should be shaken for 35 sec at maximum speed. The flask of the Braun homogenizer should be precooled in ice and/or the whole Braun homogenizer can be placed in a cold room at 4°. The cell homogenate must now be separated from the glass beads. The easiest way to achieve quantitative recovery of homogenate is to use a homemade sieve tube. The author has two sizes and uses whichever is appropriate to the volume of homogenate to be separated. The smaller is made from a standard plastic scintillation vial "insert," and the larger is made from the barrel of a plastic 60-ml syringe. In both a hot needle has been used to cut a roughly circular hole in the base, leaving a 1-

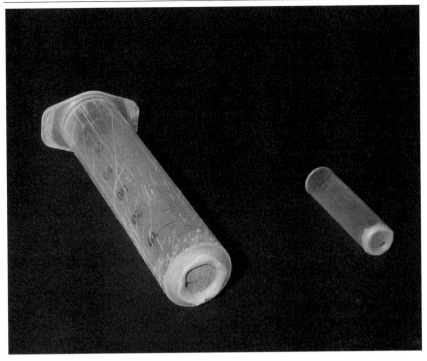

FIG. 1. Sieve tubes for the separation of glass beads from cell homogenate.

to 2-mm rim. Covering the hole is a small piece of wire gauze that has been glued in place with quick-setting epoxy resin such as Rapid Araldite (CIBA-GEIGY Plc., Stafford, UK) (see Fig. 1). A convenient source of the wire gauze is the sachet that holds the catalyst in gas generator kits for microbiologists' anaerobic jars (e.g., Oxoid gas-generating kit; Unipath Limited, Basingstoke, UK). Filtration through the sieve can be accelerated by centrifugation at low speed for about 30 sec. One of the smaller sieve tubes will sit securely in a 10-ml bench centrifuge tube; the larger version can be housed in a Sorvall (Newtown, CT) 250-ml centrifuge bottle by boring a hole in the lid. For small-scale work the crude homogenate can now be transferred to a 1.5-ml microcentrifuge (Eppendorf) tube and centrifuged at 12,000g for 15 sec. The resulting supernatant can be used immediately to initiate an assay. For larger scale work, including full purification, the homogenate should be centrifuged for 15 min to pellet large cell debris.

Purification of Yeast Branched-Chain α-Keto Acid Dehydrogenase Complex

By the dropwise addition of a 50% (w/v) stock solution, the 12,000g supernatant is made 2% (w/v) in polyethylene glycol 8000 and the mixture stirred for 10 min. After centrifugation at 12,000g, the supernatant is made 5% (w/v) in polyethylene glycol and centrifuged again. This step precipitates the pyruvate dehydrogenase and α-ketoglutarate dehydrogenase complexes. The supernatant is then slowly adjusted to 9% (w/v) in polyethylene glycol and recentrifuged as before. The resulting pellet is then resuspended in buffer B and applied to a Sephacryl S-200 column (400 × 15 mm) equilibrated with the same buffer. The branched-chain α-keto acid dehydrogenase complex elutes in peak III (Fig. 2). The appropriate fractions are then combined and concentrated with a Centricon-100 microconcentrator (Amicon, Beverly, MA) with a relative molecular weight cutoff of 100,000 at 1000g for 1 hr. There seems to be no other way to accomplish this step, because centrifugation at 170,000g results in loss of activity. The resulting solution, now occupying less than 300 μl, is applied to a DEAE-cellulose

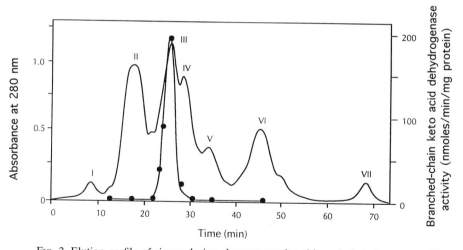

FIG. 2. Elution profile of size-exclusion chromatography of branched-chain α-keto acid dehydrogenase preparation. Partially purified, polyethylene glycol-fractionated extract was applied to a Sephacryl S-200 column (400 × 15 mm) equilibrated with buffer B. Fractions were eluted at a rate of 0.4 ml/min and assayed for branched-chain α-keto acid dehydrogenase activity (solid circles). The appropriate fractions were pooled and concentrated. [From D. A. Sinclair, I. W. Dawes, and J. R. Dickinson, *Biochem. Mol. Biol. Int.* **31**, 911 (1993).]

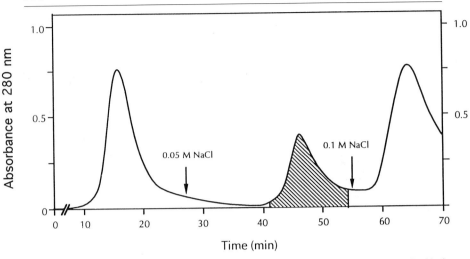

FIG. 3. Elution profile of anion-exchange chromatography of purified extract. Purified material obtained by size-exclusion chromatography and subsequently concentrated was applied to a DEAE-cellulose (100 × 0.7 mm) column equilibrated with buffer B. The column was washed with buffer B until no further proteins were eluting. The column was then washed with buffer B containing 0.05 M NaCl and the fraction containing highly purified α-keto acid dehydrogenase was collected (shaded area). [From D. A. Sinclair, I. W. Dawes, and J. R. Dickinson, *Biochem. Mol. Biol. Int.* **31**, 911 (1993).]

column (100 × 0.7 mm) equilibrated with buffer B. The application of 3 column volumes of buffer B is sufficient to elute all nonbinding proteins. The branched-chain α-keto acid dehydrogenase can now be eluted with buffer B supplemented with 0.05 M NaCl (Fig. 3). This is easily seen by eye as the passage of a brown band that is collected (approximately 2 ml) and concentrated as described previously.

The reconcentrated eluate is then applied to a Sepharose CL-2B column equilibrated with buffer B. Two main peaks are obtained. The second, larger peak elutes as brown fractions and contains branched-chain α-keto acid dehydrogenase activity. The fractions that comprise this peak are pooled, concentrated again, and frozen. The enzyme can be stored at −70° for several weeks without loss of activity. The purification described is summarized in Table I.

Properties

Sodium dodecyl sulfate–polyacrylamide gel electrophoresis (SDS–PAGE) of the enzyme complex shows that it comprises an E1α of M_r 47,000, E1β of M_r 38,000, E2 of M_r 52,000, and E3 of M_r 58,000 (Fig. 4).

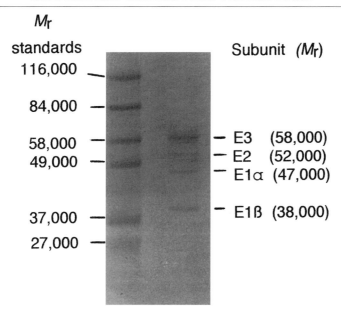

FIG. 4. SDS–polyacrylamide gel electrophoresis of purified branched-chain α-keto acid dehydrogenase complex. The protein sample (15 μg) was added to 2% SDS (w/v), 0.005% (w/v) bromphenol blue, 10% (v/v) glycerol, and TTS buffer [25 mM Tris (pH 7.5), 250 mM Tricine, 0.15 (w/v) SDS] and heated to 100° for 5 min. The gel was a linear 10 to 20% (w/v) polyacrylamide gradient gel and was buffered in TTS buffer. Electrophoresis was performed at 100 mV for 5 hr. The proteins were stained with Coomassie blue and destained with a solution comprising 5% (v/v) methanol and 5% (v/v) acetic acid. [From D. A. Sinclair, I. W. Dawes, and J. R. Dickinson, *Biochem. Mol. Biol. Int.* **31,** 911 (1993).]

TABLE I

PURIFICATION OF BRANCHED-CHAIN α-KETO ACID DEHYDROGENASE FROM YEAST[a]

Purification stage	Total protein (mg)	Specific activity (μmol/min/mg protein)	Total activity (μmol/min)	Yield (%)
Clarified homogenate[b]	9350	0.015	140	100
5% PEG pellet	6050	0.018	108	77
9% PEG pellet	630	0.130	82	59
Sepharose S-200 eluate	144	0.16	23	16
DEAE-cellulose eluate	23	0.65	15	11
Sepharose CL-2B	14	0.82	7.8	6

[a] Adapted from Sinclair *et al.*[14]
[b] Refers to supernatant after initial 12,000*g* centrifugation.

TABLE II
KINETIC CHARACTERIZATION OF BRANCHED-CHAIN α-KETO ACID DEHYDROGENASE
FROM YEAST[a]

Substrate	K_m (mM)	V_{max} (μmol/min/mg protein)	Relative V_{max}
2-Oxoisovalerate	21.3 ± 1.4[a]	19.9 ± 0.8	100
2-Oxoisocaproate	21.8 ± 1.7	7.5 ± 0.4	38
2-Oxo-β-methylvalerate	19.9 ± 0.5	7.6 ± 0.1	38
2-Oxobutyrate	7.5 ± 0.9	4.0 ± 0.4	20
4-Methylthio-2-oxobutyrate	13.3 ± 1.2	1.6 ± 0.2	8

[a] Values represent means ± SE.

Thus, the constitutent polypeptides are similar in size to the equivalent components in other eukaryotic branched-chain α-keto acid dehydrogenases. In this context retention of the E3 component, lipoamide dehydrogenase, is noteworthy because it dissociates during purification of the complex from some other eukaryotic sources including bovine kidney[10,11] and rabbit heart[12] but not bovine liver.[13] Coenzyme A appears to be extraordinarily tightly bound, because extensive dialysis against buffer B fails to render the complex coenzyme A dependent.[14] On the basis of apparent V_{max} values (Table II), α-ketoisovaleric acid is the preferred substrate among the α-keto acids derived from valine, leucine, and isoleucine, which accords with other eukaryotic branched-chain α-keto acid dehydrogenases. The yeast complex also catalyzes the decarboxylation of α-ketobutyrate and α-keto-γ-methiolbutyrate. However, it should be noted that the K_m values for the yeast complex are about 500-fold greater than those of mammalian enzymes for all substrates tested. This may be indicative that branched-chain keto acids are much more toxic to higher eukaryotes than yeast. The yeast enzyme has a pH optimum of about 7.5 and is quickly inhibited in the presence of Tris.

Acknowledgments

The author is indebted to David Sinclair for perfecting the purification of yeast branched-chain α-keto acid dehydrogenase[14]; Ian Dawes and Dean Danner for inspiration; and the Royal Society, British Council, United Distillers, and the Australian Department of Industry, Trade, and Commerce for financial support.

[10] K. G. Cook and S. J. Yeaman, *Methods Enzymol.* **166,** 303 (1988).
[11] F. H. Petit and L. J. Reed, *Methods Enzymol.* **166,** 309 (1988).
[12] R. Paxton, *Methods Enzymol.* **166,** 313 (1988).
[13] D. J. Danner and S. C. Heffelfinger, *Methods Enzymol.* **166,** 298 (1988).
[14] D. A. Sinclair, I. W. Dawes, and J. R. Dickinson, *Biochem. Mol. Biol. Int.* **31,** 911 (1993).

[37] β-Alanine Synthase, an Enzyme Involved in Catabolism of Uracil and Thymine

By THOMAS W. TRAUT

β-Alanine synthase (*N*-carbamoyl-β-alanine amidohydrolase; EC 3.5.1.6) completes the catabolism of the pyrimidine bases uracil and thymine, as shown in Fig. 1A. The enzyme specifically catalyzes the following reactions:

$$N\text{-Carbamoyl-}\beta\text{-alanine} \rightarrow \beta\text{-alanine} + CO_2 + NH_3 \qquad (1)$$
$$\text{(}\beta\text{-ureidopropionate)}$$

$$2\text{-Methyl-}N\text{-carbamoyl-}\beta\text{-alanine} \rightarrow 2\text{-methyl-}\beta\text{-alanine} + CO_2 + NH_3 \quad (2)$$
$$\text{(}\beta\text{-ureidoisobutyrate)} \qquad \text{(}\beta\text{-aminoisobutyrate)}$$

The trivial names β-ureidopropionase and β-ureidopropionate decarbamylase have also been used. The name β-alanine synthase emphasizes the physiological role of this amino acid, which is used in the synthesis of various dipeptides. These include peptides abundant in mammalian muscle such as anserine (β-alanyl-1-methylhistidine),[1] carnosine (β-alanylhistidine),[1] and balenine (β-alanyl-*N*-methylhistidine),[2] as well as β-alanylhypusine (found in cow brain),[3] various tan pigments (β-alanyldopamine) identified in plants and insects, as well as pantothenate (pantoyl-β-alanine)—the precursor to coenzyme A.[4] β-Alanine may also function as a neurotransmitter[5]; in humans disorders in the synthesis of β-alanine are accompanied by severe neurological dysfunctions and mental retardation.[4] Consistent with such a biosynthetic function, the enzyme from rat shows cooperativity for the substrate and allosteric regulation,[6,7] and is able to adopt various conformations that form different oligomeric assemblies, as shown in Fig. 1B.

In the adult rat, the pathway including β-alanine synthase appears to be active only in liver and kidney, and was not detectable in brain, lung,

[1] R. A. Clemens, J. D. Kopple, and M. A. Swendseid, *J. Nutr.* **114,** 2138 (1984).
[2] C. I. Harris and G. Milne, *Comp. Biochem. Physiol.* **86B,** 273 (1987).
[3] S. Ueno, K. Kotani, A. Sano, and Y. Kakimoto, *Biochim. Biophys. Acta* **1073,** 233 (1991).
[4] C. R. Scriver, T. L. Perry, and W. Nutzenadel, *in* "Metabolic Basis of Inherited Disease" (D. Fredrickson, J. Wyngaarden, and J. B. Stanbury, eds.), 4th Ed., p. 570. McGraw-Hill, New York, 1983.
[5] M. Sandberg and I. Jacobson, *J. Neurochem.* **37,** 1353 (1971).
[6] M. M. Matthews and T. W. Traut, *J. Biol. Chem.* **262,** 7232 (1987).
[7] M. M. Matthews, W. Liao, K. L. Kvalnes-Krick, and T. W. Traut, *Arch. Biochem. Biophys.* **293,** 254 (1992).

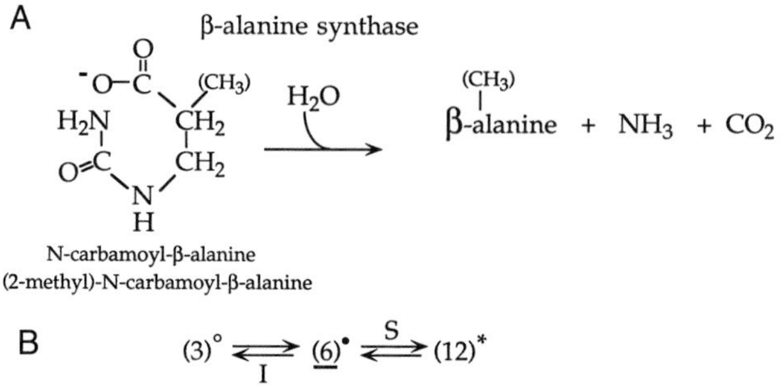

A   β-alanine synthase

N-carbamoyl-β-alanine
(2-methyl)-N-carbamoyl-β-alanine

B $(3)° \overset{}{\underset{I}{\rightleftharpoons}} \underline{(6)}^{•} \overset{S}{\rightleftharpoons} (12)^{*}$

Fig. 1. Reactions catalyzed by β-alanine synthase, and conformations of the enzyme. (A) Reduction and cleavage of the pyrimidine ring, uracil, or thymine produce the two alternate substrates for β-alanine synthase. (B) The native enzyme is a hexamer, which readily associates to an active dodecamer in the presence of substrate, or substrate analogs, and dissociates to an inactive trimer in the presence of the product, or product analogs. S, substrate; I, inhibitor.

muscle, or spleen.[6] This may reflect the ability of these two tissues to handle the ammonia produced by this reaction. The enzyme has been completely purified from the liver of the rat,[7,8] and calf,[9] as well as from *Euglena gracilis*[10] and *Pseudomonas putida*.[11] Generally the subunit mass is about 42–45 kDa,[7,9,11] but one study reported this value at 54 kDa.[8] In mammals the native enzyme forms a hexamer,[6,8,9] while the bacterial enzyme is a dimer.[11] The reaction catalyzed requires divalent metal: for the bacterial enzyme the addition of Co^{2+}, Ni^{2+}, or Mn^{2+} to the enzyme assay enhanced activity.[11] The rat enzyme contains two tightly bound Zn^{2+},[12] and removal of the zinc by chelators led to loss of activity. The cDNA sequence for the rat enzyme encodes 393 amino acids, confirming a subunit mass of 44 kDa, and suggests the position of two zinc-binding sites.[12]

Enzyme Assays

Two different experimental strategies have been employed to quantitate enzyme activity. Although isotopically labeled forms of the two substrates described above are not commercially available, it is not difficult to chemically synthesize N-carbamoyl-β-[5-^{14}C]alanine (NCβA), since it is directly

[8] N. Tamaki, N. Mizutani, M. Kikugawa, S. Fujimoto, and C. Mizota, *Eur. J. Biochem.* **169**, 21 (1987).
[9] G. Waldmann and K. D. Schnackerz, *Biol. Chem. Hoppe-Seyler* **370**, 969 (1989).
[10] C. Wasternack, G. Lippmann, and H. Reinbothe, *Biochim. Biophys. Acta* **570**, 341 (1979).
[11] J. Ogawa and S. Shimizu, *Eur. J. Biochem.* **223**, 625 (1994).
[12] K. L. Kvalnes-Krick and T. W. Traut, *J. Biol. Chem.* **268**, 5686 (1993).

formed in reasonable yields by reacting $K^{14}CNO$ with β-alanine as described.[13] The advantage of this substrate is that it is labeled in the carbon that will become the CO_2 product, which can be easily separated from the reaction and trapped on a filter. It provides a clear signal, with little background, and therefore provides a sensitive enzyme assay.

While we have used the isotope assay, an alternative procedure also provides a sensitive enzyme assay.[14] The second procedure is to isolate the product β-alanine, which in turn is derivatized with the Edman reagent phenylisothiocyanate to form phenylthiocarbamoyl-β-alanine (PTC-β-alanine), which must be isolated by high-performance liquid chromatography (HPLC) and is then easily detected. With the isotope assay, the major effort is to synthesize the [5-^{14}C]NCβA, but this is fairly stable on storage and can be used for at least one year. With the HPLC assay, the major effort is in derivatizing and purifying every assay sample before it can be quantitated.

Isotope Assay

The chemical synthesis of [5-^{14}C]NCβA is described elsewhere,[13] and only the direct enzyme assay is given below. Note that a variation of the chemical synthesis can use KCNO plus commercially available β-[^{3}H] alanine to synthesize [3-^{3}H]NCβA, which permits isolation from enzyme reactions of both substrate and product by thin-layer chromatography.[13]

Reagents

Final assay concentrations are presented.
  KPO_4 buffer, pH 7.0 at 37°, 50 mM
  Dithiothreitol (DTT), 1 mM
  N-Carbamoyl-β-[5-^{14}C] alanine, 200 μM, 0.4–2.0 Ci/mol
  NaOH, 2 M
  $HClO_4$, 4 M

Procedure. Since the product to be measured is $^{14}CO_2$, the reaction itself is done within a 1-ml plastic microcentrifuge tube. This plastic assay tube is carried, fairly upright, within a standard glass scintillation vial, since the vial can be sealed by a rubber septum stopper (obtained from Kontes, Vineland, NJ). Inserted into this septum is a plastic center well that simply carries a folded strip of filter paper. This strip of filter paper, Whatman (Clifton, NJ) 3MM paper, about 0.5×3 cm, has been moistened with 2 M NaOH and then blotted against a paper towel. As $^{14}CO_2$ is evolved it can adsorb to this filter paper and become trapped by conversion to bicarbonate.

[13] T. W. Traut and S. Loechel, *Biochemistry* **23**, 2533 (1984).
[14] G. Waldmann and B. Podshun, *Anal. Biochem.* **188**, 233 (1990).

Reaction assay tubes are prepared, containing buffer and substrate, and water as needed to adjust for a constant volume if other factors are varied. The rubber septa have their center wells filled with alkali-treated filters. To control temperature, reactions are normally done in a shaking water bath. With pure enzyme the standard reaction is done in a total volume of 50 μl, but this can readily be varied to much larger assay volumes if greater volumes of enzyme sample are required, as in the early stages of a purification. Reactions are started by the addition of enzyme, and allowed to continue for a desired time (often 10 min), when they are quenched by the addition of 100 μl of $HClO_4$ directly through the septum with a 1-ml syringe.

Enzyme can be added to the open vial, via a pipette, and the rubber septum quickly added thereafter to seal the vial. Alternatively, vials can be sealed in advance with the septum, and the enzyme is then added through the septum with a small syringe. Since CO_2 is not evolved until after the reaction is quenched by acid, either procedure for adding enzyme can be employed, with each giving similar results.

After the reactions are quenched, they are allowed to shake for at least 10 min to ensure total adsorption of the $^{14}CO^2$, after which the filters are removed with forceps and placed under a heat lamp to dry. The filters are visibly dry when they begin to turn yellow, and should then be placed into a scintillation vial with scintillation fluid. Prolonged drying causes the filters to turn brown and then black, which lowers the efficiency of scintillation counting.

Note that two syringes can be preinserted into the prepared assay vials. The first contains enzyme, the second contains $HClO_4$. This permits rapid assay times, since enzyme can be added to start the reaction, which can then be manually stopped with the second syringe at any time, but easily within 2 sec.[6]

High-Performance Liquid Chromatography Assay

Reagents

Final assay concentrations are presented.
KPO$_4$ buffer (pH 7.0), 10 mM
N-Carbamoyl-β-alanine, 200 μM
Ethanol–H$_2$O–triethylamine (2:2:1, v/v)
Ethanol–H$_2$O–triethylamine–phenylisothiocyanate (PITC) (7:1:1:1, v/v)
Buffer A: Sodium acetate (pH 6.0), 50 mM plus 15% (v/v) acetonitrile

For the published assay,[14] the enzyme is preincubated for 3 min, at which time substrate is added. The reaction is terminated by heating at 95° for 5 min. Samples are centrifuged to remove precipitated protein, and an aliquot of the supernatant is evaporated to dryness. The residue is dissolved in 20 μl of ethanol–H_2O–triethylamine, and again evaporated to dryness. After addition of 20 μl of coupling solution (ethanol–H_2O–triethylamine–PITC), mixtures are vortexed and allowed to stand at room temperature for 30 min. The samples are then dried, and dissolved in buffer A, and aliquots applied to a LiChrospher 100 C_{18} reversed-phase column. Separation is achieved by isocratic elution with buffer A, at a flow rate of 1 ml/min, while monitoring the effluent stream at 254 nm.

Purification of β-Alanine Synthase

The strategy outlined below requires at least 5 days for a complete purification. In the early stages, through isoelectric focusing, experiments are done at 4°. However, buffers are always adjusted to pH 7.5 at room temperature, and contain 1 mM dithiothreitol. This reducing agent is not absolutely essential,[8,9,11] but our results show that it improves the stability of the enzyme. Buffers used in high-performance liquid chromatography (HPLC) and fast protein liquid chromatography (FPLC) columns are filtered through 0.45-μm pore size filters and degassed before use. In the early stages, because of the sample size and the sensitivity of chromatography to background salt, the sample is dialyzed overnight with four 1-liter changes of buffer at 3- to 4-hr intervals.

Buffers

   Buffer A: 20 mM KPO_4, 0.25 M sucrose, 1 mM EDTA, plus protease inhibitors
   Buffer B1: 20 mM KPO_4 plus 10% (v/v) glycerol, plus protease inhibitors
   Buffer B2: 60 mM KPO_4 plus 10% (v/v) glycerol, plus protease inhibitors
   Buffer C: 20 mM Tris-HCl
   Buffer D: 20 mM Tris plus 50 mM NaCl
   Protease inhibitors: (phenylmethylsulfonyl fluoride (PMSF, 1 mM), leupeptin (0.5 mg/liter), antitrypsin (25 mg/liter), and aprotinin (70 units/liter)

Procedure

   The outlined procedure is for about 100 g of liver, which is homogenized in buffer A and centrifuged. The supernatant (S-20 fraction) is adjusted to 37% saturation with ammonium sulfate, stirred gently for 30 min, and centrifuged for 1 hr at 105,000g and 4°. This step accomplishes both the first salt precipitation and the removal of subcellular particles to yield a cytosolic suspension. The supernatant is brought to 47% saturation with ammonium sulfate, stirred for 30 min, and solids are pelleted at 20,000g for 20 min at 4°. The supernatant is discarded; the pellet is resuspended in buffer B1, and dialyzed overnight at 4° (four 1-liter buffer changes).

Ion-Exchange Chromatography: Step 3

The protein sample is then diluted 1:1 with buffer B1 for the QAE column, and centrifuged at 12,000g for 20 min at 4°, before application to the QAE-Sepharose column. This centrifugation is necessary to remove some undissolved protein after the dialysis. After the sample is loaded, the column is washed with buffer B1 (flow rate of 4 ml/min) until protein absorbance (280 nm) reaches a constant lower value. Elution of the enzyme is accomplished with buffer B2. Fractions of 9 ml are collected throughout the run and assayed for β-alanine synthase activity (Fig. 2). The enzyme activity elutes primarily with the 60 mM KPO$_4$ step. Fractions containing enzyme are pooled (95% of total enzyme activity) and concentrated by precipitating the protein (about 150 mg in 275 ml) at 70% ammonium sulfate. The precipitate is dissolved and dialyzed versus buffer B1 overnight at 4°.

Isoelectric Focusing: Steps 4 and 5

The sample is diluted to 50 ml with buffer B1, containing 2% (w/v) ampholytes (2.5 ml of 40% Bio-Lyte ampholytes, pH 5–7). This solution is put directly into a Rotofor isoelectric focusing cell (Bio-Rad, Hercules,

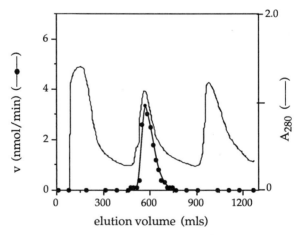

Fig. 2. Anion-exchange chromatography. The enzyme sample was applied to the QAE-Sepharose column and was eluted with 60 mM KPO$_4$ buffer.

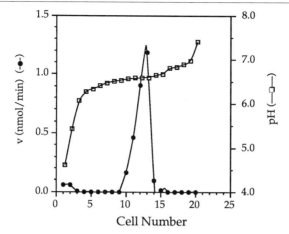

FIG. 3. Isoelectric focusing of β-alanine synthase. Enzyme from the QAE column (Fig. 2) was focused in successive runs on the Rotofor apparatus. Represented here is the second run.

CA), and focusing is carried out at 12 W constant power for 3.5 hr. The apparatus is maintained at 0–2° by recirculating a solution of 30% (v/v) ethylene glycol maintained at 0°. The Rotofor unit contains 20 separate cells, which are assayed for enzyme activity, and also for pH (to determine the pI of the enzyme). Appropriate fractions are pooled, and diluted to 50 ml with 10% (v/v) glycerol plus 1 mM DTT in distilled water, maintained at 0°. This sample is reloaded on the Rotofor (without additional ampholytes) and again subjected to a constant current of 12 W for 3 hr. Fractions are again assayed, and those with peak enzyme activity are pooled (Fig. 3) and concentrated overnight with a Centricon-30 microconcentrator (Amicon, Beverly, MA). To avoid protein precipitation during this step, the pooled sample (pH near 6.7) is adjusted to pH 7.5.

Hydrophobic Interaction Chromatography: Step 6

The sample is then adjusted to 0.6 M ammonium sulfate, and applied to a phenyl-Sepharose HPLC column (Fig. 4). The first two enzyme activity peaks are pooled and concentrated. The third enzyme peak coelutes with the major contaminating protein peak and must be discarded.

Size-Exclusion Chromatography: Step 7

The final chromatography step utilizes a Superose 6 FPLC gel-filtration column equilibrated with buffer D; 2.5 mM NCβA is included in the column buffer to promote association of β-alanine synthase to a larger polymeric

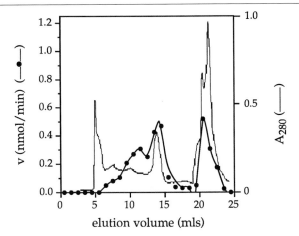

FIG. 4. Hydrophobic interaction chromatography. The concentrated sample from the isoelectric focusing step (Fig. 3) was applied to a phenyl-Sepharose HPLC column, and eluted with a decreasing gradient of ammonium sulfate (0.6–0 M), producing three peaks of enzyme activity.

species,[6] and thereby facilitate its separation from smaller contaminating proteins. The column elution profile and sodium dodecyl sulfate–polyacrylamide gel electrophoresis (SDS–PAGE) of the corresponding column fractions are shown in Fig. 5. This column produces two major UV absorbance peaks, with most of the β-alanine synthase activity eluting with the first protein peak. The corresponding SDS–polyacrylamide gel shows a single protein band that increases in intensity coincident with β-alanine synthase activity. The subunit M_r of β-alanine synthase is determined from a standard curve to be 42,000 (repeated more than three times each, with two different sets of M_r standards). The gel shows that the first protein peak is largely pure β-alanine synthase; however, a small amount of β-alanine synthase activity (with correspondingly decreased amounts of the M_r 42,000 band) continue to elute under the second major protein peak. The gel shows the M_r 42,000 band to be the major protein in the sample loaded on the Superose column (Fig. 5A, lane 1).

For the procedure outlined in Table I, a significant loss in recovered activity occurs at the original ammonium sulfate step, and again later in the final chromatographic step, in which glycerol must be omitted. Little enzyme activity is lost in the two successive isoelectric focusing steps. The pure enzyme in Table I has a specific activity of 877 nmol/min/mg protein; this represents a moderate rate of catalytic activity, with a k_{cat} of 0.6 sec^{-1}. The final specific activity of the enzyme from all sources at this time is within a factor of two.[7–9,11]

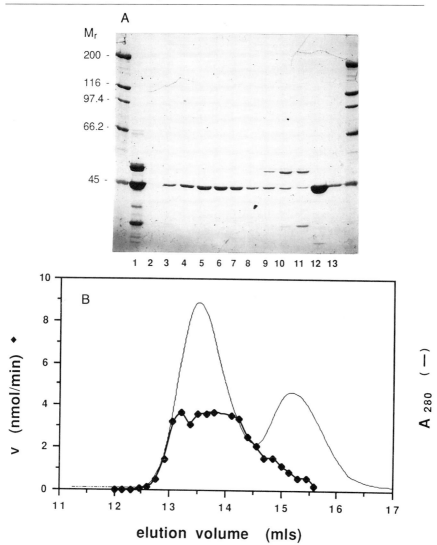

FIG. 5. Gel-filtration chromatography in the presence of NCβA. (A) Samples along the Superose 6 gel-filtration chromatography profile (shown in B) were analyzed by SDS–PAGE. Molecular weight markers (for the far left and right lanes) contained 2 μg each of myosin (200,000), β-galactosidase (116,000), phosphorylase *b* (97,400), bovine serum albumin (66,200), and ovalbumin (45,000). Lane 1 contained the enzyme sample loaded on the column; 10-μl aliquots from fractions of the column elution profile were applied to lanes 2 through 11 (at 12.6 to 15.3 ml of the elution); lane 12 contained the concentrate of pooled enzyme from 12.6 to 14.1 ml, and lane 13, the SDS wash from the tube for lane 12. (B) A 100-μl sample of the Phenyl-P-5W enzyme peak was concentrated and chromatographed on a Superose 6 gel-filtration column; both the sample and the column buffer contained 2.5 m*M* NCβA. β-Alanine synthase activity (◆) and corresponding protein absorbance (—) at 280 nm are shown.

TABLE I
PURIFICATION OF β-ALANINE SYNTHASE FROM RAT LIVER[a]

Purification step	Volume (ml)	Total protein (mg)	Specific activity (μmol/min/mg)	Recovery (%)	Purification (fold)
1. S-20	407	16,280	0.0008	100	1
2. AS 37–47%	100	2,120	0.0028	45	3.5
3. QAE-Sepharose	275	148	0.025	29	32
4. Isoelectric focusing 1	14	36.4	0.10	28	124
5. Isoelectric focusing 2	28	14.0	0.23	25	285
6. Phenyl-Sepharose HPLC	26	3.8	0.72	21	901
7. Superose FPLC	2.1	0.84	0.88	5.7	1,096

[a] Procedure shown is for about 100 g of liver.

Stability of Enzyme

Both detergents and reducing thiols help to stabilize the purified enzyme in solution, although they are not routinely used in other efforts to purify this enzyme.[8,9,11] The enzyme obtained from rat liver is hydrophobic, and adsorbs easily to surfaces. For example, a tube that had stored concentrated enzyme was emptied and washed; a subsequent second wash of this empty tube with 3% (w/v) SDS resulted in solubilizing some β-alanine synthase as shown in lane 13 (Fig. 5A).

Enzyme activity is barely detectable after storage at 4° for 26 days in buffer alone (Fig. 6). Addition of the detergent CHAPSO at 3% (v/v) results in improved recovery for 10 days, with a third of the activity still measurable after 26 days. Also shown is the importance of fresh reducing thiols; for both experiments in Fig. 6, in the early times there are increases in enzyme activity after the addition of fresh DTT to stored enzyme. In fact, detergent produces an apparent increase in activity (relative to the control) in the first few days, probably because the detergent helps resolubilize enzyme that has come out of solution. This effect of detergent is similar to the resolubilizing effect of SDS shown in lane 13 of Fig. 5A, except that detergent does not denature the enzyme. The detergent CHAPS gives results almost as good as those of CHAPSO, while digitonin is not effective.

Other laboratories[8,9] have heated the initial S-20 sample (step 1), as this is an easy batch step. Our studies show that heating destabilizes the enzyme, such that its K_m increases significantly, while the allosteric response to effectors is no longer detectable.[7]

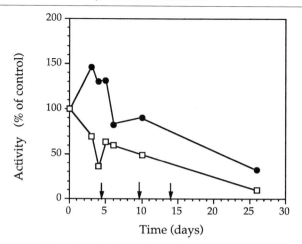

FIG. 6. Storage stability of β-alanine synthase. Pure enzyme was stored at 4° in buffer alone (□), or in the presence of detergent at 3% (●). Both solutions originally contained 1 mM DTT, and an additional 1 mM DTT was added at the times indicated by arrows.

Properties of Enzyme

The purification described yields almost 1 mg of pure enzyme from 100 g of starting tissue. Given the recovery shown in Table I, this means that liver actually contains about 15 mg of this enzyme per 100 g of tissue. This amount reflects the importance of this enzyme as well as its modest level of activity.

An interesting allosteric feature of the enzyme is its association from the hexamer to the dodecamer state in the presence of substrate, and its dissociation to the trimer state in the presence of inhibitors.[6,7] This property was used during the final purification step, and shows that the enzyme has retained its native features through the purification. Therefore the active form of the enzyme (in the presence of substrate) is as a dodecamer, while the dissociated trimer has little or no activity.

The enzyme has a K_m of 8 μM for NCβA, and 6 μM for 2-methyl-NCβA and K_i values of 1.1 mM (β-alanine), 1.6 mM (γ-aminobutyrate), and 3.9 mM (2-methyl-β-alanine). The amino group interferes with effective binding by products or product analogs, since propionate (deaminated β-alanine) has a K_i of only 90 μM.

Association and Dissociation of Oligomers

Figure 1 shows that the enzyme oligomers associate or dissociate in response to appropriate ligands. The hexamer is a fairly stable species in the absence of effectors. Since the presence of substrate (NCβA) or substrate analogs promotes the dodecamer form,[6,7] this must be the active form of the enzyme. In contrast, the presence of the product (β-alanine) or product analogs leads to dissociation of hexamers to inactive trimers. Thus the availability of the normal substrate and/or the physiological product provides allosteric regulatory effects to appropriately modulate the activity of the enzyme.

Section III

Detection and Consequences of Genetic Defects in Genes Encoding Enzymes of Branched-Chain Amino Acid Metabolism

[38] Diagnosis and Mutational Analysis of Maple Syrup Urine Disease Using Cell Cultures

By Jacinta L. Chuang and David T. Chuang

Maple syrup urine disease (MSUD), or branched-chain ketoaciduria, is caused by a deficiency in activity of the branched-chain α-keto acid dehydrogenase (BCKD) complex.[1] This metabolic block results in the accumulation of branched-chain amino acids (BCAAs) leucine, isoleucine, and valine and the corresponding branched-chain α-keto acids (BCKAs). On the basis of clinical presentation and biochemical responses to thiamine administration, MSUD patients can be divided into five phenotypes: classic, intermediate, intermittent, thiamine responsive, and dihydrolipoyl dehydrogenase (E3) deficient. The classification is based on rapidity of onset, severity of the disease, tolerance for dietary proteins, response to thiamine supplements, and enzyme analysis data.[2] The presence of alloisoleucine is diagnostic of MSUD. Activity of the BCKD complex in skin fibroblasts or lymphoblast cultures is reduced, and ranges from less than 2% of normal in the classic form to 30% of normal in variant types. The enzyme affected in MSUD, the BCKD complex, consists of three catalytic components, i.e., BCKA decarboxylase or E1, dihydrolipoyl acyltransferase or E2, and dihydrolipoyl dehydrogenase or E3. The E3-deficient MSUD presents a combined deficiency of BCKD, pyruvate dehydrogenase, and α-ketoglutarate dehydrogenase complexes.[3] Four genetic subtypes based on the affected locus of the BCKD complex have been classified.[2] Type 1A refers to mutations in the E1α gene, type 1B in the E1β gene, type II in the E2 gene, and type III in the E3 gene. Mutations in all four genetic subtypes have been identified, and some have been characterized.[2]

The BCKD complex is expressed in peripheral leukocyte preparations[4] and skin fibroblasts.[5] Thus cultured diploid lymphoblasts (from blood) or fibroblasts (from skin biopsies) derived from patients have provided a useful system for studying enzyme deficiency in this disease.

---

[1] J. Dancis, J. Hutzler, and M. Levitz, *Biochim. Biophys. Acta* **43**, 342 (1960).

[2] D. T. Chuang and V. E. Shih, *in* "Metabolic and Molecular Bases of Inherited Disease," 7th Ed., p. 1239. McGraw-Hill, New York, 1995.

[3] B. H. Robinson, J. Taylor, and W. G. Sherwood, *Pediatr. Res.* **11**, 1198 (1977).

[4] J. Dancis, J. Hutzler, and M. Levitz, *Pediatrics* **32**, 234 (1963).

[5] J. Dancis, V. Jansen, J. Hutzler, and M. Levitz, *Biochim. Biophys. Acta* **77**, 523 (1963).

Enzyme Analysis of Maple Syrup Urine Disease with Intact Cultured
  Lymphoblasts or Fibroblasts

The intact-cell assay measures the flux of BCKA and BCAA through
the BCKD complex in cultured lymphoblasts or fibroblasts. Harvested cells
are incubated with α-keto[1-^{14}C]isovalerate ([1-^{14}C]KIV) in a balanced salt
solution. The $^{14}CO_2$ evolved is trapped and counted for radioactivity. In
MSUD cells, the rate of decarboxylation of 1-^{14}C-labeled α-keto acids is
greatly reduced compared with that in normal cells and thus provides a
basis for diagnosis of MSUD using cell cultures.

Lymphoblast Cell Culture

Lymphoblasts derived from the blood of patients are preferred to fibroblasts for MSUD diagnosis, because lymphoblasts are faster growing and
easier to culture (trypsinization is not required during passage) than fibroblasts. Lymphoblasts are immortalized cells, unlike fibroblast cells, which
have a finite life span as reflected by the limited passage number. Moreover,
blood drawing required for lymphoblast culture is less traumatic than skin
biopsies needed for fibroblast culture. To establish a lymphoblast cell line,
5–10 cm^3 of blood is collected into an acid–citrate–dextrose tube and
processed within 72 hr. The blood is mixed with an equal volume of isotonic
saline (0.9%, w/v) and layered over a three-fifth volume of Ficoll-Paque
(Amersham-Pharmacia Biotech, Piscataway, NJ) in a 15-ml disposable conical sterile centrifuge tube. Samples are centrifuged at 400g for 30 min at
room temperature. The white buffy coat and the clear Ficoll layer (in the
lower half of the tube) are transferred to a new tube and an equal volume
of Earle's balanced salt solution is added. After further centrifugation at
400g for 15 min at room temperature, resulting small cell pellets from each
sample are combined and resuspended in 9–10 ml of cold RPMI 1640
medium (GIBCO-BRL, Gaithersburg, MD). Cell suspensions are centrifuged again at 400g for 15 min at room temperature. The cell pellet is
resuspended in 5 ml of RPMI 1640 medium at 37° containing 20% heat-
inactivated fetal bovine serum (incubated in a 60° water bath for 30 min)
and cells are transferred to a 25-cm^2 culture flask. Five milliliters of Epstein–
Barr virus (EBV)-containing medium (from cultured marmoset cells) and
0.1 ml of phytohemagglutinin (GIBCO-BRL) are added to each flask. Cells
are incubated in a CO_2 incubator at 37°. Established cells are maintained
in RPMI 1640 medium with 10% (w/v) heat-inactivated fetal bovine serum.
Cells are fed every 3 to 4 days, and cell feeding is accomplished by removing
half the growth medium and replacing it with fresh medium. Cells are ready
to be split when the medium turns yellow in less than 3 days. Cell splitting
is accomplished by doubling the growth medium.

Fibroblast Cell Culture

All steps are carried out under sterile conditions in a laminar flow hood. Fibroblasts are grown from explants of skin biopsies[5] derived from normal individuals and MSUD patients. Explants are immobilized with sterile silicon grease in a 35 × 10 mm culture dish, and overlaid with coverslips. To the dish is added 2 ml of Waymouth medium (GIBCO-BRL) containing 20% (v/v) fetal calf serum (Intergen, Purchase, NY) and antibiotics [penicillin (50 U/ml), streptomycin (50 μg/ml), and kanamycin (30 μg/ml)]. The dish is incubated in a CO_2 incubator at 37°. The medium is changed twice a week. Within 7–11 days, cells that have migrated out of the explant are dispersed or passaged by trypsinization as follows. The growth medium is removed and attached cells rinsed once with 4 ml of Puck's saline A without glucose [0.8% (w/v) NaCl, 0.04% (w/v) KCl, 0.085% (w/v) $NaHCO_3$, and phenol red].[6] The rinsed monolayer culture is incubated for 2 to 3 min in the same buffer containing 0.02% (w/v) ethylene diaminetetraacetic acid (EDTA). Cells are detached from the surface of the flask by incubating them in the same solution containing 0.04% (w/v) trypsin at 37° for 3–5 min. The dispersed cell suspension is transferred to a 60 × 15 cm cell culture dish, supplemented with 4 ml of Waymouth medium containing 15% (v/v) fetal calf serum. Cells are grown to confluence, with the medium changed every week. Passages are repeated until adequate numbers of cells are available from the original explant.

Intact Cell Assay

To measure rates of decarboxylation of α-keto[1-^{14}C]isovalerate (1-^{14}C]KIV), lymphoblasts or fibroblasts harvested by trypsinization are counted in a hemocytometer. Cells are collected by centrifugation at 400g at room temperature and washed with Krebs–Ringer phosphate buffer[7] [0.018 M sodium phosphate (pH 7.5), 0.68% (w/v) NaCl, 0.045% (w/v) KCl, 0.03% (w/v) $MgSO_4$, and 0.02% (w/v) $CaCl_2$]. The washed cells are suspended in the same buffer at a density of 1.5 × 10^6 cells (lymphoblasts) or 1 × 10^6 cells (fibroblasts) per 50 μl for assays.

The intact-cell assay is carried out on a 24-well tissue culture plate as described in [19] in this volume.[7a] The reaction mixture in each well contains 295 μl of Krebs buffer with 500 μg of thiamine hydrochloride and 50 μl of cell suspension. After the addition of 25 μl of 7.4 mM [1-^{14}C]KIV (specific

[6] P. I. Marcus, S. J. Cieciura, and T. T. Puck, *J. Exp. Med.* **104**, 615 (1956).

[7] H. A. Krebs, Z. *Physiol. Chem.* **217**, 191 (1933).

[7a] J. L. Chuang, J. R. Davie, R. M. Wynn, and D. T. Chuang, *Methods Enzymol.* **324**, [19], 2000 (this volume).

radioactivity, 1000 cpm/nmol) the plate, which has been kept on ice, is transferred to a shaking 37° water bath and incubated for 80 min. The $^{14}CO_2$ evolved is absorbed on an NaOH-impregnated filter paper wick located in a center cup inside each well. The reaction is terminated by the injection of 50 μl of 15% (w/v) trichloroacetic acid (TCA) directly into the reaction mixture, followed by an additional 60-min incubation at 37° to recover all $^{14}CO_2$ released. The filter wicks are transferred into scintillation cocktail and counted for radioactivity. To serve as a control for cell viability, 0.5 mM [1-^{14}C]pyruvate (specific radioactivity, 1275 cpm/nmol) is used as substrate and incubated with the cell suspension for 10 min at 37°. The remaining steps are as described, using [1-^{14}C]KIV as a substrate.

Definition of Rate of Decarboxylation

The $^{14}CO_2$ released from [1-^{14}C]KIV by intact lymphoblasts or fibroblasts is expressed as nanomoles of CO_2 released per minute per milligram of protein. To calculate the amount of CO_2 released, specific radioactivity of the prepared substrate [1-^{14}C]KIV is used without correction for the endogenous α-keto acid in cells. Protein is determined by the method of Lowry et al.,[8] using bovine serum albumin (BSA) as standard.

Remarks

The intact whole-cell assay permits an estimate of the rate of decarboxylation of BCKA in cultured lymphoblasts or fibroblasts under defined assay conditions. The close correlations between the patient phenotype and tolerance for dietary protein with the degree of enzyme deficiency determined by the intact-cell assay[2] suggest that this method approximates the situation in vivo.

The choice of [1-^{14}C]BCKA over [1-^{14}C]BCAA as substrate in the intact-cell assay avoids the interposed step of transamination. This allows a more direct analysis of enzyme deficiency in the BCKD complex with MSUD cells. [1-^{14}C]KIV is the preferred substrate because it is decarboxylated at the highest rate among the BCKAs (the other two being α-ketoisocaproate and α-keto-β-methylvalerate) tested. At 0.5 mM KIV concentration, normal fibroblasts decarboxylate [1-^{14}C]KIV at a rate of 0.05 to 0.1 nmol/min/mg of protein. The rate of decarboxylation is three- to fourfold higher in lymphoblast cells than in fibroblast cells. Classic MSUD cells exhibit no decarboxylating activity.[2] Residual activities (0.5 to 36% of the normal) are observed with intact cells from patients with variant forms of MSUD.[2]

[8] O. H. Lowry, N. J. Rosebrough, A. L. Farr, and R. J. Randall, J. Biol. Chem. **193,** 265 (1951).

Identification and Mutational Analysis of Affected Branched-Chain α-Keto Acid Dehydrogenase Subunits in Maple Syrup Urine Disease Cells

The human BCKD complex consists of three catalytic components and two regulatory enzymes that are encoded by six genetic loci. Mutations in any one of these loci could theoretically produce the MSUD phenotype. Enzyme assays for each component (E1, E2, or E3) were developed using model reactions. These component assays have allowed the detection of deficiencies in E1[9] and E3.[8,9] The assays for purified E1 and E2 components are reliable, but are difficult with cell culture homogenates because of low sensitivity.[10] In contrast, the E3 assay is sensitive for both purfied enzyme[11] and cell culture material.[3,9] Cofactor effects on the decarboxylation of BCKAs in certain MSUD cells were previously measured to determine the component affected.[12]

The availability of specific antibodies and cDNA probes has allowed the use of more direct molecular approaches to identify the mutant locus in MSUD cells. In most instances, Western blot analysis can be used to identify deficient BCKD catalytic subunits. There are exceptions, which are discussed below.

Western Blot Analysis

To identify the enzyme subunit involved in the dysfunction of BCKD complex in MSUD cells, the levels of specific protein subunits (E1α, E1β, and E2) are measured by Western blot analysis, using antibodies raised against each purified subunit. Lymphoblasts or fibroblasts suspended in 30 mM potassium phosphate buffer, pH 7.5, containing 1 mM phenylmethylsulfonyl fluoride (PMSF) and 1 mM benzamidine are homogenized by sonic oscillation. Cells are solubilized in sodium dodecyl sulfate–polyacrylamide gel electrophoresis (SDS–PAGE) sample buffer containing 50 mM Tris-HCl (pH 6.8), 10% (v/v) glycerol, 1% (w/v) SDS, 1% (v/v) 2-mercaptoethanol, and 0.02% (w/v) bromphenol blue and boiled for 5 min. Proteins (300 μg per lane) are applied to a 10% (w/v) gel (1.5 mm thick), and SDS–PAGE is performed according to the method of Laemmli.[13] Proteins separated on the gel are electrotransferred at 150 mA for 4 hr to a polyvinylidene

---

[9] C. W. Fisher, J. L. Chuang, T. A. Griffin, K. S. Lau, R. P. Cox, and D. T. Chuang, *J. Biol. Chem.* **264**, 3448 (1989).

[10] D. T. Chuang and R. P. Cox, *Methods Enzymol.* **166**, 135 (1988).

[11] Y. Sakurai, Y. Feduyoshi, M. Hamada, T. Hayakawa, and M. Koike, *J. Biol. Chem.* **245**, 4453 (1970).

[12] L. J. Elsas, B. A. Pask, F. B. Wheeler, D. P. Perl, and S. Trusler, *Metabolism* **21**, 929 (1972).

[13] U. K. Laemmli, *Nature* (*London*) **227**, 680 (1970).

difluoride (PVDF; Millipore, Marlsborough, MA) membrane in 25 mM Tris-HCl (pH unadjusted), 200 mM glycine, and 20% (v/v) methanol. Protein bands are visualized by staining the blot briefly by Coomassie blue staining, followed by destaining with 100% methanol. The blot is blocked for 3 hr with PBS (75 mM NaCl, 1.8 mM KH$_2$PO$_4$, and 57 mM Na$_2$HPO$_4$) containing 5% (w/v) nonfat dry milk. Affinity-purified anti-E1α and/or anti-E1β IgGs or E2 antiserum is added to the blocking solution, and the incubation is continued overnight at 4° on a rocking platform. Unbound antibodies are removed by washing the blot with several changes of a wash buffer [PBS containing 0.05% (v/v) Tween 20 and 0.05% (v/v) Nonidet P-40 (NP-40)]. The blot is incubated with ^{125}I-labeled protein A (Amersham-Pharmacia) [5 μCi per 100 ml of PBS with 5% (w/v) milk] at room temperature for 3 hr and rinsed with several changes of the wash buffer. Radioactivity bound to antigens is detected by either exposing the blot to X-ray film or by phosphor-imaging analysis.

Figure 1 shows the results of Western blot analysis with cell lysates from normal and MSUD fibroblasts, probed with combined anti-E1α and anti-E1β antibodies (Fig. 1A) or anti-E2 antibodies alone (Fig. 1B). As compared with normal controls, the E2 subunit is immunologically undetectable in cells derived from patients WG-34, E.C., A.L., and GM-612. E1α and E1β subunits are barely detectable in patient L.O. The E1α subunit is reduced in patient P.K., with E1β also undetectable. Further sequence analysis shows that both patients P.K. and L.O. have a mutation in the E1α subunit. The disappearance of normal E1β subunit results from its failure to form stable $\alpha_2\beta_2$ heterotetramers with the mutant E1α subunit. In most cases, affected loci can be easily identified by Western blot analysis. However, if the mutation does not grossly affect the size and abundance of the mutant polypeptide (as shown in type IA patient F.J.), Western blot analysis is not informative. In this situation, vectors carrying a normal cDNA for one of the three catalytic subunits (E1α, E1β, or E2) can be transfected into MSUD cells (see below). Cultured lymphoblasts are preferred because of the ease of handling and the faster rate of growth than fibroblasts. The restoration of decarboxylation of [1-^{14}C]KIV in transfected MSUD cells by one of the vectors will identify the mutant locus.

Isolation of RNA and Genomic DNA from Cultured Maple Syrup Urine Disease Cells

For mutational analysis, total RNA is prepared from freshly harvested lymphoblasts or fibroblasts with an RNeasy Midi kit from Qiagen (Chatsworth, CA). Typically, 500 μg of total RNA is obtained from 20 × 10^6 cells. Isolated RNA is kept frozen at −80°. Genomic DNA is prepared either

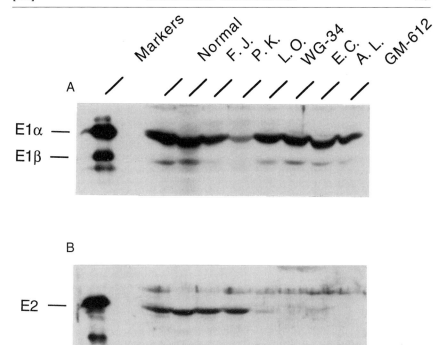

FIG. 1. Western blot analysis with cellular extracts from normal and MSUD fibroblasts. Cellular extracts from cultured fibroblasts were subjected to Western blot analysis (300 μg/ lane) with either a combination of affinity-purified anti-E1α and anti-E1β antibodies (A) or with anti-E2 antibodies, (B) as a probe. Markers of bovine E1α, E1β, and E2 subunits are shown in the far left lane. F. J., P. K. (Mennonite), L. O., E. C., A. L., and GM-612 are classic patients. WG-34 is a thiamine-responsive patient. [From C. W. Fisher, J. L. Chuang, T. A. Griffin, K. S. Lau, R. P. Cox, and D. T. Chuang, *J. Biol. Chem.* **264**, 3448 (1989), with permission.]

from freshly harvested cells or frozen cell stocks, using a standard genomic DNA purification method[14] or the Wizard genomic DNA purification system from Promega (Madison, WI). Isolated genomic DNA is kept frozen at −20°.

Northern Blot Analysis

Northern blot analysis can be used in addition to Western blot analysis to determine if dysfunction is at the mRNA level. Total RNA is isolated

[14] J. Sambrook, E. F. Fritsch, and T. Maniatis, "Molecular Cloning: A Laboratory Manual," 2nd Ed. Cold Spring Harbor Laboratory Press, Cold Spring Harbor, New York, 1989.

from cultured cells as described above. Standard procedures for formaldehyde gel electrophoresis, RNA transfer to nitrocellulose filters, prehybridization, and hybridization are used.[14] cDNA probes are prepared by nick translation, using DNA polymerase I.[14]

Reverse Transcription and Amplification of mRNA

The first-strand cDNA is reverse transcribed from 10 μg of total RNA isolated from MSUD or normal cells. The primer used is designed from the 3'-noncoding region of cDNA for the affected subunit. Reverse transcription is carried out with the avian myeloblastosis virus (AMV) reverse transcriptase (Promega). Polymerase chain reaction (PCR) amplification of the entire cDNA is accomplished by using the 5' and 3' primers flanking the coding region and *Taq* DNA polymerase.[14] Amplified cDNA is sequenced directly with a Thermo Sequenase radiolabeled terminator cycle sequencing kit from Amersham-Pharmacia.

Sequencing of Genomic DNA

Mutations detected by cDNA sequencing need to be confirmed in the corresponding genomic sequences from the patient to rule out PCR artifacts. Exons containing the putative mutations are amplified with flanking intronic primers and directly sequenced. Primer sequences designed for amplifying exons of the human E1α[15,16] and E2[17] genes have been published. When a heterozygous mutation in one allele results in unstable mRNA, RT-PCR of total RNA often leads to an exclusive amplification of the second allele.[16] Direct sequencing of the amplified cDNA may erroneously indicate a homozygous mutation. In this case, amplification and direct sequencing of exons involved from genomic DNA are essential.

Mutations involving exon deletions, or deletions/insertions in mRNA often results from intronic mutations at or near the 5' or 3' splice site.[17] However, if an intronic mutation involves a large deletion distant from the 5' and 3' splice sites, its identification becomes less than straightforward.[17] For example, the IVS4del[−3.2kb : −15] allele produces a 17-bp insertion in the E2 mRNA immediately 5' to exon 5 (Fig. 2). The insertion was initially detected in a cDNA subclone from type II (E2-deficient) MSUD patient WG-34, who was a compound heterozygote for this allele. However,

[15] J. L. Chuang, J. R. Davie, J. M. Chinsky, R. M. Wynn, R. P. Cox, and D. T. Chuang, *J. Clin. Invest.* **95**, 954 (1995).

[16] J. L. Chuang, C. R. Fisher, R. P. Cox, and D. T. Chuang, *Am. J. Hum. Genet.* **55**, 297 (1994).

[17] J. L. Chuang, R. P. Cox, and D. T. Chuang, *J. Clin, Invest.* **100**, 736 (1997).

E2 GENE

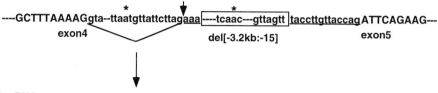

E2 mRNA

17-bp insertion
----GCTTTAAAAG AAATACCTTGTTACCAG ATTCAGAAG---
     exon4   exon5

FIG. 2. Amplification and sequencing of intron 4 of the E2 gene, showing a 3.2-kb intronic deletion in the IVS4del[−3.2kb : −15] allele, which produces aberrant splicing. The 3.2-kb deletion occurs immediately 5′ to the 14-bp intronic sequence present in the E2 transcript. The *aaa* trinucleotides that are also present in the 17-bp insert are located between the 3.2-kb deletion and the invariant *ag* dinucleotides that are used as a cryptic 3′ acceptor splice site. The lowercase letters represent intronic sequences. The uppercase letters refer to exonic or mRNA sequences. The 17-bp inserted intronic sequence in the mRNA is underlined. The original (base−29) and the new branch point based on a consensus sequence YTRAY are indicated by asterisks (∗). [From J. L. Chuang, R. P. Cox, and D. T. Chuang, *J. Clin. Invest.* **100,** 736 (1997), with permission.]

the authenticity of the 17-bp insertion allele was questioned because there were no intronic alterations when 5′ and 3′ splice site regions of exon 5 of this patient were amplified. Later, a second type II MSUD patient was identified as homozygous for the same 17-bp insertion in the E2 mRNA. The inability to amplify exon 5 from this patient led to the speculation of an intronic deletion encompassing the sense primer region. This prompted the amplification of the entire intron 4 from a normal subject (11.2 kb) and the second homozygous type II MSUD patient, using the Expand long template PCR system from Boehringer GmbH (Mannheim, Germany). Subcloning and sequencing of restriction fragments of the PCR products identified a 3.2-kb internal segment of intron 4 that was truncated in the homozygous IVS4del[−3.2kb : −15] allele of the second patient. Figure 2 depicts the mechanism for an aberrant splicing that involves the activation of a cryptic splice site in intron 4, resulting in the 17-bp insertion in the E2 mRNA. These studies demonstrate the utility of long PCR in analyzing internal intronic mutations, and emphasize the importance of homozygote availability in detecting secondary insertions/deletions caused by this type of mutations.[17]

Functional Characterization of Maple Syrup Urine Disease Mutations

Missense mutations detected by nucleotide sequencing described above need to be established functionally as the cause of the MSUD phenotype. This is accomplished by introducing a mutation-containing cDNA into MSUD cells that are deficient in that subunit. The inability of the mutation-containing cDNA to restore BCKD activity in the mutant host cells establishes the mutation as a cause of MSUD. Restoration of BCKD activity with a normal cDNA for the subunit in question serves as a positive control.[18]

The EBO-pLPP plasmid[19] is used as a vector for carrying cDNAs of different subunits into cultured lymphoblasts. This vector contains the EBNA-1 sequence, which encodes the trans-acting EBV nuclear antigen. It also harbors the HPH sequence encoding hygromycin phosphotransferase, which confers antibiotic resistance. The EBO plasmid containing a normal or mutant cDNA of a different BCKD subunit is transfected into host lymphoblasts by electroporation at 250 V and 1180 μF in a Cell-Porator (GIBCO-BRL). Transfected cells are cultured in RPMI 1640 medium with 20% (v/v) fetal bovine serum for 48 hr. Viable cells are separated from nonviable cells by a Ficoll gradient. Hygromycin selection is initiated by replacing half of the medium with that containing hygromycin (200 μg/ml) at each feeding, and is maintained during the entire culture period. Transfected cells are harvested and the intact-cell assay is performed with [1-^{14}C]KIV as substrate, as described previously.

Prenatal Diagnosis

For families with known MSUD offspring, if the mutation is not yet identified, prenatal diagnosis can be performed by intact-cell assays of cultured aminionic cells[20,21] or chorionic villus sampling (CVS) cells.[22] Culture of these cell types is similar to that described for fibroblasts. If control aminionic or CVS cells are not available, normal and MSUD fibroblasts can be used as positive and negative controls, respectively.

A DNA-based prenatal diagnosis has been performed in this laboratory

[18] C. R. Fisher, J. L. Chuang, R. P. Cox, C. W. Fisher, R. A. Star, and D. T. Chuang, *J. Clin. Invest.* **88,** 1034 (1991).

[19] R. F. Margolskee, P. Kavathas, and P. Berg, *Mol. Cell. Biol.* **8,** 2837 (1988).

[20] R. P. Cox, J. Hutzler, and J. Dancis, *Lancet* **2,** 212 (1978).

[21] U. Wendel, H. W. Rudiger, E. Passarge, and M. Mikkelsen, *Humangenetik* **19,** 127 (1973).

[22] W. J. Kleijer, D. Horsman, G. M. Mancini, A. Fois, and J. Boue, *N. Engl. J. Med.* **313,** 1608 (1985).

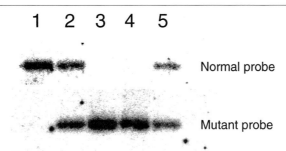

FIG. 3. DNA-based prenatal diagnosis of cultured CVS cells. Exon 6 of the human E1α gene was amplified from genomic DNA prepared from cultured CVS cells or fibroblasts. Amplified products were subjected to allele-specific probing with a normal probe or a mutant probe containing a G-to-C transversion at the 5′ splice site. Lane 1, normal subject; lane 2, the heterozygous father; lane 3, a homozygous affected sibling; lane 4, CVS 1 from a previous pregnancy; lane 5, CVS 2 from the current pregnancy of the same heterozyous mother.

for a family with a known homozygous MSUD mutation.[23] This diagnosis offers much higher sensitivity and accurracy than the above-described enzyme-based assay. Genomic DNA was prepared from cultured CVS cells (50 μg of DNA/1 × 10^6 cells) using the Wizard genomic DNA purification kit (Promega). The exon of interest is amplified by PCR from 1 μg of genomic DNA. Amplified DNA fragments are transferred to GeneScreen Plus membranes (Du Pont, Wilmington, DE) and subjected to allele-specific oligonucleotide (ASO) probing. Figure 3 shows results of the DNA-based prenatal diagnosis. The family carried a homozygous exon 6 deletion in the E1α gene (type IA MSUD) caused by a G-to-C transversion at the 5′-splice donor site. CVS cells obtained from previous pregnancy (CVS 1) and the current pregnancy (CVS 2) were diagnosed. Results obtained with a ^{32}P-labeled normal probe and a mutant probe, unequivocally show that CVS 1 (Fig. 3, lane 4) is homozygous whereas CVS 2 (lane 5) is heterozygous for the mutation. DNA samples from fibroblasts of a normal subject (lane 1), the heterozygous father (lane 2), and a homozygous-affected sibling (lane 3) were also studied for comparison. The advantage of using CVS cell culture for prenatal diagnosis is that cell samples can be obtained early in the first trimester. An important caveat, however, is that maternal contribution must be ruled out by chromosome typing the CVS cell culture.

[23] J. L. Chuang and D. T. Chuang, unpublished observations, 1999.

[39] Detection of Gene Defects in Branched-Chain Amino Acid Metabolism by Tandem Mass Spectrometry of Carnitine Esters Produced by Cultured Fibroblasts

By CHARLES R. ROE *and* DIANE S. ROE

Introduction

Many inherited defects involving the acyl-CoA dehydrogenases in the catabolic pathways of the branched-chain amino acids have been characterized.[1] An inherited disorder due to electron-transferring flavoprotein–dehydrogenase (ETF–QO) deficiency has been characterized that mimicks the biochemical effects of hypoglycin. Like Jamaican vomiting sickness, which results from toxicity of the unripened ackee (akee) fruit, this disorder affects all FAD-dependent acyl-CoA dehydrogenases and has been designated glutaric aciduria type II (GA II) or multiple acyl-CoA dehydrogenase deficiency (MADD).[2] This autosomal inherited disorder affects not only the branched-chain amino acid pathways but also mitochondrial acyl-CoA dehydrogenases (required for fatty acid degradation) and glutaryl-CoA dehydrogenase (serving the metabolism of lysine, hydroxylysine, and tryptophan).

In rat liver, there is a single acyl-CoA dehydrogenase protein that uses both isobutyryl-CoA (valine pathway) and S-2-methylbutyryl-CoA (isoleucine pathway) as equally preferred substrates and has been designated "2-methyl-branched" chain dehydrogenase.[3–5] Humans have a similar enzyme. Subsequently, deficiency of this 2-methyl-branched chain dehydrogenase, affecting the metabolism of both valine and isoleucine, has been proposed in the diagnosis of a few children.[6] The procedure described here for probing branched-chain amino acid pathways in intact fibroblasts was

[1] L. Sweetman and J. C. Williams, *in* "The Metabolic and Molecular Bases of Inherited Diseases" (C. R. Scriver, A. L. Beaudet, W. S. Sly, and D. Valle, eds.), p. 1387. McGraw-Hill, New York, 1995.

[2] F. E. Frerman and S. I. Goodman, *in* "The Metabolic and Molecular Bases of Inherited Diseases" (C. R. Scriver, A. L. Beaudet, W. S. Sly, and D. Valle, eds.), p. 1611. McGraw-Hill, New York, 1995.

[3] Y. Ikeda, C. Dabrowski, and K. Tanaka, *J. Biol. Chem.* **258,** 1066 (1983).

[4] Y. Ikeda and K. Tanaka, *J. Biol. Chem.* **258,** 1077 (1983).

[5] Y. Ikeda and K. Tanaka, *J. Biol. Chem.* **258,** 9477 (1983).

[6] A. B. Burlina, F. Zacchello, C. Dionisi-Vici, E. Bertini, G. Sabetta, M. J. Bennett, D. E. Hale, E. Schmidt-Sommerfeld, and P. Rinaldo, *Lancet* **338,** 1522 (1991).

developed to diagnose a child who appeared to have an isolated deficiency of isobutyryl-CoA dehydrogenase. Because of the overlap in substrate specificities of several enzymes using four-carbon compounds as substrates, direct enzyme assay in fibroblasts was not able to distinguish an isolated defect specific for isobutyryl-CoA. The method for probing the branched-chain pathways in intact cells proved to be useful in recognizing several inherited defects as well as distinguishing between the mild and severe clinical phenotypes of ETF–QO dehydrogenase deficiency (GA II or MADD).

Procedure

Reagents

Culture media: Two different media are used in this analysis: (1) Dulbecco's modified Eagle's medium (DMEM), low glucose (5.5 mM), and L-glutamine (4 mM), and (2) DMEM, D-glucose at 4500 mg/ liter (25 mM), and L-glutamine (4 mM), but devoid of L-leucine, L-isoleucine, and L-valine (GIBCO, Grand Island, NY). The complete medium is prepared by adding 10% (v/v) fetal bovine serum albumin (FBS) and 1% (v/v) antibiotic/antimycotic to the base media

Other materials and solutions: trypsin–EDTA (1×) [0.05% (w/v) trypsin in 0.53 mM EDTA]; phosphate-buffered saline (1×) (PBS), pH 7.2; fatty acid-free bovine serum albumin (BSA) (60 mg/ml in serum-free medium); L-carnitine (inner salt), 161 (40 mM in serum-free medium)

Labeled precursors to the fat oxidation and branched-chain amino acid metabolic pathways: [16-*methyl*-^{2}H$_3$]palmitic acid (98%, MW 259, 20 mM in absolute ethanol); L-[U-^{13}C$_6$]leucine (97–98%, MW 137, 40 mM in deionized water); L-[U-^{13}C$_6$]isoleucine (97–98%, MW 137, 40 mM in deionized water); and L-[U-^{13}C$_5$]valine (97–98%, MW 122, 40 mM in deionized water). Filter sterilize these solutions with a 0.2-μm × 25 mm syringe filter into sterile 1.5-ml aliquots. Store frozen at $-20°$

Preparation of Probe Media

A 3.5-ml volume of probe medium is used per 25-cm^2 flask.

[16-^{2}H$_3$]Palmitic Acid Probe Medium. For each culture flask, add 35 μl of 20 mM [16-^{2}H$_3$]palmitic acid to a sterile centrifuge tube and, using a sterile 2-ml pipette, dry under nitrogen until barely dry. (*Note:* Do not overdry.) Perform the remaining procedures under a laminar flow hood,

using aseptic technique. For each flask, add 350 μl of filtered BSA to the tube containing dry palmitic acid. Place the tubes in a 37° water bath for approximately 30–60 min. Pipette 3.15 ml of complete medium into each tube containing the BSA–palmitate mixture. For each flask, add 35 μl of filtered L-carnitine (40 mM) to the probe medium. Mix well. The final concentration of components in each 25-cm^2 culture flask is 2 mM BSA, 0.2 mM [16-^{2}H$_3$]palmitic acid, and 0.4 mM L-carnitine.

Labeled Branched-Chain Amino Acid Probe Medium. For each flask, add 70 μl of the desired labeled amino acid to a sterile centrifuge tube. For each flask add 3.4 ml of complete amino acid-devoid DMEM and 35 μl of L-carnitine and mix well. The final concentration of labeled amino acid in each flask is 0.8 mM with 0.4 mM L-carnitine.

Incubation of Cell Lines

Fibroblasts are grown to 80–100% confluency (0.1–0.3 mg of protein) in DMEM containing 10% (v/v) FBS and 1% (v/v) antibiotic/antimycotic at 37° in a humidified 5% CO_2–95% air (v/v) incubator. When the cells are nearly confluent, remove the old medium, rinse the cells with 3–5 ml of PBS, and add 3.5 ml of the freshly prepared probe medium to each flask. Incubate for 72 hr.

Cell Harvest

Examine each flask by microscope for signs of contamination or cell death. If cell death is less than 5%, the medium can be transferred to a 5-ml vial and stored frozen at −20°. Medium containing a large number of dead cells should be centrifuged for 5 min at 2500 rpm to remove dead cells before storage. Add 1 ml of trypsin–EDTA to detach cells. Add 5 ml of DMEM to neutralize the trypsin and ensure total detachment of cells. Transfer the detached cell mixture to a 15-ml conical centrifuge tube. Rinse each flask two times with 4 ml of PBS to collect any remaining cells. Transfer the PBS rinses to the same conical test tube. Examine each flask under a microscope to ensure total cell harvest. Repeat trypsinization if necessary. Centrifuge for 5 min at 2500 rpm. Without disturbing the cell pellet, aspirate all but 1 ml of supernatant. Rinse the cell pellet twice more with 10 ml of PBS and centrifuge as described above. Aspirate all PBS without disturbing the cell pellet. Add 300 μl of deionized water to the pellet, gently homogenize, and store frozen at or below −20° until analysis. The cell homogenate is used for protein determination as well as tandem mass spectrometry (MS–MS) analysis.

Sample Preparation for Tandem Mass Spectrometry

For MS–MS,[7] add 5 μl of internal standard containing [^{2}H$_5$]propionyl-carnitine (40 pmol), [^{2}H$_7$]butyrylcarnitine (20 pmol), [^{2}H$_6$]octanoylcarnitine (10 pmol), and [^{2}H$_6$]palmitoylcarnitine (50 pmol) to a labeled 1.5-ml micro-tube.[7] Add 67 μl of incubation medium and 33 μl of cell homogenate; vortex. Add 800 μl of 100% ethanol to precipitate proteins; cap and vortex vigorously. Centrifuge for 5 min at 14,000 rpm. Transfer the supernatant to a glass 12 × 32 mm vial and evaporate to dryness under nitrogen. Add 100 μl of 3 M methanolic HCL; cap and vortex. Heat in a dry block heater for 15 min at 50°. Decap the vial and evaporate to dryness under nitrogen. Reconstitute the derivatized (methyl esters) sample in 50 μl of octyl sodium sulfate (OSS; 1%, w/v) in methanol–glycerol (1 : 1, v/v). [*Note:* For butyl ester derivatization replace 3 N methanolic HCl with 3 N butanolic HCl, heat the samples at 65° for 15 min, and reconstitute in 0.1% (w/v) OSS matrix.]

Tandem Mass Spectrometry

Any tandem mass spectrometer capable of performing precursor (parent) scan function and equipped with the correct ion source [liquid secondary ionization mass spectrometry (LSIMS), fast atom bombardment (FAB), or electrospray ionization (ESI)] can be used to analyze the acylcarnitines produced by the *in vitro* procedures described above. Acylcarnitines have a unique structural property, a quaternary ammonium group, which forms a preformed cation amenable to the above-cited ionization processes. Esterification of the carboxyl group with methanol or butanol improves detection limits of acylcarnitines (into the low picomolar range) from all types of bodily fluids and tissue. The method detects the common fragment, m/z 99 (methyl esters) or m/z 85 (butyl esters), produced from the collision-induced dissociation of the esterified acylcarnitines. The second quadrupole is set at m/z 99 (or m/z 85) and the first quadrupole is scanned from m/z 200 to 550, generating a profile predominantly of acylcarnitines with little interference from other biological components.

Application

The status of various short-chain acyl-CoA dehydrogenases was analyzed in intact fibroblasts. Normal cells and cells deficient in butyryl-CoA dehydrogenase (SCAD), and cells presenting both the severe and mild

[7] B. Z. Yang, J. H. Ding, T. Dewese, D. Roe, G. He, J. Wilkinson, D. W. Day, F. Demaugre, D. Rabier, M. Brivet, and C. Roe, *Mol. Genet. Metab.* **64**, 229 (1998).

phenotypes of ETF–QO dehydrogenase (ETF–QO severe and ETF–QO mild), were examined after incubation with [16-^{2}H$_3$]palmitate and [U-^{13}C]-labeled leucine, valine, and isoleucine. For cells incubated with palmitate, the following labeled acylcarnitines were measured: palmitoylcarnitine (C$_{16}$), decanoylcarnitine (C$_{10}$), octanoylcarnitine (C$_8$), and butyrylcarnitine (C$_4$) (Fig. 1). None of these were significantly elevated in normal fibroblasts. SCAD-deficient cells revealed the characteristic increase of butyrylcarnitine as well as a mild elevation of the other acylcarnitines, reflecting the impact of the deficiency on the rate of β oxidation. ETF–QO severe cells demonstrate the complete absence of palmitate oxidation with the corresponding elevation of palmitoylcarnitine. In this form of the disease, there were no other labeled acylcarnitine intermediates. In contrast, in the milder phenotype (ETF–QO mild) considerably less elevation of palmitoylcarnitine was apparent, accompanied by evidence of oxidation with labeled intermediates going down to C$_8$ but no label in C$_4$ or C$_6$ compounds. This is consistent with peroxisomal degradation to the level of C$_8$, but in view of the nonfunctioning medium-chain acyl-CoA dehydrogenase (MCAD) due to the primary deficiency, mitochondrial oxidation cannot proceed. This represents an interesting cooperativity between peroxisomal and mitochondrial fat oxidation, and illustrates as well a clear distinction between the phenotypes of this disease.

Because all three acyl-CoA dehydrogenases involved in branched-chain amino acid metabolism are affected in ETF–QO dehydrogenase deficiency, these cell lines were examined with the labeled precursor to each metabolic

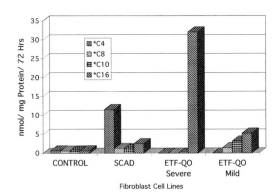

Fig. 1. Levels of acylcarnitine intermediates as methyl esters following a 72-hr incubation of fibroblasts from control, SCAD-deficient (butyryl-CoA dehydrogenase), and both severe and mild forms of ETF–QO dehydrogenase deficiency with [16-^{2}H$_3$]palmitate. *C4, [^{2}H$_3$]Butyrylcarnitine; *C8, [^{2}H$_3$]octanoylcarnitine; *C10, [^{2}H$_3$]decanoylcarnitine; *C16, [^{2}H$_3$]palmitoylcarnitine.

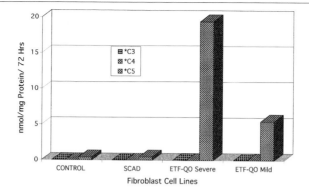

FIG. 2. Levels of acylcarnitine intermediates as methyl esters following a 72-hr incubation of fibroblasts from control, SCAD-deficient (butyryl-CoA dehydrogenase), and both severe and mild forms of ETF–QO dehydrogenase deficiency with $[^{13}C_6]$leucine. *C5, $[^{13}C_5]$Isovalerylcarnitine. No labeled C_3 or C_4 acylcarnitine intermediates were detected with $[^{13}C_6]$leucine.

pathway. Figure 2 illustrates the results with $[^{13}C_6]$leucine. Leucine oxidation can produce isovaleryl-CoA (C_5) but there are no three- or four-carbon intermediates to be observed as acylcarnitines. As expected, there was no significant difference between normal and SCAD-deficient cells in terms of the labeled isovalerylcarnitine (C_5). In the ETF–QO dehydrogenase phenotypes, the levels of labeled isovalerylcarnitine were markedly increased. It is not clear why the amount in the severe phenotype is so much greater than that observed with the mild form of the disease.

Incubation of cells with $[^{13}C_5]$valine is expected to produce $[^{13}C_4]$isobutyrylcarnitine (C_4) as well as $[^{13}C_3]$propionylcarnitine if the entire pathway is intact; however, there is no five-carbon intermediate in the pathway that appears as an acylcarnitine. This example illustrates the equivalent increase in labeled isobutyrylcarnitine in both phenotypes of ETF–QO dehydrogenase deficiency but without detection of labeled propionylcarnitine (Figure 3).

Incubation of cells with $[^{13}C_6]$isoleucine is expected to produce labeled S-2-methylbutyrylcarnitine (C_5) as well as labeled propionylcarnitine (C_3) if the pathway is intact (Fig. 4). The amounts of these two acylcarnitines were equivalent for the normal and SCAD-deficient fibroblasts. ETF–QO severe was characterized by a marked increase in S-2-methylbutyrylcarnitine (C_5) but no label appeared in propionylcarnitine (C_3). Cells representing the milder phenotype produced lower but also significant levels of S-2-methylbutyrylcarnitine (C_5) and, surprisingly, near-normal levels of $[^{13}C_3]$propionylcarnitine. This finding may contribute to the milder clinical

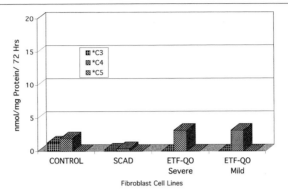

FIG. 3. Levels of acylcarnitine intermediates as methyl esters following a 72-hr incubation of fibroblasts from control, SCAD-deficient (butyryl-CoA dehydrogenase), and both severe and mild forms of ETF–QO dehydrogense deficiency with [$^{13}C_5$]valine. *C4, [$^{13}C_4$]Isobutyrylcarnitine; *C3, [$^{13}C_3$]propionylcarnitine. There was no labeled C_5 acylcarnitine.

course of this form of the disease, as it suggests that gluconeogenesis from isoleucine via propionyl-CoA is intact.

Discussion

Probing the fat oxidation pathways (mitochondrial and peroxisomal), as well as the branched-chain amino acid pathways, in intact cells, is a useful method for identifying a number of inherited biochemical disorders

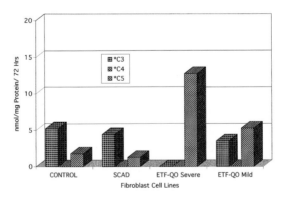

FIG. 4. Levels of acylcarnitine intermediates as methyl esters following a 72-hr incubation of fibroblasts from control, SCAD-deficient (butyryl-CoA dehydrogenase), and both severe and mild forms of ETF–QO dehydrogenase deficiency with [$^{13}C_6$]isoleucine. *C5, S-2-methyl [$^{13}C_5$]butyrylcarnitine; *C3, [$^{13}C_3$]propionylcarnitine. There was no labeled C_4 acylcarnitine.

in children. The ability to distinguish clinical phenotypes of ETF–QO dehydrogenase deficiency has real significance if applied to prenatal diagnosis of families at risk. The ability to distinguish these phenotypes by probing with palmitate and isoleucine emphasizes the potential role of alternative pathways influencing clinical severity in children with this disorder. Other disorders that have been clearly recognized by probing intact cells include propionic acidemia and the various forms of methylmalonic acidemia by the marked elevation of labeled propionylcarnitine (not shown). This method has also successfully characterized an inherited defect of isobutyryl-CoA dehydrogenase deficiency that is distinct from the 2-methyl-branched chain dehydrogenase that serves both the valine and isoleucine pathways.[8] Further development of this method to include quantification of free acid intermediates from the culture medium and cells will be of great value for critically evaluating other known and, as yet, unknown inherited defects affecting the branched-chain amino acid pathways.

Characterization of S-2-Methylbutyryl-Coenzyme A Dehydrogenase Deficiency (S-2-MBCDase)

The method described earlier was successful for the characterization of the first documented case of S-2-MBCDase, an inherited defect unique to the metabolism of L-isoleucine in humans.[9] The demonstration of an isolated deficiency of isobutyryl-CoA dehydrogenase in the L-valine pathway and the corresponding deficiency of S-2-methylbutyryl-CoA dehydrogenase in the L-isoleucine pathway indicates two separate enzymes involved in humans which have now been further characterized by molecular techniques. This methodology provides a more specific diagnostic approach for the resolution of deficiencies of short-chain acyl CoA dehydrogenases in human fibroblasts.

[8] C. R. Roe, S. D. Cederbaum, D. S. Roe, R. Mardach, A. Galindo, and L. Sweetman, *Mol. Genet. Metab.* **65,** 264 (1998).
[9] K. M. Gibson, T. Burlingame, B. Hogema, C. Jakobs, R. Schutgens, D. Millington, C. Roe, D. Roe, L. Sweetman, R. Steiner, L. Linck, P. Pohowalla, M. Sacks, D. Kiss, P. Rinaldo and J. Vockley, *Ped. Res.* **47**(6), 2000 (in press).

[40] Molecular and Enzymatic Methods for Detection of Genetic Defects in Distal Pathways of Branched-Chain Amino Acid Metabolism

By K. Michael Gibson, Magdalena Ugarte, Toshiyuki Fukao, and Grant A. Mitchell

This chapter focuses on the enzymatic and molecular characterization of inherited defects in the distal pathways of branched-chain amino acid (L-isoleucine, L-leucine, and L-valine) metabolism (Fig. 1). The disorders covered include 3-methylcrotonyl-coenzyme A (CoA) carboxylase (MCC) deficiency (EC 6.4.1.4; MIM 210200), propionyl-CoA carboxylase (PCC) deficiency (EC 6.4.1.3; MIM 232000 and 232500, types A and B), 3-methylglutaconyl-CoA (3-MG-CoA) hydratase deficiency (EC 4.2.1.18; MIM 250950), 3-hydroxy-3-methylglutaryl-CoA (HMG-CoA) lyase deficiency (EC 4.1.3.4; MIM 246450), short-chain 3-ketoacyl-CoA thiolase deficiency (or β-ketothiolase, isoleucine specific) (EC 2.3.1.9; MIM 203750), and putative methylmalonate-semialdehyde dehydrogenase (MMSDH) deficiency (EC 1.2.1.27; MIM 236795). Disorders involving branched-chain keto acid dehydrogenase (maple syrup urine disease), isovaleryl-CoA dehydrogenase (isovaleric acidemia), and methylmalonyl-CoA mutase (methylmalonic acidemia) are discussed in [24], [38], and [39] in this volume[1] (Fig. 1). In addition, those disorders involving only a single reported patient (3-hydroxyisobutyryl-CoA deacylase and putative isobutyryl-CoA dehydrogenase deficiencies) are not discussed.[1a,2]

The enzymatic and molecular techniques presented here are employed in the authors' laboratories for diagnosis of the respective disorders in routinely available human samples, including leukocytes isolated from whole blood, cultured skin fibroblasts, cultured amniocytes, and cultured/biopsied chorionic villi tissue. With the exception of β-ketothiolase, all techniques presented are radiometric. Several other spectrophotometric/tendem mass spectrometric techniques are available to assay some of the enzymes discussed in this chapter, and even in the same tissues. Many

[1] J. L. Chuang and D. T. Chuang, *Methods Enzymol.* **324**, [38], 2000 (this volume), J. Vockley, A.-W. A. Mohsen, B. Binzak, J. Willard, and A. Fauq, *Methods Enzymol.* **324**, [24], 2000 (this volume); C. R. Roe and D. S. Roe, *Methods Enzymol.* **324**, [39], 2000 (this volume).

[1a] G. K. Brown, S. M. Hunt, R. Scholem, K. Fowler, A. Grimes, J. F. B. Mercer, R. M. Truscott, R. G. H. Cotton, J. G. Rogers, and D. M. Danks, *Pediatrics* **70**, 532 (1982).

[2] C. R. Roe, S. C. Cederbaum, D. S. Roe, R. Mardach, and A. Galindo, *J. Inher. Metab. Dis.* **21**, 54 (1998).

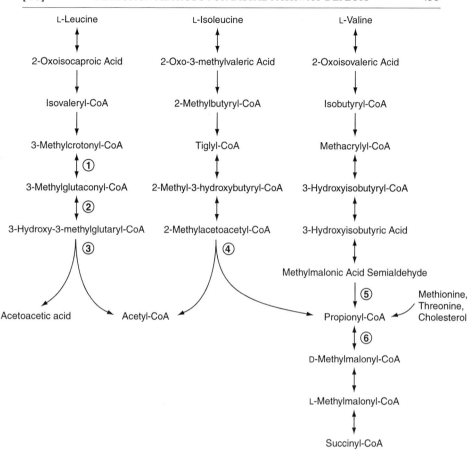

FIG. 1. Catabolic pathways for the branched-chain amino acids L-leucine, L-isoleucine, and L-valine. Specific enzymes discussed in this chapter are depicted by numbers: (1) 3-methylcrotonyl-CoA carboxylase (MCC); (2) 3-methylglutaconyl-CoA (3-MG-CoA) hydratase; (3) 3-hydroxy-3-methylglutaryl-CoA (HMG-CoA) lyase; (4) 2-methylacetoacetyl-CoA thiolase (β-ketothiolase, or short-chain 3-ketoacyl-CoA thiolase); (5) methylmalonate-semialdehyde dehydrogenase; and (6) propionyl-CoA carboxylase (PCC).

of these alternative methods are applicable to other tissues (i.e., plant, mammalian organs, and bacterial sonicates containing recombinant enzymes), and are useful for monitoring enzyme purification protocols when radiometric analyses are impractical. (see [3] and [4] in this volume[2a]).

[2a] P. Schadewaldt, *Methods Enzymol.* **324,** [3], 2000 (this volume); P. Schadewaldt, *Methods Enzymol.* **324,** [4], 2000 (this volume).

The reader should not feel that the methods described here are the only methods available.

Assay of Short-Chain 3-Ketoacyl-CoA Thiolase Ketothiolase (β-Ketothiolase) Activity

Short-chain 3-ketoacyl-CoA thiolase (β-ketothiolase; 3-oxothiolase) is an enzyme of the L-isoleucine catabolic pathway (Fig. 1). All mammalian cells have short-chain thiolase enzymes that are active in the cytosol and not stimulated by K^+ ion; conversely, β-ketothiolase is a mitochondrial enzyme that is stimulated by K^+ ion. Thus, assay of β-ketothiolase activity in the presence and absence of K^+ ion serves as an estimate of mitochondrial short-chain 3-ketoacyl-CoA thiolase activity. Because the β-ketothiolase assay relies on the ratio of activity with and without K^+ ion, it is important that the sample be free of K^+. Any washes of cells to be assayed (fibroblasts, leukocytes, amniocytes, or chorionic villi) should be carried out in 0.9% (w/v) saline. Also, when adjusting the pH of reaction buffers, potassium hydroxide should be avoided.

Reaction mixtures are monitored at 303 nm and 30° in a water-jacketed cuvette, using an Aminco SLM DW 2000 spectrophotometer with a split beam selector (Spectronic Unicam, Rochester, NY). Cell sonicate (fibroblasts or leukocytes, although in our hands the latter require at least 30 ml of whole venous blood in order to obtain sufficient leukocytes for assay) is prepared in 50 mM sodium phosphate, pH 8.0, supplemented with 0.1% (v/v) Triton X-100. The final assay includes (total volume, 1 ml): 100 mM Tris-HCl (pH 8.0), 50 mM $MgCl_2$, 50 mM KCl, 10–15 μM acetoacetyl-CoA (sufficient to produce an optical density of approximately 0.5 at 303 nm, and prepared in 10 mM sodium acetate, pH 5.2), and cell sonicate. Potassium chloride is omitted from assays monitoring activity without potassium. Linearity of reaction rates should be assessed in the respective laboratory, although analyses with potassium are linear with respect to fibroblast sonicate up to 0.28 mg of protein, and those without potassium are linear to 0.19 mg of protein. All reagents are mixed and incubated for 5 min at 30°. The baseline is then monitored for at least 1 min, after which coenzyme A is added [final concentration 65 μM, lithium salt, dissolved in 40 mM aqueous dithiothreitol (DTT)] and the change in optical density (OD) monitored for an appropriate length of time. Final enzyme-specific activity is determined by correlating the change in OD with time (minus background OD) with respect to protein content, and using a molar extinction coefficient of 21.4 mM^1 cm^{-1}.

A radiometric analysis has been used to estimate β-ketothiolase activity. In this method, tiglyl-CoA was incubated in the presence of NAD^+ and

TABLE I

β-Ketothiolase Activities in Sonicates of Fibroblasts Derived from Controls and Patients in Presence and Absence of Potassium[a]

Subject	$+K^+$	$-K^-$	Ratio $+K^+/-K^+$
Control ($n = 8$)	11.1 ± 3.0	6.3 ± 2.1	1.8 ± 0.3
	(5.6–15.9)	(2.4–10.1)	(1.3–2.3)
Patient 1	4.7	4.8	1.0
Patient 2	4.7	4.6	1.0
Patient 3	10.6	10.2	1.0
Patient 4	7.4	9.4	0.8

[a] Two of four patients have been reported and studied elsewhere [K. M. Gibson, C. F. Lee, V. Kamali, and O. Sovik, *Clin. Chim. Acta* **205,** 127 (1992)]. All enzyme activities are reported as nanomoles per minute per milligram of protein.

coenzyme A to generate propionyl-CoA *in situ* due to the action of β-ketothiolase, in conjunction with the endogenous activities of tiglyl-CoA hydratase and 2-methyl-3-hydroxybutyryl-CoA dehydrogenase in fibroblast sonicates (see Fig. 1). The amount of propionyl-CoA formed is estimated with endogenous propionyl-CoA carboxylase and fixation of radiolabeled sodium bicarbonate, as described in the preceding sections. Propionyl-CoA carboxylase activity thus serves to estimate β-ketothiolase activity.[3] However, this assay has been discontinued because a reliable commercial supplier for tiglyl-CoA is no longer available. As shown in Table I, patients with β-ketothiolase deficiency do not show stimulation with K^+ ion, while control cell lines demonstrate a measurable stimulation.

Mutation Analysis of Short-Chain 3-Ketoacyl-CoA Thiolase

General Considerations

β-Ketothiolase deficiency is a deficiency of mitochondrial acetoacetyl-CoA thiolase (T2). The human T2 cDNA has a 1281-base open reading frame, and its gene spans 27 kb and contains 12 exons.[4,5] Table III shows

[3] K. M. Gibson, C. F. Lee, V. Kamali, and O. Sovik, *Clin. Chim. Acta* **205,** 127 (1992).

[4] T. Fukao, S. Yamaguchi, M. Kano, T. Orii, Y. Fujiki, T. Osumi, and T. Hashimoto, *J. Clin. Invest.* **86,** 2086 (1990).

[5] M. Kano, T. Fukao, S. Yamaguchi, T. Orii, T. Osumi, and T. Hashimoto, *Gene* **109,** 285 (1991).

TABLE II

Oligonucleotides for Mutation Analysis of T2 Deficiency[a]

RT-PCR	Sense primer	Antisense primer	Product size (bp)
cDNA synthesis			
Specific primer		5'-^{1360}tgacccacagtagtcacac-3'	
cDNA amplification			
Fraction A	5'-$^{-40}$agtctacgcctgtggagccga-3'	5'^{646}tagcataagcgtcctgttca-3'	686
Fraction B	5'-^{583}agctgtgctgagaataca-3'	5'-^{1326}ttctggtcacatagggtt-3'	743
Additional sequence primers other than above five primers			
For fraction A	5'-^{140}ctacaagaacacccattgg-3'	5'-^{199}gcttagtggctggcagcaa-3'	
	5'-^{388}atgaaagccatcatgatggc-3'		
For fraction B	5'-^{949}gacgctgctgtagaacctat-3'	5'-^{996}agcatatacaggagcaattgg-3'	
Exon			
1	5'-agtctacgcctgtggagc-3'	5'-tagatctgagtgctgtggggacga-3'	147
2	5'-atgatcataaggatgcaggaagaaa-3'	5'-cgtctagagggaatgttgttttg-3'	401
3,4	5'-ggaagcttctgtcctcaaacct-3'	5'-ggatggtctccatctcttgacctc-3'	884
3(second)	5'-gggtaaaataccctataattt-3'	5'-ttacatacatagttaatgc-3'	245
4(second)	5'-gtgtaatgtcacctacctt-3'	5'-gctgggattacaggcatga-3'	262
5	5'-catgctctaattaagttctgcag-3'	5'-atggatccagacactcttgagca-3'	319
6	5'-acttcctgtattcactgt-3'	5'-tcaacatttatagagcccatc-3'	391
7	5'-atgatcttcctgccttggc-3'	5'-aaagtaatactcagggatctta-3'	535
7(second)	5'-cactataagttaggcaaagt-3'	5'-tgaaaagtctattcatcctt-3'	269
8	5'-atctagatgagtgtttacttgg-3'	5'-acatgagcctgatatactg-3'	438
9	5'-tgcagttagctgtgtacatg-3'	5'-atgatcatgccaccatgctcagcta-3'	441
10	5'-ttgatctaaacttggcttgtgata-3'	5'-aagatcaaccttggtataagcaaac-3'	216
11	5'-gagacagagcaagactgttg-3'	5'-gccttatgaaagatgggtagg-3'	457
12	5'-ttgatcagcaagtagtagtggcta-3'	5'-tgacccacagtagtcacac-3'	316

[a] The A of the initiator ATG codon is designated base number 1 in the cDNA sequence. Underlined nucleotides were introduced for *de novo* restriction sites. For T-A cloning or direct sequencing, the underlined sequences are not essential. The PCR conditions included 40 cycles at 94° for 1 min, 54° for 2 min, 72° for 3 min, and 72° for 7 min for all fragments, using Takara *Taq* (Takara Shuzo) and the Takara PCR thermal cycler (Takara Shuzo).

mutations identified in 19 T2-deficient patients thus far.[6–11] The mutations are distributed widely in the T2 gene and no common mutation has been detected. Hence, diagnostic applications are restricted. Screening for previously identified mutations in T2 deficiency is less useful for diagnosis than

[6] T. Fukao, S. Yamaguchi, T. Orii, R. B. H. Schutgens, T. Osumi, and T. Hashimoto, *J. Clin. Invest.* **89,** 474 (1992).

[7] T. Fukao, S. Yamaguchi, S. Wakazono, T. Orii, G. Hoganson, and T. Hashimoto, *J. Clin. Invest.* **93,** 1035 (1994).

[8] T. Fukao, S. Yamaguchi, T. Orii, and T. Hashimoto, *Hum. Mutat.* **5,** 113 (1995).

in diseases that have limited common mutations, such as medium-chain acyl-CoA dehydrogenase deficiency.[12] To identify mutations in newly identified patients, the mutation analysis is performed at the cDNA level, followed by genomic confirmation of mutations. cDNA analysis provides critical insight into mutations. For example, the Q272X mutation was identified as a nonsense mutation that also is caused by exon 8 splicing.[7] This important observation would not have been made with genomic screening alone. In addition, some mutations causing frame shifts are difficult to identify at the cDNA level.

cDNA Analysis

The human T2 cDNA has a 1281-base open reading frame, which can be amplified in overlapping fragments. Total cellular RNA is extracted from fibroblasts, lymphocytes, or Epstein–Barr virus (EBV)-transformed lymphoblasts, using an Isogen kit (Nippon Gene, Tokyo, Japan; a modified acid guanidinium thiocyanate–phenol–chloroform extraction method). The first-strand cDNA synthesis is performed with 5 μg of total RNA, the specific primer (Table II, 10 pmol), and 200 U of Moloney murine leukemia virus (Mo-MuLV) reverse transcriptase (GIBCO-BRL, Gaithersburg, MD) in a 20-μl volume at 37° for 60 min as recommended by the manufacturer. One microliter of this cDNA solution is used as a template with 30 pmol each of primers, employing conditions described in the footnotes to Table II. After electrophoresis in a 2% (w/v) agarose gel, fragments are purified with a GeneClean II kit (Bio 101, La Jolla, CA), subcloned into a pT7blue T-A cloning vector (Novagen, Madison, WI), sequenced with DNA sequencing kits (Applied Biosystems, Foster City, CA), and analyzed on an ABI 373A DNA sequencer (Applied Biosystems).

Genomic Amplification

The human T2 gene spans about 27 kb and includes 12 exons. Table II shows primer pairs to amplify each exon with 5′ and 3′ intron sequences. Using the same polymerase chain reaction (PCR) conditions for all frag-

[9] T. Fukao, X.-O. Song, S. Yamaguchi, T. Orii, R. J. A. Wanders, B. T. Poll-The, and T. Hashimoto, *Hum. Mutat.* **5,** 94 (1995).

[10] T. Fukao, X.-O. Song, S. Yamaguchi, N. Kondo, T. Orii, J. M. Jean-Marie Matthieu, C. Bachmann, and T. Orii, *Hum. Mutat.* **9,** 277 (1997).

[11] T. Fukao, H. Nakamura, X.-O. Song, K. Nakamura, Y. Kohno, M. Kano, S. Yamaguchi, T. Hashimoto, T. Orii, and N. Kondo, *Hum. Mutat.* **12,** 245 (1998).

[12] K. Tanaka, I. Yokota, P. M. Coates, A. W. Straus, D. P. Kelly, Z. Zhang, N. Gregersen, B. S. Andersen, Y. Matsubara, D. Curtis, and Y.-T. Chen, *Hum. Mutat.* **1,** 271 (1992).

TABLE III
MUTATIONS OF T2 GENE

Exon/intron	Mutation	Nucleotide changes	Nucleotide position	Residual activity[a]	Comments[b]	Cases
E1	M1K	T→A	2	—	Initiation codon mutation	GK08
E2	83del2	Del. of AT	83	ND	FS, PT	GK14
E2	Y33X	T→A	99	ND	PT	GK12
E3	149del1C	Del. of C	149	ND		GK01
E4	delE85	Del. of GAA	253–255	—		GK02
E4	N93S	A→G	278	10%		GK19
E5	K124R	A→G	371	ND		GK18
I5	435+1G→C	G→C		ND	Ex5 skip, FS, PT	GK13, 14
E6	G152A	G→C	455	—		GK02
E6	N158D	A→G	472	—	AM	GK11, 15
E6	G183R	G→A	547	—		GK04, 05
E7	754ins2	Ins. of CT	754	—	FS, PT	GK13
I7–E8	del68	Del. (731-46 to 752)		—	Ex8 skip	GK17
E8	Q272X	C→T	816	—	Partial Ex8 skip	GK07, 15
I8	828+1G→T	G→T		—	Ex8 skip	GK05, 16
E9	T297M	C→T	890	10%		GK11
E9	A301P	G→C	901	5%	AM	GK16
E9	I312T	C→T	935	10%	AM	GK17
E10	A333P	G→C	997	—	AM	GK01
I10	1005-2A→C	A→C		—	Ex11 skip, FS	GK04
I10	1005-1G→C	G→C		—	Ex11 skip, FS	GK10
E11	G379V	G→T	1136	—		GK07
E11	A380T	G→C	1138	7%		GK06
I11	1163+2T→C	T→C		—	GCAGins to cDNA, FS	GK09, 17

[a] Residual activity indicates residual activity estimated by transient expression of the cDNA.

[b] FS, causing frame shift; PT, causing premature termination; AM, mutant protein having altering mobility in SDS–PAGE.

ments, sizes less than 500 bp long are suitable for sequencing and screening mutations by a heteroduplex detection method.

Heteroduplex Detection by Mutation Detection-Enhanced Gel Electrophoresis as Mutation Screening. Heteroduplex detection is a simple and reliable method with which to detect mutations.[13] To detect homozygous mutations, heteroduplex DNA may be formed by mixing fragments from a control and a patient, heating the mixture to 95°, and slowly cooling it

[13] J. Keen, D. Lester, C. Inglehearn, A. Curtis, and S. Bhattacharya, *Trends Genet.* **7**, 5 (1991).

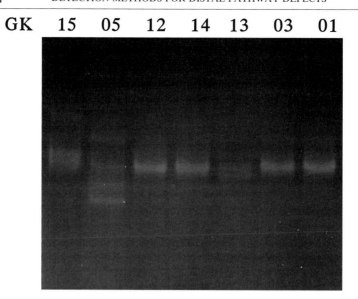

FIG. 2. Heteroduplex detection in exon 8 fragments of the human T2 gene. Fragments from seven T2-deficient patients were electrophoresed on an MDE gel. A single band was detected in patients GK12, GK14, GK03, and GK01. In patients GK15, GK05, and GK13, extra bands with different electrophoretic mobilities are evident. Sequence analysis revealed that GK15 had Q272X, GK05 had IVS8(+1)g to a, and GK13 had intronic polymorphism heterozygously in their fragments.

to room temperature. This method is sensitive enough to detect a single base substitution with the use of a mutation detection-enhanced (MDE) gel (Worthington Biochemical Corp., Lakewood, NJ). The exact protocol is available from the manufacturer. This method detected 17 of 21 gene mutations and is more sensitive than single-strand conformational polymorphism (SSCP) analysis in T2 mutation analysis at the genomic level. Typical results are shown in Fig. 2. This method also was used for familial analysis.[10,11,14,15]

Restriction Enzyme Assay for Familial Analysis. Familial analysis is important to confirm mutations and for genetic counseling. One easy way to confirm familial segregation of mutations is to perform a restriction enzyme assay. If there is no available restriction enzyme, in most cases a

[14] A. Wakazono, T. Fukao, Y. Yamaguchi, T. Hori, T. Orii, M. Lambert, G. A. Mitchell, G. W. Lee, and T. Hashimoto, *Hum. Mutat.* **5,** 34 (1995).
[15] T. Fukao, A. Wakazono, X.-O. Song, S. Yamaguchi, R. Zacharias, M. A. Donlan, and T. Orii, *Prenatal Diagn.* **15,** 363 (1995).

de novo restriction enzyme site can be introduced for either a normal or mutant allele, using modified PCR with a mismatch primer.[14,16]

Assay of 3-Methylcrotonyl-CoA Carboxylase and Propionyl-CoA Carboxylase

Both enzyme activities are determined by quantifying biotin-dependent acid-nonvolatile fixation of radiolabeled sodium bicarbonate into the appropriate substrates, propionyl-CoA or 3-methylcrotonyl-CoA.[17] The final assay (total volume, 0.1 ml) contains 100 mM Tris-HCl (pH 8.0), 0.5 mM Na$_2$EDTA, 2.5 mM reduced glutathione, 3.0 mM ATP, 50 mM KCl, 6.0 mM MgCl$_2$, 2.4 mM propionyl-CoA or 3-methylcrotonyl-CoA, 10 mM NaH^{14}CO$_3$ (final specific activity, 2–10 mCi/mmol), and clarified cell sonicate (generally prepared in 50 mM Tris-HCl, 0.025 mM Na$_2$EDTA, pH 8.0). All reactions are performed in duplicate, and blanks contain water in place of the corresponding coenzyme A ester.

Incubation is carried out in a shaking water bath for approximately 30 min at 30°. The linearity of product formation should be assessed with respect to time and protein content in individual laboratories. All manipulations with radiolabeled bicarbonate should be performed in a fume hood. Blanks without cell extract should be run to assess the level of impurity in individual lots of radiolabeled sodium bicarbonate. After incubation, the reactions are acidified in the fume hood with 0.02 ml of 15% (v/v) formic acid, the reaction contents are transferred to glass or polypropylene scintillation vials, and the reaction contents are taken to dryness under a heat lamp in the hood (usually 6 hr to overnight). The addition of small pieces of dry ice to individual vials enhances the rate of evolution of unreacted, radiolabeled carbon dioxide. Scintillation fluid is added to the samples, and acid nonvolatile radioactivity is assessed in a liquid scintillation spectrometer. Product formation is related to the protein content of cell extracts, which is quantified by a standard technique such as the method of Bradford or Lowry.

This procedure can be readily applied in sonicates of cultured fibroblasts or leukocytes isolated from whole blood. The average laboratory control value for PCC in fibroblasts is 887 ± 313 pmol/min/mg of protein ($n = 7$ cell lines; range, 503–1274 pmol/min/mg of protein) and in leukocytes is 263 ± 88 pmol/min/mg of protein ($n = 16$; range, 138–484 pmol/min/mg of protein). The values of a recently identified patient were 0 in fibroblast

---

[16] T. Fukao, S. Yamaguchi, C. R. Scriver, G. Dunbar, A. Wakazono, M. Kano, T. Orii, and T. Hashimoto, *Hum. Mutat.* **2**, 214 (1993).

[17] W. Weyler, L. Sweetman, D. C. Maggio, and W. L. Nyhan, *Clin. Chim. Acta* **76**, 321 (1977).

sonicates and 4 pmol/min/mg of protein in leukocyte extracts, while another patient's residual activity in fibroblast extracts was 8 pmol/min/mg of protein. For MCC, the average control activity in fibroblasts is 567 ± 205 pmol/min/mg of protein (n = 7 cell lines; range, 295–961 pmol/min/mg of protein) and in leukocytes is 34 ± 15 pmol/min/mg of protein (n = 14; range, 16–66 pmol/min/mg of protein). Residual leukocyte MCC activities in four adult individuals of Amish/Mennonite ancestry were <0.3 pmol/min/mg of protein.[18] Mutation analysis in human MCC has not been reported; mutation analyses for human PCC are outlined in the following section.

Mutation Analysis of *PCCB* Gene

Lack of data on the genomic structure has been a significant impediment to full characterization of PCC mutant chromosomes causing the inborn error of metabolism termed propionic acidemia. The implementation of long-distance PCR resulted in an effective method of characterizing the genomic organization of the coding region of the PCCB gene.[19] The gene consists of 15 exons. Sequences flanking splice junctions were determined. Gene sequence information provided the basis for a complete mutation analysis of the PCCB gene, using genomic DNA.

Procedure

Principle. Mutation analysis is performed on PCR products derived from the amplification of the 15 different exons of the PCCB gene from genomic DNA. Reverse transcriptase PCR (RT-PCR)-based methods are used to confirm the predicted consequences of splice mutations.

Amplification of PCCB Exons. Genomic DNA is isolated from fibroblast cell culture or peripheral blood leukocytes in accordance with the method of cell lysis and DNA isolation described by Old.[20] This DNA is then used as a template for the analysis of the PCCB gene.

The primers designed to amplify each PCCB exon are listed in Table IV. PCR is performed in a final volume of 100 μl, by use of 0.5–1 μg of

[18] K. M. Gibson, M. J. Bennett, E. W. Naylor, and D. H. Morton, *J. Pediatr.* **132,** 519 (1998).

[19] P. Rodríguez-Pombo, J. Hoenicka, S. Muro, B. Pérez, C. Pérez-Cerdá, E. Richard, L. R. Desviat, and M. Ugarte, *Am. J. Hum. Genet.* **63,** 360 (1998).

[20] J. Old, *in* "Human Genetic Diseases: A Practical Approach" (K. Davies, ed.), p. 1. IRL Press, New York, 1986.

TABLE IV
PCCB Primers used for Exon Amplification

Exon	5'-Intron primer	Distance from exon (bp)	3'-Intron primer	Distance from exon (bp)	Length of PCR (bp)	GenBank/EMBL accession no.
1	cDNA1: 5'-GCTCTAGACGCGCCGGCACAGCAA		AS1: 5'-ccggcagcagtgataggtc	65	272	AJ006487
2	S184: 5'-cttcaagccctccatgaacagc	85	AS184: 5'-ggtaggccagctgagccacacct	56	261	AJ006488
3	S304: 5'-ggcatagtggccaaactcattag	71	AS304: 5'-agtttcattcccaagatttgc	92	232	AJ006489
4	S373: 5'-cactgggttttcatcctagc	64	AS373: 5'-gtacctagctcagtgtcttgt	94	215	AJ006490
5	S430: 5'-ctgtggtattttgtgaatgtcg	74	AS430: 5'-gccaggccatggcttagagaa	65	253	AJ006491
6	S544: 5'-tatctttccacagataatgcctc	67	AS544: 5'-caacctagatctgggtcagga	91	269	AJ006492
7	S655: 5'-gaatcaactcaaggctgtgacc	70	AS655: 5'-aggctctttggcaatgccctct	65	249	AJ006493
8	S764: 5'-gaagcagcaactttggcctggc	63	AS764: 5'-cctgtcaggcagagaaaggcaa	67	251	AJ006494
9	S885: 5'-aagcccagcagggacagaactgg	128	AS885: 5'-caagtggtccagtgcttcctcac	92	302	AJ006495
10	S965: 5'-gaaatgtcattcgtagtgtcat	79	AS965: 5'-gacagaaaggataccatggaac	77	280	AJ006496
11	S1091: 5'-ggatgctgctgaggacaaa	85	AS1091: 5'-tgggaaaggcaactcctcctagcc	44	237	AJ006497
12	S1120: 5'-CGTTTGTCAGATTCTGTGATGC	317	AS1199: 5'-cactactgccttccagttccc	198	616	AJ006497
13	S1237: 5'-GCCAAGCTTCTCTACGCATTTG	442	AS1300: 5'-agctatgccttgagctctgc	72	613	AJ006497
14	S1399: 5'-cagttcctccctatgctatg	157	AS1399: 5'-cactagtcaacagattagactgc	119	376	AJ006498
15	S1499: 5'-ctaccatctctgtatcagttg	134	cDNA2: 5'-GCTCTAGACTTGGTTTCTTTTCCTTTGATTT		289	AJ006499

[a] cDNA1 and cDNA2 are exon-based primers designed by Ohura et al.[22] as complementary to the 5'- and 3'-untranslated regions of the human PCCB gene. S1120 and S1237 are exonic primers hybridizing to the preceding exon.

genomic DNA, a 1 μM concentration of each primer, a 0.25 mM concentration of each dNTP, 1.5 mM MgCl$_2$, 10% (v/v) dimethyl sulfoxide (DMSO), and 0.5 units of AmpliTaq (Perkin-Elmer, Norwalk, CT). A "hot" start is performed, followed by 30 cycles of 1 min at 94°, 1 min at 55°, and 1 min at 72°, by use of a Peltier thermal cycler PTC-200 (MJ Research, Waltham, MA). To amplify exon 1, the annealing temperature is 60°. In some cases, PCR amplifications are performed directly, using dried blood spots as the source of DNA with a 1.5× special PCR buffer [10× buffer composition: 20 mM MgCl$_2$, 0.1 M Tris-HCl (pH 8.6), 0.5 M KCl, 0.068% (v/v) 2-mercaptoethanol, and bovine serum albumin (BSA, 1 μg/ml)]. The PCR-amplified fragments are purified with the Wizard PCR Preps DNA purification system (Promega, Madison, WI) and direct sequenced with the fmol DNA sequencing kit (Promega) with fluorescently labeled primers and 100 ng of the PCR product. Sequences are analyzed on an automated DNA sequencer (ALF Express; Pharmacia, Piscataway, NJ).

Restriction Enzyme Analysis. Detection of certain mutations (Table V), as well as analysis of control chromosomes, is accomplished by restriction–digestion analysis of PCR products, either directly or after creation of the restriction site in the amplification (ACRS). Digestion conditions are in

TABLE V
RESTRICTION ENZYME ANALYSIS TO IDENTIFY MUTATIONS IN PCCB GENE[a]

Mutation	Nucleotide change sequence	Effect on coding change[b]	Restriction site
ins/del	1218del14ins12	fs and stop codon	(−)*Msp*I
c1170insT	ins of T at 1170–1174	fs and stop codon	—
E168K	502G→A	Glu168Lys	(−)*Mbo*II (ACRS)
A497V	1490C→T	Ala497Val	(−)*Bso*FI
G198D	593G→A	Gly198Asp	(+)*Ava*II
R44P	131G→C	Arg44Pro	(−)*Msp*I
S106R	318C→A	Ser106Arg	(−)*Bst*UI (ACRS)
G131R	391G→C	Gly131Arg	(−)*Mnl*I
R165W	493C→T	Arg165Trp	(+)*Nla*III
R410W	1228C→T	Arg410Trp	(−)*Msp*I
1298–1299insA	ins of A at 1297–1299	fs and stop codon	(+)*Xcm*I (ACRS)
R512C	1534C→T	Arg512Cys	(−)*Pml*I
L519P	1556T→C	Leu519Pro	(+)*Msp*I
W531X	1593G→A	Trp531Stop	(−)*Sty*I
IVS1+3GΔC	183+3G→C	Splice mutation	(−)*Bsa*I(ACRS)
IVS10-11del6	1091-11del6	Splice mutation	—

[a] Mutations are designed according to Antonarakis and the Nomenclaure Working Group (1998), except for ins/del and c1170insT, which were named previously. S. E. Antonarakis, *Hum. Mutat.* **11,** 1 (1998).
[b] fs, frameshift.

accordance with the manufacturer protocols. After digestion, fragments are electrophoresed on 4% (w/v) NuSieve gels and visualized after staining with ethidium bromide.

Analysis of cDNA. Total RNA (1 μg), isolated from cultured skin fibroblasts by a guanidinium thiocyanate–phenol–chlorophorm extraction method,[21] is used to perform the analysis on PCR products derived from reverse-transcribed RNA.

First-strand cDNA is synthesized from 1 μg of total RNA, using the GeneAmp RNA PCR kit (Perkin-Elmer Cetus) according to the instructions of the manufacturer, and with random hexamers as downstream primer. PCCB cDNA is amplified in five overlapping fragments ~400 bp in length, with primers and amplification conditions described elsewhere.[22,23] The amplified cDNA fragments are purified and sequenced as described previously.

Remarks

Until now, this procedure has effectively defined 64 among 66 mutant PCCB chromosomes (50 from Spain and 16 from Chile, Brazil, and Ecuador), with 18 different mutations detected.

Assay of 3-Methylglutaconyl-CoA Hydratase

In the preceding volume of this series devoted to branched-chain amino acid metabolism, a method for preparation of 3-methyl[5-^{14}C]glutaconyl-CoA was described, which employed 3-methylcrotonyl-CoA carboxylase purified from bovine kidney for preparation of the ^{14}C substrate.[24] The enzymatically prepared substrate was purified by reversed-phase high-performance liquid chromatography (HPLC). For assay of 3-MG-CoA hydratase in patient samples, cell sonicates were incubated with this substrate, and product separation and quantification achieved by reversed-phase HPLC.[24] Although accurate and sensitive, this method is laborious and time-consuming, and does not lend itself to rapid implementation in the clinical diagnostic laboratory. A coupled assay estimating 3-MG-CoA hydratase activity has been more recently employed in our laboratory.[25]

In the coupled assay, 3-methylcrotonyl-CoA is incubated with ATP, radiolabeled sodium bicarbonate, and tissue extract in order to generate

[21] P. Chomczynski and N. Sacchi, *Anal. Biochem.* **162,** 156 (1987).

[22] T. Ohura, K. Narisawa, and K. Tada, *J. Inher. Metab. Dis.* **16,** 863 (1993).

[23] R. A. Gravel, B. R. Akerman, A.-M. Lamhonwah, M. Loyer, A. León-del-Rio, and I. Italiano, *Am. J. Hum. Genet.* **55,** 51 (1994).

[24] K. M. Gibson, *Methods Enzymol.* **166,** 214 (1988).

[25] K. Narisawa, K. M. Gibson, L. Sweetman, and W. L. Nyhan, *Clin. Chim. Acta* **184,** 57 (1989).

radiolabeled 3-MG-CoA *in situ* through the action of endogenous MCC. The assay contents are identical to those described above for MCC, with minor modifications. Radiolabeled sodium bicarbonate is added to a final concentration of 12.5 mM, and the incubation includes 2 mM NADH and 0.25 units of 3-hydroxybutyrate dehydrogenase. In addition, the protein content of the 0.1-ml incubation mixture is higher (up to 0.25 mg of protein versus 0.02 mg of protein in carboxylase assays described above). Linearity of product formation with respect to time and protein content should be determined in individual laboratory settings. The products of the incubation include ^{14}C-labeled 3-methylglutaconyl-coenzyme A, ^{14}C-labeled 3-hydroxy-3-methylglutaryl-coenzyme A, and ^{14}C-labeled acetoacetic acid, which is converted to the more stable 3-hydroxybutyric acid by action of added NADH and 3-hydroxybutyrate dehydrogenase. Reactions are generally carried out for 100 min in our laboratory, after which 0.01 ml of 4.2 M HClO$_4$ is added to each sample in a fume hood, and the samples kept on ice to deproteinize. The addition of small amounts of dry ice to the incubation tubes enhances evolution of unreacted, radiolabeled sodium bicarbonate.

After removal of precipitated protein by centrifugation, the pH of the incubation mixtures is carefully adjusted to pH 4–7 with 6 N KOH, followed by removal of insoluble KClO$_4$. The addition of excess KOH results in hydrolysis of coenzyme A esters, which alters chromatographic resolution. An aliquot of the neutralized reaction mixture is streak-spotted on a 20 × 20 cm cellulose, thin-layer chromatography plate over a 1.5- to 2.0-cm lane, and the plate is developed in *n*-butanol–water–formic acid (77 : 13 : 10, by volume) to within 5 cm of the top of the plate. After drying, individual reaction mixture lanes are cut into 1-cm pieces and placed in scintillation vials, and reaction products quantified after addition of fluor in a liquid scintillation spectrophotometer. Using this method, products of the reaction are reported as total coenzyme A esters (CoA esters R_f 0.05–0.1) and 3-hydroxybutyric acid (3-HOB; R_f 0.8–0.9). This assay is used routinely in sonicates of cultured fibroblasts, although the method should also be applicable to isolated leukocytes.

As shown in Table VI, control cell lines accumulate considerable 3-HOB and a lesser amount of CoA esters, reflecting the large K_{eq} of HMG-CoA lyase (see Fig. 1) toward cleavage to acetoacetic acid and acetyl-CoA. However, patients with 3-MG-CoA hydratase deficiency (Table VI) accumulate small amounts of 3-HOB and considerably higher quantities of CoA esters, with a skewed ratio.[25,26] This assay is also useful to detect patients with inherited HMG-CoA lyase deficiency, as shown in Table VI.

[26] K. M. Gibson, R. S. Wappner, S. Jooste, E. Erasmus, L. J. Mienie, E. Gerlo, B. Desprechins, and L. De Meirleir, *J. Inher. Metab. Dis.* **21**, 631 (1998).

TABLE VI

3-Methylglutaconyl-CoA Hydratase Activity in Sonicates of Fibroblasts
Derived from Controls and Patients[a]

Subject	CoA esters	3-HOB	Total products	3-HOB/CoA esters
Control[b] range	105 ± 19	123 ± 53	208 ± 42	1.2 ± 0.4
	(76–120)	(65–204)	(158–263)	(0.7–1.8)
Hydratase deficiency	91	15	106	0.16
Hydratase deficiency	136	21	157	0.15
Lyase Deficiency	120	25	145	0.21
Lyase deficiency	97	22	119	0.23

[a] All activities in picomoles per minute per milligram of protein; control values are reported as means ± SD with ranges in parentheses. Values for fibroblasts derived from patients with either 3-MG-CoA hydratase deficiency or HMG-CoA lyase deficiency are single determinations.
[b] $n = 5$.

To date, a mammalian cDNA encoding 3-MG-CoA hydratase has not been reported, and thus mutation analysis in the approximate 15 patients with this disorder has not been undertaken.

Assay of 3-Hydroxy-3-Methylglutaryl-CoA Lyase

The procedure employed in our laboratory is comparable to that used for estimating 3-MG-CoA hydratase activity described above, with the exception that 3-hydroxy-3-methyl[3-^{14}C]glutaryl-CoA is the substrate and radiolabeled sodium bicarbonate is omitted. The standard assay mix (final volume, 0.1 ml) contains 8–20 μg of protein (fibroblasts, leukocytes, amniocytes, or chorionic villus tissue), 40 mM Tris-HCl (pH 8.0), 2 mM NADH, 0.25 U of 3-hydroxybutyrate dehydrogenase, and 0.16 mM DL-[3-^{14}C]HMG-CoA (final specific activity, 5 Ci/mol), as modified from previously published methods.[27,28] Blanks contained heat-denatured cell extract, and recovery of hydroxy[3-^{14}C]butyric acid is achieved by inclusion of known quantities of stock radiolabeled 3-hydroxy[^{14}C]butyric acid in blank assays (approximately 75–90% recovery). Incubation is carried out for 15 min at 37°. Reaction termination, neutralization, chromatography, and product quantification are identical to methods employed for the 3-MG-CoA hydratase assay described above. For comparison purposes, mean HMG-CoA

[27] K. M. Gibson, Methods Enzymol. 166, 219 (1988).
[28] K. M. Gibson, C. F. Lee, V. Kamall, K. Johnston, A. L. Beaudet, W. J. Craigen, B. R. Powell, R. Schwartz, M. Y. Tsai, and M. Tuchman, Clin. Chem. 36, 297 (1990).

lyase activity in sonicates of cultured fibroblasts is 4.8 ± 0.6 nmol/min/mg of protein (mean $\pm$ SD; $n = 7$ cell lines; range, 3.9–5.7). For two recently studied patients, activities in fibroblasts sonicates were 0.01 and 0.03 nmol/min/mg of protein.

3-Hydroxy-3-methylglutaryl-CoA Lyase

3-Hydroxy-3-methylglutaryl-CoA (HMG-CoA) lyase is a mitochondrial matrix homodimer expressed in all tissues studied to date. HMG-CoA lyase mediates the last step of leucine catabolism and also the last step of ketogenesis from fatty acids, an important pathway of energy transfer during fasting.[29] The HMG-CoA lyase precursor (325 residues, 34.4 kDa) contains an N-terminal targeting sequence that is cleaved on mitochondrial entry, producing a 31.6-kDa mature monomer (298 residues). Interestingly, the HMG-CoA lyase precursor contains a C-terminal type I peroxisome-targeting sequence and is imported into peroxisomes,[30] where it exists as an enzymatically active monomer[31] of unknown function.

Clinical Features. Clinical features of HMG-CoA lyase deficiency have been reviewed.[32,33] HMG-CoA lyase deficiency presents clinically as episodes of hypoketotic hypoglycemia, often with coma and acidosis. In a series of 62 HMG-CoA lyase-deficient patients collected from the literature and from collaborators,[32] about half the patients presented in the neonatal period, and presentation after the age of 2 years was exceptional. Neurological complications may be common after hypoglycemic episodes, as expected because ketone bodies are the main nonglucose energy source for the brain. The diagnosis is made by the finding of a typical urinary organic acid pattern. Some patients have a urinary organic acid pattern resembling that of HMG-CoA lyase deficiency, but normal HMG-CoA lyase activity in cultured cells. Therefore, enzymatic confirmation of the diagnosis is recommended. Molecular analysis complements the enzymatic approach and is beginning to yield insight into the structure–function relationships of HMG-CoA lyase and into common mutations in high-risk populations.

[29] G. A. Mitchell, S. Kassaovska-Bratinova, Y. Boukaftane, M.-F. Robert, S. P. Wang, L. Achmarina, M. Lambert, P. Lapierre, and E. Potier, *Clin. Invest. Med.* **18,** 193 (1995).
[30] L. I. Ashmarina, N. Rusnak, H. M. Miziorko, and G. A. Mitchell, *J. Biol. Chem.* **269,** 31929 (1994).
[31] L. I. Ashmarina, M.-F. Robert, M.-A. Elslinger, and G. A. Mitchell, *Biochem. J.* **315,** 71 (1996).
[32] G. A. Mitchell and T. Fukao, *in* "Inborn Errors of Ketone Body Metabolism" (C. R. Scriver, A. Beaudet, W. Sly, and D. Valle, eds.), Chapter 102. McGraw-Hill, New York, 1995.
[33] K. M. Gibson, J. Breuer, and W. L. Nyhan, *Eur. J. Pediatr.* **148,** 180 (1988).

Mutation Detection in HMG-CoA Lyase Deficiency

Single-Strand Conformational Polymorphism Analysis. The conditions for amplification and electrophoresis have been slightly modified from the original description.[34] Primers have been designed that provide efficient amplification of exons and surrounding splice sequences (Table VII). Amplification is performed with 25 ng of genomic DNA, 2.5 U of *Taq* DNA polymerase, 5′ and 3′ primers (Table VIII, each 1 mM), dNTPs (each 12.5 mM), 1.5 mM MgCl$_2$, 12.5 mCi of [α-^{35}S]dATP, 20 mM Tris-HCl (pH 8.4), and 50 mM KCl. After a 6-min denaturation at 94°, a hot start is performed at 80°, with 30 amplification cycles (94°, 15 sec; annealing [see below], 15 sec; 72°, 15 sec) followed by a 5-min extension at 72°. For HMG-CoA lyase (HL) exons 2, 3, 6, 7, and 8, the annealing temperature is 58°; for exons 4, 5, and 9, it is 60°. Because SSCP migration variants may be subtle, any exon for which the amplicon is suspected of abnormal migration is sequenced.

Sequencing of Genomic Amplicons. To amplify HL exons for sequencing, the same solutions and conditions as those for SSCP are used, with the exception that the concentration of dNTPs is 50 mM each and [^{35}S]dATP is absent. Sequencing is performed with the thermal Sequenase radiolabeled termination cycle sequencing kit (Amersham, Arlington Heights, IL), according to the manufacturer instructions. The amplification primers are used for sequencing, except for exon 3, where an internal 5′ primer, HLH108, is employed. Because false negatives can occur with SSCP, this technique may be replaced in future by direct sequencing of all exons.

Reverse Transcription and Polymerase Chain Reaction. Reverse transcription-PCR was used to identify the first HL mutation described,[35] and has been used to describe three premature termination mutations resulting in aberrant splicing.[36–38] cDNA sequencing is also the preferred approach for analysis of HL exon 1, for which the genomic sequence has not yet been described in humans. It is accessible by rapid amplification of cDNA

[34] S. P. Wang, M.-F. Robert, K. M. Gibson, R. J. A. Wanders, and G. A. Mitchell, *Genomics* **33,** 99 (1996).

[35] G. A. Mitchell, M.-F. Robert, P. W. Hruz, G. Fontaine, C. E. Behnke, L. M. Mende-Mueller, S. Wang, K. Shaeppert, C. Lee, K. M. Gibson, and H. Miziorko, *J. Biol. Chem.* **268,** 4376 (1993).

[36] C. Buesa, J. Pie, A. Barcelo, N. Casals, C. Mascaro, C. H. Casale, D. Haro, M. Duran, J. A. Smeitink, and F. G. Hegardt, *J. Lipid Res.* **37,** 2420 (1996).

[37] J. Pie, N. Casals, C. H. Casale, C. Buesa, C. Mascaro, A. Barcelo, M. O. Rolland, T. Zabot, D. Haro, F. Eyskens, P. Divry, and F. G. Hegardt, *Biochem. J.* **323,** 329 (1997).

[38] C. H. Casale, N. Casals, J. Pie, N. Zapater, C. Perez-Cerda, B. Merinero, M. Martinez-Pardo, J. J. Garcia-Penas, J. M. Garcia-Gonzalez, R. Lama, B. T. Poll-The, J. A. Smeitink, R.J. Wanders, M. Ugarte, and F. G. Hegardt, *Arch. Biochem. Biophys.* **349,** 129 (1998).

TABLE VII
HUMAN HMG-CoA LYASE GENE EXONS: ARRANGEMENT AND AMPLIFICATION/SEQUENCING PRIMERS

Exon no.	Length	5' Extremity	First codon	Reading frame	Primer position	Name	Restriction size (1)	Sequence (2)	Amplicon length
2	84	61	Val-21	1	5'	HLH49	EcoRI	ATG AAT TCG GTC TCC CTG GGA ATT G	206
					3'	HLH45	HindIII	TTA AGC TTC ACG TAA TAC TCA AAG CAG ATG	
3	108	145	Asn-49	1	5'	HLH47	EcoRI	ATG AAT TCT GCA TTT TGA GGC TGT TT	231
					3'	HLH48	HindIII	ATA AGC TTC AAA GGC AAA TGC AAA AC	
					5'	5HLH108	—	GTT ACT AAC TTT GCT GCC TT	
4	96	253	Met-83	1	5'	HLH40	EcoRI	ATG AAT TCT CTG CTC TTG GTG ATG ACT	213
					3'	HLH41	HindIII	ATA AGC TTC AAG ACA AGG CAG GGA C	
5	149	349	Val-117	1	5'	HLH51	EcoRI	ATG AAT TCG CAA GAC TCC ATC TCA AAC A	254
					3'	HLH52	HindIII	ATA AGC TTG AAC GGT ACA GAG GAA AGG A	
6	64	498	Gly-166	3	5'	HLH53	EcoRI	ATG AAT TCG CCC TGC CTC AGT TCT	171
					3'	HLH44	HindIII	ATA AGC TTA CCC TCA CCA AAC CCC	
7	189	562	Val-188	1	5'	HLH42	EcoRI	ATG AAT TCA TTC TGT ATC CTC CAA GG CC	262
					3'	HLH43	HindIII	TAA AGC TTC GTG ACC TTT GGG AGA AT	
8	126	751	Met-251	1	5'	HLH54	EcoRI	ATG AAT TCG GCA ACA GAC GAT TGG G	266
					3'	HLH55	HindIII	TTT AAG CTT GGC CCC CTG GTC AGT TC	
9	527	877	Gly-293	1	5'	HLH56		CCT GGT GTT GAG GGC ATA CC	205
					3'	HLH57		GTG CCC CTA TTT CCA CAT CAT C	

TABLE VIII
DELETERIOUS MUTATIONS IN HMG-CoA LYASE

Mutation[a]		DNA change	cDNA position	Exon	Functional[b] effect	Ethnicity	Ref.
Deleterious point mutations							
E37X	GAA→TAA		109	2	Truncation	Iberian, Arabic	Pie et al.[37]
R41ter	CGA→TGA		121	2	Truncation	American	Mitchell et al.[39]
R41Q	CGA→CAA		122	2	Expression	Saudi, other	Mitchell et al.[39]
D42H	GAT→CAT		124	2	Expression	African-American	Mitchell et al.[39]
D42G	GAT→GGT		125	2	Expression	Dutch	Mitchell et al.[39]
D42E	GAT → GAG		126	2	Expression	Austrian	Mitchell et al.[39]
N46fs(+1)	CAAAAT→CAAAAAT		134–137	2	Truncation	Italian	Mitchell et al.[39]
S69fs(−2)	CTCTCT→CTCT—		202–207	3	Truncation	Acadian, French-Canadian	Mitchell et al.[39]
V168fs(−2)	GTCTCC→GT—CC		504–505	6	Truncation	Spanish	Casale et al.[38]
H233R	CAG→CGC		698	7	Expression	American	Roberts et al.[40]
IVS8+1 G-C	CGgt→CGct		NA	Intron 8	Truncation	Turkish	Buesa et al.[36]
F305fs(−2)	TTT→T—		913–915	9	Truncation	Saudi	Mitchell et al.[39]
Gross deletions							
Del	Unknown		NA	3 to 6	Large deletion	Turkish	Wang et al.[34]
DEL	Unknown		NA	<2 to 7	Large deletion	English	Wang et al.[34]
Frequent polymorphisms							
L218L	CTA→CTG		562	7	NA	Many	Unpublished
IVS3+34C→T	C→T		NA	IVS 3	NA	Many	Unpublished

[a] Defined as effect on enzyme activity. In general, truncation/deletion mutations were not expressed.
[b] fs, frameshift.

ends (RACE)-PCR as described.[34] Lymphoblasts and fibroblasts provide convenient sources for RNA isolation.

Bacterial Expression of HMG-CoA Lyase

A bacterial expression system is used for missense mutations that do not obviously inactivate HMG-CoA lyase.[39] A human HMG-CoA lyase cDNA fragment corresponding to mature (mitochondrial) HMG-CoA lyase is expressed in the pTrc vector.[40] A mutation, C323S, has been introduced into the HMG-CoA lyase cDNA. HMG-CoA lyase containing C323S is more stable than wild-type HMG-CoA lyase, but the catalytic properties of the mutant peptide closely resemble those of wild-type HMG-CoA lyase.[40] Putative mutations are introduced into the vector either by site-directed mutagenesis or by replacement of restriction fragments with mutant cDNA cassettes derived from patient cells. The mutated cassette is sequenced to ensure that it contains only the desired mutation.

For expression, the conditions described by Roberts *et al.*[40] are used. A 500-ml culture of *Escherichia coli* containing the HMG-CoA lyase plasmid is grown to an optical density of 0.6 at 600 nm, and then is incubated in the presence of 1.0 mM isopropyl-1-thio-β-D-galactoside (IPTG), shaken overnight at 25°, and then centrifuged at 3000g for 45 min at 4°. The pellet is resuspended in ice-cold lysis buffer [10 mM potassium phosphate (pH 7.8), 5 mM EDTA] in the presence of 100 μM phenylmethylsulfonyl fluoride (PMSF), DNase (10 μg/ml), and RNase (10 μg/ml). The cells are sonicated on ice, five times for 30 sec each. The crude lysate is centrifuged at 100,000g at 4° for 1 hr. HMG-CoA lyase accounts for about 1% of total protein in the lysate supernatant. The expressed HMG-CoA lyase activity is assayed spectrophotometrically.

Pure recombinant HMG-CoA lyase from the lysate supernatant has been isolated from lysates[40] by Q-Sepharose anion exchange chromatography, ammonium sulfate precipitation, and reversed-phase chromatography on a phenyl-agarose column. Fractions containing HMG-CoA lyase activity are concentrated and then subjected to molecular sieve chromatography on a Superose-12 column, yielding recombinant HMG-CoA lyase with specific activity and kinetic properties similar to those of HMG-CoA lyase purified from vertebrate tissues. Measurement of HMG-CoA lyase activity in crude lysate supernatants is adequate to establish the function of most

[39] G. A. Mitchell, P. T. Ozand, M.-F. Robert, L. Ashmarina, J. Roberts, K. M. Gibson, R. J. Wanders, S. Wang, I. Chevalier, E. Ploechl, and H. Miziorko, *Am. J. Hum. Genet.* **62,** 295 (1998).

[40] J. R. Roberts, C. Narasimhan, P. W. Hruz, G. A. Mitchell, and H. M. Miziorko, *J. Biol. Chem.* **269,** 17841 (1994).

missense mutations, although studies with purified mutant HMG-CoA lyase are required for detailed mutational analysis.

Mutations Causing HMG-CoA Lyase Deficiency

Table VIII summarizes reported HL mutations. In our series, 21 of 82 (26%) mutant alleles involve the hexanucleotide CAGGAT, corresponding to codons R41 and D42. Other mutations are distributed throughout the gene. Specific mutations may be frequent in some populations. For instance, the highest known concentration of HL deficiency is in Saudi Arabia, where R41Q accounts for most mutant alleles. We (G. A. Mitchell and K. M. Gibson, unpublished, 1999) and others[40] have noted the E37X mutation in several patients of Spanish, Portuguese, and Arabic descent.

Methods for Analyses of Defects in L-Valine Catabolic Pathway

With the exception of the branched-chain keto acid dehydrogenase complex or propionyl-CoA carboxylase, the defects of L-valine catabolism are exceedingly rare (see Fig. 1). In addition, there is some confusion as to the exact sequence of reactions in L-valine catabolism.[41] Brown and colleagues published a single case of enzymatically confirmed 3-hydroxyisobutyryl-CoA deacylase deficiency in the early 1980s, and to this date, no other patient has been reported.[1a] Roe and co-workers have reported an isolated deficiency of isobutyryl-CoA dehydrogenase in a patient whose fibroblasts were studied with deuterium-labeled L-valine in conjunction with unlabeled L-carnitine, and tandem mass spectrometric analysis of acylcarnitine conjugates.[2] This patient is reviewed in [39] in this volume.[41a]

In the mid-1980s, Pollitt and co-workers reported a patient with a unique metabolic profile suggesting combined malonate and methylmalonatesemialdehyde dehydrogenase deficiencies.[42,43] Oxidation of radiolabeled L-valine in fibroblasts derived from this patient was consistent with a deficiency of methylmalonate semialdehyde dehydrogenase. Gibson and co-workers reported an additional patient whose metabolic profile was different from that of the patient of Pollitt and co-workers.[44,45] This patient

---

[41] P. Kamoun, *Trends Biochem. Sci.* **17,** 175 (1992).

[41a] C. R. Roe and D. S. Roe, *Methods Enzymol.* **324,** [39], 2000 (this volume).

[42] J. Pollitt, A. Green, and R. Smith, *J. Inher. Metab. Dis.* **8,** 75 (1985).

[43] R. G. F. Gray, R. J. Pollitt, and J. Webley, *Biochem. Med. Metab. Biol.* **38,** 121 (1987).

[44] F.-J. Ko, W. L. Nyhan, J. Wolff, B. Barshop, and L. Sweetman, *Pediatr. Res.* **30,** 322 (1991).

[45] K. M. Gibson, C. F. Lee, M. J. Bennett, B. Holmes, and W. L. Nyhan, *J. Inher. Metab. Dis.* **16,** 563 (1993).

presented with chronic lactic acidemia and 3-hydroxyisobutyric aciduria; the intact cell studies employing radiolabeled L-valine and β-alanine suggested a combined semialdehyde dehydrogenase deficiency. In both these patients, direct assessment of methylmalonate-semialdehyde dehydrogenase activity has not been feasible because of the lability of substrate methylmalonate semialdehyde, a highly unstable β-keto acid prone to spontaneous decarboxylation. Roe and colleagues reported a third patient with methylmalonic aciduria in whom *in vivo* loading studies with L-carnitine and L-valine or thymine suggested a deficiency of methylmalonate-semialdehyde dehydrogenase.[46] No confirmatory enzyme studies were reported in fibroblasts from this patient. Because most of the cDNA sequence for methylmalonate-semialdehyde dehydrogenase has been known since 1992,[47] molecular characterization of cDNA from these patients should help to clarify which of these patients have mutations in the methylmalonate semialdehyde dehydrogenase gene.

Acknowledgments

The authors gratefully acknowledge the expert technical assistance of Beth Lee in development of the methodology for determining β-ketothiolase activity. We thank the following physicians and researchers who have supplied cultured fibroblast cell lines from patients: Dr. R. S. Wappner, Dr. J. Filiano, Dr. A. Sewell, Dr. W. Lehnert, Dr. B. A. Barshop, Dr. V. Proud, and Dr. K. Grange. We also thank Shupei Wang, Marie-France Robert, and Linge Pan for invaluable laboratory assistance; Henry Miziorko for long-term collaboration; and the numerous collaborators who provided samples and clinical information.

[46] C. R. Roe, E. Struys, R. M. Kok, D. S. Roe, R. A. Harris, and C. Jakobs, *Mol. Genet. Metab.* **65**, 35 (1998).
[47] N. Y. Kedishvili, K. M. Popov, P. M. Rougraff, Y. Zhao, D. W. Crabb, and R. A. Harris, *J. Biol. Chem.* **267**, 19724 (1992).

[41] Genetic Defects in E3 Component of α-Keto Acid Dehydrogenase Complexes

By Mulchand S. Patel, Young Soo Hong, and Douglas S. Kerr

Introduction

Mammalian dihydrolipoamide dehydrogenase (E3), a homodimeric flavoprotein, is a shared component of the three mitochondrial multienzyme α-keto acid dehydrogenase complexes: the pyruvate dehydrogenase complex (PDC), α-ketoglutarate dehydrogenase complex (KDC), and branched-chain α-keto acid dehydrogenase complex (BCKDC), respectively.[1] E3 is also a component (referred to as L protein) of the mitochondrial glycine synthase (also known as the glycine cleavage system).[2] E3 catalyzes the reoxidation of dihydrolipoyl moieties covalently linked to specific lysine residues in the acyltransferase components of the three α-keto acid dehydrogenase complexes and the dihydrolipoyl moiety of the H protein of glycine synthase.[3] Each subunit of homodimeric E3 has four distinctive subdomain structures, namely the FAD-binding, NAD$^+$-binding, central, and interface domains with one noncovalently but tightly bound FAD molecule per subunit. E3 has two identical active sites composed of specific amino acid residues from both subunits. For this reason, E3 dimers are active but monomers are not. During catalysis, electrons are sequentially transferred from a reduced lipoyl moiety covalently linked to the substrate protein to the redox disulfide center in the E3 protein, the FAD cofactor, and then to NAD$^+$, generating NADH.

Because E3 is common to PDC, KDC, and BCKDC, a deficiency of E3 impairs oxidation of the α-keto acid substrates for these complexes, resulting in the accumulation of the α-keto acids as well as the respective α-hydroxy and amino acids in plasma and urine.[4] Detection of abnormal amounts of these compounds and low levels of E3 activity in cell or tissue preparations is evidence of E3 deficiency.[4] About 15 patients with E3 deficiency have been identified using enzymatic assays for E3 and the α-keto acid dehydrogenase complexes (e.g., PDC, KDC, and BCKDC).[5–11]

[1] L. J. Reed, *Acc. Chem. Res.* **7,** 40 (1974).

[2] G. Kikuchi and K. Higara, *Mol. Cell. Biochem.* **45,** 137 (1982).

[3] M. S. Patel and T. E. Roche, *FASEB J.* **4,** 3224 (1990).

[4] B. H. Robinson, *in* "Metabolic and Molecular Basis of Inherited Disease" (C. R. Scriver, A. L. Beaudet, W. S. Sly, and D. Valle, eds.), 7th Ed., p. 1479. McGraw-Hill, New York, 1995.

[5] J. C. Haworth, T. L. Perry, J. P. Blass, S. Hansen, and N. Urquhart, *Pediatrics* **58,** 564 (1976).

[6] B. H. Robinson, J. Taylor, and W. G. Sherwood, *Pediatr. Res.* **11,** 1198 (1977).

[7] A. Munnich, J. M. Saudubray, J. Taylor, C. Charpentier, C. Marsac, F. Rocchiccioli, O. Amedee-Manesme, F. X. Coude, and B. H. Robinson, *Acta Paediatr. Scand.* **71,** 167 (1982).

[8] R. Matalon, D. A. Stumpf, K. Michals, R. D. Hart, J. K. Parks, and S. I. Goodman, *J. Pediatr.* **104,** 65 (1984).

[9] S. Matuda, A. Kitano, Y. Sakaguchi, M. Yoshino, and T. Saheki, *Clin. Chim. Acta* **140,** 59 (1984).

[10] Y. Sakaguchi, M. Yoshino, S. Aramaki, I. Yoshida, F. Yamashita, T. Kuhara, I. Matsumoto, and T. Hayashi, *Eur. J. Pediatr.* **145,** 271 (1986).

[11] O. N. Epeleg, A. Shaag, J. Z. Glustein, Y. Anikster, A. Joshep, and A. Saada, *Hum. Mutat.* **10,** 256 (1997).

Residual E3 activity levels in cultured skin fibroblasts from affected patients have ranged from 3 to 20% of normal; none of the patients had complete loss of E3 activity. The composite clinical presentation from E3-deficient patients is similar to PDC deficiency, including failure to thrive, delayed neurological development, hypotonia, and seizures.[5–11] We have identified specific mutations in three E3-deficient subjects and their heterozygous parents, and have shown that prenatal diagnosis is possible. In this chapter, we describe methods for characterization of E3 deficiency using activity assays, immunological detection, and sequence analysis of specific cDNAs and genomic DNA.

Identification of Genetic Defects in E3-Deficient Patients

Reagents

Minimum essential medium (Life Technologies, Gaithersburg, MD)
Fetal bovine serum (Life Technologies, Grand Island, NY)
Dimethyl sulfoxide (DMSO)
Phosphate-buffered saline, pH 7.5 (PBS)
Tween 20 (polyoxyethylenesorbitan monolaurate)
Tris-buffered saline containing 2% (v/v) Tween 20 (TTBS)
Piperazine-N,N'-bis (2-ethanesulfonic acid) (PIPES)
Formaldehyde
Salmon sperm DNA
Sodium chloride–sodium citrate buffer (SSC; 0.15 M NaCl and 0.015 M sodium citrate, pH 7.0)
Sodium dodecyl sulfate (SDS)
NAD^+, 120 mM
Dihydrolipoamide, 120 mM [in 95% (v/v) ethanol, prepared freshly before each assay]
EDTA, 100 mM
Dichloroacetic acid (DCA)
[1-^{14}C]Pyruvate (approximately 2000 dpm/μmol)
α-Keto[1-^{14}C]isovalerate (prepared from L-[1-^{14}C]valine)
α-Keto[1-^{14}C]glutarate
Thiamine pyrophosphate (TPP)
Dithiothreitol (DTT)
Trichloroacetic acid (TCA)
CoASH
[^{32}P]dCTP

Deoxynucleotide solution (mixture of 10 mM each: dATP, dGTP, dCTP, and dTTP)

Taq DNA polymerase

Oligo(dT)$_{12-18}$ (0.5 μg/μl; Life Technologies)

Stopping solution: 25 mM NaF, 25 mM EDTA, 4 mM DTT in 40% (v/v) ethanol, pH 7.4

Assay reaction mixture (for 100 reactions of the fibroblast assay): 0.2 ml of 0.1 M MgCl$_2$, 14 mg of NAD$^+$, 0.01 ml of phosphotransacetylase (1 U/μl), 2.5 ml of 60 mM potassium oxalate, 0.25 ml of fetal calf serum, 2.5 ml of 200 mM potassium phosphate buffer (pH 8.0), and 2.44 ml of water; total volume is 7.9 ml

Solution A (for 100 blanks): 0.4 ml of 20 mM dithiothreitol and 0.6 ml of water

Solution B (for 100 assays): 5 mg of CoA, 0.4 ml of 20 mM dithiothreitol, 0.1 ml of 10 mM thiamine pyrophosphate and 0.5 ml of water

Phenylmethylsulfonyl fluoride (PMSF)

Triton X-100

Hyamine hydroxide

Stemmed center wells

Rubber serum stoppers

Whatman (Clifton, NJ) filter paper

Scintillation fluid

Anti-E3 rabbit serum

Nitrocellulose membranes

Dried skim milk

Horseradish peroxidase conjugated to goat anti-rabbit gamma globulin antibody.

Skin Fibroblast Cell Culture

Fibroblast cells of the subjects, family members (parents and siblings, depending on availability), and a control subject are cultured in minimum essential medium (MEM) with 10% (v/v) fetal calf serum (FCS; GIBCO, Grand Island, NY) plus streptomycin (100 μg/ml), penicillin (60 μg/ml), and amphotericin B (1.5 μg/ml). All cell lines are grown in 100 $\times$ 20 mm polystyrene plates in a humidified 95% air, 5% CO$_2$ environment at 37°. When cells are confluent, they are either harvested (by scraping with a cell lifter or by trypsin treatment) for immediate use or frozen in MEM–10% (v/v) fetal calf serum with 10% (v/v) dimethyl sulfoxide (DMSO) at −180° until further use. Cultured skin fibroblasts should be tested for mycoplasma contamination, which can cause low specific activities and poor replicates.

Measurements of E3 and Pyruvate Dehydrogenase Complex,
α-Ketoglutarate Dehydrogenase, Complex, and Branched-Chain
α-Keto Acid Dehydrogenase Complex Activities in Fibroblasts

E3 Activity. E3 activity measurement (forward reaction) is based on the production of NADH as shown in the following reaction. A detailed protocol for the measurement of E3 has been reported.[12]

$$\text{Dihydrolipoamide} + NAD^+ \rightleftharpoons \text{lipoamide} + NADH + H^+$$

The production of NADH is detected spectrophotometrically at 340 nm, in the initial 2–3 min of the reaction. The final reaction mixture (1 ml) contains 3 mM NAD$^+$, 3 mM dihydrolipoamide, and 1.5 mM EDTA in 100 mM potassium phosphate buffer, pH 8.0. The reaction cuvettes are equilibrated at 37°, 10–50 μl of appropriately diluted E3 preparation (0.1–2 mg/ml of crude E3) is added to start the reaction, and the formation of NADH is monitored for 2–3 min. Reactions are run in duplicate, at two protein concentrations. Blank reactions are run in parallel, under the same conditions except for omission of dihydrolipoamide. The specific activity of E3 is calculated as units per milligram of total cellular protein. The definition of 1 unit of activity is 1 μmol of NADH produced per minute at 37°.

Measurement of Activities of Pyruvate Dehydrogenase Complex, α-Keto-glutarate Dehydrogenase Complex, and Branched-Chain α-Keto Acid Dehydrogenase Complex. PDC activity is assayed by measuring $^{14}CO_2$ produced from [1-^{14}C]pyruvate, as described previously.[13] To assay the activity of PDC in the activated (dephosphorylated) state, harvested fibroblasts are treated with dichloroacetic acid (DCA) before the reaction is initiated. Fibroblasts are resuspended in PBS with protease inhibitors and aliquoted in small quantities (150 μl). Seven microliters of 125 mM DCA is added to a final concentration of 5 mM and incubated for 15 min at 37°. The treatment is stopped by addition of 30 μl of stopping mixture and 5 μl of FCS followed by two cycles of freeze–thawing before measurement of activity.

Assays are run in quadruplicate tubes (at two protein concentrations and two incubation times), and a blank reaction (without TPP or coenzyme A in the reaction mixture) is run for each condition (four tubes). The assay mixture (0.1 ml) contains final concentrations of 2 mM MgCl$_2$, 2 mM NAD$^+$, 15 mM potassium oxalate, 0.3 mM CoA, 1 mM dithiothreitol, 0.1 mM TPP, 0.5 mM [1-^{14}C]pyruvate for PDC, phosphotransacetylase (1 U/ml), and

[12] M. S. Patel and Y. S. Hong, *in* "Methods in Molecular Biology" (D. Amstrong, ed.), p. 337. Humana Press, Totowa, New Jersey, 1998.
[13] K.-F. R. Sheu, C.-W. C. Hu, and M. F. Utter, *J. Clin. Invest.* **67**, 1463 (1981).

FCS (25 μl/ml) in 50 mM potassium phosphate buffer, pH 8.0; and 5 or 10 μl of sample solution. The blank reaction mixture is identical, except that TPP and CoA are omitted. Seventy-nine microliters of the "assay reaction mixture" is placed at the bottom of a 16 $\times$ 100 mm test tube, and 10 μl of solution A or B is added. First, the test tubes containing all reagents except the sample are incubated in a 37° water bath for 3 min, and then 10 μl of the sample is added and the reaction is initiated by addition of 1 μl of [1-^{14}C]pyruvate. A rubber serum stopper fitted with a stemmed hanging center well containing a folded piece of filter paper saturated with 0.1 ml of hyamine hydroxide is immediately placed onto the tube. After 5 or 10 min, the reactions are stopped by injecting 20 μl of ice-cold 20% (w/v) trichloroacetic acid (TCA) containing 30 mM unlabeled pyruvate through the serum stopper, using a syringe. After mixing, the acidified reaction mixtures are incubated in a 37° water bath for 30 min to collect evolved $^{14}CO_2$ in the hyamine hydroxide. The radioactivity of the hyamine hydroxide-saturated filter paper is measured in a scintillation counter. Using the specific radioactivity of [1-^{14}C]pyruvate added to the reaction mixture, nanomoles of evolved CO_2 are calculated. One milliunit of enzyme activity is defined as 1 nmol of CO_2 produced per minute at 37°. Modifications of this method are available for measuring PDC activity in the inactive (dephosphorylated) or untreated states, and for assay of activated/inactivated PDC in frozen tissues.[14] For measurement of KDC, the activation (dephosphorylation) process is not required, and α-keto[1-^{14}C]glutarate is used in place of pyruvate as the substrate. For measurement of BCKDC, the activation step is carried out in the presence of 5 mM α-keto-isocaproate,[15] and α-keto[1-^{14}C]isovalerate is used as the substrate. α-Keto[1-^{14}C]isovalerate is prepared from L-[1-^{14}C]valine.[16]

Preparation of Total Cellular and Mitochondrial Proteins for Immunoblotting

For the preparation of total soluble cellular proteins, harvested skin fibroblasts are washed twice with phosphate-buffered saline (PBS) and resuspended in PBS. One-tenth volume of 10% (v/v) Triton X-100 is added to the cell suspension and mixed thoroughly by brief vortexing, and placed on ice for 1 hr with brief vortexing every 10 min. Cell suspensions are frozen by immersing the tube in an ethanol–dry ice bath and thawed in a 30° waterbath. This freeze–thaw procedure is repeated one more time and

[14] D. S. Kerr, L. Ho, C. M. Berlin, K. F. Lanoue, J. Towfighi, C. L. Hoppel, M. M. Lusk, C. M. Gondek, and M. S. Patel, *Pediatr. Res.* **22,** 312 (1987).

[15] R. Paxton and R. A. Harris, *J. Biol. Chem.* **257,** 14433 (1982).

[16] H. W. Rudiger, U. Langenbeck, and H. W. Geodde, *Biochem. J.* **126,** 445 (1972).

the final suspension containing total soluble cellular protein is kept on ice until use. For the preparation of mitochondrial proteins, cultured fibroblasts are washed with PBS, and 0.5 ml of 10 mM Tris-HCl (pH 7.4) containing 250 mM sucrose, 1 mM 2-mercaptoethanol, 0.5 mM phenymethylsulfonyl fluoride (PMSF), and leupeptin (25 μg/ml) is added and spread evenly on each plate. Plates are then tilted on ice and excess buffer is removed by pipetting, and fibroblasts are collected with a sterile cell lifter and combined into one Eppendorf tube (volume is 0.85–1 ml from four plates). Cells are homogenized with a glass homogenizer. To remove nuclei and unbroken cell debris, the suspension is centrifuged at 500g for 10 min at 4°, and the supernatant is centrifuged at 10,000g for 15 min at 4° to recover mitochondria. After discarding the supernatant, mitochondria are suspended in 30–50 μl of the same buffer as described above and stored at −80° until use.

Preparation of Total RNA from Cultured Fibroblasts

To prepare E3 cDNAs from E3-deficient subjects and their family members, total RNA is isolated from cultured skin fibroblasts. The guanidium thiocyanate method[17] or a commercially available Perfect RNA isolation kit (from 5′→3′, Boulder, CO) may be used for the isolation of total RNAs by the protocols provided by the manufacturer. Because RNA is easily susceptible to degradation by RNase, caution must be exerted to avoid RNase contamination in all containers during RNA preparation. With a microscale kit, typically 80–120 μg of pure RNA, in the range of 1.5–2.5 μg/μl, can be obtained from four confluent plates, which is sufficient for several RNA blot analyses and reverse transcription-polymerase chain reactions (RT-PCR).

Immuno- and RNA Blot Analyses

E3 deficiency can be caused by the lack of a functional E3 protein, resulting from impaired transcription or translation, as well as by functionally crippled mutant E3 proteins, which may be present in normal amounts, or by instability of expressed E3 in $vivo$. To test these possibilities, immuno- and RNA blot analyses can be performed with total soluble proteins (or mitochondrial proteins) and total RNA obtained from cultured fibroblasts.

For immunoblot analysis, appropriate amounts of total soluble protein (25 μg) or mitochondrial protein (14 μg) are loaded onto a 12% (w/v) SDS–polyacrylamide gel. The amount of protein applied can be varied depending on the sensitivity of the specific antibodies. After electrophoresis, the separated, unstained proteins are transferred onto a nitrocellulose mem-

---

[17] P. Chomczynski and N. Sacchi, *Anal. Biochem.* **162,** 156 (1987).

brane. After incubating with 10% (w/v) nonfat skim milk in Tris-buffered saline containing 2% (v/v) Tween 20 (TTBS) for 1 hr at room temperature, the membranes are washed three times with TTBS and incubated for 1 hr at room temperature with a mixture of appropriately diluted (depending on titer of each specific antibody) rabbit antisera raised against bovine kidney pyruvate dehydrogenase (E1), bovine heart dihydrolipoamide acetyltransferase (E2), and bovine heart E3. The membrane is then washed again and incubated with horseradish peroxidase-conjugated anti-rabbit goat antibody (generally 1 : 1000–2000 diluted antibody gives good results), and E1, E2, and E3 protein bands are detected by a chemiluminescence system (DuPont, Wilmington, DE). The best results are usually obtained with less than 1 min of exposure to X-ray film. The relative band ratio of E3/E1 or E3/E2 is measured with a densitometer.

For RNA blot analysis, 20 μg of total RNA isolated from cultured fibroblasts usually produces good results with a ^{32}P-labeled E3 cDNA probe. Total RNAs are separated electrophoretically by size fractionation in a 0.8% (w/v) agarose gel containing 50% (v/v) formaldehyde and subsequently transferred to a Genescreen nylon membrane (NEN-DuPont, Boston, MA). Suitable molecular weight markers such as *Escherichia coli* ribosomal RNA, globin mRNA, and tRNA are coelectrophoresed to estimate the molecular weight of the hybridizing mRNA samples. The membrane is baked at 80° for 1 hr and prehybridized in the solution containing 20 mM piperazine-N,N'-bis(2-ethanesulfonic acid) (PIPES, pH 6.4), 2 mM EDTA, 0.8 M NaCl, 50% (v/v) formaldehyde, and salmon sperm DNA (100 μg/ml) at 42° overnight. The membrane is probed with a ^{32}P-labeled E1β/E3 cDNA probe mixture overnight at 68°. The membrane is then washed twice with 2 × SSC–2% (w/v) SDS for 30 min at 65°, followed by 0.2 × SSC–0.2% (w/v) SDS twice for 30 min and exposed to X-ray film overnight. Two E3 mRNA bands, 2.4 and 2.2 kb, are detected with three bands of E1β mRNA, 5.2, 1.6, and 1.3 kb, respectively. By scanning and measuring the relative intensities of E3 mRNA bands against those of the E1β RNA bands, the quantities of E3 mRNAs relative to that of controls can be calculated. As a second internal control, labeled β-actin probe can be used.

Mutational Analysis of E3 cDNA

Reverse Transcriptase-Polymerase Chain Reaction. For the construction of E3 cDNAs from total RNA and subsequent insertion into a cloning vector for mutation analysis, two-step RT-PCR is performed with SuperScript II reverse transcriptase (Life Technology) and *Taq* DNA polymerase. Cloning is facilitated by incorporating restriction sites in the PCR primers to synthesize full-length E3 cDNA fragments with staggered ends compatible with

cloning sites in commercially available cloning vectors. The following protocol produces good results.

Total RNA purified as described above is freshly diluted to a concentration of 0.1 $\mu g/\mu l$ in diethylpyrocarbonate (DEPC)-treated distilled water.
Each reaction tube contains the following solutions: 2 μl of total RNA
(0.1 $\mu g/\mu l$), 2.5 μl of oligo (dT)$_{12-18}$ (0.5 $\mu g/\mu l$), and 7.5 μl of water. After
mixing well, the mixture is incubated at 70° for 10 min followed by placing
it on ice for 1 min prior to brief centrifugation to collect liquid drops. To
this mixture, the following solutions are added in order: 3.5 μl of 5× first-
strand cDNA buffer, 2 μl of 5 mM dNTP mixture, 2 μl of 0.1 M dithiothreitol, and 1 μl of SuperScript II RT (200 U/μl). After a brief spin, the reaction
mixture is incubated for 10 min at room temperature and the reaction is
allowed to proceed at 42° for 50 min. The reverse transcriptase is then
denatured at 70° for 15 min, and the tube is placed on ice. Once again, the
liquid drops in the tube are collected by brief centrifugation, and RNA,
which might interfere with the PCR, is digested by the addition of 1 μl of
RNase H (1–4 U/μl) and incubated at 37° for 20 min. The reaction tube,
containing single-strand cDNA without RNA, is kept on ice until used as
a template for PCR.

To synthesize double-stranded E3 cDNA with new restriction sites at
both ends, PCR is done as follows: 1 μl of single-stranded cDNA (from
above), 10 μl of 10× PCR buffer (Mg^{2+} free), 1 μl of 5′ primer (0.1 nmol,
with *Bam*HI restriction site), 1 μl of 3′ primer (0.1 nmol, with *Bam*HI
restriction site), 5 μl of 10 mM dNTPs, 10 μl of 25 mM MgCl$_2$, 0.5 μl of
Taq DNA polymerase (5 U/μl), and 71.5 μl of water. The sequences of
the 5′ and 3′ primers are 5′-GCG CGC GGA TCC GGA GGT GAA
AGT ATT GGC GG-3′ and 5′-GCG CGC GGA TCC TCA AAA GTT
GAT TGA TTT GCC-3′, respectively. The final reaction volume is 100 μl.
Two drops of light mineral oil are placed on top of the reaction mixture.
The amplification conditions consist of initial denaturation at 94° for 7 min;
followed by 35 cycles of denaturing at 94° for 1 min, annealing at 55° for
1 min, and extension at 72° for 2 min; and a final extension at 72° for 7
min. The PCR products are separated by 1% (w/v) agarose gel electrophoresis and visualized by staining with ethidium bromide, and the existence of
the appropriate product (about 1.6 kb) is confirmed. If nonspecific DNA
bands are detected, the PCR conditions can be changed, such as by elevating
the annealing temperature or by varying the Mg^{2+} concentration in the
reaction mixture. E3 cDNA is purified with a Wizard PCR preparation kit
(Promega, Madison, WI) to remove any contaminants.

Preparation of Recombinant E3 cDNA. For DNA sequencing, full-
length E3 cDNA, including the leader sequence, is inserted into a pBlue-
Script SK (+) cloning vector and the recombinant DNA is transferred into

E. coli XL1-Blue competent cells. After screening for transformed cells by antibiotic selection, the recombinant DNA is amplified by culturing overnight at 37° in 20 ml of LB liquid medium containing ampicillin (100 μg/ml), with vigorous shaking. The recombinant DNA is purified with a Wizard plasmid minipreparation kit (Promega). The insert is liberated by digestion with *Bam*HI and visualized on a 1% (w/v) agarose gel with ethidium bromide staining. If confirmation of proper orientation of the insert is required, digestion with restriction enzymes that have cutting sites in both the E3 cDNA and vector can be performed and the orientation of inserted E3 cDNA in the vector is determined by the resulting DNA fragment sizes.

DNA Sequencing. DNA sequencing is done with a Sequenase kit version 2.0 (U.S. Biochemicals, Cleveland, OH) as instructed by the manufacturer. Alternatively, an automated sequencing apparatus can be used with speed and accuracy. Primers for this step are listed in Table I. DNA polymerase and Sequenase may introduce errors (up to 1 error in every 400 bases); therefore, DNA sequencing is repeated in several E3 cDNA clones from each RT-PCR amplification to rule out any unwanted errors. Once a mutation is identified and confirmed in several E3 cDNA clones, additional cDNA clones are analyzed for the sequence in the targeted area of the known mutation. At least two mutations (one in each allele-specific cDNA) are expected in most cases of E3 deficiency. To identify new mutations, several clones must be sequenced completely. Once a new mutation is located, it can be confirmed in several additional cDNA clones, using the same strategy.

TABLE I
PRIMERS FOR E3 cDNA DNA SEQUENCING ANALYSIS

Primer	Sequence	Remark
E3-211	5'-TATGTTGCTGCTATTAAAGCT-3'	Sense
E3-346	5'-ATCTTTTCCATGGGCCATAT-3'	Antisense
E3-417	5'-ATGATGGAGCAGAAGAGTACTGCA-3'	Sense
E3-768	5'-CAGCAGTTGAACGTTTAGGTCATG-3'	Sense
E3-1088	5'-TTTAGTTTGAAATCTGGTATTGAC-3'	Antisense
E3-1263	5'-TCTTTCAACTGCTCTTCTGA-3'	Antisense
E3-1462	5'-TGGAATATGGAGCATCCTGT-3'	Sense
E3-1522	5'-TAAGGTCGGCTGTGCATGA-3'	Antisense
−40 primer	5'-GTTTTCCCAGTCACGAC-3'	pBS primer

TABLE II
BIOCHEMICAL ANALYSES OF THREE E3-DEFICIENT PATIENTS

Subject	E3 activity in fibroblasts (% of control)	PDC activity in fibroblasts (% of control)	Immunoreactive E3 (% of control)	Mutations	Ref.
1	6	20	40	K37E (AAA to GAA) P453L (CCG to CTG)	21
2	14	11	50	Y-1X (TAC to TAAC) R460G (AGA to GGA)	22
3	3	12	100	Δ101G (AGG deletion) E340K (GAA to AAA)	23

Direct sequencing of genomic DNA[18] is then used to avoid any artifacts as well as to confirm the mutations found. Total cellular DNA is isolated by a conventional method.[19,20] Two bands (two peaks for automated sequencing), generated by a wild-type nucleotide and a mutated nucleotide from each allele of the E3 gene, are detected in the same position in the case of substitution mutations. However, in the case of deletion or insertion mutations (causing a frame shift), ladder patterns with overlapping of two independent sequencing bands originating from two different alleles are observed.

Findings in E3-Deficient Subjects and Their Families

From our investigation of three unrelated E3-deficient subjects, two of the three subjects had approximately 50% of E3 protein compared with that of controls (Table II). In one of these subjects, DNA sequence analysis revealed a nonsense mutation in the last codon of the leader peptide sequence resulting in a premature termination of E3 protein. In the third subject, the level of immunoreactive E3 remained unaltered in cultured fibroblasts, suggesting a catalytically crippled protein. The relative levels of E3-specific mRNA in all three subjects were similar to that seen in controls. Therefore, E3 deficiency in these cases was not considered to be caused by impaired transcription or instability of E3

---

[18] B. R. Krishnan, R. W. Blakesky, and D. E. Berg. *Nucleic Acids Res.* **19,** 1153 (1991).

[19] M. A. Peinado, S. Malkhosyan, A. Velazquez, and M. Perucho, *Proc. Natl. Acad. Sci. U.S.A.* **89,** 10065 (1992).

[20] M. Perucho, M. Goldfarb, K. Shimizu, C. Lama, J. Fogh, and M. Wigler, *Cell* **27,** 467 (1981).

mRNA. We identified six different mutations including four substitutions, a three-nucleotide deletion, and a single-nucleotide insertion causing premature termination of the E3 protein (Table II).[21–23] DNA sequencing analysis of family members was used for the confirmation of the mode of inheritance. In two families tested, the parents showed a heterozygous state, confirming that E3 deficiency is inherited in an autosomal recessive manner. In the case of patient 2, the nonsense mutation (TAC to TAAC) was inherited from the father, and the substitution mutation (R460G, AGA to GGA) from the mother. In the case of subject 3, the substitution mutation (E340K, GAA to AAA) was inherited from the father, and the deletion mutation (Δ101G, AGG deletion) was found in the mother as well as in a younger sister.

If specific mutations that cause E3 deficiency are established in the parents, it should be possible to identify E3 gene mutations in products of conception. A chorionic villus sample was obtained from the mother of subject 3 at approximately the sixth week of a subsequent pregnancy and used for measurement of E3 and PDC activities as well as isolation of total RNA. E3 and PDC activities of chorionic villus homogenates were 45 and 93% of controls, respectively, while the activities in cultured chorionic villus fibroblasts were 64 and 73% of controls. Full-length E3 cDNAs were constructed by RT-PCR with isolated total RNA isolated from the chorionic villus sample, as described above. Also, direct nucleotide sequencing around the regions of known mutations was performed with genomic DNA isolated from cultured chorionic villus fibroblasts to confirm the mutations. We identified only one mutation, derived from the mother (AGG deletion), establishing the heterozygous state of the fetus. This conclusion was later confirmed with DNA isolated from cultured fibroblasts derived from a skin biopsy obtained after birth.

Acknowledgments

The authors acknowledge grant support from the National Institute of Health: DK42885 (to M.S.P.) and MJC009122 (to D.S.K.).

[21] T.-C. Liu, H. Kim, C. Arizmendi, A. Kitano, and M. S. Patel, *Proc. Natl. Acad. Sci. U.S.A.* **90**, 5186 (1993).

[22] Y. S. Hong, D. S. Kerr, W. J. Craigen, J. Tan, Y. Pan, M. M. Lusk, and M. S. Patel, *Hum. Mol. Genet.* **5**, 1925 (1996).

[23] Y. S. Hong, D. S. Kerr, T.-C. Liu, M. M. Lusk, B. R. Powell, and M. S. Patel, *Biochim. Biophys. Acta* **1362**, 160 (1997).

[42] Targeting E3 Component of α-Keto Acid Dehydrogenase Complexes

By MARK T. JOHNSON, HSIN-SHENG YANG, and MULCHAND S. PATEL

Introduction

Branched-chain α-keto acid dehydrogenase complex (BCKDC) is a mitochondrial multienzyme complex that catalyzes a series of reactions that form the first irreversible step in the catabolism of the essential branched-chain amino acids: leucine, isoleucine, and valine. After transamination of these amino acids to their respective α-keto acids, BCKDC catalyzes the irreversible oxidative decarboxylation and subsequent formation of acyl-CoAs. BCKDC is estimated to be 4–5 million Da in size and consists of multiple copies of three enzymatic and two regulatory components. The three enzymatic components of BCKDC are abbreviated as E1b, E2b, and E3 components, indicating their sequential roles in the enzymatic cascade. The E3 component, also referred to as dihydrolipoamide dehydrogenase (EC 1.8.1.4), is shared among two other α-keto acid dehydrogenase complexes, namely pyruvate dehydrogenase and α-ketoglutarate dehydrogenase complexes, as well as with the glycine cleavage system.[1,2] E3 functions in these complexes by oxidizing the dihydrolipoyl moiety of the transacylase components and transferring the electron pair to a recipient NAD$^+$ molecule.

Deficiency of E3[3–8] results in an inborn error of metabolism referred to as a variant form of maple syrup urine disease (MSUD). The presence of branched-chain keto acids and lactate in the urine of patients with MSUD imparts a distinctive odor. Branched-chain keto acids in the urine are

---

[1] N. N. Vettakkorumakankav and M. S. Patel, *Ind. J. Biochem. Biophys.* **33,** 168 (1996).

[2] L. J. Reed, *Acc. Chem. Res.* **7,** 40 (1976).

[3] B. H. Robinson, J. Taylor, and W. G. Sherwood, *Pediatr. Res.* **11,** 1198 (1977).

[4] T. C. Liu, H. Kim, C. Arizmendi, A. Kitano, and M. S. Patel, *Proc. Natl. Acad. Sci. U.S.A.* **90,** 5186 (1993).

[5] J. P. Bonnefont, D. Chretien, P. Rustin, B. Robinson, A. Vassault, J. Aupetit, C. Charpentier, D. Rabier, J. M. Saudubray, and A. Munnich, *J. Pediatr.* **212,** 255 (1992).

[6] N. Guffon, C. Lopez-Mediavilla, R. Dumoulin, B. Mousson, C. Godinot, H. Carrier, J. M. Collombet, P. Divry, M. Mathieu, and P. Gubaud, *J. Inher. Metab. Dis.* **16,** 821 (1993).

[7] A. Kohlschutter, A. Behbehani, U. Langenbeck, M. Albani, P. Heidemann, G. Hoffmann, J. Kleineke, W. Lehnert, and U. Wendel, *Eur. J. Pediatr.* **138,** 32 (1982).

[8] J. H. Cross, A. Connelly, D. G. Gadian, B. E. Kendall, G. K. Brown, R. M. Brown, and J. V. Leonard, *Pediatr. Neurol.* **10,** 276 (1994).

0076-6879/00 $30.00

indicative of a systemic problem; these keto acids accumulate, causing an organic acidemia and neurologic damage. There are several phenotypic categories of MSUD that, if untreated, all result in varying degrees of physical and mental retardation within the first 2 years of life.[9,10] The E3-deficient category manifests with a more complex organic acidosis, including elevated levels of lactate, and has a phenotype similar to the severe or classic form of MSUD, in which there is rapid neurologic deterioration that results in neonatal death.[11] This chapter discusses the generation of E3 knockout mice and the characterization of a null allele of the dihydrolipo-amide dehydrogenase (*Dld*) gene.

Materials and Reagents

Bacteriophage libraries: λ EMBL-3 bacteriophages containing *Mbo*I partial digests of DBA/2J genomic DNA (Clontech, Palo Alto, CA) and λ Fix II bacteriophages containing *Sau*3A partially digested 129SVJ genomic DNA (Stratagene, La Jolla, CA)

Bacterial strains: LE392 for EMBL-3, XL-1 Blue MRA for Fix II, DH5α for plasmid cloning (BRL-Life Technologies, Gaithersburg, MD)

Polymerases: Klenow fragment of DNA polymerase I, *Taq* polymerase

Zeta-Probe GT blotting membrane (Bio-Rad, Hercules, CA)

Plasmids: pUC 19 (BRL-Life Technologies), β-actin promoter-driven neomycin phosphotransferase gene and lacking a polyadenylation signal, diptheria toxin fragment minigene cassette[12]

Mouse strains: C57BL (Taconic, Germantown, NY), 129 SVJ (Jackson Laboratory, Bar Harbor, ME), MTKNeo2[13]

Buffers and Media

Church hybridization solution: 500 mM sodium phosphate (pH 6.8), 1 mM EDTA, 7% (w/v) sodium dodecyl sulfate (SDS), and bovine serum albumin (BSA; 0.5%, w/v)

TAE buffer: 40 mM Tris–acetate (pH 7.5) and 1 mM EDTA

SSC solution: 150 mM NaCl and 15 mM sodium citrate

Embryonic stem (ES) cell growth medium: Dulbecco's modified

---

[9] D. T. Chuang and V. E. Shih, *in* "The Metabolic and Molecular Base of Inherited Disease" (C. R. Scriver, A. L. Beaudet, W. S. Sly, and D. Valle, eds.), p. 1239. McGraw-Hill, New York, 1995.

[10] F. Peinemann and D. J. Danner, *J. Inher. Metab. Dis.* **17**, 3 (1994).

[11] P. A. W. Harper, P. J. Healy, and J. A. Dennis, *Acta Neuropathol.* **71**, 316 (1986).

[12] H. Tomasiewicz, K. Ono, D. Yee, C. Thompson, C. Goridis, U. Rutishauser, and T. Magnuson, *Neuron* **11**, 1163 (1993).

[13] S. J. Abbondanzo, I. Gadi, and C. L. Stewart, *Methods Enzymol.* **225**, 808 (1993).

Eagle's medium (DMEM) with 4.5 g of glucose per liter (Life Technologies) supplemented with 15% (v/v) ES cell-tested fetal calf serum (JRH Biosciences, Lenexa, KS), 0.1 mM 2-mercaptoethanol (Sigma, St. Louis, MO), penicillin (100 U/ml), streptomycin (100 U/ml) (Life Technologies), and ESGRO (1000 U/ml; Life Technologies)

Trypsin solution: 0.25% (w/v) trypsin, 0.02% (v/v) EDTA (JRH Biosciences)

Phosphate-buffered saline (PBS) buffer: 137 mM NaCl, 0.3 mM KCl, 10 mM Na$_2$HPO$_4$, and 0.18 mM KH$_2$PO$_4$, pH 7.4

Low-TE electroporation solution: 10 mM Tris-HCl (pH 7.5), 0.1 mM EDTA

G418 100× stock solution: Dissolve 180 mg of G418 in 10 ml of DMEM, filter sterilize, and store aliquots at −20°

Freezing solution (2×): 60% (v/v) DMEM, 20% (v/v) fetal calf serum, 20% (v/v) dimethyl sulfoxide (DMSO), stored at 4°

Genomic DNA extraction buffer: 10 mM Tris-HCl (pH 7.5), 10 mM EDTA, 10 mM NaCl, 0.5% (w/v) sarcosyl, and proteinase K added immediately before use to final concentration of 1 mg/ml

DNA precipitation solution: 1.5 μl of 5 M NaCl added to ice-cold 100% ethanol immediately before use

Embryo lysis buffer: 100 mM Tris-HCl (pH 8.0), 2 mM MgCl$_2$, 0.01% (w/v) gelatin, 0.45% (v/v) Tween 20, 0.45% (v/v) Nonidet P-40 (NP-40), and proteinase K (500 μg/ml)

Cloning of Mouse Dld Gene

The mouse *Dld* gene is cloned from a λ EMBL-3 mouse genomic DNA library prepared from the DBA/2J strain (Clontech).[14] The full-length human E3 cDNA[15] is random-prime labeled to incorporate approximately 1×10^8 cpm/μg[16] and used to screen approximately 1×10^6 recombinant bacteriophages. Hybridizations are performed in Church buffer overnight at 65° and then serially washed in progressively lower salt concentrations ranging from 1× SSC and 0.1% (w/v) SDS down to 0.1× SSC at 65°. The washed filters are then exposed to Kodak (Rochester, NY) XAR-5 film overnight at −80°. Positive plaques are picked with Pasteur pipettes and purified by several rounds of replating and rehybridization until a pure clonal population of hybridizing plaque-forming units (PFU) is identified.

[14] M. Johnson, H. S. Yang, G. L. Johanning, and M. S. Patel, *Genomics* **41,** 320 (1997).
[15] G. Pons, C. Raefsky-Estrin, D. J. Carothers, R. A. Pepin, A. A. Javed, B. W. Jesse, M. K. Ganapathi, D. Samols, and M. S. Patel, *Proc. Natl. Acad. Sci. U.S.A.* **85,** 1422 (1988).
[16] A. P. Feinberg and B. Vogelstein, *Anal. Biochem.* **132,** 6 (1983).

The phage DNA is then purified, digested with restriction enzymes *Eco*RI, *Bam*HI, and *Sal*I, and subjected to Southern hybridizations with the same probe and under similar hybridization conditions. The DNAs from the positive phage clones are isolated and subcloned into pUC19 (Life Technologies) or pBluescript KS⁺ (Stratagene) vectors. Restriction enzymes mapping and partial nucleotide sequencing are performed to identify the intron and exon locations of the mouse *Dld* gene.[14]

Construction of *Dld* Gene Targeting Vector

To create an isogenic construct, approximately 1.8×10^6 clones from an isogenic 129SV genomic library are screened with two labeled fragments from the DBA/2J genomic clones (a 0.7-kb *Sal*I–*Sph*I fragment containing exon 9 and a 0.8-kb *Hind*II–*Sau*3A fragment within intron 4). In our experiments one clone isolated from this screen contained a 13-kb insert extending approximately from exon 3 to exon 11. A targeting constuct is then designed by using the positive–negative selection strategy,[17] in which a positively selectable marker, a neomycin phosphotransferase driven by the β-actin promoter (*neo*), is utilized as a sequence to disrupt the reading frame and a negative marker, the diptheria toxin fragment A (DT-A),[12] is included to select against random insertions of the vector within the ES cell genome. A 2-kb *Eco*RI–*Bgl*II DT-A cassette with the β-actin promoter is inserted by blunt-ended ligation into an *Eco*RI site of the pUC19 vector (this vector is named pUC/DT), thereby destroying the *Eco*RI site. A 7-kb *Xba*I fragment including the 3′ half of the λ insert is subcloned into the *Xba*I site of the pUC/DT vector to serve as the region of homology in the targeting vector, with roughly equal-sized arms of homology on either side of an *Eco*RI site within exon 10 at codon 301 (Fig. 1). From structure–function studies, it has been shown that residues downstream of the *Eco*RI site, such as H452 and E457, are essential for function.[18] The *Dld* gene is disrupted by cloning a neomycin phosphotransferase gene driven by the β-actin promoter lacking a polyadenylation signal (*neo*) into the *Eco*RI site. *neo* is inserted in the same transcriptional orientation as the endogenous E3 in order to utilize the endogenous polyadenylation signal of the E3 gene following a homologous recombination event. The lack of a polyadenylation signal confers an additional power of selection against random insertions, as many of these events will not be conveniently located upstream of a polyadenylation signal. A *Not*I site from the λ polylinker, which is carried along with

[17] S. L. Mansour, K. R. Thomas, and M. R. Capecchi, *Nature* (*London*) **336**, 348 (1988).
[18] H. Kim and M. S. Patel, *J. Biol. Chem.* **167**, 5128 (1992).

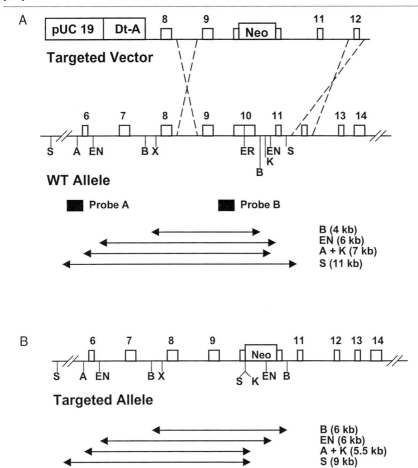

Fig. 1. Targeting of the *Dld* gene. (A) Linearized targeting vector is composed of the pUC19 vector, the β-actin–diphtheria toxin A chain gene (DT-A) negative maker, 7 kb of homology, and the β-actin–neomycin phosphotransferase gene (*neo*). The wild-type *Dld* allele is represented by numbered boxes indicating exons with restriction enzyme sites listed below. The two probes used for distinguishing the wild-type from the targeted alleles are designated by the solid boxes. The sizes of the various restriction fragments and the enzymes used are shown below the restriction map. (B) The targeted allele after a recombination event is shown with the integrated *neo* and the altered restriction map. A, *Apa*I; B, *Bst*XI; ER, *Eco*RI; EN, *Eco*NI; K, *Kpn*I; S, *Sal*I; X, *Xba*I.

the genomic fragment at its 3' end, is utilized as a site for linearizing the construct before electroporation.

Targeted Disruption of *Dld* Gene in Embryonic Stem Cells

The E14.1 embryonic stem cell line derived from the 129/Ola strain[19] is cultured under routine conditions similar to those described by Ramirez-Solis *et al.*[20] Briefly, monolayers of primary embryonic feeders are prepared from day 13 embryos,[13] using the MTKNeo2 transgenic line that carries a neomycin resistance gene. Feeder cells are maintained for up to five passages in ES cell growth medium lacking ESGRO. To prepare a monolayer of growth-arrested feeders, the primary embryonic feeders are trypsinized, γ irradiated with 1000 rads, and plated at a density of 1×10^6 cells per 60-mm dish at least 6 hr prior to ES cell passaging. ES cells are maintained on the feeder layer in a 37° incubator with 5% CO_2 with passaging every 2–3 days, in which 6×10^5 ES cells are plated per 60-mm dish after trypsinization. For electroporation, 7×10^6 ES cells and linearized targeting construct (25 μg/ml) are suspended in 750 μl of low-TE solution in 0.4-cm cuvettes and pulsed at settings of 270 V and 650 μF, using a Gene Pulser electroporator (Bio-Rad). The cells are incubated at room temperature for 10 min and then plated at 10^6 cells per 100-mm plate in ES cell growth medium for 24 hr, followed by the addition of selection medium containing G418 (125 μg/ml). After a 10-day selection period, G418-resistant clones are individually transferred to separate wells in gelatinized 96-well plates lacking feeder cells, using a stereoscope and pulled Pasteur pipettes. The frequency of production of G418-resistant colonies is quite low, with each electroporation of 7×10^6 cells producing fewer than 50 colonies surviving postselection, most likely due to the poly(A)$^-$ design of this construct. After several days of growth, clones in individual wells are split to two plates, with one plate grown to confluence for the isolation of genomic DNA while the second plate is prepared for freezing. The plate to be frozen is prepared by treating each well with 30 μl of trypsin solution followed by the addition of an equal volume of 2× freezing medium, and then placing the plate in a Styrofoam container to cool slowly to −70°.

Analysis of Targeted Embryonic Stem Cells

The initial identification of homologous recombination between the targeting DNA and the endogenous *Dld* gene is performed by Southern

[19] M. Hooper, K. Hardy, A. Handyside, S. Hunter, and M. Monk, *Nature* (*London*) **326**, 292 (1987).

[20] R. Ramirez-Solis, A. C. Davis, and A. Bradley, *Methods Enzymol.* **252**, 855 (1995).

blot analysis. The genomic DNA is isolated from colonies grown in the 96-well plates. Cells are washed twice with PBS and then 50 μl of genomic DNA extraction buffer is added per well. The plates are then sealed and incubated at 60° overnight with gentle rocking. Genomic DNA is recovered by adding 100 μl of DNA precipitation solution per well and then spinning the plate at 2500 rpm for 10 min. After careful decanting of the solution, the wells are washed twice with 70% (v/v) ethanol and allowed to air dry. The samples are resuspended in 35 μl of restriction digestion buffer (supplied by the manufacturer) containing the SacI isoschizomer Ecl136II (New England BioLabs, Beverly, MA), and incubated at 37° overnight. To identify which G418-resistant clones have undergone homologous recombination events, the digested DNA from each clone is fractionated on a 0.8% (w/v) agarose gel in TAE buffer, transferred to Zeta-Probe membranes (Bio-Rad), and probed with a HindIII–SalI fragment from DNA upstream of the region contained within the targeting vector (Fig. 1, probe A). After hybridization in Church buffer, the membranes are washed with 1× SSC and 0.1% (w/v) SDS at room temperature for three 20-min periods and then with 0.25× SSC and 0.1% (w/v) SDS at 65° for 20 min.

Genomic Southern blot analysis with Ecl136II digestion yields an 11-kb wild-type allele and a 9-kb targeted allele (Fig. 1). For further confirmation of the site-specific integration of the targeting DNA at the Dld locus, genomic DNA from putative targeted ES cells is digested by several additional restriction enzymes (ApaI plus KpnI, EcoNI, or BstXI) and analyzed by Southern blot, using a 1-kb HindIII fragment that hybridizes to sequences immediately upstream of the EcoRI site (Fig. 1, probe B).[21] The presence of predicted wild-type and targeted alleles identifies heterozygous ES cell lines. Also, none of these lines demonstrates additional, unanticipated hybridizing bands, indicating that there are no additional insertions of the targeting construct into the genome.

In addition to Southern blot analysis, the targeted ES cell lines are also analyzed for E3 enzymatic activity.[22] Cells are washed in PBS twice, resuspended in 20 mM potassium phosphate (pH 7.5) containing 0.5% (v/v) Triton X-100, and then subjected to three freeze–thaw cycles. After centrifugation, aliquots of the supernatants are added to the reaction cocktail [100 mM potassium phosphate (pH 8.0), 1.5 mM EDTA, 3 mM dihydrolipoamide, 3 mM NAD$^+$], to a final volume of 1 ml. The production of NADH is measured spectrophotometrically at a wavelength of 340 nm over

[21] M. T. Johnson, H.-S. Yang, and M. S. Patel, Proc. Natl. Acad. Sci. U.S.A. **94**, 14512 (1997).
[22] M. S. Patel, N. N. Vettakkorumakankav, and T. C. Liu, Methods Enzymol. **252**, 186 (1995).

several minutes at 37°. An approximately 50% reduction in E3 activity is observed in targeted ES cells. ES cells that are confirmed as heterozygous by molecular and biochemical analysis are karyotyped by preparing standard chromosome spreads and staining with Giemsa.

Establishment of Chimeric Mice

The targeted ($Dld^{+/-}$) ES cell lines with good karyotypes (>85% diploid) are microinjected into C57BL/6 recipient blastocysts and transferred to the uteri of CH3B16 pseudopregnant females (performed by J. Duffy at the University of Cincinnati, Cincinnati, OH). The offspring are judged to be chimeric on the basis of coat color. The recipient blastocysts are derived from a strain that is homozygous for the alleles, aCP, which results in a black coat color whereas the ES cells are derived from the 129/Ola strain that is homozygous $A^w c^{ch}p$, producing a cream-colored coat. Contribution of ES cells in the chimeras can be visually assessed by the presence of areas of agouti fur, indicating contribution from the ES cells to the epithelium, with areas of a cream color indicating a region of higher ES cell contribution. These overtly chimeric animals are initially bred to the outbred Black Swiss line so that ES cell-derived progeny are identified by the presence of agouti offspring as a result of transmission of the dominant agouti allele.

Identification of Genotype of Disrupted *Dld* Gene

The offspring with coat color are genotyped by Southern blot or polymerase chain reaction (PCR) of their genomic DNA shortly before weaning. The genomic DNA from mouse tail tissue is isolated with a genomic extraction kit (GeneMate, ISC BioExpress, Kaysville, UT) according to the manufacturer instructions. The genomic DNA is analyzed by Southern blot as previously described or by PCR, using a trio of primers that simultaneously amplify regions of the wild-type and targeted alleles. The amplification conditions consist of 35 cycles of 95° for 1 min, 55° for 1 min, and 72° for 1 min. The wild-type allele is amplified as a 0.9-kb fragment from the wild-type exon 9 to exon 10 by 5' common primer (5'-GGTGGAATTGGAATT-GACAT-3') and 3' wild-type specific primer (5'-TTATTGACTGGAATT-CTACCTTTGGGATCT-3'). Under these conditions, the targeted allele does not amplify because the insertion of *neo* disrupts the annealing of 3' wild-type specific primer. The targeted allele is amplified as a 1.1-kb fragment from the wild-type exon 9 to *neo* by 5' common primer and 3' targeted allele specific primer (5'-ACCCCCTCTCCCCTCCTTTTG-3'). More recently, the founder chimeras have been bred with the 129/J and 129/Ola inbred strains to maintain the mutant allele in a homogenous genetic background.

Analysis of $Dld^{+/-}$ Animals

In analyzing the agouti progeny, the presence of a $1:1$ ratio of $Dld^{+/+}:Dld^{+/-}$ indicates that there are no dominant effects on viability from the introduced mutation. The $Dld^{+/-}$ progeny from the chimera/Black Swiss mating are examined for any obvious phenotypic effect of carrying a mutant E3 allele by performing gross anatomic surveys and following growth curves for $Dld^{+/-}$ and $Dld^{+/+}$ littermates. As may be expected from the human data, the $Dld^{+/-}$ littermates are phenotypically normal. Liver samples from the $Dld^{+/-}$ and $Dld^{+/+}$ littermates are assayed for E3 activity as described above. To assess the effect of the heterozygous state on four of the E3-requiring multienzyme complexes, liver samples from both $Dld^{+/-}$ and $Dld^{+/+}$ littermates are assayed for pyruvate dehydrogenase complex (PDC), α-ketoglutarate dehydrogenase complex (KDC), BCKDC, and glycine synthase (GS).[23–26] These assays follow the decarboxylation of specific substrates labeled at the C-1 position. The CO_2 produced is trapped in hyamine hydroxide and counted by a liquid scintillator. For determination of PDC, KDC, and BCKDC activities, crude liver homogenates are used. To measure total PDC and BCKDC activities, the samples are activated by treatment with phospho-E1 phosphatase and λ protein phosphatase (New England BioLabs), respectively, in the presence of dichloroacetate and α-chloroisocaproic acid, respectively, for inhibition of kinase activities. For measurement of GS activity, instead of liver homogenates, liver mitochondrial extracts are used. Liver samples are minced and homogenized in a buffer (70 mM sucrose, 220 mM mannitol, 2 mM HEPES and EDTA, pH 7.4). After six or seven passes by a motor-driven Teflon homogenizer, the unbroken cells are removed by centrifugation at 650g for 10 min at 4°. The mitochondria are pelleted by centrifugation at 12,000g, washed twice with PBS, and resuspended in 20 mM Tris-HCl (pH 8.0), followed by three freeze–thaw cycles. The measurement of GS activity is determined according to a previously described protocol.[26] As a mitochondrial marker, citrate synthase activity is measured by the production of free CoASH, which can be measured spectrophotometrically by assaying the production of free CoASH after treatment with dithionitrobenzoic acid.[27]

The $Dld^{+/-}$ animals have approximately 50% of wild-type E3 activity as well as four E3-requiring complex activities (Table I), suggesting that the disruption results in a loss of function. Northern blot analysis showed

[23] D. S. Kerr, S. A. Berry, M. M. Lusk, L. L. Ho, and M. S. Patel, *Pediatr. Res.* **24,** 95 (1988).

[24] D. T. Chuang, C. W. Hu, and M. S. Patel, *Biochem. J.* **214,** 177 (1983).

[25] G. W. Goodwim, B. Zhang, R. Paxton, and R. A. Harris, *Methods Enzymol.* **166,** 189 (1988).

[26] H. Kochi, K. Hayasaka, K. Hirraga, and G. Kikuchi, *Arch. Biochem. Biophys.* **198,** 589 (1979).

[27] P. A. Srere, *Methods Enzymol.* **7,** 3 (1969).

TABLE I

COMPARISON OF E3 AND E3-REQUIRING COMPLEXES IN $Dld^{+/+}$ AND $Dld^{+/-}$ MICE[a]

Animals	Enzymatic activity (mean ± SD)[b]					
	E3	PDC	KDC	BCKDC	GS	CS[c]
$Dld^{+/+}$	21.0 ± 2.9	3.1 ± 0.3	5.1 ± 0.4	0.79 ± 0.05	0.87 ± 0.06	0.17 ± 0.01
$Dld^{+/-}$	12.4 ± 1.1	1.5 ± 0.1	2.7 ± 0.4	0.43 ± 0.04	0.42 ± 0.10	0.17 ± 0.01
%[d]	59	48	53	54	48	100

[a] $n = 4$ except for GS, where $n = 3$.
[b] Milliunits per milligram of total protein, except for GS, where entries represent milliunits per milligram of total mitochondrial protein.
[c] CS, Citrate synthase, a mitochondrial protein as control.
[d] Percentage (%) of enzymatic activity of control ($Dld^{+/+}$) animals.

that the $Dld^{+/-}$ animals have approximately 50% of wild-type levels of Dld mRNA in liver and kidney samples.[21]

Identification of the $Dld^{-/-}$ Embryos

The litters from intercrosses between heterozygotes are found to be approximately 25% smaller. Genotypic analysis of progeny reveals the presence of $Dld^{+/+}$ and $Dld^{+/-}$ animals in the expected Mendelian ratio (66 : 130), but there are no $Dld^{-/-}$ animals. The absence of $Dld^{-/-}$ mutants indicates that a recessive prenatal lethal allele has been created. To determine the time of death, embryos from various stages of postimplantation development are dissected from the decidua and genotyped by PCR. Embryos are incubated with an embryo lysis buffer for at least 2 hr at 50°. For larger embryos, 7.5 days postcoitum (dpc) and later, embryos are genotyped with a trio of primers that allow for simultaneous amplification of both the targeted and wild-type alleles, using a common primer annealing to a sequence within intron 9 (5'-CACTAAGCTCCATCTTCAGCCATGAG-3'), a wild-type allele-specific primer annealing to a sequence within intron 10 (5'-GGTCTGTTTTTATCTTTAGAGAGAGCCAAAAA-3'), and targeted allele-specific primer annealing to a sequence within the β-actin promoter of the neomycin marker (5'-CCTCCGCCCTTGTGGACA CT-3'). The thermocycling parameters consist of 35 cycles of 95° for 1 min, 55° for 1 min, and 72° for 1 min. In the case of smaller preimplantation embryos, additional primers are synthesized that are internal to the original set of primers to allow for nested PCR. The second set of primers includes a primer from exon 9 (5'-GGTGGAATTGGAATTGACATGGAGAT-3'), a wild-type allele primer (5'-GGTCTGTTTTTATCTTTAGAGAGAGC-CAAAAA-3'), and a mutant allele primer (5'-ACCCCTCTCCCCT-

CCTTTTG-3′). The final amplification product for the targeted allele is 400 bp, whereas the wild-type allele product is 455 bp. For nested PCR, a 5-μl sample from the first 50-μl reaction is added to the second amplification cocktail and the sample is cycled with cycling parameters as described above.

At the blastocyst stage, corresponding to 3.5 dpc, all embryos are similar in gross appearance, with the expected Mendelian distribution of genotypic classes: 16 : 35 : 14 for $Dld^{+/+} : Dld^{+/-} : Dld^{-/-}$. At 7.5 dpc, which corresponds to several days after uterine implantation, two types of embryos are observed. Approximately 75% of the embryos (56 of 81) appear to be normal in size and morphology at pre- to early primitive streak stages and genotype as $Dld^{+/+}$ or $Dld^{+/-}$ class.[28] The remaining 25% of embryos are much smaller and resemble normal 6.5 dpc egg cylinders, indicating that they are delayed in development. The genotypes of these embryos indicate that the majority of these embryos are $Dld^{-/-}$ class, suggesting that $Dld^{-/-}$ embryos survive at least this stage. One day later, at 8.5 dpc, the normally developing embryos have reached the head-fold stage and are all in the $Dld^{+/+}$ or $Dld^{+/-}$ genotypic class (11 : 18), whereas all the embryos genotyped as $Dld^{-/-}$ are abnormal in morphology, still appearing in size to be 6.5 dpc embryos. One day later, at 9.5 dpc, 42 decidua are collected, of which 8 contain only resorption sites with insufficient embryonic material for genotypic analysis. The rest of the embryos are genotyped as $Dld^{+/+}$ or $Dld^{+/-}$, with a ratio of 11 : 23.

Analysis of $Dld^{-/-}$ Embryos

To assess the $Dld^{-/-}$ embryonic phenotype in more detail, histologic analysis is preformed. Decidua are collected at several time points around the established time of embryonic death and fixed overnight in 4% (w/v) paraformaldehyde in PBS. The following day, the decidua are dehydrated through increasing concentrations of ethanol and embedded in paraffin. A series of 7-μm sections of the decidua are stained with hematoxylin–eosin.

At 6.5 dpc, all embryos appear to be normal egg cylinders, suggesting that the $Dld^{-/-}$ class is not significantly delayed in growth or differentiation at this stage. One day later, at 7.5 dpc, the normal embryos have initiated gastrulation with the formation of the three germ layers: mesoderm, endoderm, and ectoderm. Approximately one-quarter of the embryos are presumed to be developmentally delayed by approximately 1 day. Further analysis of these abnormal embryos at 7.5 and 8.5 dpc reveals the presence of mesoderm in most embryos, suggesting that the homozygous mutant embryos initiate gastrulation and then cannot progress further in development.

[28] J. R. Clough and D. G. Whittingham, *J. Embryol. Exp. Morphol.* **74**, 133 (1983).

The *Dld* null phenotype provides direct evidence of the importance of oxidative metabolism during the early postimplantation period. Because of the lack of PDC and KDC enzymatic activities, the $Dld^{-/-}$ embryos would be unable to metabolize glucose oxidatively. Instead, embryos must rely exclusively on glycolysis, which converts pyruvate to lactate. An *in vitro* study using embryos in the presence of radiolabeled glucose demonstrated that virtually all glucose is converted to lactate at the egg cylinder period.[29] At 6.5 dpc, both $Dld^{+/+}$ and $Dld^{-/-}$ embryos are similar in size, mophology, and phenotype, suggesting that anaerobic glucose oxidation may supply sufficient energy for this stage. However, at 7.5 dpc gastrulation begins, a period of rapid growth and differentiation that imposes a significant energy demand on the embryo. One plausible explanation for the cessation of development of the homozygous mutants at this stage is that glycolysis does not generate significant energy for $Dld^{-/-}$ embryos at this period. The generation of these animals has provided some of the first *in vivo* data indicating the importance of oxidative metabolism in early postimplantation embryogenesis.

Prospect

This chapter has described the techniques involved in generating and analyzing mice that harbor a null mutation in the *Dld* gene. The existence of these mice now opens the possibility for further studies to determine if heterozygotes manifest with metabolic differences compared with their wild-type littermates. It may be possible to subject these animals to extreme environmental conditions such as endurance exercise or diets enriched in branched-chain amino acids to bring out metabolic differences and in turn gain some insight into the roles of these metabolic pathways. Also, with the existence of an E3 null allele, it would be possible to introduce less severe mutations into this background to mimic more closely the phenotypes seen in humans to serve as a model for testing various therapeutic interventions.

Acknowledgments

We thank Dr. Clemencia Colmenares of the Cleveland Clinic for the E14.1 ES cell line, Dr. John Duffy of the University of Cincinnati for performing the blastocyst injections, and Dr. Terry Magnuson of Case Western Reserve University for helpful discussions and guidance. Mark Johnson was the recipient of a predoctoral fellowship supported by Metabolism Training Grant AM07319. This work was supported by U.S. Public Health Service Grant DK 42885. The first two authors (MTJ and HSY) made equal contributions to this article.

[29] K. M. Downs and T. Davis, *Development* **118**, 1255 (1993).

Section IV

Regulation and Expression of Enzymes of
Branched-Chain Amino Acid Metabolism

[43] Regulation of Expression of Branched-Chain α-Keto Acid Dehydrogenase Subunits in Permanent Cell Lines

By JEFFREY M. CHINSKY and PAUL A. COSTEAS

A large number of studies have demonstrated that there are many influences on the activity of the mammalian mitochondrial enzyme complex, branched-chain α-keto acid dehydrogenase (BCKAD, EC 1.2.4.4). Changes in BCKAD activity may be produced in response to endocrine factors, exercise, and nutritional state through posttranslational mechanisms of regulation, such as kinase-mediated phosphorylation (inactivation), as well as through pretranslational mechanisms that lead to increases in the RNAs that encode the BCKAD subunit proteins.[1] Examination of steady state levels of RNAs encoding BCKAD subunits (E1α, E1β, E2, and kinase) in rodent tissues obtained from animals at different postnatal developmental ages or fed diets with various protein and/or caloric contents suggested that regulation of BCKAD subunit gene expression plays an important role in the response of tissues to varying physiologic conditions.[2-5] To study the mechanisms by which these tissues regulate BCKAD promoter activity, cell lines amenable to both the influences of external agents and the expression of transfected BCKAD promoter minigenes needed to be identified. The first step was to identify cell lines that demonstrate altered levels of BCKAD subunit RNAs in response to glucocorticoids, insulin, acidosis, or state of differentiation, known effectors of BCKAD tissue activity *in vivo*. The cell lines demonstrated to satisfy these conditions included those demonstrating hepatic (H4IIEC3, Hepa 1), renal (LLC-PK1), and fibroblast inducible adipocyte (3T3-L1) characteristics or origin.[4-6] Of note is that lack of glucocorticoid effect on BCKAD expression in some cell lines may be due to lack of expression of sufficient glucocorticoid receptor, a condition rectified with stably transfected minigenes (LLC-PK1-GR101).[6] We have focused on using hepatic cell lines because mammalian liver demonstrates the highest activity per organ and the majority of the total body BCKAD

[1] M. S. Patel and R. A. Harris, *FASEB J.* **9**, 1164 (1995).
[2] Y. Zhao, S. C. Denne, and R. A. Harris, *Biochem. J.* **290**, 395 (1993).
[3] Y. Zhao, K. M. Popov, Y. Shimomura, N. Y. Kedishvili, J. Jaskiewicz, M. J. Kuntz, M. J. Kain, B. Zhang, and R. A. Harris, *Arch. Biochem. Biophys.* **308**, 446 (1994).
[4] J. M. Chinsky, L. M. Bohlen, and P. A. Costeas, *FASEB J.* **8**, 114 (1994).
[5] P. A. Costeas and J. M. Chinsky, *Biochem. J.* **318**, 85 (1996).
[6] X. Wang, C. Jurkovitz, and S. R. Price, *Miner. Electrolyte Metab.* **23**, 206 (1997).

capacity in laboratory rats.[7] A large number of studies have confirmed the response of hepatic BCKAD subunit RNA levels to physiologic states associated with known changes in levels of circulating hormones.[5,8] Our most consistent results have been obtained with the rat hepatic cell line H4IIEC3. Previous studies had demonstrated that the presence of insulin decreases levels of E1α RNA with no significant effect on E2 RNA levels observed at 24 hr, while glucocorticoids appear to demonstrate the reverse effect.[5] The effects of both glucocorticoids and insulin on BCKAD E2 gene expression are described below.

General Methodology

Cell Culture

Rat hepatoma H4IIEC3 cells are obtained from the American Type Culture Collection (Manassas, VA) and maintained in Dulbecco's modified Eagle's medium (DMEM) containing 10% (v/v) fetal calf serum (DMEM–FCS) at 37° in a humidified 5% CO_2, 95% air atmosphere. Specific conditions are indicated in the figure legends, but experiments are performed in an identical manner by plating cells in serum containing medium for 24 hr, washing them with phosphate-buffered saline (PBS) twice, feeding them with serum-free medium for 14–16 hr, and then adding dexamethasone (0.001–1.0 μM), insulin (0.01–0.1 μM), and/or dibutyryl-cAMP (0.1–1.0 mM) in fresh serum-free medium for the indicated time interval prior to harvest for BCKAD subunit RNA or reporter gene analysis. Cells are prepared for RNA analysis as previously described.[5] For transfections, cells are multiply plated from a single batch of freshly trypsinized cells at 3.5 × 10^5 cells per 20-mm well into six-well cell culture dishes (Falcon; Becton Dickinson Labware, Lincoln Park, NJ).

Murine Hepa 1 cells (ATCC) demonstrate similar responses but not to the extent observed with H4IIEC3. Human HepG2 cells (ATCC) do not demonstrate statistically significant changes in BCKAD gene expression, perhaps because of their lack of sufficient expression of the glucocorticoid receptor.

Isolation and Blot Hybridization Analysis of Steady State RNA

Total cellular RNA is isolated by the guanidine thiocyanate–phenol extraction method, separated by electrophoresis in 1% (w/v) agarose–

[7] S. Soemitro, K. P. Block, P. L. Crowell, and A. E. Harper, *J. Nutr.* **119**, 1203 (1989).

[8] A. G. Chico, S. A. Adibi, W.-Q. Liu, S. M. Morris, and H. S. Paul, *J. Biol. Chem.* **269**, 19427 (1994).

formaldehyde gels, transferred to nylon membranes (MagnaGraph; Micron Separations, Westboro, MA) and hybridized with randomly primed ^{32}P-labeled cDNA as previously described.[4,9] Ethidium bromide staining of 28S and 18S ribosomal RNA confirms equivalent loading in all lanes. Rehybridization of all blots with several cDNA probes (see below) is performed to ensure relative equality of different RNA preparations. All individual RNA preparations are retested several times to confirm any results presented.

DNA Probes

The following ^{32}P-labeled probes are used for Northern blot hybridization experiments: 1.7-kb *Eco*RI fragment from murine BCKAD E1α cDNA (GenBank accession no. 47335),[5] the 1.4-kb *Eco*RI–*Cla*I fragment from murine BCKAD E2 cDNA (GenBank accession no. L42996) containing sequences encoding only the mature preprotein,[10] and the 0.6-kb *Eco*RI–*Bam*HI fragment of CHOB (GenBank accession no. L22552), which detects a single RNA species in rodents and is considered to be the mRNA for ribosomal protein S2.[11,12] Other useful probes included cDNAs for actin and glyceraldehyde-3-phosphate dehydrogenase (GAPDH).

Minigene Plasmids

The isolation, cloning, and sequencing of the murine BCKAD E2 promoter region containing a 7.0-kb 5′ upstream genomic sequence and construction of minigene plasmids containing a downstream luciferase reporter sequence (derived from the pGL-Basic luciferase vector; Promega, Madison, WI) are described elsewhere.[13] Available restriction sites in the promoter and 5′ upstream genomic region are used in a series of restriction endonuclease and religation steps to obtain sequentially smaller BCKAD E2 gene sequences ranging from 7.0 kb to 300 bp from the full-sized 7.0-kb pGLE2-7.0 minigene plasmid. The restriction enzymes used in their creation include pGLE2-4.0, *Kpn*I; pGLE2-2.3, *Spe*I; pGLE2-0.9, *Sac*I; pGLE2-0.3, *Pst*I. The promoter activities of these minigenes demonstrate a 10-fold decrease with the loss of 5′ upstream genomic material down to 300 bp.[13] The end sequences of the larger minigenes as well as the entire sequence of the smaller minigenes are confirmed by use of the double-stranded DNA

[9] P. Chomczynski and N. Sacchi, *Anal. Biochem.* **162,** 156 (1987).
[10] P. A. Costeas, L. A. Tonelli, and J. M. Chinsky, *Biochim. Biophys. Acta* **1305,** 25 (1996).
[11] M. M. Harpold, R. M. Evans, M. Salditt-Georgieff, and J. E. Darnell, *Cell* **17,** 1025 (1979).
[12] L. T. Putowski, D. Choi, J. Mordacq, W. J. Scherzer, K. E. Mayo, E. Y. Adashi, and R. M. Rohan, *J. Soc. Gynecol. Invest.* **2,** 735 (1995).
[13] P. A. Costeas and J. M. Chinsky, *Biochim. Biophys. Acta* **1399,** 111 (1998).

(dsDNA) cycle sequencing system (BRL Life Technologies, Gaithersburg, MD) as well as cycle sequencing on an Applied Biosystems (Foster City, CA) automated sequencer (model 373A). All the BCKAD E2 sequences terminate at the *Bsp*E2 sequence (TCCGGA) 6–11 bp upstream from the initiation ATG codon in exon 1, and 3–8 bp downstream from the determined BCKAD E2 transcription initiation site.[13] pRSV-SEAP, a minigene plasmid that expresses a secretory form of alkaline phosphatase, is obtained from Tropix (Bedford, MA).

Transient Transfection of H4IIEC3 Cells

H4IIEC3 cells are plated in DMEM containing 10% (v/v) FCS, incubated overnight, washed with PBS, and replaced with DMEM without FCS. DMEM (100 μl) containing plasmid DNA (2 μg: 0.4 μg of RSV-SEAP and 1.6 μg of test BCKAD minigene) is mixed with 100 μl of medium containing LipofectAMINE (8 μl; GIBCO-BRL, Bethesda, MD) and incubated at room temperature for 30 min. The optimal amounts of Lipofect-AMINE and plasmid DNA needed to obtain highest transfection efficiency are determined by systematic titration of both parameters and determination of transfection efficiency on the basis of levels of secreted alkaline phosphatase (SEAP) in the medium (data not shown). After 800 μl of DMEM is added to the suspension, it is added to the cells and incubated for 4 hr (37°, 5% CO_2). The plates are washed twice with PBS and DMEM–10% (v/v) FCS is added. After overnight incubation, the cells are washed twice with PBS to remove dead cells and medium is replaced with fresh DMEM–10% (v/v) FCS. The cells are incubated for 8–10 hr, at which time medium is collected for SEAP analysis to assess relative transfection efficiency. The cells are washed twice with PBS and FCS-free DMEM is added. After overnight incubation, the medium is replaced with fresh FCS-free DMEM containing insulin (I), dexamethasone (DEX), and/or dibutyryl-cAMP (dBcAMP); the cells are then incubated for the indicated times (usually 24 hr) and harvested for determination of luciferase activity.

Production and Analyses of Stably Transfected Cell Lines

H4IIEC3 cells are transfected with DNA at a 1:4 ratio of a neomycin expression plasmid under the control of the Rous sarcoma virus (RSV) promoter (RSV-*neo*) to a cloned luciferase expression plasmid (pGL2 Basic; Promega) under the control of an inserted BCKAD E2 genomic promoter sequence (minigene pGLE2-7.0 or pGLE2-0.3). Cells are plated at 2×10^5 cells per 60-mm dish, incubated overnight in DMEM–10% (v/v) FCS, washed with PBS, and then covered with 5 ml of DMEM containing 8% (v/v) modified bovine serum (MBS; Stratagene, La Jolla, CA). Ten micrograms of plasmid DNA is then mixed in 450 μl of water with 50 μl of solution I (2.5 M CaPO$_4$) and 500 μl of solution II [2× N,N-bis-(2-hydroxyethyl)-2-aminoethanesulfonic acid (BES)-buffered saline] and the suspension is incubated at room temperature for 10–20 min and added to the cells dropwise. After 3 hr of incubation at 35°, 3% CO$_2$, the cells are washed three times with PBS, covered with DMEM–10% (v/v) FCS, and incubated at 37°, 5% CO$_2$.[14] After overnight incubation, the cells are washed to remove dead cells and the medium replaced with fresh DMEM–10% (v/v) FCS. The following day, the medium is supplemented with G418 (Geneticin, 400 μg/ml; GIBCO-BRL) and maintained under this selection for 3–4 weeks, with the medium replaced every 4 days. Colonies formed by neomycin-resistant cells are individually trypsinized and plated separately or in pools. On expansion, the cells are assayed for luciferase activity and frozen under liquid nitrogen in 10% (v/v) dimethyl sulfoxide (DMSO)–20% (v/v) FCS.

Previously frozen cultures of cloned transfected cells are thawed, passaged several times to ensure high viability, and then tested as a single batch for comparative analyses. Cells are plated at 3.0×10^5 cells per 20-mm well in six-well cell culture dishes, incubated overnight in DMEM–10% (v/v) FCS, washed twice with PBS, and incubated overnight (14–16 hr) with DMEM without FCS. The medium is then replaced with FCS-free DMEM containing insulin (I), dexamethasone (DEX), and/or dibutyryl-cAMP (dBcAMP); the cells are then incubated for 24 hr and harvested for determination of luciferase activity. Analyses using a singly cloned cell line containing pGLE2-7.0, H4S1, are presented in the figures, but analyses using other independently cloned cells lines are performed to ensure the results reported for H4S1.

Reporter Minigene Assays

Secreted alkaline phosphatase (SEAP) activity is measured from culture medium (100 μl) collected from the transfected cells according to the directions of the supplier of the assay kit (Tropix). This medium is mixed with 1× dilution buffer (300 μl) and heated to 65° for 30 min, and then 100 μl of heat-treated medium is mixed with 100 μl of assay buffer containing a mixture of different alkaline phosphatase inhibitors. These inhibitors allow the detection of the secreted placental alkaline phosphatase while minimizing the interference from nonheat inactivated endogenous phosphatase activities. The reaction buffer containing CSPD® (Tropix) chemiluminescent substrate (100 μl) is added to the reaction tube and after 20 min of incuba-

---

[14] F. M. Ausebel, "Current Protocols in Molecular Biology," pp. 9.13–9.14. John Wiley & Sons, New York, 1995.

tion at room temperature, the luminescence is measured with a model 20e luminometer (Turner Designs, Sunnyvale, CA).[13]

Luciferase (LUC) activity is measured according to the directions of the assay kit supplier (Promega). Transfected cells are washed twice with PBS and covered with 200 μl (per 20-mm well) of 1× reporter lysis buffer, incubated for 15 min at room temperature, and scraped into a 1.5-ml microcentrifuge tube. After vortex mixing for 10 sec and microcentrifuge centrifugation for 10 sec, 20 μl of cell extract is mixed with 100 μl of luciferase assay reagent (470 μM luciferin, 270 μM coenzyme A, and 530 μM ATP) and luminescence determined with the Turner Designs model 20e luminometer.

All assay values are measured in duplicate or triplicate to ensure proper values. All transient transfections are performed in triplicate in the same multiwell dish and, similarly, all stably transfected cells are assayed in triplicate. Luciferase activity from each sample is corrected for SEAP activity in medium collected prior to serum starvation and exposure of the cells to hormonal effectors. The corrected luciferase activities from the triplicate samples are averaged and the Student t test performed on the basis of the mean and the standard deviation (s_{n-1}).

Gel Retardation (Gel Shift) Assay

Nuclear Extract Preparation. H4IIEC3 cells are grown to confluency (5 × 150 mm dishes), washed and scraped into 20 ml of cold PBS, centrifuged for 10 min at 1000 rpm, and rewashed in PBS. They are then resuspended and incubated on ice for 10 min in 5 ml of 10 mM HEPES (pH 7.9), 1.5 mM MgCl$_2$, 10 mM KCl, 0.5 mM dithiothreitol (DTT), and 0.5 mM phenylmethylsulfonyl fluoride (PMSF) and then centrifuged, and the pellet is resuspended in fresh buffer to which 0.05% (v/v) Nonidet P-40 has been added. The pellet is homogenized (20 strokes) in a tight Dounce homogenizer and the pellet recentrifuged. The pellet is resuspended in 2 ml of 5 mM HEPES (pH 7.9), 1.5 mM MgCl$_2$, 0.2 mM EDTA, 26% (v/v) glycerol, 0.5 mM DTT, and 0.5 mM PMSF, and to this suspension concentrated NaCl is added to 420 mM and incubated on ice for 30 min. The lysate is centrifuged at 12,000 rpm for 20 min in an Sw41 Beckman (Fullerton, CA) rotor, dialyzed through 0.0025-μm pore size VS membrane (Millipore, Bedford, MA) against 5 mM HEPES (pH 7.9), 100 mM KCl, 0.2 mM EDTA, 26% (v/v) glycerol, 0.5 mM DTT, and 0.5 mM PMSF at 4°, and frozen in 50-μl aliquots in liquid nitrogen.[15]

Gel Retardation Assay. Template DNA from the upstream E2 promoter

[15] P. E. Berg, D. W. Williams, R. B. Cohen, R.-L. Qian, M. Mittelman, and A. N. Schechter *Nucleic Acids Res.* **17,** 8833 (1989).

region (-47 to -140 bp upstream from the start site) is amplified by polymerase chain reaction (PCR), purified, and eluted into Tris–EDTA (TE). The gel-purified DNA (3 pmol) is end labeled with $[\gamma\text{-}^{32}P]ATP$ by T4 polynucleotide kinase in a 10-μl reaction volume at 37° for 30 min. The reaction is terminated with 1 μl of 0.5 M EDTA and diluted in TE to 200 μl. One microliter of the end-labeled template is mixed with nuclear extract (approximately 3 μg), incubated at room temperature for 20 min, and loaded in 1× gel shift binding buffer [5× buffer: 20% (v/v) glycerol, 5 mM MgCl$_2$, 2.5 mM EDTA, 2.5 mM DTT, 250 mM NaCl, 50 mM Tris (pH 7.0), poly(dI–dC) (0.25 mg/ml)] onto a 5% (w/v) nondenaturing polyacrylamide gel containing 2.5% (v/v) glycerol. After electrophoresis at 100 V for 3 hr in 0.5× TBE, the gel is dried under vacuum and exposed to X-ray film at $-80°$.[15,16]

Results

Glucocorticoids Affecting BCKAD E2 RNA Accumulation in H4IIEC3

After exposure of H4IIEC3 cells to titrated amounts of dexamethasone (DEX), steady state levels of BCKAD E2, but not E1α, subunit RNA increased (Fig. 1). Studies were also performed with dibutyryl-cAMP (dBcAMP), an agent known to interact synergistically with dexamethasone under certain conditions. In contrast to dexamethasone, which consistently produced increased accumulation of E2 RNA, exposure of cells to dBcAMP alone had no consistent effect on the accumulation of the E2 transcript. However, low levels of dBcAMP in combination with dexamethasone caused an enhanced increase in the observed level of E2 RNA, usually threefold, by 24 hr after treatment.

Glucocorticoids and Insulin Affecting E2 Minigene
Expression in H4IIEC3

Using an E2 minigene plasmid (pGLE2-7.0) containing 7.0 kb of 5' murine genomic sequence with promoter activity directing expression of the reporter gene luciferase, stably transfected H4IIEC3 cells were isolated and tested for their responsiveness to dexamethasone (DEX), dBcAMP, and insulin (Fig. 2). Similar to the results observed with RNA levels, exposure to DEX produced twofold increases in luciferase, dBcAMP alone produced no increases, and in combination, dBcAMP appeared to augment

---

[16] F. M. Ausebel, "Current Protocols in Molecular Biology," pp. 12.03–12.2.10. John Wiley & Sons, New York, 1995.

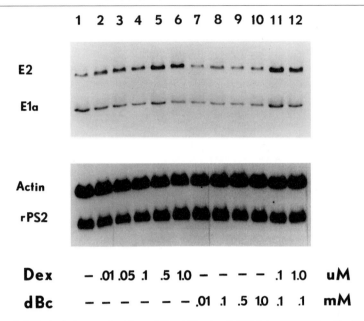

FIG. 1. Effect of glucocorticoids on BCKAD subunit RNA in H4IIEC3 cells. Northern blot hybridization analysis of 20 μg of total cellular RNA prepared from pooled samples from three individual plates of cells grown in media containing the indicated concentrations of dexamethasone (Dex) and dibutyryl-cAMP (dBc) is shown. Rehybridization with actin and rPS2 probes as control for loading is indicated. Densitometric analysis comparing lane 1 and 11 suggested a 3.8-fold increase in BCKAD E2 RNA in the combined presence of 0.1 μM Dex/0.1 mM dBc in this experiment.

the DEX-induced increase to about threefold within 24 hr. Further increases were observed at 48 hr, up to six- to sevenfold (data not shown). Titration analysis of DEX exposure of H4S1 cells demonstrated increases in relative promoter activity from 0.001 to 1.0 μM in the presence of a constant amount of dBcAMP known to augment the DEX response (data not shown). To confirm the involvement of a glucocorticoid-receptor mediated mechanism in the activation of the E2 gene, we treated H4S1 cells with a combination of dexamethasone and the glucocorticoid receptor antagonist RU486 (Roussel Uclaf, Romainville, France), which prevented the DEX-associated increase in E2 promoter activity (data not shown). This finding suggested that the glucocorticoid receptor participated in the regulation of E2 by glucocorticoids.

In contrast to glucocorticoids, exposure of transfected H4IIEC3 cells to insulin (I) resulted in decreases in luciferase activity (about twofold) at 24 hr. The presence of insulin along with DEX or DEX plus dBcAMP

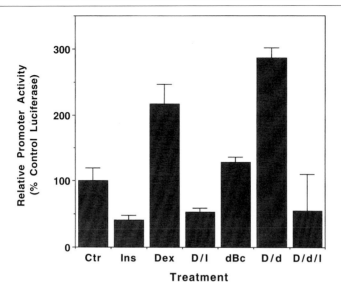

FIG. 2. Hormonal effects on H4IIEC3 cells stably transfected with BCKAD 7.0-kb E2 promoter–luciferase minigene DNA. Cloned isolate H4S1 cells derived from stably transfected H4IIEC3 cells were plated in triplicate, grown overnight in DMEM–10% FCS, washed with PBS, incubated for 16 hr in FCS-free medium, and then exposed to dexamethasone (DEX or D, 1.0 μM), insulin (INS or I, 0.1 μM), and/or dBcAMP (dBc or d, 0.5 mM), or not exposed (Ctr), for 24 hr in FCS-free medium. Cells were then harvested for luciferase assay as described in General Methodology. The relative promoter activity represents the ratio of the mean ($\pm$SD) luciferase activity per treatment group ($n = 3$) to the activity of control (Ctr) untreated cells, and is expressed as a percentage [control cells (Ctr) by definition had 100% relative promoter activity].

did not result in the othewise consistent increase in E2 promoter activity observed in association with glucocorticoid exposure alone (Fig. 2). This was further investigated by examination of H4IIEC3 cells transiently transfected with E2-7.0 minigenes and subsequently exposed to these hormones. The results of a typical experiment, showing the ratio of luciferase to secreted alkaline phosphatase activities, are shown in Table I. Using the presence of SEAP activity measured after transfection but prior to the addition of hormone, one could standardize for the relative efficiencies of individually transfected dishes. A number of experiments have confirmed that the presence of glucocorticoids increases the BCKAD E2-7.0 promoter activity by two- to fourfold in H4IIEC3 cells. In contrast, the presence of insulin appears to consistently suppress this increase, but to a variable extent. The range of insulin concentration required to produce this effect appears to be from 10 to 100 nM in these cells, based on titration experiments.

TABLE I
EFFECT OF DEXAMETHASONE, dBcAMP, AND INSULIN ON
BCKAD PROMOTER ACTIVITY IN TRANSIENTLY
TRANSFECTED H4IIEC3 CELLS

	Luciferase activity	
Additive[a]	LUC/SEAP[b]	Fold difference[c]
None	2.429 ± 0.069	1.00
DEX	6.601 ± 0.155	2.71
dBcAMP	3.041 ± 0.047	1.22
DEX/dB	8.830 ± 0.147	3.64
Insulin	2.484 ± 0.120	1.02
DEX/Ins	3.717 ± 0.327	1.55
dB/Ins	2.792 ± 0.196	1.15
DEX/dB/Ins	5.067 ± 0.149	2.09

[a] Dexamethasone (DEX), 1.0 μM; dBcAMP (dB), 0.1 mM; insulin (Ins), 0.1 μM.
[b] Values represent mean ratios ± SD of luciferase (LUC) to secreted alkaline phosphatase (SEAP) activities determined for each transfected dish ($n = 3$).
[c] Fold increase in promoter activity (LUC/SEAP) compared with control.

Regional Localization of Sequences Involved in Regulation of BCKAD E2 Promoter Activity

To identify the region that contained *cis*-acting elements responsible for the regulation of BCKAD E2 promoter activity, a series of luciferase-expressing minigenes containing decreasing amounts of E2 genomic DNA directly upstream from the transcription start site was assayed. The 0.3-kb E2-containing minigene demonstrated notable increases in response to DEX or DEX/dBcAMP, which could be affected by the presence of insulin (data not shown). This finding suggested that this proximal promoter region contained sequences sufficient for glucocorticoid-mediated activation. To narrow further the critical promoter region-containing sequences needed for this glucocorticoid effect, a series of minigenes with progressively shorter regions of the BCKAD promoter DNA (from 315 to 44 bp upstream from the transcription initiation site) was assayed by both transfection studies and linker scanning analysis. The minimum BCKAD promoter sequence that could be demonstrated contained 44 bases upstream from the transcription start site.[13] The utility of using H4IIEC3 cells for identification of potential DNA protein-binding sites in the proximal E2 promoter region was tested by preparation of nuclear extracts and DNA gel shift analysis

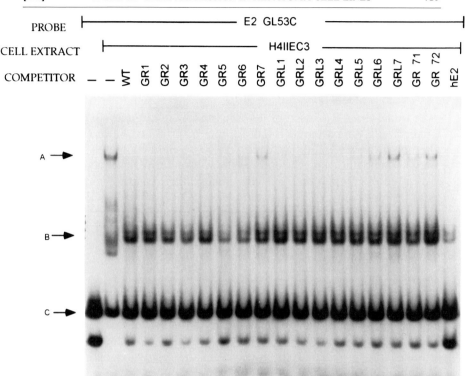

FIG. 3. Gel retardation analysis of the BCKAD E2 upstream promoter region. Oligonucleotide DNA prepared from the murine BCKAD E2 promoter region (−47 to −140) by PCR amplification was gel purified, end labeled with [γ-^{32}P]ATP, and incubated with 3 μg of H4IIEC3 nuclear extract and the indicated competitor DNA (100×) at room temperature for 20 min. The DNA–protein reaction mix was then separated by nondenaturing PAGE and the gel dried and exposed for autoradiography. (−) No competitor; WT, unaltered wild-type murine sequence; GR1–GR7, GR71, GR72, sequences with linker scanning mutations; GRL1–GRL7, series of contiguously deleted sequences; hE2, amplified DNA derived from the corresponding region of the human BCKAD E2 promoter.[16a]

of E2 DNA containing sequences from −47 to −140 bases from the start site. Using these nuclear extracts, several shifted bands were identified, indicating binding of nuclear protein to the labeled promoter sequence (see Fig. 3, bands A and B). All the shifted bands were efficiently competed with excess unlabeled probe but not with oligonucleotides containing SP1 or AP2 consensus sequences. The top, slower migrating band (band

[16a] K. S. Lau, W. J. Herring, J. L. Chuang, M. McKean, D. J. Danner, R. P. Cox, and D. T. Chuang, J. Biol. Chem. 267, 24090 (1992).

A) was easy to compete with as little as a 100× excess of unlabeled probe, but the lower, faster migrating band (band B) required at least a 200× excess.

This same promoter DNA region (-47 to -140) was PCR amplified from a series of minigenes containing contiguous linker scanning mutations or contiguously deleted sequences along this region.[13,17] Each of these prepared DNAs was used at 100× excess to determine competition for binding with the prepared H4IIEC3 nuclear proteins. As shown in Fig. 3, the upper, slowly migrating band was completely competed with all the mutant competitors as efficiently as the wild-type (WT) sequence, with the exception of competitor DNA containing linker scanning mutations in or near the reverse CAAT box site (pGR6, pGR7, pGR71, pGR72) at -78 to -83 bp upstream from the transcription start site.[13] Similarly, the competitor DNAs containing sequences whose deletions extended down to but did not include the CAAT box (pGRL1–pGRL5) competed the shifted band efficiently. The DNA with deletions termininating near or within the region of bp -78 to -83, GRL6: to -85 and GRL7: to -76, failed to compete out the shifted band. More detailed binding kinetic studies are required to establish the dynamics of DNA–protein interaction in this region. However, these studies indicate that this cell line will provide a useful model with which to identify potentially important sites of DNA protein binding that may be involved in the regulation of E2 promoter activity in hepatic cells.

Summary

The rat hepatoma cell line H4IIEC3 has demonstrated a response to both insulin and glucocorticoids in its accumulation of BCKAD subunit RNAs. It is amenable to BCKAD promoter minigene transfection analyses, demonstrating positive (glucocorticoids) and negative (insulin) regulatory effects. These cells can therefore be used as a model to identify cis-acting sites responsible for regulation of BCKAD subunit promoter activity.

Acknowledgments

These studies were supported in part by a Life and Health Insurance Medical Research Fund grant to J. M. Chinsky and presented as a portion of the doctoral thesis submitted by P. A. Costeas.

[17] F. M. Ausebel, "Current Protocols in Molecular Biology," pp. 8.5.1–8.5.9. John Wiley & Sons, New York, 1995.

[44] Expression of Murine Branched-Chain α-Keto Acid Dehydrogenase Kinase

By Christopher B. Doering and Dean J. Danner

The focus within the field of amino acid metabolism has changed significantly. During the first part of the century a wealth of knowledge was obtained with the elucidation of the metabolic pathways, but understanding how these individual pathways are regulated has proved more arduous. With the advances in molecular biology and genetic manipulation, several new approaches are proving extremely effective in advancing our understanding of metabolic regulation.[1,2] Possibly the most important of these approaches is that of gene manipulation in mice, with the ability to completely control or abolish gene expression. This technology has revitalized studies concerning both the regulation of metabolic enzymes and pathways and the role of individual tissues in maintaining overall homeostasis. To take full advantage of these technologies and utilize the mouse as a model system, initial characterization of the system of interest must be achieved. To this end the characterization of branched-chain amino acid (BCAA) metabolism has begun in mice. BCAA levels in the body are regulated by protein synthesis and/or breakdown and the irreversible catabolism of the branched-chain α-keto acids (BCKAs) by the branched-chain α-keto acid dehydrogenase complex (BCKD). BCKD is a nuclear-encoded multienzyme complex, which is present in the mitochondria of all cell types in mammals. Regulation of BCKD activity occurs through the phosphorylation of the E1α subunit by a BCKD-specific kinase (BCKDK), which renders it completely inactive.[3] Most studies to date concerning BCKDK have utilized rats and have been limited to genetic characterization and measurements of enzymatic activity in various tissues and nutritional states.[4,5] In an attempt to eventually utilize the excellent mouse genetic technologies to address more complex questions surrounding BCAA metabolism, this

[1] B. Lamothe, A. Baudry, P. Desbois, L. Lamotte, D. Bucchini, P. De Meyts, and R. L. Joshi, *Biochem. J.* **335,** 193 (1998).

[2] F. Bosch, A. Pujol, and A. Valera, *Annu. Rev. Nutr.* **18,** 207 (1998).

[3] R. Odessey, *Biochem. J.* **204,** 353 (1982).

[4] K. M. Popov, Y. Zhao, Y. Shimomura, M. J. Kuntz, and R. A. Harris, *J. Biol. Chem.* **267,** 13127 (1992).

[5] R. A. Harris, J. W. Hawes, K. M. Popov, Y. Zhao, Y. Shimomura, J. Sato, J. Jaskiewicz, and T. D. Hurley, *Adv. Enzyme Regul.* **37,** 271 (1997).

chapter outlines the available information concerning murine BCKDK (mBCKDK).

General Methods

Animals

All animal studies are done according to IACUC-approved protocols. Mice (C57BL/6) are maintained on 13-hr light/11-hr dark cycle and fed standard rodent chow and water *ad libitum*. For embryonic studies, estrous females are paired individually with stud males in the late afternoon and checked for copulation plugs the next morning (embryonic day 0.5). Pregnant females are killed by cervical dislocation at noon on the designated days and the embryos are collected as previously described.[6] Embryos are examined to ensure similar levels of development, and embryos at incorrect stages are discarded.

cDNA Cloning and Sequencing

A mouse muscle λgt10 cDNA library (Clontech, Palo Alto, CA) is screened with rat random-primed (Rediprime; Amersham, Arlington Heights, IL) ^{32}P-labeled BCKDK (kindly provided by R. A. Harris, Indiana University, Bloomington, IN) as a probe. The full-length cDNA is constructed from the two individual λ clones and subcloned into pBluescript KS⁻ (Stratagene, La Jolla, CA). The nucleotide sequence of the full-length clone is determined experimentally (Emory DNA Sequencing Core Facility, Atlanta, GA).

Antibody Preparation

The amino acid sequence (TDTHHVELARERSK) derived from an N-terminal portion of the mBCKDK protein is supplied by Research Genetics (Huntsville, AL) for the generation of rabbit antisera.

Preparation of Mitochondria

Using a Dounce homogenizer, isolated tissue is minced into 1-mm³ pieces on ice in 5 ml of specified tissue buffer: liver and DG75 lymphoblasts—220 mM manitol, 70 mM sucrose, and 2 mM HEPES, pH 7.4; brain—250 mM sucrose, 10 mM Tris, 0.5 mM EDTA, pH 7.4; kidney— 300 mM sucrose; heart and skeletal muscle—220 mM sucrose, 70 mM

---

[6] B. L. Hogan, R. S. P. Beddington, F. Costantini, and E. Lacy, "Manipulating the Mouse Embryo—a Laboratory Manual," 2nd Ed. Cold Spring Harbor Laboratory Press, Cold Spring Harbor, New York, 1994.

mannitol, and 2 mM HEPES pH 7.4. The homogenate is centrifuged at 660g (liver, brain, heart, and kidney) or 900g (skeletal muscle) for 10 min at 4°. The remaining supernatant is then transferred to a fresh tube and centrifuged again at 9500g for 10 min at 4°. The yellowish-brown mitochondrial pellet is then washed to remove any residual tissue contamination (whitish color) and the appropriate volume of buffer (220 mM mannitol, 70 mM sucrose, and 2 mM HEPES, pH 7.4) to achieve a 10- to 15-$\mu g/\mu l$ mitochondrial protein concentration is added. Quantitation of mitochondrial protein is performed with the bicinchoninic acid (BCA) protein assay system (Pierce, Rockford, IL) according to the manufacturer instructions.

In Vitro Protein Synthesis

In vitro transcription/translation reactions are performed with the Promega (Madison, WI) TNT coupled system according to the manufacturer protocol. T7 polymerase is used for the transcription portion of the reaction and proteins are labeled with [^{35}S]methionine (10 mCi/ml; Amersham). The resulting lysate is immediately used for mitochondrial import without purification.

Murine BCKD Kinase cDNA

Two overlapping murine BCKDK (mBCKDK) cDNA clones are obtained by screening a muscle cDNA library. Combined, the clones encompass 1654 bp including the entire coding region along with 106 bases of 5' untranslated region (UTR) and 309 bases of 3' UTR (GenBank accession number AF043070). The predicted protein is 412 amino acids long, including a 30-amino acid mitochondrial targeting sequence (MTS). Both at the nucleotide and amino acid levels mBCKDK shows a high identity to the rat (93 and 98%, respectively) and human (84 and 95%, respectively) sequences.

Tissue Distribution

To establish comparisons between mice and the other species for which BCKDK has been studied, we have examined the tissue distribution of mBCKDK in mice. Mitochondria from skeletal muscle (hindquarter), heart, brain, kidney, and liver are isolated as described above. Ten micrograms of mitochondrial protein is loaded per lane and resolved by 10% (w/v) sodium dodecyl sulfate–polyacrylamide gel electrophoresis (SDS–PAGE) at 200 V for 45 min. The resulting gel is transblotted to Hybond-ECL nitrocellulose (Amersham) for 2 hr at 200 mA. Membranes are incubated overnight in IM2 buffer [50 mM Tris (pH 7.5), 150 mM NaCl, 5 mM EDTA (pH 8.0), 0.25% (w/v) gelatin, 1% (v/v) Tween 20] to minimize nonspecific

liver   brain   kidney   muscle

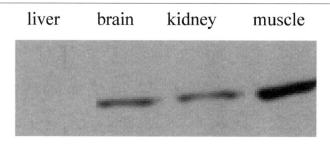

FIG. 1. Western blot of skeletal muscle, brain, kidney, and liver mitochondrial proteins, using antisera against an mBCKDK peptide (1 : 10,000 dilution). Ten micrograms of mitochondrial protein was loaded per lane and resolved by 10% (w/v) SDS–PAGE.

binding. Antibody detection of mBCKDK is performed with mBCKDK polyclonal antibody at a 1 : 10,000 dilution in IM2. Chemiluminescent visualization (ECL Western blotting system; Amersham) reveals a distribution pattern similar to that found in rats (Fig. 1).[7] Muscle mitochondria contain the highest levels of the protein, with heart, brain, and kidney levels being intermediate, and liver containing the least amount of kinase protein. Tissue distribution of mBCKDK has also been analyzed by reverse transcriptase-polymerase chain reaction (RT-PCR) with similar findings.[8] These tissue-specific kinase levels have been shown previously to correlate well with the tissue-specific activity state of BCKD complex in several species.[9,10]

Embryonic Expression

Previously it has not been addressed when in embryonic development BCKD activity is first turned on and, correspondingly, when BCKDK regulation of activity begins. Developmental regulation of BCKD activity in mice is examined by measuring mBCKDK transcript levels throughout embryogenesis by RT-PCR. Total RNA is isolated from embryonic day 6.5–18.5 (e6.5–e18.5) mouse embryos with TriReagent (Sigma, St. Louis, MO) according to the manufacturer instructions. Reverse transcriptase reactions are performed on ice in a solution containing (in 20 μl) 20 mM Tris-HCl (pH 8.4), 50 mM KCl, 5 mM MgCl$_2$, dNTPs (1 mM each), 20 units of RNasin (Promega), random primer (0.5 mg/ml), RNA (0.5

[7] K. M. Popov, Y. Zhao, Y. Shimomura, J. Jaskiewicz, N. Y. Kedishvili, J. Irwin, G. W. Goodwin, and R. A. Harris, *Arch. Biochem. Biophys.* **316,** 148 (1995).

[8] C. B. Doering, C. Coursey, W. Spangler, and D. J. Danner, *Gene* **212,** 213 (1998).

[9] A. Suryawan, J. W. Hawes, R. A. Harris, Y. Shimomura, A. E. Jenkins, and S. M. Hutson, *Am. J. Clin. Nutr.* **68,** 72 (1998).

[10] R. A. Harris, K. M. Popov, Y. Zhao, N. Y. Kedishvili, Y. Shimomura, and D. W. Crabb, *Adv. Enzyme Regul.* **35,** 147 (1995).

embryonic day 6.5 7.5 8.5 9.5 10.5 11.5 12.5 13.5 14.5 15.5 16.5 17.5 18.5

— mBCKDK

Fig. 2. RT-PCR analysis of mBCKDK embryonic expression.

mg/ml), and 250 units of Moloney murine leukemia virus (Mo-MuLV) reverse transcriptase (Amersham) and brought to volume with diethyl pyrocarbonate (Sigma)-treated water. The reactions are incubated for 10 min at room temperature followed by 30 min at 42° and 5 min at 99°. The reactions are then cooled on ice, and 20 pmol of kinase-specific primers (forward, 5'-GAAGCTTTCCTCCCGGCCATCAATGTG-3'; reverse, 5'-GCTGCCCGTTTCCCCTTCATTCCTATG-3'; amplification product of 425 bp) are added followed by thermocycling at 94° for 30 sec, 50° for 30 sec, and 74° for 30 sec for 39 cycles followed by 1 cycle of 94° for 30 sec, 50° for 30 sec, and 74° for 5 min. mBCKDK transcripts are present at varied intensity throughout embryonic development (Fig. 2). Kinase transcript is present as early as sufficient RNA can be collected to perform the RT-PCR (e6.5). Apparently low levels of mBCKDK transcript are found at e10.5 and e18.5 despite examining embryos derived from three different mothers. These changes in relative kinase transcript levels suggest embryonic regulation of mBCKDK and thus BCKD activity.

Gender-Specific Kinase Expression

It has been established that there is a sex-specific increase in the amount of BCKD kinase protein that occurs in female rats with the progression of the light cycle and that this change corresponds with a decrease in the active state of the BCKD complex.[11] To document a similar mode of regulation this experiment has been repeated with mice. Mitochondria are isolated from liver, kidney, heart, brain, and skeletal muscle as described above from 6-week-old mice killed at 10 A.M. and 4 P.M. Western blot detection of mBCKDK levels is performed as described above. Liver mitochondria isolated from male and female mice at 10 A.M. and 4 P.M. reveal a phenomenon identical to that seen in rats (Fig. 3A). Densitometric analysis of the kinase protein bands reveals a twofold increase in mBCKDK protein levels

[11] R. Kobayashi, Y. Shimomura, T. Murakami, N. Nakai, N. Fujitsuka, M. Otsuka, N. Arakawa, K. M. Popov, and R. A. Harris, *Biochem, J.* **327**, 449 (1997).

A

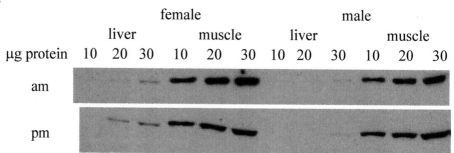

B

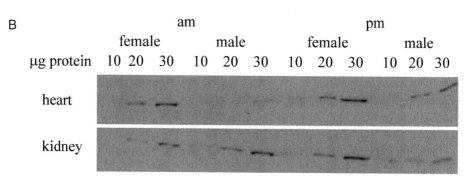

FIG. 3. Western blot of mitochondrial proteins taken from female and male rats at 10 A.M. and 4 P.M. and probed with α-mBCKDK antisera. Five, 10, and 20 μg of mitochondrial protein were loaded for quantitation. (A) Isolated mitochondrial proteins from liver and skeletal muscle. (B) Isolated mitochondrial proteins from heart and kidney.

at the end of the light cycle in female but not in male mice. No corresponding increases are found in skeletal muscle protein, but slight changes are observed in kidney and heart (Fig. 3B). The significance of this regulation is currently not understood.

Mitochondrial Import

In vitro transcription and translation of mBCKDK are performed as described above. The resulting lysate is mixed with an equal volume of mitochondria and incubated for 20 min at 30°. The mitochondria from the import reaction are then isolated by centrifugation at 10,000g for 10 min at room temperature. Mitochondrial proteins are resuspended in sample buffer containing 10% (v/v) 2-mercaptoethanol and resolved in a 10%

lysate   import

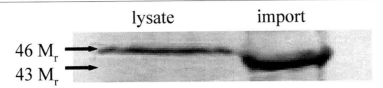

$46 \, M_r$ →

$43 \, M_r$ →

FIG. 4. *In vitro* import of mBCKDK into human DG75 lymphoblast-derived mitochondria. Two microliters of *in vitro* TNT lysate was loaded onto the lysate lane. Fifteen microliters of TNT lysate was used for import.

(w/v) SDS–PAGE as described earlier. The gel is fixed for 5 min in 10% (v/v) acetic acid–40% (v/v) methanol and rehydrated for 5 min in water, and then exposed to a 30-min soak in Fluoro-Hance (Research Products International, Mount Pleasant, IL). The gel is then dried and exposed to Hyperfilm (Amersham) for visualization of the radiolabeled proteins. Rainbow molecular weight markers (RPN56; Amersham) are used to determine relative protein mobility. *In vitro* transcription/translation of the mBCKDK sequence results in the formation of a protein with the predicted relative mobility for the mBCKDK preprotein, which includes the 30-amino acid MTS. On incubation of the lysate with freshly isolated mitochondria, there is a shift in the relative mobility of the protein (Fig. 4). This shift corresponds with the predicted alteration in protein molecular weight after cleavage of the MTS and import into the mitochondria. After a 20-min incubation with mitochondria there is no longer any remaining visible mBCKDK preprotein, suggesting efficient import.

Remarks

BCKD kinase plays an important role in the maintenance of cellular BCAA levels. Data are now available concerning kinase expression in four mammalian species. These studies have proved helpful in addressing models of whole-body BCAA metabolism, and in elucidating mechanisms for kinase gene expression. The question that cannot be addressed by current experiments in nonmurine animals is the phenotypic consequence of altered kinase expression *in vivo*. No known mutations in BCKDK have been uncovered in any species and the phenotype of this situation is unknown. The studies presented here provide the first steps toward utilizing the power of the experimental mouse to help us understand the complete role of BCKDK in regulating BCAA metabolism.

Acknowledgments

This work was supported in part by a grant from the NIH (DK38320) and by funds from the Emory University Research Committee. C.B.D. was supported in part by an NIH training grant (5T32GM08490). Special thanks to L. Jackson and W. Spangler for technical assistance.

[45] Regulation of Branched-Chain α-Keto Acid Dehydrogenase Kinase Gene Expression by Glucocorticoids in Hepatoma Cells and Rat Liver

By YI-SHUIAN HUANG *and* DAVID T. CHUANG

Introduction

The mammalian mitochondrial branched-chain α-keto acid dehydrogenase (BCKD) complex catalyzes the rate-limiting step in the catabolism of branched-chain amino acids, i.e., the oxidative decarboxylation of branched-chain α-keto acids derived from leucine, isoleucine, and valine. The activity of the BCKD complex is tightly controlled by a reversible phosphorylation (inactivation)–dephosphorylation (activation) cycle, mediated by a specific kinase[1] and a specific phosphatase,[2] respectively.

The activity state, i.e., the percentage of active dephosphorylated form, of the BCKD complex varies in tissues and is regulated by nutritional and hormonal stimuli.[3] This provides an effective mechanism to modulate the flux of branched-chain amino acids through the BCKD complex. The kinase activity in the liver of rats fed 0% protein diets was shown to increase three- to fourfold compared with rats fed a 50% protein diet.[4] The increase in kinase activity was accompanied by a concomitant elevation in levels of the kinase protein and mRNA. The same study[4] also showed that the kinase protein and mRNA levels were low in the liver and high in the kidney and heart of rats fed a normal diet, which inversely correlated with the activity state of the complex. These findings suggest that the kinase gene is subject to transcriptional regulation in response to nutritional stimuli in a tissue-specific manner. Whether changes in the complex activity state by hormones are mediated through transcriptional regulation of the kinase gene remains to be investigated.

To better understand prolonged hormonal effects on the BCKD kinase gene expression, we studied the effects of dexamethasone, dibutyryl-cAMP, and insulin on the level of the kinase mRNA in a rat hepatoma cell line

---

[1] R. Odessey, *Biochem. J.* **204**, 353 (1982).

[2] Z. Damuni, M. L. Merryfield, J. S. Humphreys, and L. J. Reed, *Proc. Natl. Acad. Sci. U.S.A.* **81**, 4335 (1984).

[3] R. A. Harris, R. Paxton, S. M. Powell, G. W. Goodwin, M. J. Kuntz, and A. C. Han, *Adv. Enzyme Regul.* **25**, 219 (1986).

[4] K. M. Popov, Y. Zhao, Y. Shimomura, J. Jaskiewicz, N. Y. Kedishvili, J. Irwin, G. W. Goodwin, and R. A. Harris, *Arch. Biochem. Biophys.* **316**, 148 (1995).

H4IIE. The transcription of phospho*enol*pyruvate carboxykinase (PEPCK) was shown to be tightly regulated by cyclic AMP (cAMP), dexamethasone, and insulin[5-7] in these cells, which therefore offer a suitable system to study the effects of dexamethasone, dibutyryl-cAMP, and insulin on the kinase gene expression. Our data show that the BCKD kinase mRNA is downregulated by dexamethasone at the transcriptional level in H4IIE cells, which results in full activation of the complex activity in dexamethasone-treated cells. Results obtained from studies with the rat indicate that the downregulation of the kinase gene by dexamethasone is liver specific. However, promoter studies indicated that the negative glucocorticoid-responsive element (GRE) is not located in the 3.0-kb kinase promoter region that was systemically investigated. In this chapter, we describe the details of our experiments and discuss possible approaches for identifying the *cis*-acting elements in the kinase promoter by which hormonal and nutritional stimuli modulate enzyme activity.

H4IIE Cell Culture

Rat hepatoma cells H4IIE obtained from the American Type Culture Collection (ATCC, Manassas, VA) are grown in Dulbecco's modified Eagle's medium (DMEM) supplemented with 5% (v/v) fetal bovine serum and 5% (v/v) calf serum. For hormonal studies, H4IIE cells are cultured in the same medium with the desired reagents or hormones, e.g., 1 μM dexamethasome, 0.5 mM dibutyryl-cAMP, 100 nM insulin, or any combination of these agents, for 24 and 48 hr. H4IIE cells cultured in a 100-mm culture dish are rinsed with phosphate-buffered saline (PBS) and dissolved in 3 ml of an ice-cold lysis buffer containing 4 M guanidinium isothiocyanate, 1 mM EDTA, 25 mM sodium acetate, and 0.5% (v/v) Sarkosyl (sodium lauroyl sarcosinate). The lysate is sheared with $27\frac{1}{2}$-gauge syringe needles, layered onto a discontinuous 1-ml CsCl (5.7 M CsCl, 1 mM EDTA, and 25 mM sodium acetate) gradient, and centrifuged at 85,000g [Beckman (Fullerton, CA) SW-50.1] for at least 15 hr at 4°. The CsCl gradient-pelleted RNA is dissolved in 400 μl of diethyl pyrocarbonate (DEPC)–H$_2$O, extracted twice with an equal volume of H$_2$O-saturated phenol–chloroform, and precipitated with ethanol. Normally 500 to 700 μg of total RNA from one 100-mm dish of confluent cells is obtained and 20 μg is used for

[5] A. Wynshaw-Boris, T. G. Lugo, J. M. Short, R. E. K. Fournier, and R. W. Hanson, *J. Biol. Chem.* **259**, 12161 (1984).
[6] R. M. O'Brien, P. C. Lucas, T. Tomoyuki, E. L. Noisin, and D. K. Granner, *J. Biol. Chem.* **269**, 30419 (1994).
[7] R. K. Hall, F. M. Sladek, and D. K. Granner, *Proc. Natl. Acad. Sci. U.S.A.* **92**, 412 (1994).

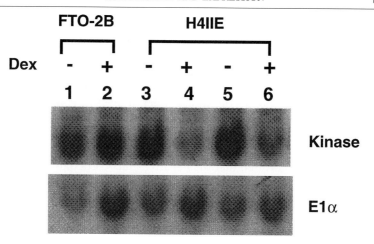

FIG. 1. Effects of dexamethasone on BCKD kinase expression in cultured H4IIE and FTO-2B cells. Cells cultured in the medium with or without 100 n*M* dexamethasone (Dex) for 30 and 48 hr were used for total RNA isolation and the BCKD complex activity assays, respectively. Northern blot shows levels of kinase and E1α mRNAs in FTO-2B (lanes 1 and 2) and H4IIE (lanes 3–6) cells cultured with (lanes 2, 4, and 6) or without (lanes 1, 3, and 5) 100 n*M* dexamethasone for 30 hr. [With permission from Y. S. Huang and D. T. Chuang, *Biochem. J.* **339,** 503 (1999).]

Northern analysis. To investigate whether serum contains trace amounts of hormones capable of interfering the study, the same treatments are applied to H4IIE cells cultured in medium containing 10% (v/v) charcoal-treated fetal bovine serum (Cocalico Biologicals, Reamstown, PA). No differences are found in the amounts of the kinase mRNA between cells cultured in the regular or charcoal-treated sera. However, we routinely culture cells in the medium containing charcoal-treated serum for hormonal studies. For Northern blotting, a 670-bp *Bam*HI-digested kinase cDNA fragment and a 720-bp *Eco*RI/*Acc*I-excised E1α cDNA fragment are used as templates for probe synthesis[8] by hexamer random priming. GeneScreen Plus membranes (Du Pont, Boston, MA) are used for Northern analyses, and can be recycled for hybridization with different probes without significantly increased background. Figure 1 shows that dexamethasone causes a marked decrease in the kinase mRNA levels (lanes 4 and 6) compared with the control (lanes 3 and 5). The decrease in the kinase mRNA content associated with the dexamethasone treatment is not observed in FTO-2B (rat hepatoma) cells, which do not express a functional glucocorticoid receptor (Fig. 1, lanes 1 and 2). Dibutyryl-cAMP and insulin have no effect on

[8] Y. S. Huang and D. T. Chuang, *Biochem. J.* **339,** 503 (1999).

the kinase mRNA level with or without the presence of dexamethasone (data not shown).

By titrating dexamethasone used in the H4IIE cell culture, we have found that the dexamethasone effect on the kinase mRNA is sensitive, as it is observed at a hormonal concentration as low as 10 nM, which is below the physiological concentration of glucocorticoid.[8] Moreover, the apparent half-lives of the kinase mRNA are estimated to be 16 and 15 hr in cells cultured with or without dexamethasone, respectively. This is accomplished by measuring the mRNA abundance as a function of time after addition of a transcription inhibitor, actinomycin D (1 μg/ml), to H4IIE cells cultured in the presence or absence of 100 nM dexamethasone. The kinase mRNA in H4IIE cells is stable, which explains why the downregulation of the kinase mRNA by dexamethasone is observed only after long-term (>24 hr) treatments.[8] The results also establish that the reduced steady state level of the kinase mRNA is not caused by a decrease in the kinase mRNA stability, which raises the possibility of regulation at the level of gene transcription.

To examine whether the reduced kinase mRNA level results in an elevated activity state of the BCKD complex, we measure the actual and total activities of the complex in H4IIE cells cultured with or without dexamethasone for 48 hr. In the presence of the kinase inhibitor α-chloroisocaproate (α-CIC), the complex is completely activated (dephosphorylated) by endogenous BCKD phosphatase, which gives rise to the total (fully activated) activity of the complex. (α-CIC is a generous gift from R. Simpson, Sandoz, East Hanover, NJ.) Cultured H4IIE cells are harvested and washed twice with Krebs buffer [0.7% (w/v) NaCl, 0.046% (w/v) KCl, 1.24 mM MgSO$_4$ · 7H$_2$O, 18 mM sodium phosphate (pH 7.4), and 0.0195% (w/v) CaCl$_2$]. Washed cells are suspended in Krebs buffer containing 1 mM phenylmethylsulfonyl fluoride (PMSF) and 1 mM benzamidine at a density of 2 × 10^6 cells/0.1 ml. To assay actual activity, cells are rapidly frozen at $-70°$ after harvesting. To assay total activity, cell suspensions are incubated in the presence of 1 mM α-CIC at 37° for 15 min and then rapidly frozen at $-70°$.

The enzyme activity of the BCKD complex is assayed in a modified 24-well plate system developed in this laboratory (see [19] in this volume[9]). The assay mixture contains in 295 μl: 62.7 mM Tris-HCl (pH 7.5), 0.25 mM EDTA, 0.25 mM coenzyme A, 0.44 mM MgCl$_2$, 0.25 mM NAD$^+$, 0.25 mM thiamine pyrophosphate (TPP), and 1.76% (v/v) fetal bovine serum. The BCKD complex activity is determined by measuring ^{14}CO$_2$, released from

---

[9] J. L. Chuang, J. R. Davie, R. M. Wynn, and D. T. Chuang, *Methods Enzymol.* **324**, Chap. 19, 2000 (this volume).

TABLE I
EFFECTS OF DEXAMETHASONE ON ACTIVITY AND ACTIVITY STATE OF BCKD
COMPLEX IN H4IIE CELLS[a]

Medium	Activity (nmol $^{14}CO_2$/min/mg protein)		Activity state (%)
	Total	Actual	
DMEM	0.228 ± 0.036	0.110 ± 0.020	48.2 ± 2.1
DMEM + Dex	0.389 ± 0.071	0.373 ± 0.067	95.9 ± 3.3

[a] The actual and total activities of the BCKD complex in H4IIE cells cultured in the presence or absence of 100 nM dexamethasone for 48 hr were measured as described in text. The activity state is 100% of actual activity divided by total activity. The data are expressed as means $\pm$ SEM ($n = 3$). [From Y. S. Huang and D. T. Chuang, *Biochem J.* **339**, 503 (1999). With permission.]

substrate α-keto[1-^{14}C]isovalerate. The specific activity of the BCKD complex is expressed as nanomoles of $^{14}CO_2$ released per minute per milligram of protein. The protein concentration is determined by the method of Lowry *et al.*,[10] using bovine serum albumin as a standard. The activity state is expressed as percent actual activity divided by total activity of the BCKD complex. In Table I, the dexamethasone-induced decrease in the kinase mRNA level is accompanied by a twofold increase in the activity (dephosphorylation) state of the BCKD complex, when H4IIE cells cultured for 48 hr in the presence of the hormone are compared with untreated cells. The total activity of the BCKD complex in dexamethasone-treated H4IIE cells also increases 70% over untreated cells. The induction of BCKD total activity correlated with a slight increase in the E1α mRNA level (Fig. 1, lanes 3 and 5 versus lanes 4 and 6). The combined effects on the kinase and the E1α mRNA contents result in a threefold increase in the actual BCKD activity in dexamethasone-treated H4IIE cells compared with the untreated cells.

Animal Model

Because the amount of the hepatic kinase mRNA is low under normal nutritional condition, the dexamethasone down-regulatory effect would be difficult to observe in chow-diet fed rats. To address whether the dexamethasone effect on the kinase gene expression is physiologically significant, male Sprague-Dawley rats, weighing approximately 200–220 g, are accli-

[10] O. H. Lowry, N. J. Rosebrough, A. L. Farrmethod, and R. J. Randall, *J. Biol. Chem.* **193**, 265 (1951).

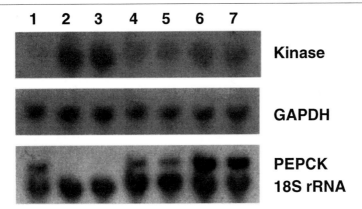

Fig. 2. Effects of dexamethasone administration on the kinase mRNA level in rats fed a 0% protein diet. Rats were fed 0% protein diets for 14 days to induce the hepatic kinase mRNA level, and then intraperitoneally injected with dexamethasone or a diluent control. Northern blot analyses of the kinase, PEPCK, and GAPDH mRNAs and 18S rRNA were carried out. The total RNA (20 μg each) of liver was isolated from a chow diet-fed rat (lane 1); 0% protein diet-fed rats injected with diluents (lanes 2 and 3, duplicates from one rat), or identically fed rats injected with 1 mg of dexamethasone/100 g body weight/day (lanes 4 and 5 isolated from one rat, lanes 6 and 7 from another rat). [Taken with permission from Y. S. Huang and D. T. Chuang, *Biochem. J.* **339,** 503 (1999).]

mated to a 12-hr day–light cycle and fed *ad libitum* a no-protein diet[4] for 14 days to induce the hepatic kinase mRNA. Rats are intraperitoneally injected with 1 mg of dexamethasone per 100 g of body weight per day at 10:00 A.M. for 3 days. During the period of injection, rats are still under the no-protein diet. The average weight of rats fed a no-protein diet for 14 days is about 160 g, and thus 1.5 mg of dexamethasone daily is injected per rat. Vehicle (saline)-injected rats serve as controls. The prolonged hormonal treatment *in vivo* takes into consideration the relatively long half-life of the kinase mRNA measured *in vitro*. Rats are killed 2 hr after the final injection, and the liver and other tissues are collected by freeze-clamping in liquid nitrogen and stored at $-70°$. Frozen tissues are ground to fine powders and dissolved in 4 M guanidinium isothiocyanate lysis buffer, followed by total RNA isolation as described previously.[11] As shown in Fig. 2, the hepatic kinase mRNA level is high in no-protein diet-fed rats (lanes 2 and 3) compared with that in chow diet-fed rats (lane 1), and markedly declines in rats injected with dexamethasone (lanes 4–7), even when the rats are maintained on a protein-free diet. As a positive control, the hepatic PEPCK mRNA is undetectable in rats fed no-protein diets

[11] P. Chomczynski and N. Sacchi, *Anal. Biochem.* **162,** 156 (1987).

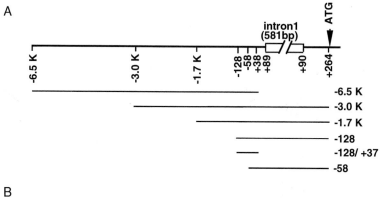

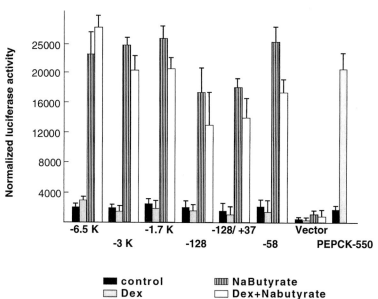

Fig. 3. Promoter activity of the BCKD kinase gene in H4IIE cells with or without dexamethasone treatment. DNA fragments representing various lengths of the promoter-regulatory region of the rat kinase gene were ligated into the pBST-Luc PA plasmid containing the luciferase reporter cDNA. The promoter–reporter constructs were ndividually electroporated into H4IIE cells. (A) Constructs are designated according to the position of the 5′ terminal base in the DNA fragment. Except for constructs −6.5 k and −128/+37 with the 3′ end at base +37 in both clones, the 3′ end of all constructs is at base +264. The broken open box depicts intron 1 (581 bp), which interrupts the 5′-untranslated region (264 bp). (B) Electroporated cells were divided and plated in four 60-mm dishes. Transfected cells were cultured in regular medium (solid bars) or the same medium containing 1 μM dexamethasone (stippled bars), 2 mM sodium butyrate (lined bars), or both dexamethasone and sodium

(Fig. 2, lanes 2 and 3), and sharply increases after dexamethasone injections (lanes 4–7). Although the glucocorticoid receptor is ubiquitously expressed in tissues, glucocorticoid-mediated reduction of the kinase mRNA is observed only in the liver (Fig. 2), and not in the heart, kidney, spleen, or skeletal muscle from rats fed no-protein diets.[8] The intact animal study confirms the downregulation of the kinase mRNA level by dexamethasone observed in H4IIE hepatoma cells. More significantly, the present data suggest the dexamethasone effect on kinase gene expression is mediated through a liver-specific transcription factor(s). Transcriptional repression of the kinase gene is not due to a global downregulation of basal transcription factors, because the dexamethasone treatment does not alter the glyceraldehyde-3-phosphate dehydrogenase (GAPDH) mRNA level in the liver, which is not regulated by dexamethasone (Fig. 2).

Promoter Studies of BCKD Kinase Gene

The finding that the dexamethasone treatment does not affect the half-life of the kinase mRNA strongly suggests that the expression of the kinase gene is downregulated at the transcriptional level. To identify the location of the putative negative glucocorticoid-responsive *cis* element (GRE) in the kinase promoter, an approximately 7-kb segment upstream of the translation start site, including the 6.5-kb promoter region, 581-bp intron 1, and 264-bp 5'-untranslated region, is analyzed by promoter–reporter assays in H4IIE cells cultured in the presence or absence of dexamethasone.

The different lengths of the kinase promoter are fused to a luciferase reporter plasmid. The kinase promoter region used in the dexamethasone study is illustrated in Fig. 3A and preparations of these constructs have been described previously.[8] H4IIE cells are harvested by trypsinization and transfected with the promoter–luciferase vector by electroporation. About 2×10^7 H4IIE cells in 1 ml of serum-free DMEM are mixed with approximately 2.5 pmol (about 10 to 15 μg) of various promoter–luciferase vectors, 5 μg of cytomegalovirus (CMV)-β-galactosidase plasmid, and supplemented with salmon sperm DNA to a total of 100 μg of DNA in each transfection mixture. Cells are electroporated at 1980 μF, 250 V. Electropor-

butyrate (open bars). To correct for variation in transfection efficiency, luciferase activity was normalized to cotransfected β-galactosidase. Luciferase activity was measured in cell extracts and expressed as relative light units/OD at 570 nm (β-galactosidase activity). pPEPCK-550 and pBST. Luc PA (vector) served as positive and negative control, respectively. Error bars represent standard errors of the mean ($n = 3$ or 4). [Taken with permission from Y. S. Huang and D. T. Chuang, *Biochem. J.* **339**, 503 (1999).]

ated cells are cultured on 60-mm dishes in a culture medium containing 10% (v/v) charcoal-treated fetal bovine serum and desired reagents (1 μM dexamethasone, 2 mM sodium butyrate, or both) for 48 hr before harvesting. Under these conditions, some cells that survive after electroporation do not carry plasmid cDNAs. Variations in transfection efficiency are corrected by normalization with cotransfected β-galactosidase activity. Cells are harvested 48 hr after transfection, washed with PBS, and resuspended in 0.25 ml of buffer [0.1 M potassium phosphate (pH 7.5), 1 mM PMSF, 1 mM dithiothreitol (DTT), 1 mM benzamidine]. Cells are disrupted by freezing and thawing three times, followed by microcentrifugation to remove cell debris (5 min, 4°. For the luciferase assay, 50 μl of cell lysate is mixed with 250 μl of assay buffer [0.1 M potassium phosphate (pH 7.8), 1% (v/v) bovine serum albumin (BSA), 25 mM glycylglycine, 2.5 mM ATP, 15 mM MgSO$_4$, 1 mM DTT] and the luciferase activity is measured with a luminometer for 0.2 sec immediately after injecting 100 μl of luciferin substrate (0.5 mM luciferin in 50 mM potassium phosphate, pH 7.8). For the β-galactosidase assay, 50 μl of cell lysate is mixed with 800 μl of CPRG reagent [0.05 mM sodium phosphate (pH 7.5), 2 mM chlorophenol red β-D-galactopyranoside, 9 mM MgCl$_2$, 90 mM α-mercaptoethanol] and incubated at 37° for 1 hr. The OD$_{570nm}$ is measured with 50 μl of lysis buffer as reference. Luciferase activity is assayed and normalized by β-galactosidase activity.

As indicated in Fig. 3B, there is no obvious change in kinase promoter activities in H4IIE cells with (stippled bars) or without (solid bars) dexamethasone present. Likewise, dexamethasone is without effect on promoter activities of additional constructs −449, −225, and −185, all of which contain intron 1 and terminate at position +264 (data not shown). The absence of dexamethasone response is observed in the transient transfection assay even when the ratios of plasmid DNA to the cell number are varied (data not shown). These constructs are also transfected into H4IIE cells pretreated with dexamethasone for 2 days, in which the kinase mRNA level is low and presumably the dexamethasone downregulatory mechanism is active. No differences in promoter activities are found in the presence or absence of dexamethasone. As a positive control, the PEPCK-550 promoter construct, which contains the GRE,[7] shows a 10-fold induction of promoter activity in the presence of dexamethasone (Fig. 3B). Results of the transient promoter assay indicate the absence of GRE in the −3. 0-kb region of the kinase gene promoter. Although dexamethasone does not affect promoter activity of the −6.5 kb construct, deletion constructs between −3.0 and −6.5 kb need to be studied to establish that GRE is not located in this region.

The inefficient DNA transfection into H4IIE cells is depicted by low luciferase activity of 1400 to 1800 light units with the kinase promoter constructs (Fig. 3B). It is therefore speculated that the repression of the kinase gene by dexamethasone may not have been readily observed. Several other DNA transfection methods, such as N-[1-(2,3-dioleoyloxy)propyl]-N,N,N-trimethylammonium methyl sulfate (DOTAP) (Boehringer Mannheim, Indianapolis, IN), LipofectAMINE (GIBCO, Gaithersburg, MD), and calcium phosphate precipitation, have been attempted, but none improves the efficiency of transfection of the kinase promoter constructs into H4IIE cells. To promote transcription of the reporter gene, sodium butyrate is added to the medium. This results in about 10-fold increases in the kinase promoter activities compared with those measured in the absence of sodium butyrate. However, the luciferase activities in dexamethasone-treated cells (Fig. 3B, open bars) or untreated cells (Fig. 3B, lined bars) cultured in the presence of sodium butyrate are again similar. It is speculated that the inclusion of intron 1 in the promoter constructs may comlicate the promoter analysis. Both repressor and enhancer elements have been shown to be present in introns. However, there are no significant differences in promoter activities between constructs $-128/+264$ and $-128/+37$, with and without intron 1, respectively (Fig. 3B). The data indicate that cis-acting elements are absent from intron 1 located in the kinase promoter, and that inclusion of intron 1 does not affect the promoter assay. On the other hand, all constructs with or without intron 1 exhibit luciferase activity. We interpret the data to indicate that intron 1 is correctly spliced out. This is confirmed by the absence of a 581-bp (the size of intron 1)-longer kinase mRNA in H4IIE cells.

The transient promoter assays described above are rapid and free from interference by the chromosome positional effect; however, the results are complicated by the low transfection efficiency. We therefore have established H4IIE cell lines containing stably integrated kinase promoter sequences (-3, -449, and -58 kb). There are no significant differences in luciferase activity in stably transfected cells cultured either with or without dexamethasone,[8] similar to the results obtained with transient transfection. The stable transfection study confirms that the negative GRE in the kinase gene is located either further upstream or downstream of the 3-kb region.

Primary Hepatocytes

The induction of the E1α mRNA was previously observed in primary rat hepatocytes cultured in the presence of dexamethasone and a cAMP analog, 8-(4-chlorophenylthio)-cAMP.[12] The same study also shows that the activity state of the BCKD complex in cultured hepatocytes is not

---

[12] A. G. Chicco, S. A. Adibi, W.-Q. Liu, S. M. Morris, and H. S. Paul, *J. Biol. Chem.* **269**, 19427 (1994).

affected by dexamethasone and the cAMP analog, suggesting a lack of change at the kinase level. To understand the conflicts between our data and previous findings, the same hormonal treatments are applied to primary hepatocytes isolated from rats fed chow and no-protein diets. Isolation of hepatocytes from rats is based on the method described by Seglen[13] with modifications. Livers from male adult rats (Sprague-Dawley), initially weighing about 200–220 g, fed *ad libitum* with a chow diet or no-protein diet for 14 days, are used for hepatocyte isolation. Rats are anesthetized by inhalation of methoxyflurane. The portal vein is cut in half and immediately perfused with 1× Hanks' solution, pH 7.4–7.5 [0.8% (w/v) NaCl, 0.04% (w/v) KCl, 0.006% (w/v) KH_2PO_4, 0.035% (w/v) $NaHCO_3$, 0.009% (w/v) $Na_2HPO_4 \cdot 7H_2O$, 0.1% (w/v) D-glucose, and 0.001% (w/v) phenol red] at the rate of 40 ml/min for 5 min. The liver is then removed from the rat and placed in a bottle at 37° and recirculated with 150 ml of 1× Hanks' solution containing collagenase (100 units/ml) (Worthington, Freehold, NJ) and 5 mM $CaCl_2$ at the rate of 40 ml/min for 10 min. The collagenase-perfused liver is cut into small pieces and suspended in RPMI 1640 medium to which penicillin G sodium (100 units/ml), streptomycin sulfate (100 μg/ml), and kanamycin sulfate (0.2 μg/ml) are added. The suspension is passed through 310-mm Spectra/Mesh (Spectrum, Houston, TX), followed by centrifugation at 30g for 2 min. Pelleted cells are resuspended in the RPMI medium and sequentially passed through sieves of 310, 104, and 75 mm mesh, followed by centrifugation as described above. The pelleted cells are resuspended with the RPMI medium and centrifuged several times until the supernatant is clear. The viability of isolated hepatocytes is examined by the trypan blue exclusion method. Hepatocytes isolated from rats fed either a low-protein or chow diet have similar viability, about 70–80%. To culture primary hepatocytes, approximately 3×10^6 viable cells are plated on a 100-mm culture dish coated with collagen type I from rat tail (Falcon; Becton Dickinson Labware, Lincoln Park, NJ). Isolated hepatocytes attach more firmly on collagen-coated dishes, which significantly improves the survival of hepatocytes in culture. Cells are maintained in DMEM containing 10% (v/v) charcoal-treated serum and desired agents. After 24 or 48 hr of incubation, total RNAs from primary hepatocytes are purified with an RNeasy® total RNA system (Qiagen, Chatsworth, CA). About 6×10^6 hepatocytes from two 100-mm dishes are dissolved in 700 μl of lysis buffer and 60 to 80 μg of total RNA is recovered by mini-RNeasy spin column, following the manufacturer protocol. The RNeasy system is chosen for total RNA isolation from hepatocytes because many samples can be processed in a short time at a reasonable cost.

---

[13] P. O. Seglen, *Methods Cell Biol.* **13**, 29 (1976).

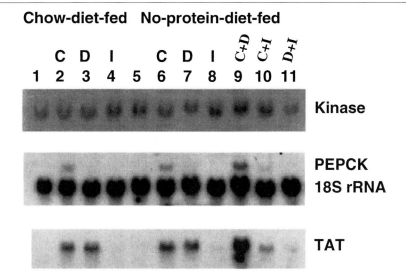

Chow-diet-fed   No-protein-diet-fed

FIG. 4. Effects of dexamethasone, cAMP, and insulin on kinase expression in rat hepatocytes. Total RNAs were isolated from hepatocytes cultured in medium containing 10% charcoal-treated serum and with various reagents for 48 hr. Lanes 1–4 and lanes 5–11 contained 10 μg of total RNA harvested from hepatocytes isolated from chow diet-fed or no-protein diet-fed rats, respectively. Lanes 1 and 5 are from the control culture. D, 1 μM Dexamethasone; C, 0.5 mM dibutyryl-cAMP; I, 100 nM insulin. The Northern blot was probed with the kinase, PEPCK, or TAT cDNAs, or with 18S rRNA probe.

Primary hepatocytes isolated from chow-fed (Fig. 4, lanes 1–4) or no-protein diet-fed (lanes 5–11) rats are used as model systems for evaluating the effects of metabolic regulators on the kinase gene expression. The level of the kinase mRNA does not change after incubation with dibutyryl-cAMP (Fig. 4, lanes 2 and 6), insulin (lanes 4 and 8), dexamethasone (lanes 3 and 7), or any combination of these regulators (lanes 9–11). As controls to evaluate the quality of the hepatocyte preparation, mRNA levels of PEPCK and tyrosine aminotranferase (TAT) are examined. PEPCK and TAT mRNAs are induced by dexamethasone and dibutyryl-cAMP (Fig. 4, lanes 2, 3, 6, and 7), and the induced mRNAs are downregulated in the presence of insulin (Fig. 4, lanes 10 and 11). Moreover, synergistic induction of the PEPCK and TAT mRNAs by dibutyryl-cAMP and dexamethasone is also observed (Fig. 4, lane 9). It is noteworthy that dexamethasone only slightly increases the PEPCK mRNA level (Fig. 4, lanes 3 and 7, longer exposure; and data not shown). In contrast, hepatic PEPCK mRNA is dramatically induced in rats injected with dexamethasone (Fig. 2, lanes 4–7). Studies in hepatocytes grown in different culture systems have shown that these cells rapidly dedifferentiate.[14] Several liver-specific genes, including those encoding liver-enriched transcription factors, are expressed at a higher or lower level in hepatocytes than in the liver.[15,16] The bases for these conflicting results obtained with hepatoma cells (Table I) and cultured hepatocytes[12] are not clear, but it may be that transcription factors required for mediating dexamethasone effects on PEPCK and kinase are defective in hepatocytes during culture. Thus, it is concluded that the primary hepatocyte culture is not an ideal system for studying hormonal regulation of the kinase gene expression.

The kinase mRNA level is relatively high in H4IIE hepatoma cells and the liver of rats fed no-protein diets, but low in the liver of rats fed a normal chow diet. Moreover, the amount of kinase mRNA in hepatocytes isolated from chow diet-fed rats is as abundant as that from no-protein diet-fed rats after 48 hr of culture (Fig. 4, lanes 1 and 5). This has led to a hypothesis that kinase transcription in the liver is repressed under normal nutritional conditions and that kinase mRNA induced by protein malnutrition is the result of derepression of kinase transcription. To approach this question, hepatocytes isolated from rats fed a no-protein or chow diet are cultured in DMEM without hormonal supplements. Cells are harvested for total RNA isolation and the BCKD complex activity assay as described above at the time points of 2, 4, 8, 24, and 48 hr. It is found that in hepatocytes isolated from rats fed chow diets, the kinase mRNA level gradually increases, whereas PEPCK and TAT mRNAs decrease as the culturing time is prolonged, up to 48 hr. In hepatocytes isolated from rats fed no-protein diets, the relatively abundant kinase mRNA remains constant during the 48-hr culture, whereas PEPCK and TAT mRNAs are undetectable by Northern analysis, similar to the results obtained with liver.[8] The activity state of the complex correlates inversely with the level of kinase mRNA. The activity state of the BCKD complex in hepatocytes isolated from chow-diet rats decreases from 97.4 to 37.4% when the culturing time is increased from 2 to 24 hr. In contrast, the activity state of the complex in hepatocytes from rats fed no-protein diets remains consistently low (34.2 to 37%) during the same culture period.[8]

Remarks

The increased activity state of the BCKD complex was previously observed in skeletal muscle 2 hr after injection of synthetic glucocorti-

[14] E. Knop, A. Bader, K. Boker, R. Pichlmayr, and K. F. Sewing, *Anat. Rec.* **242**, 337 (1995).
[15] M. Nagaki, Y. Shidoji, Y. Yamada, A. Sugiyama, M. Tanka, T. Akaike, H. Ohnishi, H. Moriwaki, and Y. Muto, *Biochem. Biophys. Res. Commun.* **210**, 38 (1995).
[16] V. L. Nebes and S. M. Morris, *Mol. Endocrinol.* **2**, 444 (1988).

coids.[17] In contrast, in the present study, the kinase mRNA level in the same tissue did not change after a 3-day treatment with the hormone. Wang et al.[18] have reported that glucocorticoids increase the BCKD complex activity state from 67 to 82% in LLC-PK-GR101 cells. Our Northern blot analysis in collaboration with Price's group showed, however, that the kinase mRNA level in these cells was not affected in the presence of up to 200 nM dexamethasone (data not shown). The results imply that the hormone may exert its effects in the skeletal muscle and the renal cells through a posttranslational inactivation of the kinase or an up-regulation of the phosphatase.

The differential kinase mRNA level in cultured hepatocyte isolated from rats fed chow diets strongly suggests the existence of in vivo factors in the liver that regulate the kinase gene expression, and these factors are absent in the culture medium and other tissues. The trace amount of hepatic kinase mRNA under normal nutritional conditions is likely caused by constitutive repression present in the liver. On the other hand, the "induced" hepatic kinase mRNA in no-protein diet-fed rats may occur through inhibition of the repression cascade. We propose that the kinase gene is regulated in the liver by a repression/derepression cycle imparted by glucocorticoids and nutritional stimuli. In support of this notion is the observation that DNA-binding activities and the levels of several liver-specific or liver-enriched transcription factors decrease under protein-restricted conditions.[19] In addition, we suggest that primary hepatocytes not be used for studying the effect of dexamethasone on kinase gene expression. Apparently, the cell lost part of its hepatic characteristics, e.g., lack of dexamethasone induction of PEPCK mRNA.

Glucocorticoids decrease the steady state of the hepatic BCKD kinase mRNA, presumably by downregulating kinase transcription. Although H4IIE cells appear to be a good model system for studying dexamethasone effects on kinase regulation, the approach of transient transfection experiments was unable to identify the hormone-responsive element, which is likely to be located in the far upstream promoter region. A transgenic approach is necessary to locate GRE, liver-specific repressive elements and the elements that are regulated by nutritional stimuli in the kinase promoter. It will be interesting to demonstrate the cross-talk between these three elements in regulating BCKD kinase gene expression under various physiological conditions.

[17] K. P. Block, W. B. Richmond, W. B., Mehard, and M. G. Buse, Am. J. Physiol. **252,** E396 (1987).
[18] X. Wang, C. Jurkovitz, and S. R. Price, Am. J. Physiol. **272,** 2031 (1997).
[19] N. W. Marten, F. M. Sladek, and D. S. Straus, Biochem. J. **317,** 361 (1996).

Author Index

Numbers in parentheses are footnote reference numbers and indicate that an author's work is referred to although the name is not cited in the text.

A

Abbondanzo, S. J., 466, 470(13)
Abumrad, N. N., 40
Achenbach, S., 289
Achmarina, L., 447
Adams, M. D., 129
Adashi, E. Y., 481
Adelmeyer, F., 24, 26(11), 27(11), 30(12), 31(11, 12), 32(11, 12)
Adibi, S. A., 48, 480, 507, 510(12)
Adlington, R. M., 354, 355(32)
Aebersold, R. H., 175
Aftring, R. P., 48, 49, 61(15)
Agbayani, R., Jr., 123
Ahih, V. E., 33, 37(4)
Aitken, A., 200
Akabaysahi, A., 24
Akaike, T., 510
Akamatsu, Y., 114
Akanuma, S., 314, 322
Akerman, B. R., 444
Akers, H. A., 377
Akutsu, N., 309, 312, 314, 316(12), 319(20), 322(20)
Alban, C., 282, 286(18), 289(18), 291(18), 292
Albani, M., 465
Allen, R. H., 239, 240(13, 14)
Allen, S. H. G., 284
Allison, R. D., 32
Altieri, F., 153
Amanuma, H., 123, 124
Ambudkar, S. V., 123, 126
Amedee-Manesme, O., 454, 455(7)
Ames, B. N., 14
Ames, G. F.-L., 123
Andersen, B. S., 437
Anderson, B., 251, 253, 255

Anderson, D. H., 139, 147(3), 153, 282, 283(8), 292(8)
Anderson, K. P., 284
Anderson, V. E., 79, 141, 147(10)
Andreasen, P. H., 344
Andres, R., 47
Anikster, Y., 454, 455(11)
Anraku, Y., 121, 123, 124, 125, 311, 313(16), 316(16)
Ansari, A. A., 344, 346(11), 352(11), 353(11)
Antholine, W. E., 146
Antonucci, T. K., 129
Aoki, M., 314, 316, 322
Aoshima, M., 314, 316, 322
Aoyama, H., 380
Aoyama, T., 241
Aparicio, C., 260
Arad, S., 13
Arakawa, N., 48, 59, 59(5), 61(6), 162, 495
Aramaki, S., 454, 455(10)
Arizmendi, C., 463(21), 464, 465
Arnoldi, L., 66, 71(15)
Arro, M., 260
Asano, S., 108, 109, 110
Aschner, M., 356
Ashmarina, L. I., 151, 447, 450(39), 451
Atalay, A., 63
Atkinson, M. R., 323, 329(5)
Audhya, T. K., 293
Aulabaugh, A., 12, 21, 23
Aupetit, J., 465
Ausebel, F. M., 483, 485, 490
Aussel, C., 63
Austen, B. M., 27
Ævarsson, A., 130, 131(6), 179
Awata, H., 344
Axiotis, S., 282, 286(18), 289(18), 291(18), 292

Subject Index

A

Acetoacetyl-CoA thiolase, *see* Short-chain 3-ketoacyl-CoA thiolase

Acetohydroxy acids, intracellular concentration determination in bacteria
- calculations, 19–20
- extraction, 19
- gas chromatography, 19
- principle, 19
- reagents, 19
- sensitivity, 20

Acetohydroxy acid synthase
- inhibitors, 12, 22
- isozymes and inhibitor sensitivity, 20–22
- quaternary structure and lability, 95, 103
- reconstitution of subunits from isozyme III
  - assay, 99–100
  - research applications, 102–103
  - titration data analysis, 101
  - valine inhibition, 100
- subunit isolation from isozyme III of recombinant *Escherichia coli*
  - bacterial strains and growth, 97
  - expression system, 96
  - large subunit isolation, 97–98
  - reagents, 96
  - small subunit isolation, 98–99

AHAS, *see* Acetohydroxy acid synthase

β-Alanine synthase
- assays
  - isotope assay, 401–402
  - overview, 400–401
  - reversed-phase high-performance liquid chromatography assay, 402–403
- functions, 399–400
- properties of rat liver enzyme
  - kinetic parameters, 409
  - oligomer association and dissociation, 409–410
  - stability, 408
- purification from rat liver

ammonium sulfate fractionation, 403–404
- buffers, 403
- homogenate, 403
- hydrophobic interaction chromatography, 405
- ion-exchange chromatography, 404
- isoelectric focusing, 404–405
- overview, 403
- size-exclusion chromatography, 405–406
- yield, 406, 408–409
- quaternary structure, 400
- tissue distribution, 399–400
- zinc binding, 400

β-Aminoisobutyrate, interconversion between D- and L- enantiomers
- assays
  - conversion from D- to L-enantiomer, 385–387
  - conversion from L- to D-enantiomer, 387
- clinical significance, 385

β-Aminoisobutyrate aminotransferases
- D-β-aminoisobutyrate aminotransferase properties
  - absorption spectra, 382
  - antibodies, 384
  - inhibitors, 383
  - isoelectric point, 382
  - kinetic parameters, 382–383
  - pH optimum, 382
  - size, 382
  - stability, 382
  - substrate specificity, 382–383
- L-β-aminoisobutyrate aminotransferase properties
  - absorption spectra, 384
  - antibodies, 385
  - inhibitors, 384–385
  - isoelectric point, 384
  - kinetic parameters, 384
  - pH optimum, 384
  - size, 384